Managing Water Resources

To Tanja, Dijana and Damjan
from the banks of the Danube to the Medway Creek

Managing Water Resources

Methods and Tools for a Systems Approach

Slobodan P. Simonović

UNESCO Publishing

United Nations
Educational, Scientific and
Cultural Organization

publishing for a sustainable future

London • Sterling, VA

Published jointly by the United Nations Educational, Scientific and Cultural Organization (UNESCO), 7, place de Fontenoy, 75007 Paris, France, and Earthscan, Dunstan House, 14a St Cross Street, London EC1N 8XA, United Kingdom in 2009

Copyright © UNESCO, 2009

No part of this publication may be reproduced in any form or by any means without the written permission of UNESCO.

All rights reserved

ISBN: 978-92-3-104078-8 UNESCO paperback
ISBN: 978-1-84407-554-6 Earthscan paperback
ISBN: 978-1-84407-553-9 Earthscan hardback

The designations employed and the presentation of material throughout this publication do not imply the expression of any opinion whatsoever on the part of UNESCO concerning the legal status of any country, territory, city or area, or of its authorities, or the delimitation of its frontiers or boundaries.

The author is responsible for the choice and presentation of the facts contained in this book and on any complementary website that he maintains and for the opinions expressed therein, which are not necessarily those of UNESCO and do not commit the Organization.

Typeset by FiSH Books, Enfield, Middx.
Printed and bound in the UK by MPG Books Ltd, Bodmin
Cover design by Eric Frogé

UNESCO Publishing
7, place de Fontenoy, 75352 Paris 07 SP, France
Tel: +33 (0)1 4568 4687
Fax: +33 (0)1 4568 5739
Email: publishing.promotion@unesco.org
Web: http://publishing.unesco.org

For a full list of publications please contact:

Earthscan
Dunstan House, 14a St Cross Street, London EC1N 8XA, UK
Tel: +44 (0)20 7841 1930
Fax: +44 (0)20 7242 1474
Email: earthinfo@earthscan.co.uk
Web: www.earthscan.co.uk

22883 Quicksilver Drive, Sterling, VA 20166-2012, USA

Earthscan publishes in association with the International Institute for Environment and Development

A catalogue record for this book is available from the British Library

Library of Congress Cataloging-in-Publication Data
Simonovic, Slobodan P.
  Managing water resources : methods and tools for a systems approach / Slobodan P. Simonovic.
     p. cm.
  ISBN: 978-1-84407-553-9 (hardback)
  ISBN: 978-1-84407-554-6 (pbk.)
  1. Watershed management. 2. Water resources development. 3. Water-supply. I. Title.
TC405.S56 2008
363.6'1–dc22

2008001741

The paper used for this book is FSC-certified and totally chlorine-free. FSC (the Forest Stewardship Council) is an international network to promote responsible management of the world's forests.

The highest good is like that of water.
                              Tao te Ching, chapter VIII

Contents

List of Figures, Tables and Boxes	*xiv*
About the Author	*xxiv*
Preface	*xxvi*
Foreword	*xxviii*
Introduction	*xxx*
List of Acronyms and Abbreviations	*xxxviii*

Part I   Setting the Stage   1

1   An Overview   3
  1.1   Water resources management issues – some personal experience   5
    1.1.1 An integrated water resources model for Egypt   5
    1.1.2 Sihu basin water management in China   9
    1.1.3 Red River flooding, Manitoba, Canada   12
  1.2   Tools for water resources systems management – two new paradigms   17
    1.2.1 The complexity paradigm   20
    1.2.2 The uncertainty paradigm   24
  1.3   Conclusions   26
  1.4   References   27
  1.5   Exercises   29

2   Changing Water Resources Management Practice   31
  2.1   The history of water resources engineering   32
  2.2   Early water resources engineering science   37
  2.3   Water resources management science   38
  2.4   Characteristics of water resources management   40
    2.4.1 Phase one – rapid development   41
    2.4.2 Phase two – slower development   42

		2.4.3 Phase three – operation, preventive maintenance and rehabilitation	42
		2.4.4 Technological development	43
		2.4.5 The relationship between changing practice and technological development	45
	2.5	The future direction for water resources systems management – a critical path	47
	2.6	References	48
	2.7	Exercises	49
3	**An Introduction to Water Resources Systems Management**		**50**
	3.1	Water resources systems management is human	51
	3.2	What is involved in water resources systems management?	54
	3.3	How is water resources systems management done?	56
	3.4	References	60
	3.5	Exercises	61

Part II Applied Systems Analysis   **63**

4 General Systems Theory   **65**

	4.1	Some system definitions	66
		4.1.1 What is a system?	66
		4.1.2 Systems thinking	67
		4.1.3 Systems analysis	72
		4.1.4 The systems approach	73
		4.1.5 Systems engineering	74
		4.1.6 Mathematical modelling	76
		4.1.7 A classification of systems	78
		4.1.8 A classification of mathematical models and optimization techniques	81
	4.2	Some systems concepts in elementary mathematical consideration	82
		4.2.1 The system definition in mathematical form	83
		4.2.2 Growth	86
		4.2.3 Competition	89
		4.2.4 Wholeness, the sum, mechanization and centralization	90
	4.3	Feedback	92
		4.3.1 The feedback loop	95
		4.3.2 Positive or reinforcing feedback	96
		4.3.3 Negative or balancing feedback	96
	4.4	System formulation examples	97
		4.4.1 City water supply	97

		4.4.2 Operation of a multi-purpose reservoir	99
		4.4.3 Wastewater treatment	102
		4.4.4 Irrigation flow control	105
		4.4.5 Eutrophication of a lake	108
	4.5	References	110
	4.6	Exercises	111

Part III Water Resources Systems Management   113

5 Introduction to Methods of Water Resources Systems Management   115

    5.1 Simulation   115
    5.2 System dynamics simulation   117
    5.3 Optimization   120
    5.4 Multi-objective analysis   123
    5.5 References   126
    5.6 Exercises   127

6 Water Resources Systems Management Under Uncertainty – a Fuzzy Set Approach   129

    6.1 Sources of uncertainty in water resources systems management   130
    6.2 The scope for uncertainty in the water resources systems management process   133
    6.3 Conceptual risk definitions   135
    6.4 Changing the paradigm, from probability to fuzziness   139
        6.4.1 A brief discussion of probability   139
        6.4.2 Problems with probability   141
        6.4.3 Fuzziness and probability   142
    6.5 Introduction to fuzzy set theory   144
        6.5.1 Basic definitions   144
        6.5.2 Set-theoretic operations for fuzzy sets   148
        6.5.3 Fuzzy arithmetic operations on fuzzy numbers   152
        6.5.4 Comparison operations on fuzzy sets   160
    6.6 Derivation of a fuzzy membership function for water resources systems management under uncertainty   166
        6.6.1 Theoretical discussion of system decomposition techniques and aggregation operators   169
    6.7 Derivation of a fuzzy membership function for flood control   178
        6.7.1 Definition of 'flood control' and its decomposition   178
        6.7.2 Questionnaire preparation and definition of the extremes   180
        6.7.3 Evaluation   180
        6.7.4 Results   192

		6.7.5 Additional observations	194

	6.8	Fuzzy measures for the assessment of water resources systems performance	196
		6.8.1 Key definitions	197
		6.8.2 Combined fuzzy reliability–vulnerability measure	202
		6.8.3 Fuzzy robustness measure	204
		6.8.4 Fuzzy resiliency measure	204
		6.8.5 Fuzzy performance measures for multi-component water resources systems	205
		6.8.6 An illustrative example	208
	6.9	The application of fuzzy performance measures to a regional water supply system	212
		6.9.1 System description	213
		6.9.2 Lake Huron Primary Water Supply System (LHPWSS)	214
		6.9.3 Elgin Area Primary Water Supply System (EAPWSS)	216
		6.9.4 Methodology for the application of fuzzy performance measures	218
		6.9.5 Results of the fuzzy reliability analyses	223
		6.9.6 Utility of the fuzzy performance measures	224
	6.10	References	226
	6.11	Exercises	229
7	**Water Resources Systems Management for Sustainable Development**		**233**
	7.1	Principles of sustainable water resources decision-making	233
		7.1.1 Enhancement of the decision-making process	236
		7.1.2 Multi-objective decision-making	237
		7.1.3 Sustainability and other objectives	238
		7.1.4 A methodology for applying sustainability criteria	242
	7.2	Fairness as a sustainability criterion in water resources decision-making	243
		7.2.1 Distance-based fairness measures	244
		7.2.2 Temporal distribution-based fairness measures	244
		7.2.3 Practical considerations for intra-generational and inter-generational fairness measures	246
	7.3	Risk as a sustainability criterion in water resources decision-making	248
		7.3.1 Operational risk definition	248
	7.4	Reversibility as a sustainability criterion in water resources decision-making	250
		7.4.1 Reversibility criterion algorithm	250
		7.4.2 The sensitivity analysis	252

7.5	Consensus as a sustainability criterion in water resources decision-making	252
	7.5.1 Definition of consensus	253
	7.5.2 Development of a consensus measure	253
	7.5.3 Discussion	256
7.6	The SUSTAINPRO computer program	257
7.7	Management of the Assiniboine delta groundwater aquifer (Manitoba, Canada): case study	257
	7.7.1 Case study objectives	258
	7.7.2 Data needs	259
	7.7.3 Alternative policy scenarios	262
	7.7.4 Quantitative description of scenarios	265
7.8	Application of sustainability criteria	267
	7.8.1 Fairness	267
	7.8.2 Risk	271
	7.8.3 Reversibility	280
	7.8.4 Consensus	286
	7.8.5 Discussion	288
7.9	References	289
7.10	Exercises	291

Part IV Implementation of Water Resources Systems Management Tools 295

8 Simulation 297

8.1	Definitions	297
8.2	System dynamics simulation	299
	8.2.1 Introduction	299
	8.2.2 System structure and patterns of behaviour	300
	8.2.3 Causal loop diagram	310
	8.2.4 Stocks and flows	314
	8.2.5 Introduction to system dynamics simulation	323
	8.2.6 Formulating and analysing a system dynamics simulation model	329
	8.2.7 Developing more complex system dynamics models for water resources management	337
8.3	Simulation under uncertainty	346
	8.3.1 Fuzzy simulation	347
8.4	Examples of water resources simulation	351
	8.4.1 Shellmouth Reservoir simulation model	353
	8.4.2 Global water resources assessment simulation model	366
	8.4.3 Flood evacuation simulation model	388

		8.4.4 Hydrological simulation model for predicting floods from snowmelt	399
		8.4.5 A simulation model for resolution of water sharing conflicts	413
	8.5	References	421
	8.6	Exercises	424

9 Optimization — 429

- 9.1 Linear programming — 430
  - 9.1.1 Formulation of linear optimization models — 430
  - 9.1.2 Algebraic representations of linear optimization models — 432
  - 9.1.3 Geometric interpretation of linear optimization models — 435
  - 9.1.4 Simplex method of solution — 439
  - 9.1.5 Completeness of the simplex algorithm — 443
  - 9.1.6 Duality in linear programming — 446
  - 9.1.7 Sensitivity analysis — 451
  - 9.1.8 Summary — 455
- 9.2 Fuzzy optimization — 455
  - 9.2.1 Fuzzy linear programming — 457
- 9.3 Evolutionary optimization — 464
  - 9.3.1 Introduction — 466
  - 9.3.2 Selection — 469
  - 9.3.3 Recombination – crossover — 472
  - 9.3.4 Mutation — 476
  - 9.3.5 Reinsertion — 477
  - 9.3.6 An example of evolutionary optimization — 479
  - 9.3.7 The Evolpro computer program — 481
- 9.4 Examples of water resources optimizations — 481
  - 9.4.1 An improved linear programming algorithm for hydropower optimization — 482
  - 9.4.2 Fuzzy optimization of a multi-purpose reservoir — 491
  - 9.4.3 An evolutionary algorithm for minimization of pumping cost — 505
- 9.5 References — 519
- 9.6 Exercises — 520

10 Multi-objective Analysis — 527

- 10.1 Multi-objective analysis methodology — 531
  - 10.1.1 Change of concept — 531
  - 10.1.2 Non-dominated solutions — 533
  - 10.1.3 Participation of decision-makers — 534
  - 10.1.4 Classification of multi-objective techniques — 537

	10.1.5 Water resources management applications	541
10.2	The weighting method	544
10.3	The Compromise programming method	549
	10.3.1 Compromise programming	550
	10.3.2 Some practical recommendations	557
	10.3.3 The Compro computer program	558
10.4	Fuzzy multi-objective analysis	558
	10.4.1 Fuzzy Compromise programming	559
	10.4.2 Properties of fuzzy distance metrics	561
	10.4.3 Comparing fuzzy distance metrics	566
	10.4.4 An example of a fuzzy Compromise programming application	568
	10.4.5 The FuzzyCompro computer program	572
10.5	Fuzzy multi-participant multi-objective decision-making	573
	10.5.1 Fuzzy Compromise programming for multi-participant decision-making	573
	10.5.2 An example of fuzzy Compromise programming for a group decision-making application	577
	10.5.3 The FuzzyComproGDM computer program	580
10.6	Examples of water resources multi-objective analyses	581
	10.6.1 Multi-objective analysis for water resources master planning	582
	10.6.2 Participatory planning for sustainable floodplain management	593
10.7	References	608
10.8	Exercises	610

Part V Water for Our Children — 615

11 The Future of Water Resources Systems Management — 617

11.1	Emerging issues	617
	11.1.1 Climate variability and change	618
	11.1.2 Water as a social and economic good	619
	11.1.3 Urbanization	620
	11.1.4 Participatory water resources management	622
11.2	Emerging sciences	622
	11.2.1 Nanowater	623
	11.2.2 Quantum computing	624
	11.2.3 Closing comments	625
11.3	References	626

List of Figures, Tables and Boxes

FIGURES

1.1	An old barrage on the Damietta branch of the River Nile	6
1.2	Main drainage canal in the Sihu basin	10
1.3	Main interface of the Sihu flood management decision support system	11
1.4	Red River Floodway inlet, 4 May 1997	13
1.5	LIDAR data showing the Winnipeg Floodway inlet on the Red River	16
1.6	A prototype of the Red River basin flood decision support system	17
1.7	Schematic illustration of the complexity paradigm	21
1.8	Schematic illustration of the uncertainty paradigm	25
2.1	The parastatic water clock of Ktesibios	34
2.2	Schematic presentation of the development process	41
4.1	Schematic presentation of a system definition	67
4.2	Looking for problem solution (high leverage)	72
4.3	Potential solutions for a system consisting of two types of elements	86
4.4	Exponential growth and decay	87
4.5	Logistic curve	88
4.6	Schematic presentation of (a) an open and (b) a closed system	93
4.7	A feedback loop	96
4.8	Layout of the waste facilities	103
4.9	Irrigation canal intake	106
4.10	Flow diagram of the irrigation system (a) without the float and (b) with the float	107
4.11	A causal influence diagram of the irrigation canal intake system	107
4.12	Lake eutrophication loop diagram	109

6.1	Sources of uncertainty	132
6.2	Ordinary set classification	144
6.3	Ordinary and fuzzy set representation	145
6.4	'Safe levee' fuzzy set	147
6.5	Triangular, trapezoid, Gaussian and sigmoid membership functions	149
6.6	Credibility level (α-cut) and support of a fuzzy set	149
6.7	Fuzzy intersection (a) and union (b) of two fuzzy sets	151
6.8	Fuzzy sets of the 'safe levee' $\tilde{A}$ and 'available budget' $\tilde{B}$	151
6.9	Addition of two triangular fuzzy numbers (Example 5)	154
6.10	Subtraction of two fuzzy numbers (Example 6)	156
6.11	Multiplication of two fuzzy numbers (Example 7)	157
6.12	Division of two fuzzy numbers (Example 8)	159
6.13	Graphical illustration of Chen's method for $r = 1$	166
6.14	Composition under pseudomeasure for $\eta(B_1) = 0.2$ and $\eta(B_2) = 0.4$	177
6.15	Polynomial composition under pseudomeasure for $\eta(B_1) = 0.2$ and $\eta(B_2) = 0.4$	177
6.16	Concept hierarchy of flood control	181
6.17	Comparison of membership functions obtained using different aggregation methods	195
6.18	Variable system performance	197
6.19	Fuzzy representation of the acceptable failure region	199
6.20	Compliance between the system state and the acceptable level of performance	201
6.21	Overlap analysis	202
6.22	Compatibility of the system state with different levels of the performance membership function	203
6.23	Recovery times for different types of failure	204
6.24	Serial and parallel system configurations: (a) a serial system configuration of N components; (b) a parallel system configuration of M components	206
6.25	Schematic representation of the hypothetical case studies	208
6.26	System states for cases I and III with the predefined acceptable levels of performance	211
6.27	The City of London, Ontario, regional water supply system	213
6.28	Schematic representation of the LHPWSS treatment process	216
6.29	Schematic representation of the EAPWSS treatment process	218
6.30	Typical water supply system layout	221
6.31	System integrated layout for the fuzzy reliability analysis	221
6.32	The LHPWSS system-state membership function change for different system components	225

7.1	The Assiniboine delta aquifer, Manitoba, Canada	258
8.1	(a) Positive feedback loop; (b) System behaviour	301
8.2	(a) Negative feedback loop; (b) System behaviour	302
8.3	Irrigation water intake	303
8.4	Constant flow rate	303
8.5	Quantity time graph	304
8.6	Quantity time graph	305
8.7	Flow rate water volume graph	305
8.8	Water volume and flow rate graph	307
8.9	General patterns of system behaviour	308
8.10	Casual loop diagram for 'filling a storage tank with water'	310
8.11	A causal diagram of the irrigation canal intake system	313
8.12	Lake eutrophication causal loop diagram	314
8.13	Stock and flow diagramming notation	315
8.14	Example flow diagrams	319
8.15	City growth	320
8.16	Dreamdisintegration arms race	321
8.17	John's performance in class	321
8.18	Construction site noise pollution flow diagram	322
8.19	Thermostat temperature control flow diagram	322
8.20	Solid waste generation flow diagram	323
8.21	Hydraulic metaphor for stock and flow diagram	324
8.22	Groundwater aquifer production causal diagram	327
8.23	Groundwater aquifer production flow diagram	327
8.24	Vensim equations for the groundwater production model	329
8.25	Vensim output time histories for groundwater production	330
8.26	Simple bathtub model flow diagram	332
8.27	Modified bathtub model structure	333
8.28	Complete bathtub model structure	333
8.29	Vensim screen shot of the *Drain function*	334
8.30	Vensim equations of bathtub model	335
8.31	Change of water volume in the tub over time	336
8.32	Results of bathtub simulation	336
8.33	A simple flow diagram of the Red Reservoir problem	338
8.34	A weighted-average reservoir decision model	339
8.35	Vensim equations for the Red Reservoir weighted-average operating rule	343
8.36	Dynamics of the Red Reservoir system with the weighted-average operating rule	344
8.37	Multipliers	344

8.38	A multiplicative rule reservoir decision model	345
8.39	Vensim equations for the Red Reservoir multiplicative operating rule	346
8.40	Dynamics of the Red Reservoir system with the multiplicative operating rule	347
8.41	Typical example of a fuzzy regression model $\tilde{Y} = \tilde{C}x$	349
8.42	Inferred fuzzy output $\tilde{Y}_i = \mathbf{A}^*\mathbf{x}_i$ (×: observed value)	351
8.43	Schematic diagram of the Shellmouth Reservoir	354
8.44	Shellmouth Reservoir area	355
8.45	System dynamics model development life cycle	358
8.46	Main control screen of the Shellmouth Reservoir model	359
8.47	Water levels in the Shellmouth Reservoir for flood years 1974, 1975, 1976 and 1979	361
8.48	Existing and revised rule curves with natural spill and gated spill	363
8.49	Comparative graph of water levels in the Shellmouth Reservoir (flood year 1995) with natural spill and gated spill	364
8.50	Reservoir levels by varying the initial reservoir level at the start of the simulation	365
8.51	Reservoir levels by varying the initial reservoir level at day 90	365
8.52	Reservoir levels by varying discharges through the conduit	366
8.53	Water scenarios: projected and actual global water withdrawals	370
8.54	WorldWater model causal diagram	374
8.55	State of the World: 'Standard run' results of World3 (a) and WorldWater (b) models	381
8.56	State of the World: 'Double run' results of World3 (a) and WorldWater (b) models	382
8.57	State of the World: 'Stable run' results of World3 (a) and WorldWater (b) models	383
8.58	Use of water: 'Standard run' results of WorldWater	385
8.59	Conceptual framework of a behavioural model for evacuation planning	391
8.60	Causal diagram of a behavioural model for evacuation planning	392
8.61	A graphical relationship between *Flooding_factor* and *Upstream_community_flooded*	396
8.62	Evacuation model interface	399
8.63	A schematic representation of the vertical water balance	401
8.64	Basic dynamic hypothesis of hydrological dynamics in a watershed	404
8.65	Study area and locations	407
8.66	Simulated and measured streamflow in the Assiniboine River basin for 1995 (calibration)	409
8.67	Simulated and measured streamflow in the Assiniboine River basin for 1979 (verification)	409

8.68	Simulated and measured streamflow in the Red River basin for 1996 (calibration): (a) simulated and measured streamflow at Grand Forks; (b) simulated and measured streamflow at Emerson; (c) simulated and measured streamflow at Ste Agathe	410
8.69	Simulated and measured streamflow in the Red River basin for 1997 (verification): (a) simulated and measured streamflow at Grand Forks; (b) simulated and measured streamflow at Emerson; (c) simulated and measured streamflow at Ste Agathe	412
8.70	A schematic diagram of the hypothetical water resource system	415
8.71	A causal loop diagram representing the water sharing conflict between the two users	416
8.72	A system dynamic model of the water sharing conflict	418
8.73	Variation of dissatisfactions and water allocations with time	420
9.1	Solution space	436
9.2	Alternative optimal solutions	437
9.3	Unbounded solution	438
9.4	Infeasible solution	438
9.5	A fuzzy decision	457
9.6	A linear membership function	460
9.7	Schematic presentation of the evolutionary algorithm	466
9.8	Roulette-wheel selection	471
9.9	Stochastic universal sampling	471
9.10	Possible positions of the offspring after discrete recombination	473
9.11	Loss function representing deviations in the penalty zones and coefficients	493
9.12	Piecewise linearized loss function for the release	494
9.13	Membership functions for the deviations and the objective function	497
9.14	Release and storage variations due to the reduction in the release zone	501
9.15	Release and storage values for the reduction in the release and storage zones	502
9.16	Storage variations based on the reduction in the release zone	503
9.17	H–Q–η curves	506
9.18	Power vs. head for fixed flow rate	507
9.19	Schematic of a serial pipeline	507
9.20	Graph representation of pipeline operation	508
9.21	Head vs. cost function for serial pumps	516
10.1	The Danube River – the place where everything started	528
10.2	Feasible region of a multi-objective problem presented in the objective space	533
10.3	Classification of feasible multi-objective alternative solutions	535

10.4	The feasible region and the non-dominated set in the decision space	548
10.5	The feasible region and the non-dominated set in the objective space	549
10.6	Illustration of compromise solutions	550
10.7	The decision space of the reservoir allocation problem	555
10.8	Typical fuzzy input shapes	561
10.9	Fuzzy criterion values and weights	562
10.10	Range of valid criterion values as defined by fuzzy positive and negative ideals	563
10.11	Fuzzy distance metric exponent	564
10.12	Fuzzy distance metrics for different fuzzy definitions of $\tilde{p}$	564
10.13	Fuzzy distance metrics for different multimodal criterion values and weights	565
10.14	Fuzzy distance metric comparison for two alternatives (A and B)	567
10.15	Fuzzy subjective criteria interpretation for the Tisza River problem	570
10.16	Fuzzy input for the Tisza River problem	570
10.17	Distance metrics for the Tisza River problem	571
10.18	Participant 1 fuzzy distance metrics	579
10.19	Participant 2 fuzzy distance metrics	579
10.20	Participant 3 fuzzy distance metrics	579
10.21	Participant 4 fuzzy distance metrics	579
10.22	A geometric interpretation of the FEV	599
10.23	The distance metric fuzzy membership functions	607

TABLES

1.1	Regional distribution of renewable water availability and population	4
2.1	Timeline of water resources engineering activities	33
4.1	Cost, supply and quality of available water resources	98
4.2	Available data for an illustrative example	100
6.1	Sources of uncertainty in the water management process structure	134
6.2	Examples of loads and resistance	138
6.3	Interpretation of fuzziness for various problems in water resources	169
6.4	Hierarchical approach for construction of a fuzzy set	170
6.5	Example of the query form given to experts for evaluation – bottom level	182
6.6	Example of the query form given to experts for evaluation – top level	183
6.7	The results of experts' evaluations for the first set of sample data	184
6.8	The results of Expert 1's evaluation for the second set of sample data	188
6.9	Results of the application of the OWA operator	193

6.10	Results of the application of P-CUP aggregation	194
6.11	Summary of test cases	210
6.12	Summary results of suggested reliability and robustness measures	212
6.13	Summary list of LHPWSS and EAPWSS system components	215
6.14	The LHPWSS and EAPWSS systems' fuzzy performance measures for different membership function shapes	224
6.15	Change in system fuzzy performance measures due to an improvement in the PAC transfer pump capacity	226
6.16	Change in system fuzzy performance measures due to a change in the supply capacity of the PAC transfer pump	226
7.1	Payoff table	243
7.2	Distance-based fairness measures	245
7.3	Rural municipality areas included in the ADA region	260
7.4	Data requirements and sources for application of sustainability criteria in the ADA region	261
7.5	Population of ADA region by RM and township	262
7.6	Irrigated crop areas within the ADA region	263
7.7	Livestock populations by RM	263
7.8	Projected agricultural and population statistics under three policy scenarios	266
7.9	Projected water budgets under three policy scenarios	266
7.10	Number of users in each year in two sub-basins under different policy scenarios	268
7.11	Intra- and inter-generational fairness calculations for access to irrigation licensing for two sub-basins of the ADA	269
7.12	Intra-generational equity calculations for two sub-basins of the ADA under different policy scenarios	270
7.13	Possible risks associated with the three policy scenarios	271
7.14	Average risk preferences for three stakeholder groups in the ADA	275
7.15	Average sustainable development categories weights as indicated by three stakeholder groups in the ADA	276
7.16	Scaled risk preferences for three stakeholder groups in the ADA	276
7.17	Qualitative scale and quantitative equivalent used for estimating risk probabilities in the ADA	276
7.18	Summary of risk probability estimates for three policy scenarios in the ADA	277
7.19	Risk estimates for status quo scenario in the ADA	277
7.20	Risk estimates for development scenario in the ADA	278
7.21	Risk estimates for the conservation scenario in the ADA	278
7.22	Average risk preferences for three stakeholder groups in the ADA	279

7.23	Breakdown of positive and negative risk estimates	279
7.24	Reversibility criteria: possible impacts	280
7.25	Reversibility impacts and weights for three policy scenarios in the ADA	284
7.26	Reversibility R-metric results	284
7.27	Results of R-metric weight sensitivity analysis for the ADA case study	285
7.28	Results of the impact sensitivity analysis for the ADA case study	286
7.29	ADA groundwater management problem	287
7.30	Ranking of alternative scenarios	287
7.31	Degree of consensus measures for ADA case study	288
8.1	Flow rate and water volume calculation	307
8.2	Input/output data for prefabricated water tanks	350
8.3	Possibility $\mu^*_{Y_i}(y_i)$, lower bounds, centres, upper bounds of 0-level sets for inferred fuzzy output Y_i^*	352
8.4	Flood management with revised operating rules for selected flood years	362
8.5	Impacts on flooding by changing reservoir levels (a) start of year; (b) start of flooding season for 1976 flood year without using gated spillway	362
8.6	Impacts on flooding by changing flow through the conduit	363
8.7	Assessment of water use in the world (in km^3/year) by sector of economic activity	368
8.8	Comparison of standard projections for 2025 with the results of WorldWater (km^3/year)	386
9.1	Simplex tableau for the Example 5 problem	442
9.2	Relationship between primal and dual problems	448
9.3	Water supply problem solutions	464
9.4	Example data	470
9.5	Flow elevation data	480
9.6	EMSLP results for different ending storage values	489
9.7	EMSLP results for different system loads	490
9.8	EMSLP results for varying release limits	490
9.9	Storage and release zones and corresponding penalties for the winter season	495
9.10	Calculation of pipeline pressure losses	517
9.11	Cost function parameters	517
9.12	Solutions for each generation using the proposed evolutionary optimization	518
10.1	Payoff matrix	539
10.2	Available data for an illustrative example	546
10.3	Pairs of weights and associated non-dominated solutions	549

10.4	Non-dominated solutions	555
10.5	Scaled non-dominated solutions	556
10.6	Final compromise solution	556
10.7	Original values used in David and Duckstein (1976)	569
10.8	Tisza River alternative rankings	571
10.9	Preferences of participants	578
10.10	Participants' ranking values, using the Chang and Lee method with $\chi_I = 0.5$	580
10.11	Individual preference relation matrices	581
10.12	Activities, modelling purpose and suggested modelling techniques at the first step of the planning process	586
10.13	Activities, modelling purpose and suggested modelling techniques at the second step of the planning process	587
10.14	Activities, modelling purpose and suggested modelling techniques at the fourth step of the planning process	589
10.15	Input payoff matrix for system S1	591
10.16	Input payoff matrix for system S2	592
10.17	Alternative weight sets	592
10.18	Ranking results for system S1	592
10.19	Ranking results for system S2	593
10.20	Flood management payoff (decision) matrix	602
10.21	Three response types as they appear in the survey	605
10.22	Resultant FEVs	606
10.23	Final rank of flood management alternatives	606

BOXES

0.1	From the banks of the Danube...	xxiv
2.1	Katmandu Internet	39
3.1	The river doctor	51
3.2	Did you know?	55
4.1	The Three Gorges Dam	69
5.1	River life	119
6.1	Water quality	136
6.2	Flooding	179
7.1	Water and religion	234
7.2	How much water does it take?	241
8.1	Water use	299
8.2	Water in Canada	369

9.1	The largest power plant in the world	431
9.2	Fish catcher	465
10.1	Fishing on the Yangtze River	557
10.2	Emerson flooding	594
10.3	Super dike	599
11.1	To the Medway Creek	625

About the Author

> **Box 0.1** From the banks of the Danube . . .
>
> The surface of the Danube was visible through our balcony window. From the one room, all that we had, my mind would quite often wander to the banks of the big river.
>
> There, my Dad and I were sitting in the soft sand with our fishing tackle in the water. We were watching together the muddy waters moving slowly, tired from the long run through seven countries. They had two more to go before a rest in the Black Sea. The other bank was hardly visible. Passing boats would leave small waves and shake the pictures of gothic cathedrals, ancient forts, opera houses, beautiful bridges and so on that were floating on the surface.
>
> The conversation was usually sparse but the bond was very strong. Our presence was real. Real as the river. Eternal as the river. A fish would bite and my Dad would take it out, sometimes put it back and sometimes take it home.
>
> (A memory from 1977)

Slobodan P. Simonovic was born and raised in Belgrade, Yugoslavia. He obtained his undergraduate degree in civil engineering (water resources division) from the University of Belgrade in 1974. By joining the interdisciplinary Master's degree programme at the University of Belgrade he was able to direct his education further into the application of formal systems theory to water resources systems management. His MSc included training in formal systems theory from the Department of Electrical Engineering and in water resources engineering from the Department of Civil Engineering. He graduated in 1976.

From 1974 until 1978 he worked as a researcher for the Jaroslav Cerni Institute for Water Resources Development in Belgrade. In 1978 he continued his graduate education at the University of California in Davis, where he obtained his PhD in engineering in 1981. Until 1986 he worked as consulting engineer for the large international consulting company Energoproject in Belgrade. In 1986 he moved to Canada and joined the Department of Civil Engineering at the University of Manitoba in Winnipeg. He was the professor in the department until 1996, when he became the Director of the Natural Resources Institute, an interdisciplinary graduate programme in natural resources management at the University of Manitoba.

In 2000 Dr Simonovic moved to London, Ontario where he became professor at the Department of Civil and Environmental Engineering and an engineering research chair of the Institute for Catastrophic Loss Reduction at the University of Western Ontario.

Dr Simonovic teaches courses in civil engineering and water resources systems. He still actively works for national and international professional organizations (the Canadian Society of Civil Engineers, Canadian Water Resources Association, International Association of Hydrological Sciences and International Hydrologic Program of United Nations Educational, Scientific and Cultural Organization (UNESCO)). He has received a number of awards for excellence in teaching, research and outreach, and has been invited to present special courses for practising water resources engineers in many countries. He currently serves as associate editor of two water resources journals, and participates actively in the organization of national and international meetings. He has over 300 professional publications.

Dr Simonovic's primary research interest focuses on the application of a systems approach to, and the development of decision support tools for, the management of complex water and environmental systems. Most of his work is related to the integration of risk, reliability, uncertainty, simulation and optimization in hydrology and water resources management. He has undertaken applied research projects that integrate mathematical modelling, database management, geographical information systems and intelligent interface development into decision support tools for water resources decision-makers. Most of his research is conducted through the Facility for Intelligent Decision Support (FIDS) at the University of Western Ontario.

His subject expertise focuses on systems modelling, risk and reliability, water resources and environmental systems analysis, computer-based decision support systems development, and water resources education and training. Particular topical areas of expertise are reservoirs, flood control, hydropower energy, operational hydrology and climate change.

Preface

The two editions of the United Nations World Water Development Report[1] have clearly shown that the world is facing a serious and increasingly complex water crisis that can be overcome only by a strong improvement in water governance. Such a critical state of affairs has a multidimensional nature, which encompasses lack of access to safe drinking water and basic sanitation, malnutrition, water-related disasters, illnesses and environmental degradation. Its impacts are heterogeneously distributed around the globe and across societies, being inextricably linked with poverty and human development. The solution relies heavily upon the improvement of the way water is governed, including, but not limited to, holistic approaches that take into consideration different stakeholders, groundwater and the transboundary nature of many freshwater resources.

As Professor Slobodan Simonovic stresses in this book, the management of water resources systems is based on human interventions and highlights the need for our collective and individual action. During the last decades, through successive phases of the International Hydrological Programme (IHP), UNESCO has contributed to placing water in a clear and central position on the international environmental agenda, providing needed policy-relevant science. In addition, IHP-led initiatives have provided a framework for bringing the different audiences of the water constituency together, including scientists, managers and policy-makers, to address locally defined water challenges. These efforts are at the core of the current 7th phase (2008–2013) of IHP, which intends to address water dependencies, with an emphasis on systems under stress and societal responses.

The growing complexity and uncertainty currently faced by many water managers are to a great extent due to global changes, from population changes through land use change, migration, urbanization all the way up to climate variability and change. The adaptation to the impacts of global changes on river basins and aquifer systems is therefore an issue of prime concern to IHP and constitutes one of its main working themes. The big issue here is uncertainty and risk both in the nature of global drivers and in their impacts. In this book, Professor Simonovic presents a comprehensive fuzzy set approach to deal with uncertainty and risk.

By co-publishing this valuable volume, UNESCO intends to improve the information resources available on system methods and on the existing tools for a better management of water resources. We sincerely hope that practitioners and students find this book particularly useful.

Last but not least I would like to express our sincere thanks to Professor Simonovic for all the efforts he has put into this volume throughout the past years as well as his constant support of and contributions to IHP.

András Szöllösi-Nagy
Deputy Assistant Director-General for Natural Sciences
Secretary of the International Hydrological Programme

NOTE

1 UN-WWAP (United Nations World Water Assessment Programme) (2003), *UN World Water Development Report 1: Water for People, Water for Life*, UNESCO and Berghahn Books, Paris, New York and Oxford. UN-WWAP (2006), *UN World Water Development Report 2: Water, a Shared Responsibility*, UNESCO and Berghahn Books, Paris, New York and Oxford.

Foreword

The planning and management of water resources systems is an activity that is becoming increasingly important in almost all regions of this planet. It becomes vitally important in those regions where the water available is insufficient to meet the region's human, agricultural, industrial and environmental and ecological needs. We all are told that an average one out of six persons in the world does not have access to safe drinking water. Half of us living today do not have adequate sanitary facilities. These are average values. What this means is that some regions have a much higher proportion of their population in stress. There are countries where over half of the citizens do not have safe water to drink, and are sick as a result. All of us, and especially water resources planners and managers, are challenged to reduce this statistic. It is going to be a sizeable and difficult job. Planning is a first step. But even if we are practising our profession in more developed regions where other issues may carry greater importance, planning is needed to determine how best to address existing water management issues without creating further problems. How can we most effectively meet the demands for water of sufficient quantities and qualities at the times needed, both for humans and the environment, and at reasonable costs? How can we identify the management and operating policies that best meet these needs? And finally, what will the economic, ecological and social impacts be of such policies, and will they be acceptable?

    Modelling is a way of estimating those economic, ecological and social impacts prior to making and implementing what are often expensive decisions. Some types of models are designed to help us identify what decisions we should make to best satisfy a specified set of objectives or goals. This book describes some of the mathematical computer modelling methods that have been developed and used to aid in formulating plans and defining operating policies. The book title is appropriate: *Managing Water Resources: Methods and Tools for a Systems Approach*. These modelling methods and tools aid decision-making; they are not substitutes for it.

They inform and enhance our judgement; they do not replace it. Models are full of assumptions and uncertainties. Part of any thorough modelling endeavour is the exploration of the impacts of such assumptions and uncertainties.

Professor Slobodan Simonovic has given us a comprehensive description of many of the commonly used models for water resources planning and management. In addition he has given us the benefit of some of his considerable experience in using models to address water resources management issues in different regions of the world. Students learning this trade often focus on the mathematical aspects (which are important and, indeed, most of the book is devoted to this), only to appreciate later how much of an art this discipline is. Considerable judgement is required to successfully develop, implement and complete a modelling exercise that must fit within the political environment in which decisions are made. Planning objectives change, data change, people change, and even political administrations can change during the planning process. Water resources systems analysts and planners who do not continually interact closely with their clients will be disappointed in the interest their clients may have in their results. Modelling success in practice can be judged on the extent to which the results enter and influence the political debate, in other words, what decisions to make.

One feature that distinguishes this book from the many addressing this subject is its focus on the use of fuzzy modelling. Readers will appreciate the detail contained in many of the book's chapters showing how less than precise and qualitative performance measures can be described by and modelled using fuzzy membership functions. To me it seems appropriate that Professor Simonovic has chosen to emphasize the use of fuzzy sets in modelling. Early in my career I noticed how reluctantly those in the west seemed to accept the concepts of fuzzy modelling, whereas many in the east were actively developing and successfully using this science. Professor Simonovic is originally from the east, namely Belgrade, Yugoslavia.

It was in Belgrade where, some 30 years ago, I first met 'Simon' and his young family. Little did either of us know where our careers would take us. Nevertheless I'm sure he will agree that it has been an adventurous, rewarding and fun trip. I'm honoured to have been asked by Slobodan Simonovic to write this foreword to a book I consider well worth reading and having as a text and reference. I can attest that writing a book as comprehensive as this one takes time, and the rate of return is not measurable in economic terms. We can only hope our efforts will be considered of value to many who may use books like this one to aid them in this journey we are all taking towards improving the management and use of our valued water resources worldwide.

Daniel P. Loucks
Cornell University
Ithaca, NY 14853 USA
May 2007

Introduction

I am one of the lucky few who have the opportunity to work all their professional life in a field that they enjoy. My work has brought me into contact with many great people, responsible technicians, talented engineers, capable managers and dedicated politicians. In my capacity as an academic I have also had an opportunity to work with the abundant young talent that continues to feed the water resources management workforce. I learned a lot from all these people. I learned many things about the profession, I learned a lot about different cultures, and most importantly I learned about life. Thank you.

My work has taken me all over the world. I have had an opportunity to see the water problems in the developed and developing world, in small villages and large urban centres. Projects I have been involved with range in scale from the local to the international. I have discussed the issues with farmers of the Sihu area in China as well as with the Minister for Irrigation and Water Resources of Egypt. I hope that my professional expertise continues to contribute to the solution of some of these problems. It definitely inspires me to continue to work, with greater effort and more dedication.

For nearly 30 years of personal research, consulting, teaching, involvement in policy, implementation of projects and presentation of experiences through the pages of many professional journals, I have worked hard to raise awareness of many issues relevant to the development, management and use of freshwater resources. I have accumulated tremendous experience over the years. Writing this book offered me a moment of reflection, and it elaborates on lessons learned from the past to develop ideas for the future.

During the past four decades we have witnessed a tremendous evolution in water resources systems management. Three of the characteristics of this evolution should be noted in particular. First, the application of the systems approach to complex water management problems has been established as one of the most important advances in the field of water resources management. The systems approach draws on the fields of operations research and economics to create skills in engineering problem solving. The field of operations research evolved from its origins during the Second World War, and the area known as mathematical programming found wide application as a means to simulate and optimize complex design and operational problems in water and environmental engineering. A primary emphasis of systems analysis in water resources management is on providing an improved basis for decision-making. A large number of analytical, computer-based tools, from simulation and optimization to multi-objective analysis, are available for formulating, analysing and solving water resources planning, design and operational problems.

Second, the past four decades have brought a remarkable transformation of attitude by the water resources management community towards environmental concerns, and of action to deal with them. Water, together with land and air, is under significant pressure from a growing population, and the associated needs for food production and rapid urbanization. The civil engineering field has evolved into civil and environmental engineering through an increasing emphasis on water and air quality management, solid and hazardous waste management, environmental planning for electric utilities and the siting of environmental facilities, among other important environmental issues. There are many examples of initiatives taken for environmental assessment and planning, as well as considerable investment in environmental technologies and new processes designed to recover or eliminate pollutants.

Water resources management has faced an uphill battle with the regulatory approaches that are used in many countries around the world. They have not been conducive to the systems approach that is inherent in simulation and optimization management models. Fortunately, recent trends in regulation include consideration of the entire river basin system, explicit consideration of all costs and benefits, elaboration of a large number of alternatives that reduce the amount of pollution generated, and the greater participation of all stakeholders in decision-making. Systems approaches based on simulation, optimization and multi-objective analyses have great potential for providing appropriate support for effective management in this emerging context.

In 1987, with the publication of the Brundtland Commission's report *Our Common Future*, decision-making in many fields began to be influenced by a sustainability paradigm. It can safely be assumed that sustainability is now the major unifying concept promoted, accepted and discussed by governments throughout most of the world. The original report introduced the concept of sustainable development as 'the ability to meet the needs of the present without compromising the needs of future generations'. This concept was the third main evolutionary step to affect water resources systems management in the last four decades.

Applying the principles of sustainability to water resources decision-making requires major changes in the objectives on which decisions are based, and an understanding of the complicated inter-relationships between existing ecological, economic and social factors. The broadest objectives for achieving sustainability are environmental integrity, economic efficiency and equity. In addition, sustainable decision-making regarding water resources faces the challenge of time: that is, it must identify and account for long-term consequences. We are failing to meet the basic water needs of more than 1 billion people today, and therefore are not at a base level in terms of dealing with the needs of future generations.

To make decisions designed to produce sustainable water resources also calls for a change in procedural policies and implementation. If the choice is to select projects with this outcome, it will require major changes in both substantive and procedural policies. Sustainability is an integrating process. It encompasses technology, ecology and the social and political infrastructure of society. It is not a state that may ever be reached completely. It is, however, one for which water resources planners and decision-makers strive.

The evolution of water resources systems management is occurring in the context of rapid technological change. In the same period that brought us the water resources systems approach, environmental awareness and sustainability, we were exposed to the dynamic development of computer hardware and software systems. The power of the large mainframe computers of the early 1970s is now exceeded many times over by the average laptop computer. The computer has moved out of data processing, through the user's office and into knowledge processing. Whether it takes the form of a laptop personal computer or a desktop multi-processing work station is not important. The important point is that the computer acts as a partner for more effective decision-making.

Systems can be defined as a collection of various structural and non-structural elements that are connected and organized in such a way as to achieve some specific objective through the control and distribution of material resources, energy and information. The systems approach is a paradigm concerned with systems and inter-relationships among their components. Today, more than ever, we face the need for appropriate tools that can assist in dealing with difficulties introduced by the increase in the complexity of water resources problems, consideration of environmental impacts and the introduction of the principles of sustainability. The systems approach is one such tool. It uses rigorous methods to help determine the preferred plans and designs for complex, often large-scale, systems. It combines knowledge of the available analytic tools, an understanding of when each is appropriate, and a skill in applying them to practical problems. It is both mathematical and intuitive, as is all water resources planning, design and operation.

Water resources systems management practice is changing. There is a clear need to redefine the education of water resource engineers and increase their abilities to:

- work in an interdisciplinary environment;
- develop a new framework for the design, planning and management of water infrastructure that will take into consideration current complex socio-economic conditions;
- provide the context for water management in conditions of uncertainty.

The main objectives of this book are to introduce the systems approach as the theoretical background for modern water resources management, and to focus on three main sets of tools: simulation, optimization and multi-objective analysis. At the same time this book will allow me to reflect on the past 30 years of practising and teaching water resources systems management. The process of reflection unlocks theory from practice, brings to the surface insights gained from experience, and offers a framework for uncovering many hidden aspects of applying a theoretical approach in the search for a solution for practical problems. Insights gained from reflection can then be used to elaborate and present a theoretical approach in a different way, which I hope will prove more understandable to the students of the discipline and more acceptable to the practising professionals. Therefore, my sincere hope is that this book will be able to serve multiple communities: as a text for teaching water resources systems analysis, and as a guide for the application of a systems approach to water resources management.

The text presented in the book is supported by a number of computer programs that can be used in applying the theory presented here to the solution of real-world water problems.

THE ORGANIZATION OF THE BOOK

The text is organized into five parts and 11 chapters. Part I provides an introductory discussion and sets the scene. In Chapter 1 there is a brief overview of my personal experience, which provided my motivation for writing this book. I then look at how the water engineering profession is changing in Chapter 2. Chapter 3 defines the main terms used in water resources systems management, and looks at its links with human nature.

Part II is devoted to general systems theory, mathematical formalization and classification methods. The material presented in this section should be of practical relevance during the process of selecting an appropriate tool for the solution of a problem. There is only one chapter in this section, Chapter 4, which defines systems terms and looks at how they are applied in water engineering processes.

Part III looks at engineering activities from the broader societal point of view. Chapter 5 formally defines water resources systems management, and introduces the three main approaches that the rest of the book elaborates: simulation, optimization and multi-objective analysis. Chapter 6 introduces the issue of uncertainty, and looks at the main theoretical concepts behind probabilistic and fuzzy set theoretical approaches. Until recently the probabilistic approach was the only approach for water resources systems reliability analyses. However, it fails to address the problem of uncertainty, which is inherent to the field because of the human input, subjectivity, lack of history and records. There is a real need to convert to new approaches that can compensate for the ambiguity or uncertainty of human perception.

One unique aspect of this book is that it considers in detail how fuzzy set theory can be used to address various uncertainties in water resources systems management. Many other books deal with the application of the probabilistic approach, but to my knowledge this is the first time that an attempt has been made to provide water resources specialists with detailed insight into fuzzy set theoretical approaches.

Once Chapter 6 has introduced the basic concepts of fuzzy sets and fuzzy arithmetic, Chapter 7 gives a detailed presentation of how water resources systems management can be used for sustainable development. There is a discussion of sustainable development and its implementation in water resources engineering, which takes a highly pragmatic view of what can be a rather idealistic field. It identifies and introduces four sustainability criteria that may be used in practice. The example of the management of a groundwater aquifer is used to illustrate the implementation of these criteria.

Part IV is technical in nature. Chapter 8 concerns the simulation approach. It provides a detailed description of system dynamics simulation, then goes on to look at five areas of application in detail: multi-purpose reservoir simulation, global water resources assessment modelling, flood evacuation simulation, a hydrological simulation for the prediction of floods from snowmelt, and simulation of water-sharing conflicts. The system simulation approach is presented in both deterministic and fuzzy contexts.

Optimization is addressed in Chapter 9, with a focus on two techniques: linear programming (LP) and evolutionary optimization in deterministic and fuzzy setups. Three areas of application are discussed as examples of the optimization approach to management of water resources systems: hydropower optimization, multi-purpose reservoir system analysis and optimization of pumping costs in water distribution systems.

Chapter 10 focuses on multi-objective analysis. A very practical approach is taken to the material in this chapter. Because it approaches multi-objective analysis from an application point of view, it deals with a number of important issues in addition to the selection of an appropriate technique. Deterministic and fuzzy multi-objective analysis techniques are presented for single and group decision-making. The methods described in this chapter are illustrated through the examples of water resources master planning and sustainable floodplain management.

The book ends with the presentation of my vision for the future of water resources engineering. In Part V, Chapter 11 presents this view. This section also provides additional references for readers with a deeper interest in some of the concepts discussed.

SOFTWARE CD-ROM

The application of methodologies introduced in the book is supported through a set of computer programs contained on the accompanying CD-ROM. The state-of-the-art simulation software Vensim PLE (Personal Learning Edition) is enclosed for the implementation of system dynamics simulation. This program was developed by Ventana Systems, which has kindly given permission for its use in this context.

The CD-ROM includes seven more original computer programs developed in the user-friendly Windows™ environment, for the illustration and implementation of the methods outlined in this book. They are:

- LINPRO, an LP optimization tool;
- FUZZYLINPRO, a program for the implementation of fuzzy LP optimization;
- EVOLPRO, for the implementation of evolutionary optimization;
- COMPRO, for the implementation of the deterministic multi-objective analysis tool of Compromise programming;
- FUZZYCOMPRO, which implements fuzzy Compromise programming, for multi-objective analysis under uncertainty;
- FUZZYCOMPROGDM, for the support of group decision-making using fuzzy Compromise programming under uncertainty;
- SUSTAINPRO, a package of four programs for the computation of fairness, risk, reversibility and consensus sustainability criteria.

Each program is presented in the same way on the CD-ROM, with:

- a read.me file with installation instructions;
- a folder containing the main program files;
- a folder containing all the examples discussed in the text.

Vensim PLE is accompanied by a short tutorial developed by Craig W. Kirkwood of Arizona State University. I am grateful to the author for permission to provide it here. The other seven programs have very extensive help manuals, which are integrated into the Windows environment. These provide detailed instructions on program use, data preparation, data import and interpretation of the results. This software component of the book is not intended as a commercial product. It has been developed to illustrate the application of the methodological approaches presented in the book, and to allow the solution of real water resources systems management problems. However, the responsibility for its appropriate use is in the hands of the user.

USE OF THE BOOK

This text and the accompanying CD-ROM have four main purposes:

1. They provide material for an undergraduate course in water resources systems management. A course might be based on Chapters 1 through 4, and possibly the more deterministic parts of Chapters 8, 9 and 10. Chapter 7 might also be used, depending on the background of the participants in the class.
2. They also provide support for a graduate course in water resources systems management, with an emphasis on water resources systems management under uncertainty. Such a course might draw on Chapters 1 through 6, and possibly the fuzzy theory in Chapters 8, 9 and 10. Both undergraduate and graduate courses could use the computer programs provided on the CD-ROM.
3. Water resources practitioners should find the focus on the application of the methodologies presented to be particularly helpful, and could use the programs for the solution of real water resources systems management problems. There is discussion of a large number of specific applications in Chapters 6, 7, 8, 9 and 10 that may be of assistance.
4. Specific parts of the book can be used as a tool for specialized short courses for practitioners. For example, material from Chapter 8 and parts of Chapter 4 could support a short course on: 'System dynamics for water resources systems simulation'. A course on 'System analysis for hydropower optimization' could be based on Chapters 4, 5 and parts of Chapter 9. Similarly, material from Chapter 10 and parts of Chapter 4 could be used for a short course on 'Multi-objective analysis of water resources systems'.

My plan is to maintain an active website for the book (www.slobodansimonovic.com/waterbook.html), which will provide additional exercises for each chapter as well as suggested solutions. I will maintain an active software component of the website as a platform for:

- the improvement of the enclosed computer programs through exchange of experience;

- collecting a larger number of different applications that can be shared among the users of the book.

I and the individuals involved in publishing the book have done our best to make it error free, but it is almost inevitable that there will be some mistakes. I take responsibility for any errors of fact, judgement or science that may be contained in this book. I would be most grateful if readers would contact me to point out any mistakes or make suggestions for improving the book.

Publishing this book was made possible through the contribution of many people. I would like to start by acknowledging the publication support provided by the International Hydrologic Programme of United Nations Educational, Scientific and Cultural Organization (UNESCO), and the Division of Water Science team, including Drs András Szöllösi-Nagy, J. Alberto Tejada-Guibert, José Luis Martin-Bordes and Miguel de França Doria. Most of the knowledge contained in this book came from my numerous interactions with teachers, students and colleagues throughout the world. They taught me all I know. I would like particularly to thank the students whose work is discussed in this text. In order of appearance in the text, they are Dr Ozren Despic (Chapter 6), Dr Ibrahim El-Baroudi (Chapter 6), Dr Andrew McLaren (Chapter 7), Dr Sajjad Ahmad (Chapter 8), Dr Lanhai Li (Chapter 8), Dr Nesa Ilich (Chapter 9), Dr Ramesh Teegavarapu (Chapter 9), Mr Karl Reznicek (Chapter 9), Dr Mike Bender (Chapter 6, Section 10), Dr Pat Prodanovic (Chapter 10) and Dr Taslima Akter (Chapter 10). A special thank you goes to Dr Veerakcudy Rajasekaram, who is the developer of all the computer programs. His attention to detail, love of computer programming and analytical mind are highly appreciated.

The support of my family, Dijana, Damjan and Tanja, was of the utmost importance in the development of this book. They provide a very large part of my motivation, my goals, my energy and my spirit. Without the endless encouragement, criticism, advice and support of my wife Tanja this book would never have been completed.

Slobodan P. Simonovic
London,
March 2007

List of Acronyms and Abbreviations

ADA	Assiniboine delta aquifer
ADAAB	Assiniboine Delta Aquifer Advisory Board
CAP	Common Agricultural Policy (EU)
CUP	composition under pseudomeasures
DP	dynamic programming
EAPWSS	Elgin Area Primary Water Supply System
EMMA	energy management and maintenance analysis
EMSLP	energy management by successive linear programming
EP	evolutionary programming
ES	evolution strategies
EU	European Union
FAO	Food and Agriculture Organization of the United Nations
FEV	fuzzy expected value
FIDS	Facility for Intelligent Decision Support
GA	genetic algorithms
GDP	gross domestic product
GIS	geographic information system
GNP	gross national product
GRDC	Global Run-off Data Centre
GWP	Global Water Partnership
IHP	International Hydrological Programme of UNESCO
IJC	International Joint Commission
IPCC	Intergovernmental Panel on Climate Change
IWRME	Integrated Water Resources Model for Egypt

LHPWSS	Lake Huron Primary Water Supply System
LP	linear programming
MCDC	Manitoba Crop Diversification Centre
MEMO	Manitoba Emergency Management Organization
MINLP	mixed-integer nonlinear program
MODP	multi-objective dynamic programming
NGOs	non-governmental organizations
NWRC	National Water Research Centre of Egypt
NWSRU	Nile Water Strategic Research Unit
OERI	Overall Existence Ranking Index
OWA	ordered weighted averaging
PAC	powdered activated carbon
P-CUP	polynomial composition under pseudomeasure
PLE	Personal Learning Edition
PUC	Public Utility Commission (Walkerton, Canada)
RM	rural municipality
RNPD	River Nile Protection and Development Project
RRBDIN	Red River Basin Decision Information Network
SCADA	Supervisor Control and Data Acquisition
SLP	successive linear programming
SWT	surrogate worth trade-off
TENs-T	Trans-European Networks for Transport (EU)
TGP	Three Gorges Project
UN	United Nations
UNCED	United Nations Conference on the Environment and Development
UNESCO	United Nations Educational, Scientific and Cultural Organization
UNIDO	United Nations Industrial Development Organization
UN-WWAP	United Nations World Water Assessment Programme
WCoG	weighted centre of gravity
WMO	World Meteorological Organization
WRB	Water Resources Branch (Manitoba, Canada)
WRMS	Water Resources Master Plan for Serbia

PART I
Setting the Stage

1
An Overview

Freshwater is scarce. It is a fundamental resource, part of all social and environmental processes. Freshwater sustains life. Yet freshwater systems are imperiled, and this threatens both human well-being and the health of ecological systems.

Although water is the most widely occurring substance on Earth, only 2.53 per cent is freshwater, while the remainder is saltwater. Some two-thirds of this freshwater is locked up in glaciers and permanent snow cover. In addition to the accessible freshwater in lakes, rivers and aquifers, human-made storage in reservoirs adds a further 8000 cubic kilometres (km^3). Water resources are renewable, except for fossil groundwater. There are huge differences in availability in different parts of the world, and wide variations in seasonal and annual precipitation in many places.

Precipitation is the main source of water for all human uses and for ecosystems. Precipitation is taken up by plants and soils, evaporates into the atmosphere – in what is known as *evapotranspiration* – collects in rivers, lakes and wetlands, and runs off to the sea. The water of evapotranspiration (that is, the precipitation taken up by plants and soil) supports forests, cultivated and grazing lands, and ecosystems. Humans withdraw (for all uses including agriculture) 8 per cent of the total annual renewable freshwater, and appropriate 26 per cent of annual evapotranspiration and 54 per cent of accessible runoff. Humankind's control of runoff is now global, and forms an important part of the hydrological cycle. Per capita use is increasing and the global population is growing. Together with spatial and temporal variations in available water, this leads to the consequence that water for all human uses is scarce. On a global scale, there is a freshwater crisis (UNESCO – WWAP, 2003).

The available freshwater is distributed regionally as shown in Table 1.1.

Table 1.1 *Regional distribution of renewable water availability and population*

	North and Central America	South America	Europe	Africa	Asia	Australia and Oceania
Percentage of world's population	8	6	13	13	60	<1
Percentage of world's freshwater resources	15	26	8	11	36	5

Source: UN-WWAP (2003)

Freshwater resources are further reduced by pollution. Some 2 million tonnes of waste per day are deposited in bodies of water, including industrial wastes and chemicals, human waste and agricultural wastes (fertilizers, pesticides and pesticide residues). Although reliable data on the extent and severity of pollution are incomplete, one estimate of global wastewater production is about 1500 km^3 (UNESCO – WWAP, 2003). Assuming that 1 litre of wastewater pollutes 8 litres of freshwater, the present load of pollution may be up to 12,000 km^3 worldwide. The poor are the worst affected, with 50 per cent of the population of developing countries being exposed to polluted water sources.

The precise impact of climate change on water resources is as yet uncertain. Precipitation will probably increase above latitudes 30°N and 30°S, but many tropical and sub-tropical regions will probably receive a lower and more unevenly distributed rainfall. Since there is a visible trend towards more frequent extreme weather conditions, it is likely that floods, droughts, mudslides, typhoons and cyclones will increase. Streamflows at low-flow periods may well decrease. Water quality will undoubtedly worsen because of increased pollution loads and concentrations, and higher water temperatures (Kunzewicz et al, 2007).

Good progress has been made in understanding the nature of water's interaction with the biotic and abiotic environment. Better estimates of climate change impacts on water resources are available. Over the years, the understanding of hydrological processes has enabled humans to harvest water resources for their needs, reducing the risk of extreme situations. But pressures on the water system are increasing with population growth and economic development. Critical challenges lie ahead in coping with progressive water shortages and water pollution. According to the First World Water Development Report (UN-WWAP) (2003), by the middle of this century the number of people suffering from water shortages will be at worst 7 billion people in 60 countries, and even at best 2 billion people in 48 countries.

Recent estimates suggest that climate change will account for about 20 per cent of the increase in global water scarcity.

1.1 WATER RESOURCES MANAGEMENT ISSUES – SOME PERSONAL EXPERIENCE

The many large-scale water resources projects I have been associated with have provided a rich source of knowledge and experience. Here are three examples, with a note of some of the lessons learned.

1.1.1 An integrated water resources model for Egypt

In the past, Egyptian water policies were formulated under the premise of the continued availability of ample surface freshwater. At that time, the clear policy choice was to develop water resources to the maximum extent possible. Financial and technological constraints were seen as the only limitation to such development. Economic feasibility was the main criterion for the approval of water resources projects. This meant that the analysis process used for policy formulation had well-defined aims. However, in order to meet the increasing demand for socio-economic development, the supply of surface water from the Nile had to be augmented with marginal-quality water and high-cost groundwater.

As Egyptian society strives to achieve a higher rate of economic growth, redistribution of the congested population in the Nile delta and valley, and environmental reclamation and protection, non-traditional strategies such as reallocation of water among the different uses, desalination of seawater, use of brackish water, mining of non-renewable groundwater and pollution control have had to be considered. Demand management strategies have become one of the main features of recent water policies.

To deal with integrated water management issues in Egypt, a new research group named the Nile Water Strategic Research Unit (NWSRU) has been established within the National Water Research Centre of Egypt (NWRC), through the second phase of the River Nile Protection and Development Project (RNPD-II, 1994). The NWSRU focuses on answering critical water resources development questions at all planning levels, with particular attention being given to future needs. Thus, a major task of the NWSRU is to apply a dynamic, interdisciplinary and multi-sectoral approach to modelling Egypt's complex water resources.

Egypt's share of the Nile's water is 55.5 billion cubic metres per year (m^3/yr). When this is distributed among the population, it barely reaches the water poverty threshold. To alleviate water poverty, other water resources have been made available through efforts such as the recycling of agricultural drainage. It is estimated that a volume of 4 billion m³/yr has been reclaimed through the reuse of agricultural drainage water in the Nile delta. Present extraction from the Nile aquifer is 4.8 billion m³/yr. While the agricultural sector consumes more than 80 per cent of the total water use, its contribution to gross domestic product (GDP) is only 20 per cent, which is very low when compared with the economic value of water in the industrial sector. Nevertheless, employment in the agriculture sector accounts for 40 per cent of the national labour power. Both sectors contribute to overall environmental degradation, and unlike agriculture, industry is a point source of water pollution and a major source of air pollution (see Figure 1.1, Plate 1).

The domestic demand for water accounts for less than 5 per cent of total water use, and about 20 per cent of the population have no access to safe drinking water. Losses in the distribution network are estimated to be around 50 per cent. Raising

Note: A barrage is a hydraulic structure, a gated dam. Barrages were typically built on the River Nile or its branches with the purpose of elevating upstream water levels so that all the intakes into the branching irrigation canals were fed gravitationally. This old barrage on the Damietta branch is very close to the delta apex. It was built in 1863 to guarantee perennial irrigation of the delta without pumping. See Plate 1 for a colour version.

Source: photo courtesy of Dr Hussam Fahmy

Figure 1.1 *An old barrage on the Damietta branch of the River Nile*

the distribution efficiency is expected to partially cover the predicted increase in domestic demand created by an increase in living standards and a growing population. Hydropower generation used to be considered as a consumptive use, but it is not seen in this light any longer, since water that generates power as it passes through the High Aswan Dam is not lost and can then be applied to other purposes. Navigation is another non-consumptive use that makes a significant contribution to the pollution of the Nile.

Egyptian planners now consider sustainability and the environment in their planning, thereby increasing planning complexity beyond economic and engineering assessments. An Integrated Water Resources Model for Egypt (IWRME) has been developed, which uses the systems approach to analyse various policies and their long-term effects (Simonovic and Fahmy, 1999). The model relates various development plans in the different socio-economic sectors to water as a natural resource at the national (strategic) level. Agriculture, industry, domestic use, power generation and navigation are the five socio-economic sectors that depend directly on water. The model comprises six sectors: the five socio-economic sectors that depend on water, and the water sector itself.

The main objective behind the development of IWRME is to evaluate water policies formulated to satisfy long-term socio-economic plans at the national level in the five sectors. The time horizon of most socio-economic plans is from 25 to 30 years. Seven conventional and non-conventional water sources are modelled. Most of the water sources are conceptualized in the model as reservoirs with no maximum storage capacity. Based on the storage available in each one of them, there is a constraint on the level of withdrawal. The available storage depends mainly on the inflow to these reservoirs. In the case of desalination the inflow is infinite. In the case of surface water resources the inflow is finite, and is based on the releases from the High Aswan Dam, return flow from agriculture in Upper Egypt, and industrial and domestic effluent. On the timescale, a one-year model time increment has been chosen. No geographical distribution is assumed: that is, Egypt is modelled as a single geographical unit.

Because of the complexity of the model, gathering all the information from different sectors of the economy was not a simple process. To obtain the necessary information, it was necessary to exchange planning ideas about how to balance water demands with available resources. Discussions between the affected parties and the model developers enabled conflicting demands to be addressed and more realistic plans to be developed.

Accordingly, one of the activities of the NWSRU was to organize a workshop as a forum for cooperative planning and evaluation of the modelling approach (Abu-Zeid et al, 1996). Participants endorsed the new modelling approach for complex macro-scale water resources planning in Egypt. The workshop also provided numerous ideas for model modifications, from issues raised by participating water resources stakeholders and decision-makers. One of the major recommendations was to extend the model use to a simulation of water scarcity and water quality deterioration.

The use of the IWRME can be illustrated by simulation of a simple policy alternative (A1) formulated by pushing only four policy variables (the new reclaimed area for agriculture, the treatment capacity of industrial effluent, the capacity of sewage water treatment and the industrial growth rate) away from the initial state of the system. The simple policy alternative (A1) is analysed and compared with the 'status quo' alternative (A0) in terms of number of indicators representing the different evaluation domains.

Considering the effect of the quantity of treated industrial effluent, and of the industrial growth rate, on the industrial wastewater quality index, the model shows an improvement from the status quo in the quality index from 1997 until 2010, because of an increase in the volume of treated waste. After 2010 the effect of industrial growth outweighs the effect of efforts to clean the environment, and a clear trend towards a deterioration in the quality index starts to become apparent. Similar trends can be seen when the impact of two policy variables, volume of treated sewage and population growth, on the sewage water quality index is analysed.

The economic impact of volumes of treated sewage and industrial effluent in A1, compared with A0, is a significant deterioration from 2000 to 2010. The original state cannot be recovered even by the end of the planning period.

The social impact of the A1 alternative can be evaluated in terms of the national employment indicator. This shows a drop in national employment levels during the period from 2014 and 2021, although the area of reclaimed land is projected to reach its maximum during this period.

The country will suffer from severe water shortage, and therefore not all the land that is made available by reclamation will be cultivated. The main cause of the shortage is probably natural drought in the upper Nile, since in both the A1 and A0 alternatives, the development of additional water resources is not considered.

Lessons learned

The demand for water is growing, and increases in population create serious problems. Agriculture, industry, domestic use, power generation and navigation are the five socio-economic sectors that depend directly on water. The only solution to the complex water resources problems of Egypt is through integrated planning and management based on the systems approach. There is a need to consider non-traditional sources of water supply (desalination, use of brackish water, groundwater mining, etc.). Systems modelling tools have an application in water policy analysis in Egypt. The participation of all stakeholders is essential for the development of a policy acceptable to all. Institutional change, education, training and cooperation are necessary in order to address the water problems of the 21st century.

1.1.2 Sihu basin water management in China

This example presents experiences from a water resources management study for the Sihu area in Hubei Province, China. The main objective of the Sihu drainage system is to provide flood protection and prevent the waterlogging of agricultural lands. The objective of the study (Hubei Water Resources Bureau et al, 1995) was to improve the existing Sihu water management technology through the development of an advanced decision support system which included a hydrological model, a hydraulics model, an operation planning model, a system simulation model and a real-time operation model.

The Sihu basin is situated in the south-central part of Hubei Province. The total drainage area of the Sihu basin is 11,547 square kilometres (km^2). The inner polder area protected by dykes is 10,375 km^2 and the total area of agricultural land is 4327 km^2. The population is about 4 million. The area suffers from surface flooding caused by the surrounding rivers and by rainstorms within the polder, and from surface waterlogging of agricultural lands caused by poor drainage of the runoff. Other problems include subsurface waterlogging and droughts.

The Sihu water resources management system includes two large subsystems developed in the last four decades. The first is an extensive dyke system for protection against flooding caused by the surrounding Yangtze, Hanjiang and Dongjinghe rivers. The second is a large drainage system with such engineering facilities as large storage lakes, canals, sluices and pumping stations (see Figure 1.2, Plate 2). The Sihu basin can be divided into upper, middle and lower zones according to the drainage system layout. The drainage system has six main canals, two large storage lakes (Changhu and Honghu), seven main sluices with a total design discharge of 1725 cubic metres per second (m^3/s), Seventeen first-stage pumping stations with a total design capacity of 101.6 MW and a total design discharge of 1162 m^3/s, and several hundred second-stage pumping stations with a total design discharge of 1990 m^3/s.

The existing drainage facilities are in need of repair, rehabilitation, upgrading and modernization. New facilities are also required to improve the drainage conditions in the Sihu area. The Hubei Water Resources Bureau was planning to use a World Bank loan for improving the engineering infrastructure and operations facilities. In addition, the Bureau initiated improvements in system management as an initial step in the development of a comprehensive plan for sustainable development and management of the water resources in the Sihu basin.

The Jingzhou Prefecture Flood Protection Office manages the Sihu drainage system. The Sihu Engineering Management Commission in the Jingzhou Prefecture

Note: See Plate 2 for a colour version.
Source: photo courtesy of Dr Dejiang Long

Figure 1.2 *Main drainage canal in the Sihu basin*

is the basin water resources management unit in charge of engineering works and their maintenance. The required information for decision-making includes recorded rainfall, water levels and discharges; climate and runoff forecasts; operating conditions of drainage facilities; instructions issued by the provincial flood protection office; and feedback suggestions provided by the operators. The information is sent to the Prefecture Flood Protection Office and its Hydrological Information Group by phone and/or electronically. The group processes the information for distribution to decision-makers and the operations dispatch group. The dispatch group proposes options for operations based on available climatic and hydrological data, established operating rules and the intentions of the decision-makers. The final operations decision is usually made in a regular meeting, and is based on the proposed options and other considerations. The decision-makers or the prefecture administrative officials chair the meeting. The final operations decisions are sent back to the county and city flood protection offices as well as the Sihu Engineering Management Commission for implementation.

The ultimate objective of the Sihu management study was to develop an advanced decision support system for guiding the real-time operations of the Sihu

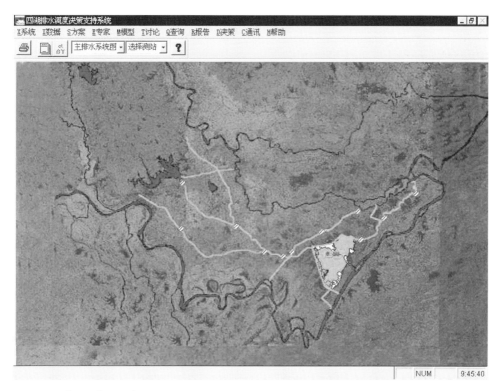

Note: See Plate 3 for a colour version.

Figure 1.3 *Main interface of the Sihu flood management decision support system*

drainage system (Figure 1.3, Plate 3). The main part of the proposed approach included the development of a planning model and a real-time operation model (Ou et al, 1995). The planning model is designed to generate optimal system operating rules. Historic hydrological input data are used to characterize and capture the long-term variability of the hydrological input for the drainage system operations. Operating rules derived from the planning model are used as either constraints or goal variables for the real-time operations model. The real-time operation model is designed to generate optimal operating decisions based on the operating rules derived from the planning model, the current state of the drainage system, and a hydrological forecast that is used to characterize and capture the short-term variability of the hydrological input. The operating decisions produced from the real-time operation model are considered by the drainage system manager for implementation in real time. A variety of models, necessary data and communication needs of different decision-making groups are incorporated in the computerized decision support system (Hubei Water Resources Bureau et al, 1998).

Lessons learned

The efficient and optimal management of complex drainage systems is important. Decision support tools including system optimization models are used for operational application. Complex decision-making processes require technical support. The involvement of decision-makers at all levels improves management.

The minimizing of flood damage is the most important objective in the Sihu basin of China. Training and institutional development are essential components in the practical application of optimal management strategies. In order to build an efficient decision-making environment, knowledge transfer and the collaboration of local institutions are required.

1.1.3 Red River flooding, Manitoba, Canada

The Red River flood of 1997 was the worst on record in many locations; it caused widespread damage throughout the Red River Valley. The governments of Canada and the US have agreed that steps must be taken to reduce the impact of future flooding. In June 1997 they asked the International Joint Commission (IJC) to analyse the causes and effects of the Red River flood of that year. The IJC appointed the International Red River Basin Task Force to examine a range of alternatives to prevent or reduce future flood damage.

The Task Force's studies (International Red River Basin Task Force, 1997, 2000) provide insights and advice for decision-makers on reducing or preventing devastation such as occurred during the 1997 flood (see Figure 1.4, Plate 4). The Task Force's work also provided useful data and tools for those who plan, design and implement flood reduction policies, programmes and projects. These data and tools provide those with operational responsibilities with a much greater ability to forecast flood events and to carry out efficiently emergency measures to save lives and property.

The Task Force considered what collaborative and integrated problem-solving mechanisms were required in the Red River basin. The aim was to enhance coordination and cooperation throughout the entire basin long after the Task Force had finished its assignment. In summary, the Task Force defined specific objectives for its investigations, to:

- develop and recommend a range of alternatives to prevent or reduce future flood damages;
- improve tools for planning and decision-making;
- facilitate integrated flood management in the basin.

An Overview 13

The 116,500 km² Red River basin slopes northward from the US Great Plains to Lake Winnipeg. The basin includes portions of South Dakota, North Dakota, Minnesota and Manitoba. The primary focus for the work was the Red River and its major tributaries. Of particular importance are those areas of the basin that were flooded in 1997.

Upstream level: 235.1 m – near peak. Downstream level: 232 m. Floodway flow: 1841 m³ per second. Flow in the Red River (above Floodway) 3905 m³ per second. Flow in the Red River (below Floodway) 2066 m³ per second. See Plate 4 for a colour version.

Source: photo courtesy of the Province of Manitoba, 1997

Figure 1.4 *Red River Floodway inlet, 4 May 1997*

The International Red River Basin Task Force managed the study. It defined required projects, coordinated the funding and scheduling, exercised quality control, provided oversight of subgroups, synthesized the findings, and prepared the recommendations for the final report to the IJC. The Task Force established three subgroups – Database, Tools and Strategies – to conduct or direct much of the data collection, model development, programme evaluation, and to prepare preliminary recommendations. Each subgroup included experts from both the US and Canada.

The concept for accomplishing the study included three main activities: database development, modelling, and the development of damage reduction strategies. A coordinated database was found to be fundamental as it supports the development of models and flood damage reduction strategies. Each of these working topics ended up as a key element in the decision support system.

The Task Force's final report (International Red River Basin Task Force, 2000), drew together the findings of the subgroups and made recommendations on policy, operations and research issues. The IJC used the final report as the basis for public hearings in the basin prior to the submission of its report to the governments.

Public participation was an important part of the process. Following the distribution of the Interim Report, the IJC and the Task Force conducted a series of public meetings throughout the basin in February and October 1998. The results from these meetings were incorporated into the study plan. Efforts were made to keep people in the basin informed throughout the study using the Internet, news releases and other means of contact. Public and technical inputs were invited throughout the study period.

The fact that this study involved two countries implied two different ways of doing business, two political systems, two or more ways of collecting, analysing and storing data, and many other political dichotomies. These dichotomies created a unique challenge for this study, but the reality that floodwaters do not recognize an international border made a basin-wide approach to flood management an imperative. Although this study did not develop a comprehensive basin-wide water management plan, the work of the Data, Tools and Strategies Groups contributed to more effective and efficient floodplain management, facilitated integrated flood emergency management in the basin, and fostered improved international cooperation and communication.

In investigating what can be done about flooding in the Red River basin, the Task Force examined the issue of storage – through reservoirs, wetlands, small impoundments or micro-storage – and drainage management. The conclusions (International Red River Basin Task Force, 2000) include:

- It would be difficult if not impossible to develop enough economically and environmentally acceptable large reservoir storage to reduce substantially the flood peaks for major floods.

- Wetland storage may be a valued component of the prairie ecosystem but it plays an insignificant hydrologic role in reducing peaks of large floods on the main stem of the Red River.

Since the Task Force concluded that storage options provide only modest reductions in peak flows for major floods, a mix of structural and non-structural options were examined. Winnipeg, the largest urban area within the basin, remains at risk. The city survived the 1997 flood relatively unharmed, but it cannot afford to be complacent. If it had not been favoured with fair weather during late April 1997, it could have suffered the fate of its southern neighbours. The Task Force made a number of recommendations to address the city's vulnerabilities and better prepare it for large floods in the future. To achieve the level of protection sufficient to defend against the 1826 or larger floods, major structural measures on a scale equal to the original Floodway project are needed to protect the city. Two options were suggested: expansion of the Floodway, and construction of a water detention structure near Ste Agathe to control floodwaters for floods larger than 1997. After detailed feasibility studies and a federal–provincial–city agreement, the Floodway expansion project began construction (it was the largest infrastructure investment in Canada in 2005).

Structural protection measures are only part of the response to living with major floods. The Task Force looked at a wide range of floodplain management issues to see how governments and residents might establish regulatory and other initiatives to mitigate the effects of major floods and to make communities more resilient to the consequences of those floods. It made a number of recommendations on defining the floodplain, adopting and developing building codes appropriate to the conditions in the Red River basin, education and enforcement.

In an effort to gain a better understanding of the flooding issues, and in recognition of weaknesses in technological infrastructure within the basin, the Task Force devoted much of its energy and resources to data issues and computer modelling. On reviewing current data availability, the Task Force concluded that further improvement and maintenance of the Red River floodplain management database was required. Federal, state and provincial governments and local authorities needed to maintain a high level of involvement in further database development, and in improving data accessibility.

The Red River Basin Decision Information Network (RRBDIN) (RRBDIN, 2005) now provides information about water management within the basin, and links to other relevant resources. While RRBDIN concentrates on information and activities on the US side, the government of Manitoba has been involved in collecting and disseminating flood information from the Canadian side (Province of Manitoba, 2005). Information from RRBDIN includes databases, references, technical tools, communication tools and geographic information system (GIS) data, as

16 Setting the Stage

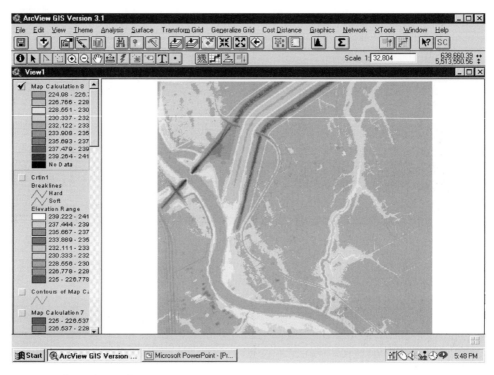

Note: See Plate 5 for a colour version.

Figure 1.5 *LIDAR data showing the Winnipeg Floodway inlet on the Red River*

well as the most up-to-date information available on weather and flood forecasting. Some of the data collected by the Task Force are shown in Figure 1.5 (Plate 5).

The Task Force found difficulty in securing public access from Canadian agencies to data and other flood-management related information. The Task Force recommended that Canadian data be made available at no cost and with no restrictions for flood management, emergency response and regional or basin-wide modelling activities. The website of the government of Manitoba now provides up-to-date reports on daily flood conditions, in the form of maps and reports, along with miscellaneous information on flood management. A prototype version of the real-time flood decision support for the Red River basin is operational (Province of Manitoba, 2004). One screen of the decision support system is shown in Figure 1.6 (Plate 6).

Lessons learned

There is a need for assistance to decision-makers in order to reduce the impacts and devastation caused by floods. Multiple alternatives must be considered in flood management. Improved systems tools for planning and decision-making are

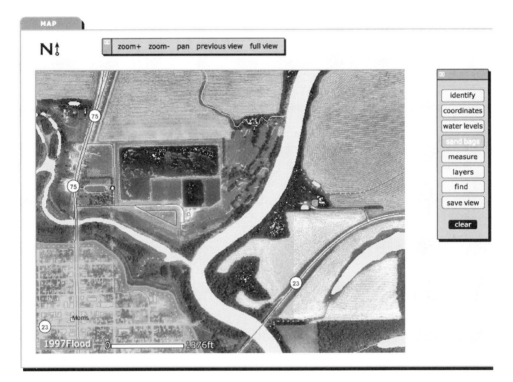

Note: See Plate 6 for a colour version.
Source: Manitoba water stewardship, http://www.geoapp.gov.mb.ca/website/rrvft (last accessed 11 June 2006)

Figure 1.6 *A prototype of the Red River basin flood decision support system*

necessary, together with a well coordinated database. Integrated flood management in an international river basin is a necessity. The public must be involved in the process of flood management. Data and communication links must be effective. Interdisciplinary studies are required to solve flood-related problems.

1.2 TOOLS FOR WATER RESOURCES SYSTEMS MANAGEMENT – TWO NEW PARADIGMS

In the context of this book any empirical, analytical or numeric procedure used for water resources planning, operations and management is referred to as a 'tool'. Tools can take the form of a simple empirical relationship. For example, the rational equation is the simplest method to determine peak discharge from drainage basin runoff. It is not as sophisticated as other methods, but is the most common method used for sizing sewer systems. The basic equation has the form:

$$Q = C_j C i A \quad [L^3 T^{-1}] \tag{1.1}$$

where:

Q = the peak rate flow
C_j = the frequency factor
C = the runoff coefficient
i = the intensity of precipitation for a duration equal to time of concentration, tc, and a return period, T
A = the drainage area.

Tools can also be expressed in analytical form. An excellent example is Bernoulli's equation, a tool that provides great insight into the balance between pressure, velocity and elevation:

$$p + \frac{1}{2}\rho V^2 + \rho g h = const \quad [L] \tag{1.2}$$

where:

p = pressure
ρ = density
V = velocity
g = gravitational acceleration
h = elevation.

The use of numerical tools can be illustrated using the example of a linear programming (LP) model for optimal allocation of, for example, scarce water resources:

$$Max(orMin) \quad x_o \sum_{j=1}^{n} = c_j x_j \tag{1.3}$$

subject to:

$$\sum_{j=1}^{n} = a_{ij} x_j = b_i \text{ for } i = 1,2,...,m \tag{1.4}$$

$$x_j \geq 0 \text{ for } j = 1,2,...,n \tag{1.5}$$

where:

c_j = objective function coefficient
a_{ij} = technological coefficient

b_i = right-hand-side coefficient
x_0 = objective function
x_j = decision variable
m = total number of constraints
n = total number of decision variables.

The application of various tools for water resources management over the last 50 years shows a pattern of change. Some of the lessons summarized by Simonovic (2000) are noted below.

Domain-specific lessons

1. Population increase creates serious water management problems.
2. Agriculture (including fisheries), industry, domestic use, power generation, navigation and recreation are the six socio-economic sectors that depend directly on water.
3. Demand for water is growing.
4. The solution of water management problems must take into consideration the water needs of ecosystems.
5. An interdisciplinary approach is required for solving water resources management problems.
6. The public must be involved in the management of water resources.
7. Institutional change, education, training and cooperation are necessary in order to address the water problems of the future.

Technical lessons

1. Integrated planning and management based on the use of systems analysis is a very efficient approach to finding solutions for complex water resources problems.
2. Mathematical modelling tools have an application in water policy analysis.
3. Decision support tools including optimization models can be considered for operational application.
4. Improved tools for planning and decision-making are necessary, together with well-coordinated databases.
5. Complex water decision-making processes require technical support.
6. Training and institutional development play an important role in the practical application of optimal management strategies.

Two paradigms were identified that will shape future tools for water resources systems management. The first focuses on the complexity of the water resources domain, and the complexity of the modelling tools, in an environment characterized by continuous, rapid technological development. The second deals with water-related data availability, and the natural variability of domain variables in time and space that affect the uncertainty of water resources decision-making.

1.2.1 The complexity paradigm

The first component of the *complexity paradigm* is that water problems in the future will be more complex. Domain complexity is increasing (Figure 1.7). Further population growth, climate variability and regulatory requirements are factors that increase the complexity of water resources problems. Water resources management schemes are planned over longer temporal scales in order to take into consideration the needs of future generations. Planning over longer time horizons extends the spatial scale. Matching increasing needs for water requires consideration of available water resources over the larger space. Meeting the water demands of people for life support, food production and industrial development calls for the integrated management of surface and groundwater. If a balance cannot be met within the watershed boundaries, water transfer from neighbouring watersheds should be considered.

The extension of temporal and spatial scales leads to an increase in the complexity of the decision-making process. Large-scale water problems affect numerous stakeholders. The environmental and social impacts of complex water management solutions must be given serious consideration. The equitable distribution of water and protection of water quality are regulated by a large number of agencies. The public interest is usually represented by non-governmental organizations (NGOs).

The second component of the complexity paradigm is the rapid increase in the processing power of computers (Figure 1.7). Since the 1950s, the use of computers in water resources management has grown steadily. Computers have moved from data processing, through the user's office and into information and knowledge processing. Whether the resource takes the form of a laptop personal computer or a desktop multi-processing workstation is not important any more. It is important that the computer is used as a partner in more effective water resources decision-making (Simonovic, 1996a, 1996b). The main factor responsible for involving computers in the decision-making process is the treatment of information as the sixth economic resource (besides people, machines, money, materials and management).

The third component of the complexity paradigm is the reduction in the complexity of tools used in water management (again, see Figure 1.7). The most important advance made in the field of water management in the last century was the introduction of systems analysis (Hall and Dracup, 1970; Friedman et al, 1984; Yeh, 1985; Rogers and Fiering, 1986; Wurbs, 1998). Systems analysis is defined

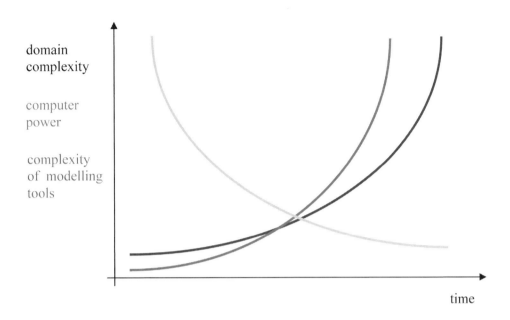

Figure 1.7 *Schematic illustration of the complexity paradigm*

here as an approach for representing water-related problems using a set of mathematical planning and design techniques. Theoretical solutions to the problems can then be found using a computer. In the context of this book systems analysis techniques, often called *operations research*, *management science* and *cybernetics*, include simulation and optimization techniques that are used to analyse the quantity and quality aspects of watershed runoff and streamflow processes, reservoir system operations, groundwater development and protection, water distribution systems, water use, and various other hydrological processes and management activities. Systems analysis is particularly promising when scarce resources must be used effectively. Resource allocation problems are very common in the field of water management, and affect both developed and developing countries, which today face increasing pressure to make efficient use of their resources.

Simulation models play an important role in water resources assessment, development and management. They are widely accepted within the water resources community and are usually designed to predict the response of a system under a particular set of conditions. Early simulation models were constructed by a relatively small number of highly trained individuals. Many generalized, well-known simulation models were developed primarily in the FORTRAN language. These models include, among many others, SSARR (streamflow synthesis and reservoir regulation – US Army Corps of Engineers, North Pacific Division), RAS (river analysis system – Hydrologic Engineering Center); QUAL (stream water quality model – Environmental Protection Agency), HEC-5 (simulation of flood control and conservation systems – Hydrologic Engineering Center), SUTRA (saturated–unsaturated transport model – US Geological Survey), and KYPIPE (pipe network analysis – University of Kentucky).

These models are quite complex, however, and their main characteristics are not readily understood by non-specialists. Also, they are inflexible and difficult to modify to accommodate site-specific conditions or planning objectives that were not included in the original model. The most restrictive factor in the use of simulation tools is that there is often a large number of feasible solutions to investigate. Even when combined with efficient techniques for selecting the values of each variable, quite substantial computational effort may lead to a solution that is still far from the best possible.

Advances made during the last decade in computer software have brought considerable simplification to the development of simulation models (High Performance Systems, 1992; Lyneis et al, 1994; Powersim Corp., 1996; Ventana Systems, 1996). Simulation models can be easily and quickly developed using these software tools, which produce models that are easy to modify, easy to understand, and that present results clearly to a wide audience of users. They are able to address water management problems with highly nonlinear relationships and constraints.

Numerous optimization techniques are used in water management too. Most water resources allocation problems are addressed using LP solvers introduced in the 1950s (Dantzig, 1963). LP is applied to problems that are formulated in terms of separable objective functions and linear constraints, as for example shown by equations (1.3) to (1.5). The objective is usually to find the best possible water allocation (for water supply, hydropower generation, irrigation, etc.) within a given time period in complex water systems. However, neither objective functions nor constraints are in a linear form in most practical water management applications. Many modifications have been used in real applications in order to convert nonlinear problems for the use of LP solvers. Examples include different schemes for the linearization of nonlinear relationships and constraints, and the use of successive approximations.

Nonlinear programming is an optimization approach used to solve problems when the objective function and the constraints are not all in the linear form. In general, the solution to a nonlinear problem is a vector of decision variables which optimizes a nonlinear objective function subject to a set of nonlinear constraints. No algorithm exists that will solve every specific problem fitting this description. However, substantial progress has been made for some important special cases by making various assumptions about these functions. Successful applications are available for special classes of nonlinear programming problems such as unconstrained problems, linearly constrained problems, quadratic problems, convex problems, separable problems, non-convex problems and geometric problems.

The main limitation in applying nonlinear programming to water management problems is in the fact that nonlinear programming algorithms generally are unable to distinguish between a local optimum and a global optimum (except by finding another better local optimum). In recent years there has been a strong emphasis on developing high-quality, reliable software tools for general use such as MINOS (Murtagh and Saunders, 1995) and GAMS (Brooke et al, 1996). These packages are widely used in the water resources field for solving complex problems, including hydropower generation problems and water network distribution problems. However, the main problem of global optimality remains an obstacle in the practical application of nonlinear programming.

Dynamic programming (DP) offers advantages over other optimization tools since the shape of the objective function and constraints do not affect it, and as a result it has been used frequently in water management. DP requires discretization of the problem into a finite set of stages. At every stage a number of possible conditions of the system (states) are identified, and an optimal solution is identified at each individual stage, given that the optimal solution for the next stage is available. An increase in the number of discretizations and/or state variables would increase the number of evaluations of the objective function and core memory requirement per stage. This problem of rapid growth of computer time and memory requirement associated with multiple-state-variable DP problems is known as 'the curse of dimensionality'. Some modifications used in the field of water management in order to overcome this limitation of DP include discrete differential DP (an iterative DP procedure) and differential DP (a method for discrete-time optimal control problems).

In the very recent past, most researchers have been looking for new approaches that combine efficiency and ability to find the global optimum. One group of techniques, known as *evolutionary algorithms*, seems to have a high potential. Evolutionary techniques are based on similarities with the biological evolutionary process. In this concept, a population of individuals, each representing a search point in the space of feasible solutions, is exposed to a collective learning process, which proceeds from generation to generation. The population is arbitrarily initialized and subjected to the process of selection, recombination and mutation through stages known as *generations*, such that newly created generations evolve towards more favourable regions of the search space. In short, the progress in the search is achieved by evaluating the fitness of all individuals in the population, selecting the individuals with the highest fitness value, and combining them to create new individuals with increased likelihood of improved fitness. The entire process resembles the Darwinian rule known as 'the survival of the fittest'. This group of algorithms includes among others, evolution strategies (ES) (Back et al, 1991), evolutionary programming (EP) (Fogel et al, 1966), genetic algorithms (GA) (Holland, 1975), simulated annealing (Kirkpatrick et al, 1983), and scatter search (Glover, 1999). Evolutionary algorithms are becoming more prominent in the water management field. Significant advantages of evolutionary algorithms include:

- no need for an initial solution;
- easy application to nonlinear problems and to complex systems;
- production of acceptable results over longer time horizons;
- the generation of several solutions that are very close to the optimum (which gives added flexibility to water resources decision-makers).

During the evolution of systems analysis in water management, it has become apparent that more complex analytical optimization algorithms are being replaced by simpler and more robust search tools. Advances in computer software have also led to considerable simplification in the development of simulation models.

1.2.2 The uncertainty paradigm

The first component of the *uncertainty paradigm* is the increase in all elements of uncertainty in time and space (Figure 1.8). Uncertainty in water management can be divided into two basic forms: uncertainty caused by inherent hydrologic variability, and uncertainty caused by a fundamental lack of knowledge. Awareness of the distinction between these two forms is integral to understanding uncertainty. The first form is described as *variability*, and the second one as *uncertainty*. Uncertainty caused by variability is a result of inherent fluctuations in the quantity of interest (that is, hydrological variables). The three major sources of variability are temporal, spatial and individual heterogeneity. Temporal variability occurs when values fluctuate over time. Values affected by spatial variability are dependent upon the location of an area. The third category effectively covers all other sources of variability. In water resources management, variability is mainly associated with the spatial and temporal variation of hydrological variables (precipitation, river flow, water quality, etc.). The more elusive type of uncertainty is caused by a fundamental lack of knowledge. It occurs when the particular values that are of interest cannot be assessed with complete confidence because of a lack of understanding or limitation of knowledge.

The second component of the uncertainty paradigm is the decrease in water data availability (Figure 1.8). Hydrological information on water levels, discharge, sediment and water quality is necessary for water management. Examples of water projects for which hydrological information is indispensable are water engineering infrastructure projects (the design of dams, reservoirs, spillways, canals, diversions,

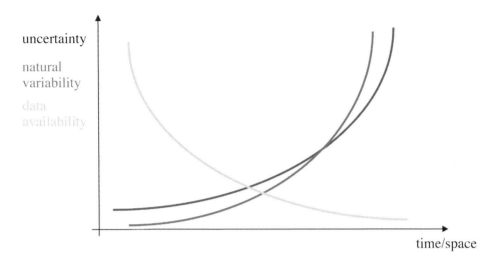

Figure 1.8 *Schematic illustration of the uncertainty paradigm*

hydropower, etc.) and projects in the area of water quality and protection from water (zoning, insurance, standards, legislation, etc.).

The numbers of hydrological stations in operation worldwide, as reported by World Meteorological Organization (WMO), are very impressive. The *INFOHYDRO Manual* (WMO, 1995) estimates that there are nearly 200,000 precipitation gauges operating worldwide and over 12,000 evaporation stations. Monitoring is taking place at over 64,000 stations for discharge, at nearly 38,000 for water level, at 18,500 for sediment, at over 100,000 for water quality, and at over 330,000 for groundwater characteristics. Despite the apparently high global numbers, the stations are not uniformly distributed, and there is a shortage over large areas.

The financial constraints of government agencies that are responsible for the collection of hydrometric data have resulted in reductions in the data collection programme in many countries. In Canada, for example, budgetary cutbacks and shifts in government priorities have led to a dramatic reduction in the hydrometric network (Pilon et al, 1996; Houghton-Carr and Matt, 2006; Thieme et al, 2007). In many countries hydrological data collection activities are very fragmented. A similar fragmentation is observed at the international level. Of particular concern are the gaps in the existing data relative to the informational requirements. Many authors agree that current data collection networks are inadequate for providing the information required to understand and explain changes in natural systems. Given the reductions in the funding of data collection activities, it is clear that a change in the approach to data collection activities is essential. The Global Run-off Data Centre

(GRDC) is working under the auspices of WMO to collect, store and disseminate discharge data for the most relevant rivers of the world. GRDC will also lead the process of establishing regional and global databases, and a global network for discharge monitoring.

The third component of the uncertainty paradigm is the increase in natural variability of water availability (again, see Figure 1.8). Water flow exhibits both temporal (between years and seasons) and spatial variation. This variation, which can be crucial for water availability to domestic, agricultural or industrial use, is not detected if the selected timescale for water balance analyses is longer than the periods of such fluctuation. The water flow from the basin is the integrated result of all physical processes in the basin. The topography, the spatial distribution of geological phenomena and land use are the main causes of spatial variability of flow.

Observed natural variability may even be affected in the future by potential climate change. One of the most important aspects of studying the hydrological consequences of potential global warming is estimating possible changes in the extreme characteristics of maximum and minimum river discharges. Through analysis of empirical data and through modelling studies, it can be shown, reasonably reliably, that potential global warming would lead to more changes in runoff extremes than in mean annual and seasonal flows, especially for small and medium-sized basins. On the one hand, an increase in maximum floods can be expected, and on the other, so can a more frequent occurrence of severe droughts. Both could have major economic and ecological consequences.

1.3 CONCLUSIONS

In the past, stakeholders not actively involved in the development of a model tended to mistrust the results of the model. Computer power has increased and costs have fallen to the point that all stakeholders in the resource can play a very important role in water resources systems management. Technology is already a facilitating force in political decision-making, and will be more so in the future. Spatial decision support systems using object-oriented programming algorithms are integrating transparent tools that will be easy to use and understand.

National and international databases, both static and dynamic, now provide much of the necessary information in digital form. The trend will continue for providing public access to all water-related data at reasonable cost and in a user-friendly format, and this will play an important role in supporting tools for water decision-making.

The speed with which data and ideas can be communicated has historically been a control mechanism of scientific progress. The Internet began in 1968 by connecting four hosts. As of December 2005, just over 1 billion hosts were connected to multiple computer networks according to the Internet Usage and World Population Statistics (http://www.internetworldstats.com/stats.htm, accessed January 2006). Virtual libraries, virtual databases, virtual forums and bulletin boards, web-enabled software packages, and the use of 'write once – run anywhere' languages (such as Java by Sun Microsystems) will create new opportunities for water managers.

The future of water resources management will be difficult in both the developing and developed world. My hope is that the tools discussed in this book, supported by good data communicated through powerful networks, will empower people to make wise decisions on how to make best use of limited water resources.

1.4 REFERENCES

Abu-Zeid, K., K. Kheireldin and S.P. Simonovic (1996), 'Stakeholder Participation in Modelling Water Resources Planning in Egypt: A Workshop Experience', in *Proceedings of the 49th Canadian Water Resources Association Annual Conference – Rational and Sustainable Development of Water Resources*, 1(6): 406–414

Back, T., F. Hoffmeister and H.P. Schewel (1991), 'A Survey of Evolution Strategies', in *Proceedings of the Fourth International Conference on Genetic Algorithms*, Morgan Kaufmann, San Mateo, CA

Brooke, A., D. Kendrik and A. Meeraus (1996), *GAMS: A User's Guide*, Scientific Press, Redwood City, CA

Dantzig, G.B. (1963), *Linear Programming and Extension*, Princeton University Press, Princeton, NJ

Fogel, L.J., A.J. Owens and M.J. Walsh (1966), *Artificial Intelligence Through Simulated Evolution*, John Wiley, Chichester

Friedman, R., C. Ansell, S. Diamond and Y.Y. Haimes (1984), 'The Use of Models for Water Resources Management, Planning and Policy', *Water Resources Research*, 20(7): 793–802

Glover, F. (1999), 'Scatter Search and Path Relinking', in *New Methods in Optimization*, D. Corne, M. Dorigo and F. Glover, eds., McGraw-Hill, New York

Hall, W.A. and J.A. Dracup (1970) *Water Resources Systems Engineering*, McGraw-Hill, New York

High Performance Systems (1992), *Stella II: An Introduction to Systems Thinking*, High Performance Systems, Inc., Nahover, NH

Holland, J.H. (1975), *Adaptation in Natural and Artificial Systems*, University of Michigan Press, Ann Arbor, MI

Houghton-Carr, H. and F.R.Y. Matt (2006), 'The Decline of Hydrological Data Collection for Development of Integrated Water Resource Management Tool in South Africa', *IAHS Publication*, 308: 51–55

Hubei Water Resources Bureau, Golder Associates Ltd and Mikuniya Corporation (1998), *Development and Implementation of the Decision Support System and Operational Models for Optimal Operations of the Sihu Water Management System, Terms of Reference*, Wuhan, China

Hubei Water Resources Bureau, Wuhan University of Hydraulic and Electric Engineering, and AGRA Earth & Environment Ltd (1995), *Optimal Operations of the Sihu Drainage System, Stage I – Phase I Study Report Volumes I to IV*, Wuhan, China

International Red River Basin Task Force (1997), *Red River Flooding: Short-Term Measures*, Interim report to the International Joint Commission, Ottawa, WA

International Red River Basin Task Force (2000), *The Next Flood: Getting Prepared*, Final report to the International Joint Commission, Ottawa, WA

Kirkpatrick, S., C.D. Gelatt, Jr. and M.P. Vecchi (1983), 'Optimization by Simulated Annealing', *Science*, 220, 4598: 671–680

Kundzewicz, Z.W., L.J. Mata, N.W. Arnell, P. Döll, P. Kabat, B. Jiménez, K.A. Miller, T. Oki, Z. Sen and I.A. Shiklomanov (2007) 'Freshwater resources and their management', *Climate Change 2007: Impacts, Adaptation and Vulnerability. Contribution of Working Group II to the Fourth Assessment Report of the Intergovernmental Panel on Climate Change*, M.L. Parry, O.F. Canziani, J.P. Palutikof, P.J. van der Linden and C.E. Hanson (eds), Cambridge University Press, Cambridge, 173–210

Lyneis, J., R. Kimberly and S. Todd (1994), 'Professional Dynamo: Simulation Software to Facilitate Management Learning and Decision Making', in *Modelling for Learning Organizations*, J. Morecroft and J. Sterman, eds., Pegasus Communications, Waltham, MA

Murtagh, B.A. and M.A. Saunders (1995), *MINOS 5.4 User's Guide, Technical report SOL 83-20R*, Systems Optimization Laboratory, Department of Operations Research, Stanford University, Stanford, CA

Ou, G., X. Bai, Z. Long, Z. Guo, Z. Huang, D. Long, Y. Yang and S.P. Simonovic (1995), 'A Modelling Approach for Sustainable Management of the Large Sihu Drainage System in China', *Proceedings of the 48th Annual Conference of Canadian Water Resources Association*, 721–734

Pilon, P.J., T.J. Day, T.R. Yuzyk and R.A. Hale (1996), 'Challenges Facing Surface Water Monitoring in Canada', *Canadian Water Resources Journal*, 21: 157–164.

Powersim Corporation (1996), *Powersim 2.5 Reference Manual*, Powersim Corporation Inc., Herndon, VI

Province of Manitoba (2004), 'Red River Valley Protection', http://geoapp.gov.mb.ca/website/rrvfp/ (accessed 24 April 2005)

Province of Manitoba (2005), 'Flood Information', http://www.gov.mb.ca/flood.html (accessed 28 March 2005)

Rogers, P.P. and M.B. Fiering (1986), 'Use of Systems Analysis in Water Management', *Water Resources Research*, 22(9): 146s–158s

RRBDIN (Red River Basin Decision Information Network) (2005), 'Red River Basin Decision Information Network', http://www.rrbdin.org (accessed 28 March 2005)

RNPD-II (River Nile Protection and Development Project) (1994), *River Nile Protection and Development Project*, Inception Report, National Water Research Centre, Cairo, Egypt

Simonovic, S.P. (1996a), 'Decision Support Systems for Sustainable Management of Water Resources 1. General Principles', *Water International*, 21(4): 223–232

Simonovic, S.P. (1996b), 'Decision Support Systems for Sustainable Management of Water Resources 2. Case Studies', *Water International* 21(4): 233–244

Simonovic, S.P. (2000), 'Tools for Water Management: One View of the Future', *Water International*, 25(1): 76–88

Simonovic, S.P. and H. Fahmy (1999), 'A New Modeling Approach for Water Resources Policy Analysis', *Water Resources Research*, 35(1): 295–304

Thieme, M., B. Lehner and R. Abell (2007), 'Freshwater Conservation Planning in Data-poor Areas: An Example From a Remote Amazonian Basin (Madre de Dios River, Peru and Bolivia)', *Biological Conservation*, 135(4): 484–501

UNESCO (UN Educational Scientific and Cultural Organization) – WWAP (World Water Assessment Programme) (2003), *Water for People, Water for Life – UN World Water Development Report (WWDR)*, UNESCO Publishing, Paris, France, co-published with Berghahm Books, UK

Ventana Systems (1996), *Vensim User's Guide*, Ventana Systems Inc., Belmont, MA

WMO (World Metrological Organization) (1995), *INFOHYDRO Manual*, Hydrological Information Referral Service, Operational Hydrology Report No. 28, WMO-No. 683, Geneva, Switzerland

Wurbs, R.A. (1998), 'Dissemination of Generalized Water Resources Models in the United States', *Water International*, 23(3): 190–198

Yeh, W.W.-G. (1985), 'Reservoir Management and Operations Models: A State-of-the-Art Review', *Water Resources Research*, 21(12): 1797–1818

1.5 EXERCISES

1. Describe the largest river in your region.
    a. What are its physical characteristics?
    b. Who is involved in the management of the basin?
    c. What is the water from this river used for?
    d. What is, in your opinion, the most important water resources problem in the basin?
    e. Give some examples of the water resources engineering works in the basin.
    f. What lessons can be learned from the past management of the river basin?

g. What are the most important principles you would apply in future management of the river basin?
2. Review the literature and find a definition of integrated water resources management.
   a. Discuss the three examples presented in this chapter in the context of this definition.
   b. What would you do, in addition to what has been done, in these three cases to make water resources management decisions sustainable?
3. Discuss characteristics of the river basin from Exercise 1 in the context of two paradigms presented in Section 1.2.
   a. What are the complexities of the river basin in Exercise 1?
   b. Identify some uncertainties in the basin.
   c. Can you find some data to illustrate the natural variability of river conditions?
   d. How difficult is to find the data? Why?
4. For the river basin in Exercise 1 identify the factors that will provide for sustainable water management decisions. What are the spatial and temporal scales to be considered?

2
Changing Water Resources Management Practice

Civilization as we view it today can be partially seen as the consequence of engineering activities (Sprague de Camp, 1963). In the context of this book, engineering is seen as intrinsic human added value. It involves, as it has done for many centuries, exploitation of the properties of matter and sources of power for the benefit of humanity. The story of civilization is inseparable from the story of engineering. The ideas of engineering are part of human nature and experience. By an organized, rational effort to use the material world around them, engineers responded to a myriad of problems and devised ways of providing food, shelter, comfort and convenience for human beings. The first engineers were irrigators, architects and military engineers. The same people were usually expected to be experts at all three kinds of work.

Engineering tends to be defined today in a much narrower sense. For example, *Webster's Dictionary* defines engineering as the application of science and mathematics by which the properties of matter and the sources of energy in nature are made useful to people. This definition rather excludes activities prior to the development of modern science and mathematics, but I would argue that they form part of a continuum with modern engineering. The Greeks were the first to show the connection between engineering and science, if we take science to be knowledge of general truths or the operation of general laws, as obtained and tested through scientific methods. However, they borrowed many ideas from the Egyptians, the Babylonians and the Phoenicians (Sprague de Camp, 1963).

Today we make fine distinctions between the meaning of, for example, craftsperson, engineer, technician and inventor. In antiquity, however, such delicate differences had no relevance. Every time an ancient craftsperson made something that was not a copy of a previous article, it was in effect an invention. In practice most ancient engineers were inventors, while most ancient inventors, at least after the beginning of civilization, may also be classified as engineers.

2.1 THE HISTORY OF WATER RESOURCES ENGINEERING

The development of the earliest agricultural settlements (6000–7000 years ago) can be seen as the start of a major human preoccupation with water issues, such as protecting people against floods, and ensuring an adequate and consistent supply of usable water. Farmers learned to produce more food than they personally needed, and this led to a diversified economy in which not everyone was a farmer, but others had the opportunity to make things that would be useful to their civilization. This was the beginning of water resources engineering. People began to solve problems related to the transportation of water and its management for irrigation. Table 2.1 presents a timeline of important events in water resources engineering.

Although the history and dates of early civilizations are not certain, most scientists believe today that the civilization of the Euphrates Valley was among the first, if not the first, complex civilization. The enormous flat plain between the Tigris and Euphrates rivers, which today lies in Iraq, was then known as Mesopotamia. This land is barren except when it is irrigated using water from the rivers. Irrigation was vital to Mesopotamia, and early irrigation works developed very quickly. They were soon followed by the first laws dealing with irrigation canals and water rights.

Flood problems were quite common in Mesopotamia. Both rivers were at their highest in the spring: an awkward time, since it is too late for winter crops and too early for summer crops. Therefore, the water needs to be stored so that it can be used at the right times to raise good crops. Mesopotamian irrigation was of the basin type. Basins do not have mechanical gates or sluices; they are opened by digging a gap in the embankment and closed by filling that gap with soil and mud.

The monuments of early Egypt are far better preserved than those of Mesopotamia, because the abundant supplies of limestone and granite meant they tended to be made of these materials and not of mud. Egyptian civilization developed around the River Nile. Irrigation canals were major endeavours of the Pharaohs. These canals were used to flood large tracts of the country during high water, which occurs in the autumn as a result of summer rainfall in the lands south of Egypt. The land to be flooded was divided into small basins. When the basins were full, the dykes were closed and the water was kept standing until the ground was completely soaked. Then the surplus was drained into the canals.

When the Nile's flow varied most from the norm, this spelled disaster. High waters washed away the dykes and villages, while low waters failed to flood the tracts and no crops grew. Perhaps as a result, as well as building canals the Egyptians

Table 2.1 *Timeline of water resources engineering activities*

Year	Activity
4000 BC	Irrigation projects in Egypt and Mesopotamia
2750 BC	Indus Valley water supply and drainage
2200 BC	Water works in China
1750 BC	Water code of Hammurabi
714 BC	Discovery of qanats (well and aqueduct)
312 BC	Rome's aqueduct
270 BC	Ktesibios pump; hydraulic pipe organ; the water clock
260 BC	Archimedes investigates hydrostatics and buoyancy
50 AD	Hero of Alexandria investigates discharge measuring
1450	Machu Picchu water supply and drainage works
1452–1519	Leonardo da Vinci's contribution: the continuity principle, velocity distribution, book on water
1608–1680	Pierre Perrault contributes to the measurement of rainfall and runoff
1623–1662	Blaise Pascal develops principles of the barometer and the hydraulic press
1642–1727	Isaac Newton: fluid resistance
18th century	Bernoulli equation; de Pitot piezometer; Chezy formula; Euler fluid pressure
19th century	Dalton evaporation; Mulvaney rational method; Darcy porous media flow; Rippl reservoir storage requirements; Manning open channel flow; Navier motion equation; de Saint-Venant equations; work of Weisbach, Froude, Stokes, Lord Kelvin, Pelton, Reynolds and others
20th century	Hazen frequency analysis; Richards unsaturated flow equation; Sherman unit hydrograph; Horton infiltration theory; Gumbel extreme value law for hydrology; Harvard water programme systems analysis

acquired the knowledge of how to build dams. Possibly as long ago as 3000 BC they built the first one in the Wadi Garawi, south-east of Cairo, to store water for the use of the workers in the nearby quarries (Sprague de Camp, 1963).

The Assyrians also developed extensive water works. After the invasion of Armenia around 714 BC they discovered *qanats*. These are tunnels used to bring water from an underground source in the hills down to the foothills.

The Greeks were the first to show the connection between engineering and science. The era from the early 500s BC to the late 400s BC is called the Golden Age of Greece, and was characterized by extraordinary advances in art, literature, engineering, science, philosophy and democratic government. During the later Hellenistic period, around 260 BC, Ktesibios contributed several inventions to water management: the force pump, the hydraulic pipe organ, the metal spring and the water clock. The water clock is one of his less widely known inventions, but it had significant importance in the practical application of feedback (which is discussed later in the book).

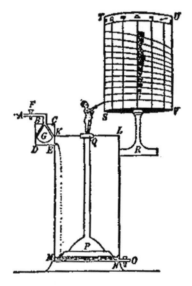

Source: www.perseus.tufts.edu/GreekScience/Students/Jesse/CLOCK1A.html (last accessed 12 June 2006).

Figure 2.1 *The parastatic water clock of Ktesibios*

Figure 2.1 shows Ktesibios' parastatic clock. To keep the rate of flow constant, Ktesibios used three chambers. The first one empties into the second one and the second into the third. The first chamber was kept full. The second one had an outlet in the bottom and an overflow outlet part-way up the side, like the overflow outlet in a contemporary bathtub. Since water rose in the second chamber only up to the overflow hole, the water in this chamber stood at a constant depth, and therefore the rate of flow out of it and into the third chamber was constant. In order to use this device as a clock, neither the first nor the second chamber could be allowed to run dry. In the third chamber, a drum-shaped float or cork floated on the water. As the third chamber filled, the float rose. By means of a rack and pinion gear, the rising float turned a shaft. A staff rising from the centre of the float bore on its top a figure holding a pointer. This pointed to the hours, which were marked on a pillar.

Archimedes of Syracuse (287 BC–212 BC) was probably the most important Hellenistic engineer, and indeed one of the greatest intellects of the classic period. He founded the science of hydrostatics, which was developed to answer the question, why do certain bodies float while others sink? (Levi, 1995). *Archimedes' law* says that a body partly or wholly immersed in a fluid loses weight equal to the weight of the fluid displaced. He realized that pressure is the fundamental physical characteristic of a fluid, and postulated two properties of pressure for a continuous and uniform fluid. First, if there is a pressure difference between two adjacent parts, the one with the higher pressure will push the other forward; and second, each fluid particle is subjected to the pressure of the fluid directly above (in the vertical direction).

Archimedes also stated, as the basis of his theory, that the surface of any fluid at rest is the surface of a sphere whose centre is the same as that of Earth. From this he proceeded to prove that a solid whose density is equal to that of a resting fluid will not move if placed within that fluid. A solid lighter than the fluid cannot sink completely, because it must rise enough above the surface for the weight of the solid to be equal to the weight of the displaced fluid. If a solid lighter than a fluid is forcibly immersed in it, it will be driven upwards by a force equal to the difference between its weight and the weight of the displaced fluid. Finally, a solid immersed in a fluid that it is heavier than will descend to the bottom of the fluid. On the other hand, when immersed, such a solid will become lighter than its true weight by the weight of the displaced fluid.

The Roman civilization was the next to make major contributions to water resources management. Roman engineering was generally based on the intensive application of simple principles. Contemporary writers often point out that the Romans contributed little to pure science. However they made a significant contribution in the applied sciences, since they were very active architects and engineers. They devoted much time to public works projects: roads, harbours, aqueducts, temples, forums, arenas, baths and sewers. Early Roman houses tended to be built on the Etruscan plan (Sprague de Camp, 1963). The *domus* of a prosperous early Roman bourgeois had about a dozen rooms, ranged around a partly roofed courtyard. There was a square hole in the roof to let in the rain and a cistern beneath to catch it. This courtyard was called an *atrium*.

Although the Romans were not the first to build *aqueducts*, the aqueducts of Rome have survived well, and have been so extensively pictured that the term 'ancient Rome' leads many people to think immediately of a row of arches. Roman aqueducts were distinguished from earlier ones by their size and number. The arcades – aqueduct bridges – were all built on a simple pattern. A series of small round arches linked a row of tall stone or brick piers. Above these lay the actual water channel, which was made of concrete, with an arched or gabled roof above it.

When an aqueduct crossed an exceptionally deep gorge, two or three rows of arches were erected on top of each other. Roman aqueducts had elevated structures because the water depended on gravity to move it from the source to the point of distribution, often flowing in open channels. Therefore the channel needed a slight and fairly constant slope, of approximately 30 cm per kilometre.

After the fall of the Western Roman Empire, some progress was made in water engineering in this part of the world by the Byzantine Empire in the early Middle Ages. It contributed the first horizontal-arch dam for flood protection, but this invention seems to have quickly been forgotten, until it was revived in modern times. Like the Byzantines, the Persians mainly maintained and improved ancient irrigation works, especially the great canal network of Babylonia.

Further east, the Indus civilization in this period did have water management systems that provided irrigation and helped with flood control. Dams and canals were built throughout the region.

Although China was connected with what is sometimes called the Main Civilized Belt by the Tarim Basin or Silk route, the Chinese tended to be isolated because of the difficulty of travelling on the route. Hence it is not surprising that Chinese civilization developed quite independently from the civilizations of the Near and Middle East. Although it is not easy to separate fact from fiction, there are stories of a great flood that caused the waters of the Yellow River to mingle with those of the Yangtze, creating a great inland sea. Many engineers are said to have worked hard for years to bring the flood under control, but we do not know what techniques they used.

In addition, for thousands of years China has been a land of canals, which were used for navigation and irrigation. As a result of about 3000 years of construction works there are about 360,000 km of canals in China, the foremost canal of which is the Grand Canal, which runs about 2000 km in a north–south direction through the eastern part of the land. It is the result of engineering labour over more than a thousand years. In the context of this book, the pre-Columbian civilizations and their extensive water works can only be mentioned. Many other examples from all over the world could have been mentioned to illustrate the universality of water engineering.

2.2 EARLY WATER RESOURCES ENGINEERING SCIENCE

Since we are not aware today of any significant advances in water sciences and engineering made during the early Middle Ages, we can take up the story of Western technology in the high Middle Ages. The *watermill* became a common source of power in this period: initially using the flow from rivers and streams and, from the

11th century, also making use of tidal waters. For this technique basins were built in bays and estuaries. Water was allowed to run into these basins at high tide and out again at low tide, turning water wheels in both directions. However it was a long time before municipal water works again equalled, let alone surpassed, the achievements of Roman engineers.

The same is true of sewerage. The people of Europe were not only far behind the Romans in the early medieval period: they have only recently begun to catch them up. In another branch of water engineering, however, medieval Europeans soon advanced beyond the Romans. This was the building of *canals*. Small irrigation canals seem to have first appeared in the Po Valley. Many of them were later enlarged to a navigable size. The first canal lock originates from the late 1300s or early 1400s, although similar developments took place in the Netherlands and in the territories of modern Italy, and it is not clear which was first.

Around the same time, a vast *land reclamation* project was taking shape on the shores of the North Sea. The coastal region of the Netherlands was originally a tangle of marshes, shallow lakes, dunes, tidal flats and low islands which were half-submerged at high tide. Beyond this a zone of dunes formed a natural dyke, but during the 13th century a series of floods occurred that broke through the dunes. Little by little the Dutch strengthened their sea defences, then they began dyking, draining and pumping to turn the watery region back into land. Over the centuries they perfected very efficient methods of doing this. Land reclamation became especially active after 1400, when the Dutch applied windmill-driven pumps to the task.

During the Renaissance there was a major shift: while the earlier developments had largely taken the form of practical engineering, a phase of development in observational water science now strengthened the theoretical basis. Leonardo da Vinci (1452–1519), one of the most creative of geniuses, made the first systematic studies of velocity distribution in streams. The French scientist Bernard Palissy showed that rivers and springs originate from rainfall. Another Frenchmen, Pierre Perrault, measured runoff and found it to be only a fraction of rainfall. He concluded that the remainder of precipitation is lost by transpiration and evaporation. The work of many subsequent scientists also contributed to a better understanding of the main processes affecting water resources systems. Blaise Pascal finalized the formulation of the principles of the barometer, hydraulic press and pressure transmissibility. Isaac Newton worked on various aspect of fluid resistance and jet contraction. His contributions span areas of inertial fluid, viscous fluid and wave mechanics.

The 18th century brought Bernoulli's equation, Chezy's formula and further improvement in measurements of precipitation and flow velocity. Euler explained the role of pressure in fluid flow and formulated the equation of motion.

In the 19th century the science of hydrology advanced, with Dalton's work on evaporation, Hagen's investigations of capillary flow, the Rational Method proposed by Mulvaney, Darcy's law of porous media flow, Rippl's diagram for determining reservoir storage and Manning's formula for open channel flow. Among further theoretical hydraulics research were Navier's extension of the equations of motion to include molecular forces, Saint-Venant's equations, Darcy's work on filtration and pipe resistance, the development of the Froude number, the invention of the Pelton wheel and the formulation of the Reynolds number.

In the 20th century quantitative hydrology continued to use empirical approaches to solve practical hydrological problems, and slowly started to replace empiricism with the analysis of observed data. Some of the major contributions are Green and Ampt's physically based infiltration model, Hazen's frequency analysis of extreme flows, Richards' governing equation for unsaturated flow, Sherman's unit hydrograph, Horton's infiltration theory and the extreme value law of Gumbel.

2.3 WATER RESOURCES MANAGEMENT SCIENCE

Since the advent of the Industrial Revolution, the world has seen remarkable growth in the size and complexity of water resources projects. The local small-scale projects of earlier eras have evolved into the billion-dollar projects of today. Water resources engineering has evolved in the direction of increasing temporal and spatial scales – from small catchments over large river basins to global systems; from storm events over seasonal cycles to climatic trends. An integral part of this revolutionary change has been a tremendous increase in water resources engineering activities, from the development (planning, design and construction) to the management of water resources systems. These activities, and the increase in the complexity of the water resources systems being handled, have led to a requirement for a fundamentally different science.

Systems analysis came to the rescue. It is a relatively new field whose development parallels that of the computer, since the increase in computational power enabled the analysis of complex relationships, involving many variables, at reasonable cost. Most of the techniques of systems analysis depend on the use of computers for practical applications. The roots of systems science include the development of the feedback concept, applied systems analysis and operations research.

The feedback concept was developed through a period of ten years (1943–1953), and is discussed in detail later in the book (Chapter 4). Applied systems analysis involves the use of rigorous methods to help determine preferred plans and designs for complex, often large-scale, systems. It combines knowledge of the available analytic tools, understanding of when the use of each is most appropriate, and skill in applying them to practical problems. Applied systems analysis covers much of the same material as operations research, but the two fields differ substantially in direction.

> **Box 2.1** Katmandu Internet
>
> During a visit to Nepal in 1997, I had the pleasure of visiting an old friend in Katmandu. He took me to his new house, built just outside old Katmandu, where his family hosted a dinner for me. We were having a very pleasant time together, and when the time came for dinner, he took me outside to wash my hands. He used the water collected in a big barrel beside the house. In that moment, I realized that the house did not have running water.
>
> After the dinner our conversation went in many directions, and at one moment he asked for my help with setting up his computer for efficient access to the Internet. He brought his laptop from the other room, took the Internet cable and plugged it into the wall outlet. I was quite surprised to realize that the house was connected to the telecommunications network, so the Internet was widely and easily accessible, whereas the essential needs of water supply would probably not be met for a very long time in the future.
>
> Unfortunately pipes that carry water cost much more than the cables that carry billions of bits of information, and that difference creates quite an interesting reality.
>
> (A memory from 1997)

Operations research tends to be interested in specific techniques and their mathematical properties, while applied systems analysis focuses on the use of the methods (Wagner, 1969).

The Second World War, and the military applications of systems analysis in allocating scarce resources to different operations and to the activities within each operation, mobilized a large number of scientists to apply a scientific approach to dealing with this and other strategic and tactical problems. It has been stated by many (Maass et al, 1962; Buras, 1972; Loucks et al, 1981; Yeh, 1985; Simonovic, 1992a, among others) that the application of systems analysis to water systems design and management has been established as one of the most important advances in the field of civil engineering since the Second World War. One important milestone in the application of systems analysis to water resources engineering was the establishment of the Harvard Water Programme in the early 1960s.

A primary emphasis of systems analysis is on providing an improved basis for decision-making. It has been concluded that a gap still exists between research studies and the application of the systems approach in practice. The objective of this book is to reduce the existing gap as much as possible. Let us move on now to discuss water resources systems management as a broader set of activities that encompasses early water resources engineering.

2.4 CHARACTERISTICS OF WATER RESOURCES MANAGEMENT

In order to set the stage for our discussion of contemporary water resources systems management, only the period since the Second World War will be considered from this point on. The role of water engineers has expanded beyond the traditional concept of design and synthesis, to a larger multidisciplinary function serving a broad social environment. A key concept in the vision of the profession is the twofold role of professional engineers: first, a technical expert role, and second, the role of generalist. Engineers need to be skilled in managing technology within a social, cultural, political, environmental and economic context (Simonovic, 1992b).

For a historical overview of water management practice since the Second World War, I shall divide the developments into three chronological phases (see Figure 2.2):

1. rapid development, with an emphasis on design and construction;
2. slower development: the consideration of more complex projects, with an emphasis on optimal planning and design;
3. the utilization of existing projects, with an emphasis on operation, preventive maintenance and rehabilitation.

In a broad sense, these three phases apply to any development conditions. Some developed countries are already in phase three, while some developing countries are in phase one.

The chronological order of these phases obviously follows the requirements of a social development process. Each phase is characterized by a certain level of technological knowledge. Analytical tools and numerical procedures are logical choices for the first phase. Systems analysis techniques, optimization and simulation are powerful tools to support the planning phase. For the further development phase, expert systems, neural networks, EP and other emerging technologies seem to be the right technological choice.

The water resources management profession is involved in seeking solutions to problems which have a complex impact on society. The range of solutions must be determined and evaluated in terms of life improvements, resource commitments, public health and safety. The solutions to such problems require the application of scientific principles and an understanding of the social, political and economic conditions in which these problems exist.

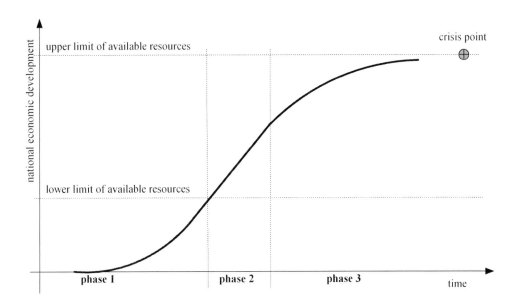

Figure 2.2 *Schematic presentation of the development process*

2.4.1 Phase one – rapid development

Initially, as society proceeds through the rapid development phase, the emphasis is on construction and design. The major activity in this phase involves determining the specific form of the end product (dam, irrigation canal, gate, etc.), its size, shape, properties and so on. This phase includes the mental activities of idea generation and evaluation, and the physical activity of creation, presentation and communication of the workable design. A trial and error procedure is the basic approach used in the design process. If an idea works, it is accepted and possibly improved through more trial and error development. If an idea fails, it is usually discarded with little thought or analysis of why. The design process is defined as the process of devising a system, component or process to meet desired needs. It is a decision-making process, in which basic science, mathematics and engineering are applied to convert resources optimally to meet stated objectives. Among the fundamental elements of the design process are the establishment of objectives and criteria, synthesis, analysis, construction, testing and evaluation.

2.4.2 Phase two – slower development

With the increasing demand and complexity of water resources systems, the need for more complex solutions grows. These more complex solutions require an increase in innovation. One design initiates several more designs. The increasing complexity of design requires a more systematic approach. That is how the need for systems analysis evolves.

In this phase of development the emphasis is on the role of planning and its close relationship to design. Planning is involved to some extent in each and every task attempted by a water manager. The planning process closely follows the systems approach, and usually involves the use of sophisticated optimization, simulation and computer tools and active participation of stakeholders (Simonovic and Akter, 2006; Soncini-Sessa and Castelletti, 2006).

The scope of planning is different from the scope of the design process. Planning is defined as the formulation of goals and objectives that are consistent with political, social, environmental, economic, technological and aesthetic constraints, and the general procedures designed to meet those goals and objectives. It is important to realize that a good plan brings together different ideas, concepts and factors, and combines them into a coherent structure.

Many water resources management activities are related to the use of natural resources. Natural resources become an economic asset when the supply begins to be insufficient for the demand. As an economic asset, natural resources must be governed by the laws of economics. Natural resources are also a common asset of the society, which means that they must be used for its benefit. While considering this, alternative solutions must be analysed carefully in order to identify a solution that is beneficial to all (Snelder and Hughey, 2005). As an awareness of the limits of resources increases, the slope of the 'development curve' changes, as shown in Figure 2.2. When the development level reaches the lower critical limit of the available resources, planning becomes a necessity.

2.4.3 Phase three – operation, preventive maintenance and rehabilitation

As the upper limit of the available resources approaches (Figure 2.2), the major water management tasks change considerably. To avoid the point when the actual development curve intersects the upper limit of available resources, shown in Figure 2.2 as a crisis point, the emphasis of the water resources management profession shifts from design and planning to optimal operation, preventive maintenance and rehabilitation of existing projects.

The problems now include finding ways to operate the complex systems and get the maximum functional benefits while considering political, social, environmental, economic and technological constraints. The management role emphasizes the need to examine all possible alternatives that could lead to more efficient use of available resources, protecting the environment while meeting the sometimes conflicting demands of the public. Water resources knowledge becomes only one facet of the interdisciplinary approach required to solve complex operational problems. In this phase the requirement exists for interdisciplinary engineering knowledge, systems analysis, experience, engineering intuition, judgement and common sense.

At this development stage preventive maintenance (of water intakes, conveyance systems, dams, water treatment systems, etc.) becomes a primary engineering task. Because long-neglected infrastructure can deteriorate rapidly, this seems to indicate that spending money to prevent deterioration can be less expensive in the long run than failing to maintain systems. The procedures required for preventive maintenance are different from the procedures used in design and operation. They incorporate design knowledge, working information and the new techniques, materials and procedures available.

The third phase, as described here, is also the stage at which the rehabilitation of ageing infrastructure becomes one of the main water management activities. Replacement of the infrastructure requires the development of new techniques for repair and renovation of existing systems that are structurally or functionally sound. In some countries considerable effort has been invested in inspecting water projects and collecting information regarding their current functional and structural performance. For efficient and successful rehabilitation, a combination of engineering knowledge, data and experience in design and operation is required.

2.4.4 Technological development

These three phases are essentially phases of socio-economic development, but they are all also characterized by certain levels of technological development. In the context of this discussion, technological development is defined as the process that leads to more effective production and delivery of new or significantly modified goods and/or services. This process also creates a body of concepts, techniques and data.

The first phase is characterized by a need for the analytical and numerical procedures required to support design activities. The second phase requires a more complex approach for the consideration of numerous alternatives. Systems analysis, offering a wide range of optimization and simulation algorithms, is the logical choice. The third phase, bounded by the upper limit of available resources, is characterized by the application of all the available engineering knowledge, experience, judgement and common sense. Systems analysis combined with the emerging technologies of artificial intelligence, such as expert or knowledge-based systems, neural networks and evolutionary optimization, provides a sound approach for solving the problems at this development stage.

Analytical and numerical procedures

Modern water resources systems management is an applied field of the natural sciences concerned primarily with the practical application of scientific knowledge. As such, water resources management relies heavily on fundamental knowledge derived from a wide range of fields. The conservative view of water resources management education considers mathematics, physics and chemistry to be the scientific basis of engineering. These subjects provide a strong background for design. The knowledge required during this phase is primarily concerned with a description of the properties and relations of quantities and magnitudes.

Systems analysis

Entrance into the second phase of development changes the emphasis of water management. Problems are more complex and increasingly interdisciplinary. The number of alternative solutions containing attractive features increases. The impacts of solutions are more important, and resources are now seen as having finite limits. As a result of these many factors, systems analysis is becoming an increasingly popular and at times necessary means of addressing engineering projects. The increasing use of computers is a contributing factor in the widespread application of systems analysis.

A structure (or theory) is essential if we are to effectively relate and interpret engineering observations. The first phase of rapid development may be described as a state of somewhat unrelated facts. Separate and often conflicting engineering impressions have not yet been brought into focus by being assembled into a unified structure. During the second phase, this basic structure of principles is developed through a unification of the diverse manifestations of social, environmental and economic processes.

In the context of this discussion, systems analysis is defined as the integration of analytical operations research techniques (optimization and simulation) and systematic approaches to the definition and solution of large-scale, complex, water resources systems. It is important to note that this definition describes the generic function of systems analysis as a way of thinking about and dealing with the complex physical realities that are the trademark of large-scale water resources problems.

Emerging technologies

The third development phase adheres to the slower development which is characteristic of the planning phase. Operation, preventive maintenance and rehabilitation activities require the integration of all available knowledge, experience and engineering common sense. Learning from experience is one very important aspect introduced during this phase. Without a structure (developed in the second phase) to interrelate facts and observations, it is difficult to learn from experience. It is difficult to use the past to educate for the future.

For the application of systems analysis, computers have been introduced into everyday engineering practice. Emerging technologies, specifically expert systems and neural networks, are extending the role of both systems analysis and computers (Ahmad and Simonovic, 2006). They have become a source of knowledge and problem-solving advice. The usefulness of expert systems is evident when a great deal of private knowledge, gained through experience, is needed to supplement the available public knowledge found in engineering textbooks and manuals. Private knowledge is not mere data, and generally cannot be adequately represented using mathematical relationships. It is also very dynamic in reflecting the heuristic nature of the learning process. Knowledge-based systems are very useful in organizing and structuring this type of knowledge. In addition to this, neural networks can deal with approximate data; learn automatically from a database of examples; learn incrementally, adapting to a changing environment; generalize to situations they have not encountered before; and execute decisions very quickly.

2.4.5 The relationship between changing practice and technological development

We can begin this discussion with a question. Is technology the impetus for change in the water resources systems management practice, or is it the changes in water resources systems management practice that have driven professionals to use more innovative methods and therefore more highly developed technology? However, perhaps no answer is necessary. It is indisputable that a strong relationship exists between the changes in water resources management and those in technology, even if it is not clear which was the driver of change (Claeys et al, 2006). There have been times when water resources management was in need of more adequate technological support. At other times, however, technology has offered our profession more than it could actually use.

The use of systems analysis combined with tools from emerging technologies is suitable for the following basic water resources management tasks.

Design

An expert system combined with either an optimization or simulation tool may help in selecting initial design requirements and in supporting design decisions related to equipment and physical facilities (Zhelev, 2007). In construction, experience plays a major role. Experience-based decisions to be made in this area are numerous: the configuration of crews, selection of equipment types, design of transportation facilities and so on.

Planning

Multi-objective analysis integrated with a knowledge-based system and its necessary data banks can help in planning projects where social and environmental issues, as well as public opinion, must be considered. Neural networks and data mining can be used to learn from data generated by the optimization and simulation models and expert knowledge. A combined system can then provide advice in the planning process.

Operation

System dynamics simulation combined with neural networks and expert systems might be very helpful in supporting the rehabilitation, repair and maintenance of water resources infrastructure, where the traditionally distinct design and construction roles of participants have become merged. A computer-based decision support system can be developed to help monitor a system where there is time-dependent behaviour. A neural network can be used to adapt to a changing environment and to help in diagnosing and/or predicting structural failures in a structure where a failure could have catastrophic results. A multi-objective tool can be used to evaluate alternative plans for rehabilitation using public input and all other available data.

Water resources management is entering an important stage, where the acceptance of available technology will have a major impact on the changing practice. On the educational side we face a dilemma. Should water resources professionals 'stick to their knitting' at least as far as the curriculum is concerned, or should educators in the field further diversify and develop courses in new areas of technology? This discussion has been going on for a while. It is the main objective of this book to provide some additional arguments in support of diversification and development of courses in new areas of technology. The choice of tools presented in the book, and its methodological basis, reflect the relationship between the development process, characteristics of water resources systems management, and technological development.

2.5 THE FUTURE DIRECTION FOR WATER RESOURCES SYSTEMS MANAGEMENT – A CRITICAL PATH

In the period from the end of the Second World War to the present, there has been an intensive international emphasis on water. The activities, the preparations that preceded them and the discussions that followed have strengthened general perceptions of the world water crisis, and understanding of the needed responses. The United Nations (UN) sponsored the International Hydrologic Decade (1965–1974), which provided support for international cooperation to effectively use transnational water resources and collect hydrological data. The Mar del Plata conference of 1977 initiated a series of global activities related to water. Of these, the International Drinking Water and Sanitation Decade (1981–1990) brought a valuable extension of basic services to the poor. These experiences show the magnitude of the present task, of providing the huge expansion in basic water supply and sanitation services needed today and in the years to come.

The International Conference on Water and the Environment in Dublin in 1992 set out the four Dublin Principles that are still relevant today:

1 Freshwater is a finite and vulnerable resource, essential to sustain life, development and the environment.
2 Water development and management should be based on a participatory approach, involving users, planners and policy-makers at all levels.
3 Women play a central part in the provision, management and safeguarding of water.
4 Water has an economic value in all its competing uses and should be recognized as an economic good.

The UN Conference on the Environment and Development (UNCED) in 1992 produced Agenda 21, which, with its seven programme areas for action in freshwater, helped to mobilize change and heralded the beginning of the still very slow evolution in water management practices. Both of these conferences were seminal in that they placed water at the centre of the sustainable development debate.

These and many other important meetings set targets for improvements in water management, very few of which have been met. However, of all the major target-setting events of recent years, the UN Summit of 2000, which set the Millennium Development Goals for 2015, remains the most influential. Among the targets set forth, the following are the most relevant to water:

1 To halve the proportion of people living on less than $1 per day.
2 To halve the proportion of people suffering from hunger.

3   To halve the proportion of people without access to safe drinking water and sanitation.
4   To ensure that all children, boys and girls equally, can complete a course of primary education.
5   To reduce maternal mortality by 75 per cent and under-five mortality by two-thirds.
6   To halt and reverse the spread of HIV/AIDS, malaria and the other major diseases.
7   To provide special assistance to children orphaned by HIV/AIDS.

All of this needs to be achieved while protecting the environment from further degradation. The UN recognized that these aims, which focus on poverty, education and health, cannot be achieved without adequate and equitable access to resources, and the most fundamental of these are water and energy. The effort continues. World Water Day 2005 marked the start of the International Decade for Action, 'Water for Life' (2005–2015), proclaimed by the United Nations General Assembly.

2.6 REFERENCES

Ahmad, S. and S.P. Simonovic (2006), 'An Intelligent Decision Support System for Management of Floods', *Water Resources Management*, 20(3): 391–410

Buras, N. (1972), *Scientific Allocation of Water Resources*, Elsevier, New York.

Claeys, F., M. Chtepen and L. Benedetti (2006), 'Distributed Virtual Experiments in Water Quality Management', *Water Science and Technology*, (53)1: 297–305

Levi, E. (1995), *The Science of Water: The Foundation of Modern Hydraulics*, ASCE Press, New York

Loucks, D.P., J.R. Stedinger and D.A. Haith (1981), *Water Resource Systems Planning and Analysis*, Prentice-Hall, Englewood Cliffs, NJ

Maass, A., M.M. Hufschmidt, R. Dorfman, H.A. Thomas, Jr., S.A. Marglin and G.M. Fair (1962), *Design of Water-Resource Systems: New Techniques for Relating Economic Objectives, Engineering Analysis and Governmental Planning*, Harvard University Press, Cambridge, MA

Simonovic, S.P. (1992a), 'Reservoir Systems Analysis: Closing Gap between Theory and Practice', *ASCE Water Resources Planning and Management Division*, 118(3): 262–280

Simonovic, S.P. (1992b), 'Challenges of the Changing Profession', *ASCE Journal of Professional Issues in Engineering*, 118(1): 1–9

Simonovic, S.P. and T. Akter (2006), 'Participatory Planning in the Red River Basin, Canada', *IFAC Annual Reviews in Control*, 30(2): 183–192

Snelder, T.H. and K.F.D. Hughey (2005), 'The Use of an Ecological Clasification to Improve

Water Resoure Planning in New Zealand', *Environmental Management*, 36(5): 741–756

Soncini-Sessa, R. and A. Castelletti (2006), 'A Procedural Approach to Strengthening Integration and Participation in Water Resource Planning', *Environmental Modelling and Software*, 21(10): 1455–1470

Sprague de Camp, L. (1963), *The Ancient Engineers*, Ballantine Books, New York

Wagner, H. (1969), *Principles of Operations Research*, Prentice-Hall, Englewood Cliffs, NJ

Yeh, W.W.-G. (1985), 'Reservoir Management and Operations Models: A State-of-the-Art Review', *Water Resources Research*, 21(12): 1797–1818

Zhelev, T. (2007), 'The Conceptual Design Approach – A Process Integration Approach on the Move', *Resources Conservation and Recycling*, 50(2): 143–157

2.7 EXERCISES

1. Research the history of water resources systems management in your country. Construct a timeline of water resources engineering activities – similar to Table 2.1 – for your country. Choose one of the water resources sectors below and study its history in your country:
   a. Municipal water supply.
   b. Flood control.
   c. Irrigation water supply.
   d. Hydropower generation.
2. Identify some common water management issues that are found in the region where you live. What management alternatives might effectively reduce some of the problems and provide additional economic, environmental or social benefits?
3. Position the region where you live on the development curve in Figure 2.2. Support your finding using two water resources engineering examples. Describe these two examples in the context of the curve in Figure 2.2.
4. What tools are available for contemporary water resources management?
5. Is the water resources management in your country an example of participatory management or not? Why?

3

An Introduction to Water Resources Systems Management

It is time to provide a clear definition of water resources systems management. To manage, in everyday language, is to handle or direct with a degree of skill. It can also be seen as exercising executive, administrative and supervisory duties. Managing can involve altering situations by manipulation, or succeeding in accomplishing something. Management is the act or art of managing. These definitions extend the definition of engineering provided earlier: the application of science and mathematics by which the properties of matter and the sources of energy in nature are made useful to people.

Water resources systems management is an iterative process of integrated decision-making regarding the uses and modifications of waters and related lands within a geographic region. This process provides a chance for users to balance their diverse needs and uses of water as an environmental resource, and to consider how their cumulative actions may affect the long-term sustainability of water and related land resources. The guiding principles of the process are a systems view, partnerships, uncertainty, a geographic focus and a reliance on strong science and reliable data.

This gives us a definition of water resources systems management which includes the traditional activities of water resources engineering: planning, design, maintenance and operation of the water-related infrastructure. It is more comprehensive, and integrates all these activities in an approach to support the decision-making process based on the engineering, natural, social and other sciences.

In this chapter we focus on this broad area of water resources systems management.

> **Box 3.1** The river doctor
>
> The Arakawa River in Yamanashi Prefecture, Japan, was slowly flowing below our feet. Dr Takeuchi was discussing some recent works on the river banks with a group of Japanese officials responsible for the management of the river.
>
> He is the *river doctor*. For every river in Japan the regional governments appoint an expert to serve as the river doctor. His/her role is to provide technical assistance and advice to those who are responsible for managing the river basin.
>
> (A memory from 1996)

3.1 WATER RESOURCES SYSTEMS MANAGEMENT IS HUMAN

To manage water resources systems is a human activity. That may seem obvious, but it has a deep relevance, in my view. Different people will have a different sense of its importance (Petroski, 1992), so let me provide some of my thoughts on the subject.

To be human is *to survive*. The basic instincts that characterize the human race ensure that we survive – through major droughts, major floods, earthquakes, wars, ups and downs of the economy – and continue our existence on this planet the best way we know. Water resources systems management is necessary for basic survival. About 83 per cent of our blood is made up of water. Water helps us digest our food, transport waste, and control our body temperature. We need water for drinking and to perform basic hygiene tasks. We need water for growing food and food preparation. Water is not equally distributed in time and space, but the management of water resources systems enables a redistribution of water resources in time and space.

To be human is *to participate actively in life*. Our effort is required for survival. There are everyday tasks that we can handle alone, and there are others that require a number of people, a variety of skills and extensive knowledge of different disciplines. Many water resources systems management activities exceed the capacity of a single person, and therefore we work with others. We create groups to tackle larger tasks. In groups, humans behave in various ways. Some of us are born leaders and some are followers. Some are great in seeing a broad picture and some are impressed with details. The successful management of water resources systems depends on every individual who may be affected by the system, or who can affect the system, identifying his or her role and participating actively.

To be human is *to learn from the past, act in the present and affect the future*. Maintaining the temporal continuity of human existence is one of the conditions for survival. Most water resources system management decisions are made in the present, but how these decisions are made shapes the system behaviour and affects the future states of the system. Information about past system performance (system states as a consequence of different exogenous/external and endogenous/internal conditions) proves to be invaluable in making new decisions. Historical information on river flows provides the range between the maximum and minimum flow that may be expected to occur in the future. Delays that occur in certain impacts as a consequence of past decisions may help in understanding better the long-term processes within the water resources system.

To be human is *to care about other humans*. We are social beings who are organized in families that form societies, and we function as complex systems in interaction with other living beings and the surrounding environment. The management of water resources systems penetrates deeply into the social structure at every level. Many water-related activities require a demonstration of responsibility by each member of different social organizations. In a family that lives in a water-deprived region of the world, there is a responsibility for carrying water over long distances in order to provide for basic needs of other family members. In organized communities there is collective responsibility, usually exercised by different levels of government, to provide safe water to meet the needs of community members. When natural disasters strike, one of the first needs of survivors is for clean water.

To be human is *to care about the environment*. Human beings are usually blamed for the careless destruction of the environment in order to support the relentless growth of the economy and the development of society, and it is right that they should be. However, ignorance of the feedback relationships between the environment and humans – as elements of the environment – cannot continue forever. There are many cases where ignorance of the environment cannot be tolerated any more, and wealth accumulated through developmental processes is being used to correct the damage done and reinstate the environment to its original form (Weiskel et al, 2007).

One common example of the management of water resources systems is the process of river diking. Dikes provide control of the water depth (helping navigation) and the extent of flooding (helping with changes of land use in floodplains). In many developed countries today, dikes are being removed to allow rivers to flood their floodplains more frequently, and support the diversity of life forms in these transitional zones. Many water resources systems are being designed and operated today to provide for multiple functions: both to satisfy human needs and to maintain the integrity of the environment.

To be human is *to care about other living beings*. Water has an important role in the maintenance of biodiversity. The environmental sector is an important water user, and one that often finds itself at the bottom of the list of priorities when supplies become scarce. One aspect of water resources systems management looks at how the needs of other ecosystems can coexist with parallel human water demands. Water maintains natural ecosystems, which sustain biodiversity, help to regulate the hydrological cycle, and bring value to people in the form of goods and services derived from activities in these ecosystems. It is increasingly being recognized that one of the costs of large-scale water resources management systems is the draining of wetlands, or reduced river flows that starve wetlands of their water. In other cases the inflow of drainage into wetlands changes the water quality and water levels, negatively affecting plant and animal life. Currently, water resources systems management decisions are being taken that do not support any reduction of the flow to wetlands, and to provide for the protection of their water quality through either natural or human-made wastewater treatment processes.

To be human is *to make mistakes*. Recent studies of Canada's infrastructure – the water supply and sewer systems that we take for granted – conclude that it has been so badly neglected in many areas of the country that it would take billions of dollars to put things back in shape. This condition resulted in part from maintenance being put off to save money during years when energy and personnel costs were taking larger portions of municipal budgets. Some water pipes in large Canadian cities are 100 or more years old, and they were neither designed nor expected to last forever. Ideally, such pipes should be replaced on an ongoing basis to keep the whole water supply system in a reasonably sound condition. Since humans are fallible, so are their water resources systems. Thus the history of water resources systems management may be told in its failures as well as in its successes. Success may be impressive, but mistakes can often teach us more.

To be human is *to be destructive*. Humans are warriors and architects of much destruction, from the fall of the Roman Empire to the hundreds of wars going on today in various parts of the world. Many destructive activities create damage to vulnerable water resources systems. At the same time, these destructions result in a more vigorous set of activities in order to bring about a return to normal life.

To be human is *to balance the positive and negative sides of our nature*. Systems theory, and especially the concept of feedback, has gained significantly from the biological notion of homeostasis. Homeostasis refers to the amazing capacity of higher organisms to maintain physiological stability in the face of dramatically varying external and internal conditions. At the centre of the idea is the notion that living organisms can apparently react automatically to counter disturbances from the preferred or normal state. The emphasis is on interactions between the negative and positive sides of our nature, because it is these interactions that embody homeostatic equilibrium. Treating every successful water resources systems management decision, and every failure, as an opportunity to test hypotheses, whether they are embodied in novel designs or in theories about the nature and process of water resources management itself, makes every case study relevant for increasing our understanding of ourselves and the world around us.

3.2 WHAT IS INVOLVED IN WATER RESOURCES SYSTEMS MANAGEMENT?

If we use the definition that water resources systems management is an iterative process of integrated decision-making regarding the uses and modifications of waters and related lands within a geographic region, this leads to a number of issues related to the management of water resources systems.

A systems view

We have inherited both natural water resources systems and many generations of human-made systems. Only recently have we come to understand the underlying structure and characteristics of natural and human-made systems in a scientific sense. The switch to thinking not in terms of single functions but in terms of 'systems' is still in progress. Nor are water resources systems isolated: they interrelate with human and physical systems, and this leads to innumerable financial, economic, social and political considerations.

Partnerships

Water resources systems management requires use of the engineering, social, natural, ecological and economic sciences. Common goals for water and land resources must be developed among people of diverse social backgrounds and values. An understanding of the structure and function – historical and current – of the water resources system is required, so that the various effects of alternative actions can be considered. The decision process must also consider the economic benefits and costs of alternative actions, and blend current economic conditions with considerations of the long-term sustainability of the ecosystem.

Uncertainty

Human modifications of waters and related lands directly alter the delivery of water, sediments and nutrients, and thus fundamentally alter aquatic systems. These alterations are made using imperfect information about many processes involved, and this brings multiple objective uncertainties into the decision-making process. People have varying goals and values related to uses of local water and related land

> **Box 3.2** Did you know?
>
> - Because 70 per cent of the Earth is covered by water, it is called the 'Blue planet'. Yet only 2.5 per cent of the world's water is freshwater, while 97.5 per cent is saltwater in the oceans. Only 0.3 per cent of the world's freshwater is available from rivers, lakes and reservoirs; 30 per cent is groundwater, while the rest is stored in glaciers, ice sheets and mountainous areas: all places that we can barely access.
> - Raindrops are not tear-shaped. Scientists using high-speed cameras have discovered that raindrops resemble the shape of a small hamburger bun.
> - About two-thirds of the human body is made up of water. Some parts of the body contain more water than others. For example, 70 per cent of your skin is made up of water.
> - You can survive about a month without food, but only from five to seven days without water.
> - Each day humans must replace 2.4 litres of water, some through drinking and the rest taken by the body from the foods eaten.
> - Most of our food is made up of water: for example tomatoes (95 per cent), spinach (91 per cent), milk (90 per cent), apples (85 per cent), potatoes (80 per cent), beef (61 per cent), hot dogs (56 per cent).
> - More than half of the world's animal and plant species live in an aquatic environment.
> - Each year 3–4 million people die of waterborne diseases, including 2 million children who die of diarrhoea.
> - In the developing countries, 80 per cent of illnesses are water-related.
>
> *Source*: UNESCO water portal, www.unesco.org/water (last accessed December 2005)

resources. These form subjective uncertainties for inclusion in the decision-making process.

Space

As a form of ecosystem management, water resources systems management encompasses the entire watershed system, from uplands and headwaters to floodplain wetlands and river channels. It focuses on the processing of energy and materials (water, sediments, nutrients and toxins) downstream through this system. Of principal concern is the management of the basin's water budget: that is, the transformation of precipitation through the processes of evaporation, infiltration and overland flow. This transformation of groundwater and overland flow defines the delivery patterns to particular streams, lakes and wetlands, and to a great extent shapes the nature of these aquatic systems.

Science and data

Like water itself, the science of water resources systems management flows in all directions: to hydrology, hydraulics, geology, meteorology, oceanography, environmental science, engineering, law, economics and so on. Water resources management decision-making requires information on both specific locations and general principles. To provide appropriate water resources management decisions requires an integrated approach and reliable data (Flugel, 2007).

Economic efficiency

With growing water scarcity and increasing competition across water-using sectors, the need for water savings and more efficient water use has increased in importance in water resources management. An improvement in the physical efficiency of water use is related to water conservation, through increasing the fraction of water beneficially used over water applied. Enhancing economic efficiency is a related but broader concept, which involves seeking the highest economic value of water use through both physical and management measures (Boland, 2007).

3.3 HOW IS WATER RESOURCES SYSTEMS MANAGEMENT DONE?

Water resources systems management provides a framework for integrated decision-making. Within this framework managers strive to:

- assess the nature and status of the water system;
- define short-term and long-term goals for the system;
- determine the objectives and actions needed to achieve the selected goals;
- assess both the benefits and the costs of each action;
- implement the desired actions;
- evaluate the effects of actions and progress towards goals;
- re-evaluate goals and objectives as part of an iterative process.

This framework is implemented through the use of various methods such as simulation, optimization and multi-objective analysis, and applied to the solution of a variety of water resources planning, design, operation and maintenance problems (Olsson and Andersson, 2007).

A series of practical steps for implementing water resources systems management were adopted from Jewell (1986). They include:

1   definition of the problem;
2   gathering data;
3   development of criteria for evaluating alternatives;
4   formulation of alternatives;
5   evaluation of alternatives;
6   choosing the best alternative;
7   final design/plan implementation.

Often several steps in the water resources systems management approach are considered simultaneously, facilitating feedback and allowing a natural progression in the problem-solving process. The water resources systems management approach has several defining characteristics. It is a repetitive process, with feedback allowed from any step to any previous step. Frequently, because water resources systems analysis takes such a broad approach to problem solving, interdisciplinary teams must be called in. Coordination and commonality of technique among the disciplines is sometimes hard to achieve. However, if applied with ingenuity and flexibility, the systems approach can provide a common basis for understanding among specialists from seemingly unrelated fields and disciplines. Close communication among the parties involved in applying the systems approach is essential if this understanding is to be achieved.

Definition of the problem

Problem definition may require iteration and careful investigation, because symptoms may mask the true cause of the problem. A key step in problem definition is the identification of any systems and subsystems that are part of the problem, or related in some way to it. This set of systems and interrelationships is called the *environment of the problem*. This environment sets the limit on factors that will be considered when analysing the problem. Any factors that cannot be included in the problem environment must be included as inputs to, or outputs from, the problem environment. When defining the problem and its environment, the best approach is to make the definition as general as possible. The largest problem over which there is a reasonable chance of maintaining control should be the problem defined.

Gathering data

There are several stages at which it may be necessary to gather data to assist in water resources systems management. Some background data will have to be gathered at the problem definition stage, and data gathering and analysis will continue through the final plan/design and implementation stage. As the process continues and more data are gathered, they will help to identify when feedback to a previous step is required.

Data will be required at the problem definition stage to evaluate whether a problem really exists, to establish what components, subsystems and elements can be reasonably included in the delineation of the problem environment, and to define the interactions between components and subsystems. Data will be needed during later steps to establish constraints on the problem and the systems involved in it, to increase the set of quantifiable variables and parameters (constants) through statistical observation or development of measuring techniques, to suggest what mathematical models might contribute effectively to the analysis, to estimate values for coefficients and parameters used in any mathematical models of the system, and to check the validity of any estimated system outputs. When feedback is required, the data previously acquired can assist in redefining the problem, systems or system models.

Development of criteria for evaluating alternatives

Criteria must be developed for measuring the degree of attainment of system objectives. This will facilitate a rational choice of a particular set of actions (from among a large number of feasible alternatives) which will best accomplish the established objectives. Sometimes the decision can be made on the basis of absolute values of these criteria, such as the cost of producing one unit of a particular product. On other occasions only relative values are available. In these circumstances the individual preferences that produce those values may dictate the ranking of alternatives. Economic comparisons such as cost–benefit analysis can be used in this process.

In most complex real-world water problems, more than one objective can be identified. A quantitative or qualitative analysis of the trade-offs between the objectives must be made. It may be possible to restate some of the objectives as required levels of performance, which then become system constraints. Then the system can be designed to perform optimally in terms of the remaining objectives. For many water problems, cost effectiveness is the primary objective. Cost effectiveness can be defined as the lowest possible cost for a set level of control of a water system, or the highest level of water system control for a set cost.

Formulation of alternatives

The formulation of alternatives essentially involves the development of system models that will be used in later analysis and decision-making, in conjunction with criteria for evaluation of the outcomes. If at all possible, these models should be mathematical in nature. However, it should not be assumed that mathematical model-building and optimization techniques are either required or sufficient for application of the systems approach. Many problems contain unquantifiable variables and parameters which would render results generated by even the most elegant mathematical model meaningless. If it is not practical to develop mathematical models, subjective models that describe the problem environment and systems included can be constructed. Models allow a more explicit description of the problem and its systems and facilitate the rapid examination of alternatives. The primary emphasis in this book is on problems that can at least partially be represented by mathematical models. However, it is made clear where these models have limitations, and it is appropriate to apply subjective models as part of the decision-making process.

Effective model building is a combination of art and science. The science includes the technical principles of mathematics, physics, engineering and other sciences. The art is the creative application of these principles to describe physical or social phenomena. Practice is the best way to learn the art of model building, but this practice must be based on a thorough understanding of the science. Although the art of modelling cannot be taught in a single course, a course that introduces the student to the water resources systems management approach will lay the foundation for further development.

Evaluation of alternatives

To evaluate the alternatives that have been developed, some form of analysis procedure must be used. Numerous mathematical techniques are available, as will be discussed later in the book. They include the simplex method for linear programming (LP) models, the various methods for solving ordinary and partial differential equations or systems of differential equations, matrix algebra, various economic analyses and deterministic or stochastic computer simulation. Subjective analysis techniques may be used for multi-objective analysis, or the subjective analysis of intangibles.

The appropriate analysis procedures for a particular problem will generate a set of solutions for the alternatives, which can be tested according to the established evaluative criteria. In addition, these solution procedures should allow efficient utilization of human and computational resources.

As part of the analysis stage, the importance of each variable should be checked. This is called *sensitivity analysis*, and it involves testing how much the model output will change given changes in the values of the decision variables and model parameters.

Choosing the best alternative

A choice of the best alternative from among those analysed must be made in the context of the objectives and evaluative criteria previously established. It must also take into account non-quantifiable aspects of the problem, such as aesthetic and political considerations. The chosen alternative will greatly influence the development of the final plan/design, and will determine in large part the implementability of the suggested solution.

Preferably the best alternative is chosen from the mathematical optimization within feasibility constraints. Near-optimum solutions can still be useful, especially if sensitivity analysis has shown that the solution is not sensitive to changes in the main variables near the optimum point.

Final plan/design/operation strategy implementation

The actual final planning, design and operations are primarily technical matters, and are conducted within the constraints and specifications developed in the earlier stages of the water resources systems management process. One of the end products of final planning, design and operation is a report which describes the recommendations made. To be effective, this report must also include information on the approach taken to the problem. The report should be written in the context of the audience for which it is intended. A well-written non-technical report can go a long way towards developing public support for the recommended problem solution, whereas a well-written technical report given to the same audience may be intimidating and actually reduce support for the recommendations.

3.4 REFERENCES

Boland, J.J. (2007), 'The Business of Water', *ASCE Journal of Water Resources Planning and Management*, 133(3): 189–191

Flugel, W.-A. (2007), 'The Adaptive Integrated Data Information System (AIDIS) for Global Water Research', *Water Resources Management*, 21(1): 199–210

Jewell, T.K. (1986), *A Systems Approach to Civil Engineering Planning and Design*, Harper and Row, New York

Olsson, J.A. and L. Andersson (2007), 'Possibilities and Problems with the Use of Models as a Communication Tool in Water Resources Management', *Water Resources Management*, 21(1): 97–110

Petroski, H. (1992), *To Engineer is Human*, Vintage Books – Random House, Inc., New York

Weiskel, P.K., R.M. Vogel and P.A. Steeves (2007) 'Water Use Regimes: Characterizing Direct Human Interaction with Hydrologic Systems', *Water Resources Research*, 43(4), Art. No. W04402

3.5 EXERCISES

1. Water resources systems management in the province of Ontario, Canada is conducted by the Conservation Authorities. Investigate how the water resources systems management is done by these agencies. As the main source of information use the Conservation Ontario website (http://conservation-ontario.on.ca last accessed 10 January 2007).
2. Define water stewardship.
3. (For those outside Canada). Give an example of water resources systems management agency in your country. Describe the management principles and institutional organization in place.
4. Present your understanding of the practical steps for implementing water resources systems management discussed in Section 3.3 for the following problems. In answering the question, be as real as possible (e.g. when listing alternatives provide real options; when mentioning data collection indicate all real data that you think is necessary; etc.).
    a. Long-term (next 50 years) water supply for the city of London.
    b. Flood protection for the (fictional) community of Riverton, which experiences flooding every two to three years.
    c. Water quality management and regulation for the State of Indiana, US.

PART II
Applied Systems Analysis

4

General Systems Theory

Systems are as pervasive as the universe around us. At one extreme, they are as large as the universe itself, while at the other, they are as small as the atom. Systems first existed in natural forms, but since the appearance on Earth of human beings, a variety of human-made systems have come into existence. Only recently have we come to understand the underlying structure and characteristics of natural and human-made systems in a scientific sense. The concept of systems thinking today plays a dominant role in a wide range of fields, from engineering to different topics of pure science. Professions and jobs have appeared in the last 40 years that go by names such as systems design, systems analysis and systems engineering.

The roots of this development are complex. One thread is the development from power engineering to control engineering, which has led to computers and automation. Self-controlling devices have appeared, from home thermostats to the self-steering missiles of the Second World War and the improved missiles of today. We have come to see technology in terms not of single functions but of 'systems'. Understanding the workings of a centrifugal pump, a spillway gate and a thermostat is within the competence of any expert trained in the relevant specialist field, but the same is not true of, for example, a modern wastewater treatment plant. It encompasses heterogeneous technologies, with components that are hydraulic, mechanical, electronic, chemical, biological and so on. It is also interrelated with wider human and physical systems, and this leads to innumerable financial, economic, social and political considerations.

Thus a systems approach based on systems analysis becomes necessary. In order to find the best (or even a good) way to realize a complex project, a system specialist (or team of specialists) must consider alternative solutions, and choose one that promises the optimal outcomes at maximum efficiency in a very complex network of interactions. This requires elaborate techniques and computing power for solving problems which are far beyond the unaided capacity of a single individual. Both the hardware of computing machines, automation and cybernetics, and the software of systems science represent a new technology which has been called the Second Industrial Revolution. The application of this technology is not limited to the industrial complex. The systems approach is also, for example, applied to pressing problems such as water pollution, traffic congestion and city planning.

It should not be taken that systems analysis is merely one of many changes in our contemporary technological society. Rather it involves a change in our basic thought processes. In one way or another, we are forced to deal with complexities, with wholes or systems, in all fields of knowledge. This implies a basic reorientation in scientific thinking across almost all disciplines, from subatomic physics to history.

4.1 SOME SYSTEM DEFINITIONS

4.1.1 What is a system?

Some kind of system is inherent in all but the most trivial engineering planning and design problems. To understand a problem, the analyst must be able to recognize and understand the system that surrounds and includes it. This has not always been done effectively in the past. Some reasons for poor system definition are poor communications, lack of knowledge of interrelationships, politics, limited objectives and transformation difficulties.

What then is a system? There are many variations in definitions of a system: for example, one dictionary alone provides no less than 15 ways to define the word. However, all of them share common traits. In the most general sense a system is a collection of various structural and non-structural elements that are connected and organized in such a way as to achieve some specific objective through the control and distribution of material resources, energy and information.

A more formal definition of a system can be stated as:

$$S: X \to Y \tag{4.1}$$

where X is an input vector and Y is an output vector. To put this differently, a system is a set of operations that transforms input vector X into output vector Y.

A schematic representation of the system definition is shown in Figure 4.1. This sees the system in terms of input, output, a transformation process, feedback and a restriction. *Input* energizes the operation of a given *transformation process*. The final state of the process is known as the *output*. *Feedback* is the name given to a number of operations which compare the actual output with an objective, and identify the discrepancies between them.

A more comprehensive definition may look at a system as an assemblage or combination of elements or parts forming a complex or unitary whole, such as a river system or a transportation system; any assemblage or set of correlated members, such as a system of currency; an ordered and comprehensive assemblage of facts, principles or doctrines in a particular field of knowledge or thought, such as a system of philosophy; a coordinated body of methods or a complex scheme or plan of procedure, such as a system of organization and management; or any regular or special method or plan of procedure, such as a system of marking, numbering or measuring (Blanchard and Fabrycky, 1990).

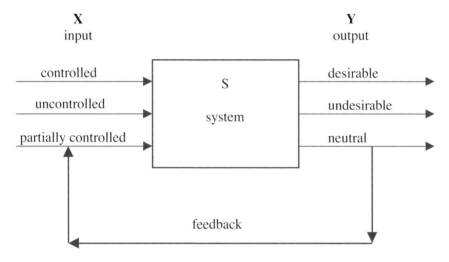

Figure 4.1 *Schematic presentation of a system definition*

4.1.2 Systems thinking

The problems that we currently face in water resources management have been stubbornly resistant to solution, particularly with a one-sided solution (that is, one that looks at the problem in a narrow linear way and 'solves' it). As we are discovering, there is no way to completely solve the problem of providing a safe water supply, and the lack of one is steadily affecting a rising number of people around the globe. The problems of water sanitation and climate change (the build-up of carbon dioxide in the atmosphere, ozone depletion, etc.) also fall into the category of 'resistant to simple solutions'.

It has probably never been possible to solve problems such as these, but the situation was less critical when it was possible to shift them out of the area of immediate concern. In an era when the connections among the various subsystems were less tight, it was possible to score a temporary victory by essentially pushing a problem into the future or into 'someone else's backyard' (Richmond, 1993). Unfortunately, there is less and less space to do that in our modern world, and temporary victories of this nature do not add up to a viable long-term strategy. In an interconnected world we do not have places that we can treat as holders for our 'garbage', and it is necessary to face the consequences of our present decisions.

Human beings are quick problem solvers. From an evolutionary standpoint, quick problem solvers were the ones who survived. We can often quickly determine a cause for a situation that we identify as a problem. For example, if a river overflows its banks, we might conclude that this is because it was raining a lot during the previous couple of days. This approach works well for simple problems, but it works less well as the problems get more complex.

If we accept the argument that the primary source of the growing intractability of problems in water resources management is a tightening of the links between the various physical and social subsystems that make up our water systems reality, we will agree that systems thinking provides us with tools for a better understanding of these difficult water resources management problems. The fundamental tools of the systems approach have been in use for over 30 years (Forrester, 1990) and are now well established. However, they require a shift in the way we think about the performance of a water resources system. In particular, they require that we move away from looking at isolated situations and their causes, and start to look at the water resources system as a system made up of interacting parts. Systems thinkers use diagramming languages to visually depict the feedback structures of these systems. They then use simulation to play out the associated dynamics. These tools give us the ability to see into a 'neighbour's backyard', even if that backyard is thousands of kilometres away. They also confer the ability to experience the consequences of our decisions, even if they are somewhere in the future.

The important question is, how can the framework, the process and the technologies of systems thinking be transferred to future water managers in a reasonable amount of time? According to Richmond (1993), if we view systems thinking within the broader context of critical thinking skills, and recognize the multidimensional nature of the thinking skills involved in systems thinking, we can greatly reduce the time it takes to pick up this framework. As this framework increasingly becomes the context within which we think, we shall gain much greater leverage in addressing the pressing water management issues that await us in the future. A switch must occur from teacher-directed learning to learner-directed learning. Open classrooms, computer-aided instruction and offering interdisciplinary courses are but a few of

Box 4.1 The Three Gorges Dam

The day we visited, explosions were blasting the rock where the future ship locks would be located. A large portion of the dam had already been erected, and we were standing on the side close to the location of the ship lift trying to comprehend, with great interest, the scale of the project and its importance for the people of China.

Figures were flying into my face. The Three Gorges Project (TGP) is designed as a concrete gravity dam with a crest elevation of 185 m and maximum height of 175 m. The dam axis is 2309.4 m long. The 483 m spillway, located in the middle of the main dam, has 23 bottom outlets 7 m wide and 9 m high. Each of its 22 surface sluice gates is 8 m wide. The maximum discharge capacity of the TGP is 116,000 cubic metres (m^3), the biggest in the world. Two powerhouses, flanking the spillway, accommodate 26 sets of turbine-generators altogether. They are the biggest units ever made: each has a generating capacity of 700 MW. With all these huge generators, the TGP is designed to generate 84.7 TW/h of electricity. The TGP ship lift is also the world's largest. With a one-step vertical hoisting mechanism, the ship lock is capable of carrying a 3000-ton passenger liner or cargo boat.

This is:

- a project to create the biggest flood control capability in the world. The total storage capacity of the reservoir is 39.3 billion m^3, of which 22.15 billion m^3 is for flood control;
- a project to build the world's largest hydropower station. The total installed capacity of the Three Gorges Hydropower Station will be 18,200 MW, with an annual electricity output of 84.68 billion KW/h;
- a water project with the world's biggest project workload. The main structure of the project calls for the excavation of 134 million m^3 of earth and rock, and the concrete placement is 27.94 million m^3;
- a project to create the highest flood discharge capability in the world, of 102.5 thousand cubic metres per second (m^3/s);
- a project to build the biggest and the most complicated ship lift in the world;
- a project which calls for the greatest ever human resettlement: 1.13 million people will have been moved by the time it is completed

On the left from us a child was smiling and showing us a small collection of pebbles which, he claimed, came from the bottom of the river, exactly where the dam was now standing. He was willing to sell them to us for a small amount of Yuan. I left with a beautiful orange-yellow stone, which is now on my work desk at home.

(A memory from 1998)

the initiatives in the right direction. It has also become apparent to me that good systems thinking means operating on multiple thought tracks simultaneously. This would be difficult even if these tracks comprised familiar ways of thinking.

Familiarity with the following aspects could be of assistance.

Dynamic thinking

Dynamic thinking involves acquiring the ability to see behaviour patterns rather than focusing on, and seeking to predict, individual events or situations. It means thinking about phenomena as resulting from ongoing circular processes unfolding through time, rather than in terms of events and causes. Dynamic thinking skills are based on the ability to trace out patterns of behaviour that change over time. They call for thinking through the underlying closed-loop processes that cycle around to produce particular situations.

Closed-loop thinking

The second type of thinking process, closed-loop thinking, is closely linked to dynamic thinking. When we think in terms of closed loops, we see the problem as a set of ongoing, interdependent processes rather than as a list of one-way relations between a group of causes and another of effects. In addition when exercising closed-loop thinking, we look to the loops themselves (i.e. the circular cause–effect relations) as being responsible for generating the behaviour patterns exhibited by a system. This is in contrast to holding a set of external forces responsible. In this model, external forces tend to be viewed as precipitators rather than as causes.

Generic thinking

Just as most of us are captivated by events, we are generally locked into thinking in terms of specifics. The notion of thinking generically rather than specifically can be applied to water resources systems. For example, it is useful to appreciate the similarities in the underlying feedback loop relations that generate a hydrological cycle, a flood–drought swing or an oscillation in water quality.

Structural thinking

Structural thinking is one of the most disciplined of the strands of systems thinking. Here we must think in terms of units of measure or dimensions. The laws of physical conservation are rigorously adhered to in this domain.

Operational thinking

Operational thinking goes hand in hand with structural thinking. Thinking operationally means thinking in terms of how things really work: not how they should work in theory, or how a model can be created by manipulating a bit of algebra and generating some convincing-looking output.

Continuum thinking

Continuum thinking is usually present when we work with simulation models that have been built using a continuous, as opposed to discrete, modelling approach. Discrete models are distinguished by their containing many 'if, then, else'-type equations. In such models, for example, we might find that water consumption is governed by some logic of the form 'IF Available Water >0 THEN Normal Water Consumption ELSE 0'. In contrast, the continuous version of this relation would begin with an operational specification of the water consumption process (e.g. Water consumption = Population × Water per person). Water per person (per year) would then be a continuous function of Available Water. Unlike its discrete analogue, the continuous formulation indicates that water consumption would be continuously affected as Available Water became depleted. That is, it allows for measures such as rationing, increases in water prices or moratoriums on new construction coming into play as it becomes apparent that there are less than adequate supplies of water. The discrete formulation, by contrast, implies 'business as usual' right up to the point where Available Water falls to zero. At that point, consumption is zero. Although from a mechanical standpoint the differences between the continuous and discrete formulations may seem unimportant, the implications for thought processes are quite dramatic.

Scientific thinking

The final component of systems thinking that Richmond (1993) identified is scientific thinking. Thinking scientifically means being rigorous about testing hypotheses. This process begins by always ensuring that you do in fact have a hypothesis to test. If there is no hypothesis, the experimentation process can easily degenerate into a game. The hypothesis-testing process itself also needs to be informed by scientific thinking. When we think scientifically we modify only one thing at a time and hold all else constant. We also test our models from a steady state using idealized inputs.

We defined the term *system* to mean an interdependent group of items forming a ' unified pattern. Since our interest here is in water management, we shall focus on systems of people and technology that are intended to plan, design, construct and operate water infrastructure. Almost everything that goes on in water resources management is part of one or more such systems. As noted above, when we face a management problem we tend to assume that some external event caused it. With a systems view, we take an alternative viewpoint: namely that the internal structure of the system is often more important than external events in generating the problem. This is illustrated in Figure 4.2. Many people try to explain aspects of performance by showing how one set of events causes another, or when they study a problem in depth, how a particular set of events is part of a longer-term pattern of behaviour. The difficulty with this cause–effect orientation is that it does not lead to very powerful ways to alter the undesirable performance. We can continue this process almost forever, and thus it is difficult to determine what to do to improve performance.

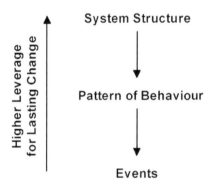

Figure 4.2 *Looking for problem solution (high leverage)*

If we shift from this event orientation to focusing on the internal system structure, we improve the possibility of finding the causes of the problem. This is because the system structure is often the underlying source of the difficulty. Unless the deficiencies in the system structure are corrected, it is likely that the problem will resurface, or be replaced by an even more difficult problem.

4.1.3 Systems analysis

Systems analysis involves the use of rigorous methods to help determine preferred plans and designs for complex, often large-scale, systems. It combines knowledge of the available analytical tools, understanding of when each is most appropriate, and skill in applying them to practical problems. It is both mathematical and intuitive, as is all planning and design (de Neufville, 1990).

It is a relatively new field. Its development parallels that of the computer, the computational power of which enables us to analyse complex relationships, involving many variables, at reasonable cost. Most of its techniques depend on the use of the computer for practical application. Systems analysis may be thought of as the set of computer-based methods essential for the planning of major projects. It is thus central to the education of water resources professionals.

Systems analysis covers much of the same material as operations research, in particular linear and dynamic programming (LP and DP) and decision analysis. The two fields differ substantially in direction, however. Operations research tends to be interested in specific techniques and their mathematical properties. Systems analysis focuses on the use of the methods.

Systems analysis includes the topics of engineering economy, but goes far beyond them in its depth of concept and the scope of its coverage. Now that both personal computers and efficient financial calculators are available, there is little need for professionals to spend much time on detailed calculations. It is more appropriate to understand the concepts and their relationship to the range of techniques available to deal with complex problems.

This approach emphasizes the kinds of real problems to be solved; considers the relevant range of useful techniques, including many besides those of operations research; and concentrates on the help they can provide in improving plans and designs. The use of systems analysis instead of the more traditional set of tools generally leads to substantial improvements in design and reductions in cost. Gains of 30 per cent are not uncommon. These translate into an enormous advantage when we consider projects that cost tens and hundreds of millions of dollars.

4.1.4 The systems approach

The systems approach is a general problem-solving technique that brings more objectivity to the engineering planning and design processes. It is, in essence, concerned with good design: a logical and systematic approach to problem solving in which assumptions, goals, objectives and criteria are clearly defined and specified. Emphasis is placed on relating system performance to specified goals. A hierarchy of systems is identified, and this makes it possible to handle a complex system by looking at its component parts or subsystems. Quantifiable and non-quantifiable aspects of the problem are identified, and the immediate and long-range implications of suggested alternatives are evaluated.

The systems approach establishes the proper order of inquiry and helps in the selection of the best course of action that will accomplish a prescribed goal, by broadening the information base of the decision-maker; by providing a better understanding of the system, and the interrelatedness of the complete system and its component subsystems; and by facilitating the prediction of the consequences of several alternative courses of action.

The systems approach is a framework for water resources analysis and decision-making. It does not solve problems, but does allow the decision-maker to undertake resolution of a problem in a logical, rational manner. While there is some art involved in the efficient application of the systems approach, other factors play equally important roles. The magnitude and complexity of decision processes requires the most effective use possible of scientific (quantitative) methods of systems analysis. However, we should be careful not to rely too heavily on the methods of systems analysis. Outputs from simplified analyses have a tendency to take on a false validity because of their complexity and technical elegance.

4.1.5 Systems engineering

Systems engineering may be defined as the art and science of selecting from a large number of feasible alternatives, involving substantial engineering content, the particular set of actions that will best accomplish the overall objectives of the decision-makers, within the constraints of law, morality, economics, resources, politics, social life, nature, physics and so on.

Another definition sees systems engineering as a set of methodologies for studying and analysing the various aspects of a system (structural and non-structural) and its environment by using mathematical and/or physical models.

Systems engineering is currently the popular name for the engineering processes of planning and design used in the creation of a system or project of considerable complexity.

Design

The design of a system represents a decision about how resources should be transformed to achieve some objectives. The final design is a choice of a particular combination of resources and a way to use them; it is selected from other combinations that might accomplish the same objectives. For example, the design of a building to provide 100 apartments involves decisions about the number of floors, the spacing of the columns, the type of materials used and so on; the same result could be achieved in many different ways.

A design must satisfy a number of technical considerations. It must conform to the laws of the natural sciences; only some things are possible. To continue with the example of the building, there are limits to the available strength of either steel or concrete, and this constrains what can be built using either material. The creation of a good design for a system thus requires solid technical competence in the matter at hand. Engineers may take this fact to be self-evident, but it often needs to be stressed to industrial or political leaders, who are motivated by their hopes for what a proposed system might accomplish.

Economics and values must also be taken into account in the choice of design; the best design cannot be determined by technical considerations alone. Moreover, these issues tend to dominate the final choice between many possible designs, each of which appears equally effective technically. The selection of a design is then determined by the costs and relative values associated with the different possibilities. The choice between constructing a building of steel or concrete is generally a question of cost, as both can be essentially equivalent technically. For more complex systems, political or other values may be more important than costs. In planning a reservoir for a city water supply for instance, it is usually the case that several sites can be made to perform technically; the final choice hinges on societal decisions about, for example, the relative importance of ease of access and the environmental impacts of the reservoir, in addition to its cost.

Planning

Planning and design are so closely related that it is difficult to separate one from the other. The planning process closely follows the systems approach, and may involve the use of sophisticated analysis and computer tools. However, the scope of the problems addressed by planning is different from the scope of design problems. Basically, planning is the formulation of goals and objectives that are consistent with political, social, environmental, economic, technological and aesthetic constraints, and the general definition of procedures designed to meet those goals and objectives. Goals are the desirable end states that are sought. They may be influenced by the actions or desires of government bodies, such as legislatures or courts, of special interest groups, or of administrators. Goals may change as the interests of the groups concerned change.

Objectives relate to ways in which the goals can be reached. Planning should be involved in all aspects of a water resources project, including preliminary investigations, feasibility studies, detailed analysis and specifications for implementation and/or construction, and monitoring and maintenance. A good plan will bring together diverse ideas, forces or factors, and combine them into a coherent, consistent structure that when implemented will improve target conditions without affecting non-target conditions. Effective use of the systems approach will help to ensure that planning studies address the true problem at hand. Planning studies that do not do this could not, if implemented, produce useful and desirable changes.

4.1.6 Mathematical modelling

In general, to obtain a way to control or manage a physical system, we use a mathematical model which closely represents the physical system. The mathematical model is solved and its solution is applied to the physical system. Models, or idealized representations, are an integral part of everyday life. Common examples of models include model aeroplanes, portraits and cartographic globes. Similarly, models play an important role in science and business, as illustrated by models of the atom, models of genetic structure, mathematical equations describing physical laws of motion or chemical reactions, graphs, organization charts and industrial accounting systems. Such models are invaluable for abstracting the essence of the subject of enquiry, showing interrelationships and facilitating analysis (Hillier and Lieberman, 1990).

Mathematical models are also idealized representations, but they are expressed in terms of mathematical symbols and expressions. Such laws of physics as $F = ma$ and $E = mc^2$ are familiar examples. Similarly, a mathematical model of a business problem is a system of equations and related mathematical expressions that describe the essence of the problem. Thus, if there are n related quantifiable decisions to be made, they are represented as *decision variables* (e.g. $x_1, x_2, ..., x_n$) whose values are to be determined. The appropriate measure of performance (e.g. profit) is then expressed as a mathematical function of these decision variables (e.g. $P = 3x_1 + 2x_2 + ... + 5x_n$). This function is called the *objective function*. Any restrictions on the values that can be assigned to these decision variables are also expressed mathematically, typically by means of inequalities or equations (e.g. $x_1 + 3x_1x_2 + 2x_2 \leq 10$). Such mathematical expressions for the restrictions are often called *constraints*. The constants (e.g. coefficients or right-hand sides) in the constraints and the objective function are called the *parameters* of the model. We might then say that the problem in using a mathematical model is to choose the values of the decision variables so as to maximize the objective function, subject to the specified constraints. Such a model, and minor variations of it, typify the models used in systems analysis.

Mathematical models have many advantages over a verbal description of the problem. One obvious advantage is that a mathematical model describes a problem much more concisely. This tends to make the overall structure of the problem more comprehensible, and it helps to reveal important cause-and-effect relationships. In this way, mathematical modelling indicates more clearly what additional data are relevant to the analysis. It also facilitates dealing with the problem in its entirety and considering all its interrelationships simultaneously. Finally, a mathematical model forms a bridge to the use of high-powered mathematical techniques and computers to analyse the problem. Indeed, packaged software for both microcomputers and mainframe computers is becoming widely available for many mathematical models.

The procedure of selecting the set of decision variables that maximizes/minimizes the objective function, subject to the systems constraints, is called the *optimization procedure*. The following is a general optimization problem.

Select the set of decision variables $x^*_1, x^*_2, \ldots, x^*_n$ such that

Min or Max $f(x_1, x_2, \ldots, x_n)$

subject to:

$$\begin{aligned} g_1(x_1, x_2, \ldots, x_n) &\leq b_1 \\ g_2(x_1, x_2, \ldots, x_n) &\leq b_2 \\ g_m(x_1, x_2, \ldots, x_n) &\leq b_m \\ x_i &\geq 0, \text{ for } i = 1, 2 \ldots n \end{aligned} \quad (4.2)$$

where $b_1, b_2, \ldots, b_m$ are known values.

If we use the matrix notation, (4.2) can be rewritten as:

Min or Max $f(\mathbf{x})$ \hfill (4.3)

subject to:

$$g_j(\mathbf{x}) \leq b_j \quad j = 1, 2, \ldots, m$$
$$\mathbf{x} \geq 0$$

When optimization fails, because of the system complexity or a computational difficulty, a reasonable attempt at a solution may often be obtained by *simulation*. Apart from facilitating trial and error design, simulation is a valuable technique for studying the sensitivity of system performance to changes in design parameters or operating procedure. Simulation is presented later in the book (Chapter 8).

According to equations (4.2) and (4.3), our main goal is the search for an optimal, or best, solution. Some of the techniques developed for finding such solutions are discussed in this book. However, it needs to be recognized that these solutions are optimal only with respect to the model being used. Since the model necessarily is an idealized rather than an exact representation of the real problem, there cannot be any utopian guarantee that the optimal solution for the model will prove to be the best possible solution that could have been implemented for the real problem. There just are too many uncertainties associated with real problems. However, if the model is well formulated and tested, the resulting solution should tend to be a good approximation to the ideal course of action for the real problem. Therefore, rather than demanding the impossible, the test of the practical success of a systems analysis study should be whether it provides a better guide for action than can be obtained by other means.

The eminent management scientist and Nobel Laureate in Economics, Herbert Simon, points out that *satisficing* is much more prevalent than optimizing in actual practice. In coining the term 'satisficing' as a combination of the words 'satisfactory' and 'optimizing', Simon described the tendency of analysts to seek a solution that is 'good enough' for the problem at hand. Rather than trying to develop various desirable objectives, a more pragmatic approach may be used. Goals may be set to establish minimum satisfactory levels of performance in various areas, based perhaps on past levels of performance or on what is expected to be achieved. If a solution is found that enables all of these goals to be met, it is likely to be adopted without further ado. Such is the nature of satisficing. The distinction between optimizing and satisficing reflects the difference between theory and the realities frequently faced in trying to implement that theory in practice.

4.1.7 A classification of systems

Systems may be classified, for convenience and to provide insight into their wide range. There are several classification systems that focus on aspects of the similarities and dissimilarities of different systems (Blanchard and Fabrycky, 1990). This section looks at some common dichotomies: between natural and human-made systems, physical and conceptual systems, static and dynamic systems, and closed and open systems.

Natural and human-made systems

The origin of systems gives a most important classification opportunity. Natural systems are those that came into being through natural processes. Human-made systems are those in which human beings have intervened by introducing or shaping components, attributes or relationships. However, once they have been brought into being, all human-made systems are embedded in the natural world. Important interfaces often exist between human-made systems and natural systems. Each affects the other in some way. The effect of human-made systems on the natural world has only recently become a keen subject for study by concerned people, especially in those instances where the effect is undesirable.

Natural systems exhibit a high degree of order and equilibrium. This is evidenced in the seasons, the food chain, the water cycle and so on. Organisms and plant life adapt themselves to maintain equilibrium with the environment. Every event in nature is accompanied by an appropriate adaptation, one of the most important being that material flows are cyclic. In the natural environment there are no dead ends and no wastes, only continual recirculation. Natural systems adapt to change of any intensity by changing the composition of the system (that is, eliminating, adding or rearranging elements as necessary to re-establish equilibrium). This evolution is the price paid for maintaining stability and system integrity. Natural systems adapt (evolve) to survive; systems that fail to adapt will become extinct.

Only recently have significant human-made systems appeared. These systems make up the human-made world. The rapid evolution of human beings is not adequately understood, but their coming upon the scene has significantly affected the natural world, often in undesirable ways. Primitive beings had little impact on the natural world, for they had not yet developed a potent and pervasive technology, but the impact of humanity has steadily increased over time.

A good example of the impact of human-made systems on natural systems is the set of problems that arose from building the High Aswan Dam on the Nile River in Egypt. Construction of this massive dam ensured that the Nile would never flood again, and provided much needed electrical energy. However, several new problems arose. The food chain was broken in the eastern Mediterranean, thereby reducing the fishing industry. Rapid erosion of the Nile delta took place, introducing soil salinity into Upper Egypt. The population of bilharzia (a waterborne snail parasite) was no longer limited by periods of drought, and this led to an epidemic of intestinal disease along the Nile. These side-effects were not adequately considered in the planning phase of the project. A systems view encompassing both natural and human-made elements might have led to a better solution to the problems of flooding and the need for power sources.

Physical and conceptual systems

Physical systems are those that manifest themselves in physical terms. They are composed of real components, and may be contrasted with conceptual systems, where symbols represent the attributes of components. Ideas, plans, concepts and hypotheses are examples of conceptual systems. A physical system consumes physical space, whereas conceptual systems are organizations of ideas. One type of conceptual system is the set of plans and specifications for a physical system before it is actually brought into being.

A proposed physical system may be simulated in the abstract by a mathematical or other conceptual model. Conceptual systems often play an essential role in the operation of physical systems in the real world. The system of elements encompassed by all components, attributes and relationships focused on a given objective employs a process in guiding the state of a system. A process may be mental (thinking, planning, learning), mental-motor (writing, drawing, testing) or mechanical (operating, functioning, producing). Processes exist equally in physical and conceptual systems.

Static and dynamic systems

Another system dichotomy is the distinction between static and dynamic systems. A static system has a structure but no activity: a bridge, for example, is a static system. A dynamic system combines structural components with activity. An example is a school, combining a building, students, teachers, books and curricula. For centuries we have viewed the universe of phenomena as unchanging. A mental habit of dealing with certainties and constants developed. The substitution of a process-oriented description for the static description of the world is one of the major characteristics separating modern science from earlier thinking.

A dynamic conception of the world has become a necessity, yet a general definition of a system as an ongoing process is incomplete. Many systems would not be included under this broad definition because they lack motion in the usual sense. A highway system is static, yet contains the system elements of components, attributes and relationships.

It should be recognized that any system can be seen as static only in a limited frame of reference. A bridge is constructed over a period of time, and this is a dynamic process. It is then maintained and perhaps altered to serve its intended purpose more fully. These are clearly dynamic aspects, which would need consideration if the field of reference was the bridge over a long period of time.

Systems may be characterized as having random properties. In almost all systems in both the natural and human-made categories, the inputs, process and output can only be described in statistical terms. Uncertainty often occurs in both the number of inputs and the distribution of these inputs over time. For example, it is difficult to predict exactly the peak flow that will arrive at a particular location in a river basin, or the exact time it will arrive. However, each of these factors can be described in terms of probability distributions, and system operation is then said to be probabilistic.

Closed and open systems

A closed system is one that does not interact significantly with its environment. The environment only provides a context for the system. Closed systems exhibit the characteristic of equilibrium resulting from internal rigidity, which maintains the system in spite of influences from the environment. An example is the chemical equilibrium eventually reached in a closed vessel when various chemicals are mixed together. The reaction can be predicted from a set of initial conditions. Closed systems involve deterministic interactions, with a one-to-one correspondence between initial and final states.

An open system allows information, energy and matter to cross its boundaries. Open systems interact with their wider environment. Examples are plants, ecological systems and business organizations. They may exhibit the characteristics of steady state, wherein a dynamic interaction of system elements adjusts to changes in the environment. Because of this steady state, open systems can be self-regulatory and are often self-adaptive.

It is not always easy to classify a system as either open or closed. Systems that have come into being through natural processes are typically open. Human-made systems have characteristics of both open and closed systems. They may reproduce natural conditions not manageable in the natural world. They are closed when designed for invariant input and statistically predictable output, as in the case of an aircraft in flight.

Both closed and open systems exhibit the property of *entropy*. Entropy is defined here as the degree of disorganization in a system, and is analogous to the use of the term in thermodynamics. In the thermodynamic usage, entropy is the energy unavailable for work resulting from energy transformation from one form to another. In systems, increased entropy means increased disorganization. A decrease in entropy takes place as order occurs. Life represents a transition from disorder to order. Atoms of carbon, hydrogen, oxygen and other elements become arranged in a complex and orderly fashion to produce a living organism. A conscious decrease in entropy must occur to create a human-made system. All human-made systems, from the most primitive to the most complex, consume entropy during the creation of more orderly states from less orderly states.

4.1.8 A classification of mathematical models and optimization techniques

Now that we have outlined some common classifications of systems, we can classify mathematical models according to the nature of the objective function and the constraints.

Linear and nonlinear models

If the objective function and all the constraints are linear in terms of the decision variables, a mathematical model is considered to be linear. Similarly, if some or all of the constraints and/or the objective function are nonlinear, a mathematical model is described as nonlinear.

Deterministic and probabilistic models

If each variable and parameter can be assigned a definite fixed number or a series of fixed numbers for any given set of conditions, a model is a deterministic one. If it contains variables, the value of which are subject to some measure of randomness or uncertainty, it is called probabilistic or stochastic.

Static and dynamic models

Models that do not explicitly take time into account are static, and those that involve time-dependent interactions are dynamic.

Distributed and lumped parameter models

Models that take into account detailed variations in behaviour from point to point throughout the system space are called *distributed parameter models*. In contrast, models that ignore the variations, in which the parameters and dependent variables can be considered to be homogeneous throughout the entire system space, are known as *lumped parameter models*.

Various techniques of optimization are available to solve different optimization problems. For static systems we can use calculus, LP, nonlinear programming (direct search, gradient search, complex method, geometric programming, etc.), DP and other techniques. Dynamic systems can be solved using queuing theory, game theory, network theory, the calculus of variation, the maximum principle, quasi-linearization and other techniques.

All the models and techniques mentioned up to now deal with a single-objective function (equations (4.2) and (4.3). If problems require more than one objective function, in mathematical terms we are dealing with *vector optimization* or multi-objective analysis, which is presented later in the book.

4.2 SOME SYSTEMS CONCEPTS IN ELEMENTARY MATHEMATICAL CONSIDERATION

Many of the concepts used in the systems sciences derive originally from work done in the mid-20th century on cybernetics (Wiener, 1948), on information theory (Shannon, 1949) and general systems theory (von Bertalanffy, 1969). These were all attempts to quantify in a rigorous way the treatment of systems as an interdisciplinary science. In other words they were a break from the old views that specialist subjects required specialist ideas.

We shall now go on to look at systems concepts in mathematical form, considering, for example, growth, competition and wholeness. These concepts have often been considered to describe only characteristics of living beings. However, they are actually general properties of systems.

4.2.1 The system definition in mathematical form

When we deal with systems as complexes of 'elements', they may be distinguished in three different ways:

1 according to their number;
2 according to their types;
3 according to the relations of the elements.

The characteristics of the first kind may be called *summative*, and of the second kind *constitutive*. We can also say that the summative characteristics of an element are those that are the same within and outside the complex. They may therefore be obtained by summing the characteristics and behaviour of isolated elements. Constitutive characteristics are those that are dependent on the specific relations within the complex. Therefore, in order to understand such characteristics, we must know not only the parts but also the relations between them.

The meaning of the expression, 'the whole is more than the sum of parts', is simply that constitutive characteristics are not explainable from the characteristics of isolated parts. The characteristics of the complex therefore appear as 'new' or 'emergent' compared with those of the elements. If, however, we know the total of parts contained in a system and the relations between them, the behaviour of the system may be derived from the behaviour of its parts.

In rigorous development, general system theory would be of an axiomatic nature. From the definition of 'system' and a suitable set of axioms, propositions expressing system properties and principles would be deduced. The considerations discussed here are much more modest (von Bertalanffy, 1969). They merely illustrate some system principles, using formulations that are simple and intuitively accessible, without any attempt at mathematical rigour and generality.

A system (as defined earlier) is a set of elements standing in interrelationship to each other. By 'interrelation' we mean that elements of type p stand in relations of type R, so that the behaviour of an element p in R is different from its behaviour in another relation, $R´$. If the behaviours in R and $R´$ are not different, there is no interaction, and the elements behave independently with respect to the relations R and $R´$.

A system can be defined mathematically in various ways. For illustration, let us choose a system of simultaneous differential equations. If we denote some measure of elements p_i ($i = 1, 2,... n$) by Q_i, then for a finite number of elements and in the simplest case, these will be of the form:

$$\frac{dQ_1}{dt} = f_1(Q_1, Q_2, ..., Q_n)$$

$$\frac{dQ_2}{dt} = f_2(Q_1, Q_2, \ldots, Q_n) \tag{4.4}$$

....

$$\frac{dQ_{n1}}{dt} = f_1(Q_1, Q_2, \ldots, Q_n)$$

A change in any measure Q_i is therefore a function of all Qs, from Q_1 to Q_n; conversely, a change in any Q_i entails a change in all other measures and in the system as a whole.

Systems of equations of this kind can, for example, be found in describing demographic problems. The equations for biological systems are special cases of equation (4.4). The definition of a system expressed by (4.4) is, of course, by no means general. It does not take into consideration spatial and temporal conditions, which would be described by partial differential equations. It also does not capture any possible dependence on the previous history of the system ('hysteresis' in a broad sense). Consideration of the previous history would require description of the system using integro-differential equations with a definite meaning – the system under consideration would be not only a spatial, but also a temporal whole.

Despite these restrictions, equation (4.4) can be used for discussing several general system properties. Although nothing is said about the nature of the measures Q_i or the functions f_i (i.e. about the relations or interactions within the system), certain general principles can be presented.

First is a condition of stationary state, characterized by the disappearance of the changes dQ_i/dt:

$$f_1 = f_2 = \ldots = f_n = 0 \tag{4.5}$$

By equating to zero we obtain n equations with n variables, and by solving them we obtain:

$$Q_1 = Q_1^*$$

$$Q_2 = Q_2^* \tag{4.6}$$

....

$$Q_n = Q_n^*$$

These values are constants, since in the system, as assumed, the changes disappear. In general, there will be a number of stationary states, some stable, some unstable. We may introduce new variables:

$$Q_i = Q_i^* - Q_i' \tag{4.7}$$

and rewrite the system (4.4):

$$\frac{dQ_1'}{dt} = f_1'(Q_1', Q_2', \ldots, Q_n') \tag{4.8}$$

$$\frac{dQ_2'}{dt} = f_2'(Q_1', Q_2', \ldots, Q_n')$$

$$\ldots\ldots\ldots\ldots\ldots\ldots\ldots\ldots$$

$$\frac{dQ_n'}{dt} = f_n'(Q_1', Q_2', \ldots, Q_n')$$

Let us develop the system in Taylor series:

$$\frac{dQ_1'}{dt} = a_{11}Q_1' + a_{12}Q_2' + \ldots + a_{1n}Q_n' + a_{111}Q_1'^2 + a_{112}Q_1'Q_2' + a_{122}Q_2'^2 + \ldots$$

$$\frac{dQ_2'}{dt} = a_{21}Q_1' + a_{22}Q_2' + \ldots + a_{2n}Q_n' + a_{211}Q_1'^2 + a_{212}Q_1'Q_2' + a_{222}Q_2'^2 + \ldots \tag{4.9}$$

$$\ldots\ldots\ldots\ldots\ldots\ldots\ldots\ldots\ldots\ldots\ldots\ldots\ldots\ldots\ldots\ldots\ldots\ldots\ldots\ldots\ldots\ldots$$

$$\frac{dQ_n'}{dt} = a_{n1}Q_1' + a_{n2}Q_2' + \ldots + a_{nn}Q_n' + a_{n11}Q_1'^2 + a_{n12}Q_1'Q_2' + a_{n22}Q_2'^2 + \ldots$$

A general solution of this system of equations is:

$$Q_1' = G_{11}e^{\lambda_1 t} + G_{12}e^{\lambda_2 t} + \ldots + G_{1n}e^{\lambda_n t} + G_{111}e^{2\lambda_1 t} + \ldots$$

$$Q_2' = G_{21}e^{\lambda_1 t} + G_{22}e^{\lambda_2 t} + \ldots + G_{2n}e^{\lambda_n t} + G_{211}e^{2\lambda_1 t} + \ldots \tag{4.10}$$

$$\ldots\ldots\ldots\ldots\ldots\ldots\ldots\ldots\ldots\ldots\ldots\ldots\ldots\ldots$$

$$Q_n' = G_{n1}e^{\lambda_1 t} + G_{n2}e^{\lambda_2 t} + \ldots + G_{nn}e^{\lambda_n t} + G_{n11}e^{2\lambda_1 t} + \ldots$$

where the G are constants and the λ are roots of the characteristic equation:

$$\begin{vmatrix} a_{11} - \lambda & a_{12} & \ldots & a_{1n} \\ a_{21} & a_{22} - \lambda & \ldots & a_{2n} \\ \ldots & \ldots & \ldots & \ldots \\ a_{n1} & a_{n2} & \ldots & a_{nn} - \lambda \end{vmatrix} = 0 \tag{4.11}$$

The roots λ may be real or imaginary. By inspection of equations (4.10) we find that if all λ are real and negative (or, if complex, negative in their real parts), Q_i', with increasing time, approaches 0 because $e^{-\infty} = 0$; since, however, according to (4.7), the Q_i thereby obtain the stationary values Q^*_i. In this case the equilibrium is *stable*, since over a sufficient period of time the system will come as close to the stationary state as possible. For the simplest case, $n = 2$, the system will approach a stable stationary state, a *node*, as illustrated in Figure 4.3a.

However, if one of the λ is positive or 0, the equilibrium is *unstable*: that is, the system will move away from equilibrium. For the simplest case the solution is represented with a spiral curve, a *loop*, as shown in Figure 4.3b.

If, finally, some λ are positive and complex, the system contains periodic terms since the exponential function for complex exponents takes the form:

$$e^{(a-ib)t} = e^{at}(\cos bt - i \sin bt) \quad (4.12)$$

In this case there will be *periodic fluctuations*, which are generally damped. Again, in the case of the simplest system there will be oscillations, *cycles*, around the stationary value as illustrated in Figure 4.3c.

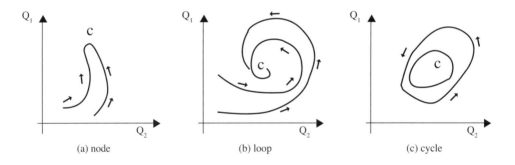

(a) node   (b) loop   (c) cycle

Source: von Bertalanffy (1969)

Figure 4.3 *Potential solutions for a system consisting of two types of elements*

4.2.2 Growth

Equations of this type are found in a variety of fields, and we can use system (4.4) to illustrate the formal identity of system laws in different fields: in other words, to demonstrate the existence of a general system theory.

Let us consider the simplest case: the system consisting of elements of only one kind. Then the system of equations (4.4) is reduced to the single equation:

$$\frac{dQ}{dt} = f(Q) \tag{4.13}$$

which can be developed into a Taylor series:

$$\frac{dQ}{dt} = a_1 Q + a_{11} Q^2 + \ldots \tag{4.14}$$

This series does not contain an absolute term when there is no 'spontaneous generation' of elements. Then dQ/dt must disappear for $Q = 0$, which is possible only if the absolute term is equal to zero.

The simplest possibility is described when we retain only the first term of the series:

$$\frac{dQ}{dt} = a_1 Q \tag{4.15}$$

This signifies that the growth of the system is directly proportional to the number of elements present. Depending on whether the constant a_1 is positive or negative, the growth of the system is positive or negative, and the system increases or decreases. The solution is:

$$Q = Q_0 e^{a_1 t} \tag{4.16}$$

Q_0 here indicates the number of elements at $t = 0$. This is the *exponential law* found in many fields (see Figure 4.4).

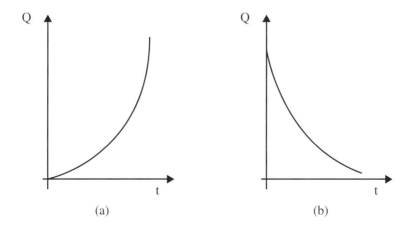

Figure 4.4 *Exponential growth and decay*

In mathematics the exponential law is called the *law of natural growth*, and with ($a_1 > 0$) is valid for the growth of a population whose birth rate is higher than its death rate. It also describes the growth of human knowledge as measured by the number of textbook pages devoted to scientific discoveries. With the constant negative ($a_1 < 0$), the exponential law applies to the rate of extinction of a population in which the death rate is higher than the birth rate, and similar examples.

If we keep two terms in equation (4.14):

$$\frac{dQ}{dt} = a_1 Q + a_{11} Q^2 \qquad (4.17)$$

a solution becomes:

$$q = \frac{a_1 C e^{a_1 t}}{1 - a_{11} C e^{a_1 t}} \qquad (4.18)$$

Keeping the second term has an important consequence. The simple exponential (4.16) shows an infinite increase; taking into account the second term, we obtain a curve which is sigmoid and attains a limiting value. This curve is known as the *logistic curve* (Figure 4.5), and is also of wide application. This curve, for example, describes the growth of populations with limited resources (food, water, space, etc.).

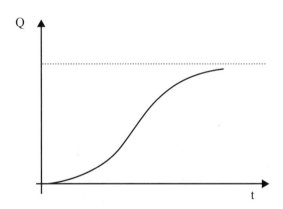

Figure 4.5 *Logistic curve*

Mathematically simple as these examples are, they illustrate something of interest to us at this point: that certain laws of nature can be arrived at not only on the basis of experience, but also in a formal way. The equations discussed signify no more than the rather general system of equation (4.4), its development into a Taylor series and the application of suitable conditions have been applied. In this sense such laws are a priori independent from their physical interpretation. In other words, this shows the existence of a general system theory which deals with formal characteristics of systems. It could be said too that such examples show a formal uniformity of nature.

4.2.3 Competition

Our system of equations may also indicate competition between elements of the system. The simplest possible case is, again, that all coefficients $(a_{j \neq i}) = 0$: that is, that the increase in each element depends only on this element itself. Then we have, for two elements:

$$\frac{dQ_1}{dt} = a_1 Q_1$$

$$\frac{dQ_2}{dt} = a_2 Q_2 \tag{4.19}$$

or

$$Q_1 = c_1 e^{a_1 t}$$

$$Q_2 = c_2 e^{a_2 t} \tag{4.20}$$

Eliminating time we obtain:

$$t = \frac{\ln Q_1 - \ln c_1}{a_1} = \frac{\ln Q_2 - \ln c_2}{a_1} \tag{4.21}$$

and

$$Q_1 = b Q_2^\alpha \tag{4.22}$$

where $\alpha = a_1/a_2$ and $b = c_1/c_2^\alpha$.

This equation applies to a wide range of problem domains. It means that a certain characteristic, Q_1, can be expressed as a power function of another characteristic, Q_2. For example, the expression in question is Pareto's law of the distribution of resource among a number of users, whereby Q_1 = number of users gaining a certain portion of the resource, Q_2 = amount of the available resource, and b and a are constants.

The situation becomes more complex if interactions between the parts of the system are assumed, if $(a_{ji}) \neq 0$. Then we come to systems of equations such as those describing competition among users of a resource, for example, and correspondingly, competition among different elements within a system. It is an interesting consequence that, in equations describing more complex interactions, the competition of two users for the same resource is, in a way, more fatal than a power relation where one element has more strength than the other. Competition eventually leads to the extermination of the elements with the smaller growth capacity; a power relation only leads to periodic oscillation of the numbers of the systems around a mean value.

Another point of philosophical interest may also be noted. If we are speaking of systems, we mean wholes or unities. Therefore it seems paradoxical that, with respect to a whole, the concept of competition between its parts is introduced. However, these apparently contradictory statements both describe essential characteristics of systems. Every whole is based upon the competition of its elements, and assumes the 'struggle between parts'. The latter is a general principle of organization in many systems around us.

4.2.4 Wholeness, the sum, mechanization and centralization

First, let us assume that equation (4.4) can be developed into a Taylor series:

$$\frac{dQ_1}{dt} = a_{11}Q_1 + a_{12}Q_2 + \ldots + a_{1n}Q_n + a_{111}Q_1^2 + \quad (4.23)$$

We see that any change in some quantity, Q_1, is a function of the quantities of all the elements Q_1 to Q_n. On the other hand, a change in a certain Q_i causes a change in all other elements and in the total system. The system therefore behaves as a *whole*: the changes in every element depend on all the others.

Now let the coefficients of the variables Q_j ($j \neq i$) become zero. The system of equations degenerates into:

$$\frac{dQ_i}{dt} = a_{ii}Q_i + a_{i11}Q_1^2 + \ldots \quad (4.24)$$

This means that a change in each element depends only on that element itself. Each element can therefore be considered independent of the others. The variation of the total complex is the (physical) sum of the variations of its elements. Such behaviour is called *physical summativity* or *independence*.

Summativity in the mathematical sense means that the change in the total system obeys an equation of the same form as the equations for the parts. This is possible only when the functions on the right-hand side of the equation contain linear terms only.

There is a further case which appears to be unusual in physical systems, but is common and basic in biological systems. This is the case where interactions between the elements decrease with time. In terms of the basic model equation (4.4), this means that the coefficients of Q_i are not constant, but decrease with time. The simplest case is:

$$\lim_{t \to \infty} a_{ij} = 0 \tag{4.25}$$

In this case the system passes from a state of wholeness to a state of independence of the elements. This is called *progressive segregation*.

The organization of physical wholes results from the union of pre-existing elements. In contrast, the organization of biological wholes is built up by the differentiation of an original whole which segregates into parts. The reason for the predominance of segregation in living nature seems to be that segregation into subsystems implies an increase of complexity in the system. Such a transition towards a higher order assumes a supply of energy, and energy is delivered continuously into the system only if it is an open system, taking energy from its environment.

In the state of wholeness, a disturbance of the system leads to the introduction of a new state of equilibrium. Increasing *mechanization* means the increasing determination of elements to function in ways that are only dependent on themselves, and a consequent loss of regulatory ability. The smaller the interaction coefficients become, the more the respective terms Q_i can be neglected, and the more 'machine-like' is the system (like a sum of independent parts).

In the contrast between wholeness and sum lies the tension in many evolutions. Progress is possible only by passing from a state of undifferentiated wholeness to differentiation of parts. This implies that the parts become fixed with respect to a certain action. Therefore, progressive segregation also means progressive mechanization. Progressive mechanization, as mentioned earlier, implies a loss of regulatory ability. As long as a system is a unitary whole, a disturbance will be followed by the attainment of a new stationary state (adaptation), because of the interactions among the elements within the system. The system is self-regulating. Progress is possible only by the subdivision of an initially unitary action into actions of specialized parts. Behaviour as a whole and summative behaviour are usually regarded as being antitheses, but it is frequently found that there is no opposition between them. Rather, there is a gradual transition from one to the other.

Connected with this is another principle. Suppose that the coefficients of one element, p_s, are large in all equations, while the coefficients of the other elements are considerably smaller or even equal to zero. In this case the system, with only linear terms, might look like this:

$$\frac{dQ_1}{dt} = a_{11}Q_1 + a_{1s}Q_s + \ldots$$

$$\frac{dQ_s}{dt} = a_{s1}Q_s + \ldots \qquad (4.26)$$

$$\frac{dQ_n}{dt} = a_{ns}Q_s + a_{n1}Q_n + \ldots$$

Then relationships are given which can be expressed in several ways. We might call the element p_s a *leading part*, or say that the system is *centred* around p_s. If the coefficients a_{is} of p_s in some or all equations are large while the coefficients in the equation of p_s itself are small, a small change in p_s will cause a considerable change in the total system, and p_s may be then called a *trigger*. A small change in p_s will be 'amplified' in the total system. The principle of *centralization* is especially important in natural systems. Progressive segregation is often connected with progressive centralization, the expression of which is the time-dependent evolution of a leading part.

The system definition and concept outlined through this mathematical structure calls for an important addition. Systems are frequently structured in such a way that their individual members are also systems at the next lower level. Hence each of the elements denoted by $Q_1, Q_2, \ldots, Q_n$, is a system of elements $O_{i1}, O_{i2}, \ldots, O_i$, in which each system O is defined by equations similar to those of (4.4):

$$\frac{dO_{ii}}{dt} = f_{ii}(O_{i1}, O_{i2}, \ldots, O_{in}). \qquad (4.27)$$

Such superposition of systems is called *hierarchical order*. For its individual levels, again the aspects of wholeness and summativity, progressive mechanization, centralization and so on apply. Such a hierarchical structure and combination into systems of ever-higher order are characteristic of reality as a whole, and are of fundamental importance, especially in natural and social systems.

4.3 FEEDBACK

The classification of systems in Section 4.1.7 provides a basic differentiation between open and closed systems. An open system is one characterized by outputs that respond to inputs, but where the outputs are isolated from and have no influence on the inputs. An open system is not aware of its own performance. In an open system, past action does not control future action (Forrester, 1990). However, a closed system (also known as a feedback system) is influenced by its own past behaviour. A feedback system has a closed-loop structure which brings results from past action of the system back to control future action. A broad purpose may imply a feedback system with many components, but each component can itself be a feedback system in terms of some subordinate purpose. We can then recognize a hierarchy of feedback structures, where the broadest purpose of interest determines the scope of the pertinent system. There is a simplified graphical representation of these two types of system in Figure 4.6.

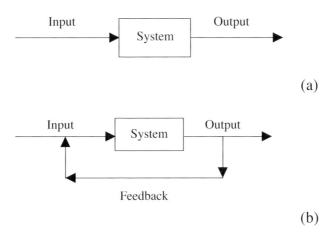

Figure 4.6 *Schematic presentation of (a) an open and (b) a closed system*

A basic example of a feedback system is a simple thermostat and the maintenance of constant temperature. The thermostat senses a difference between desired and actual room temperature, and activates the heating unit. The addition of heat eventually raises the room temperature to the desired level. Then the thermostat automatically shuts off the heater. The system description used for the thermostat applies equally well to many systems: the electric eye of a camera, a thermostatically controlled oven, the automatic pilot of an airplane and the speed governor of a turbine all follow this pattern. Although these are all mechanically controlled systems, there are also equivalents in the biological world. The human body contains numerous self-regulating physiological processes that enable it to maintain a relatively constant internal environment. This self-regulation, called *homeostasis*, maintains, for example, a normal body temperature through the continual alteration of metabolic activities and blood flow rates. Goal-directed action is fundamental to human social behaviour too.

The feedback concept is one of the most important aspects of general system theory. The basic model is a circular process (as shown in Figure 4.6b), where part of the output is monitored back into the input, as information on the preliminary outcome of the response, thus making the system self-regulating. This self-regulation can involve either the maintenance of the value of certain variables or steering towards a desired goal.

Feedback systems and homeostatic control are a significant but special class of self-regulating systems and phenomena of adaptation. The following appear to be essential criteria of feedback control systems:

- Regulation is based on a pre-established system structure.
- Causal chains within the feedback system are linear and unidirectional.
- Typical feedback or homeostatic phenomena are 'open' with respect to incoming information, but 'closed' with respect to matter and energy.

If we compare the flow diagrams of feedback (Figure 4.6b) and open systems (Figure 4.6a), the difference should be apparent. Thus dynamics in open systems and feedback mechanisms are two different model concepts, each of which has a place in its proper sphere. The open-system model is basically non-mechanistic, and supports not only conventional thermodynamics, but also one-way causality, as is basic in conventional physical theory. The cybernetic approach retains the Cartesian machine model of the organism, unidirectional causality and closed systems; its novelty lies in the introduction of concepts that transcend conventional physics, especially those of information theory.

Unfortunately, conventional water resources management education tends to pay very limited attention to feedback thought. By 'feedback thought' I mean a powerful way of thinking, linking the concepts of control and self-reinforcement, stability and instability, structure and behaviour, mutual causality, interdependence, and many more of the deepest ideas in the natural, social and behavioural sciences.

Usually implicitly, but sometimes explicitly, feedback thought is embedded in the foundations of much of engineering science and systems theory. It is a building block. The literature shows that feedback thought is both old and new. It shows that in the modern era the concept of feedback moved into prominence in the 1940s and 1950s, and has since appeared to wane. Some consider it an outmoded idea, a metaphor for the engineering sciences that has had its day and has been replaced by new metaphors. Others, perhaps as a consequence, have not encountered it at all. Still others see it not as a metaphor, but as a natural and crucial property of engineering systems. Here we look at feedback as the most important property of water resources systems.

4.3.1 The feedback loop

Another basic concept is the feedback loop. The feedback loop is a closed path connecting in sequence a decision that controls action, the level (a state or condition) of the system, and information about the level of the system (see Figure 4.7). The single loop structure is the simplest form of feedback system. There may be additional delays and distortions appearing sequentially in the loop. There may be many loops which interconnect. When reading a feedback loop diagram, the main skill is to see the 'story' that the diagram tells: how the structure creates a particular pattern of behaviour, and how that pattern might be influenced.

The feedback concept provides a basis for viewing water management problems in a different way. Water resources systems belong to a class of complex systems. They have several important characteristics (Forrester, 1990):

- Cause and effect are often separated in terms of both time and space.
- Problem resolutions that improve a situation in the short term often create bigger problems in the long term.
- The subsystems and parts of the system interact using multiple, nonlinear feedback loops. The complex flow of interactions often results in counter-intuitive behaviour.
- Because of the time delays between cause and effect, system managers tend to reduce their goals and objectives to accommodate what was originally considered as an unacceptable situation.

Note that two of the characteristics focus primarily on time, and two focus on complex interactions. The feedback concept is a valuable tool for increasing our understanding of interactions. It enables us to see interrelationships rather than linear cause–effect chains, and to see processes of change rather than snapshots of before and after changes.

To illustrate the shift from linear to feedback or system thinking, let us consider a very simple system: filling a glass of water (Senge, 1990). From the linear point of view, I am filling a glass of water. However, as I fill the glass: (i) I am watching the water level; (ii) I am monitoring the gap between the level and my goal, the desired water level; and (iii) I am adjusting the flow of water. In fact, when I fill the glass I operate in a 'water regulation' system involving five variables: *Desired Water Level*; *Current Water Level*; the *Gap* between the two; the *Tap (Faucet) Position*, and the *Water Flow*. These variables are organized in a circle or feedback loop of cause–effect relationships.

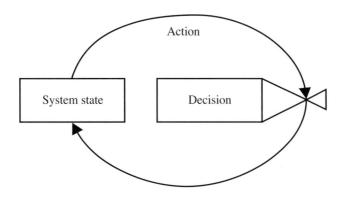

Figure 4.7 *A feedback loop*

4.3.2 Positive or reinforcing feedback

A positive, or reinforcing, feedback reinforces change with even more change. This can lead to rapid growth at an ever-increasing rate. This type of growth pattern is often referred to as exponential growth. Note that in the early stages of the growth, it seems to be slow, but then it speeds up. Thus, the nature of the growth in a water resources system that has a positive feedback loop can be deceptive. In the early stages of an exponential growth process, something that will become a major problem can seem minor because it is growing slowly. By the time the growth speeds up, it may be too late to solve whatever problem this growth is creating.

Sometimes positive feedback loops are called vicious or virtuous cycles, depending on the nature of the change that is occurring. Other terms used to describe this type of behaviour include bandwagon effects and snowballing. In summary, it is in the positive feedback form of a system structure that one finds the forces of growth.

4.3.3 Negative or balancing feedback

A negative, or balancing, feedback seeks a goal. If the current level of the variable of interest is above the goal, then the loop structure pushes its value down, while if the current level is below the goal, the loop structure pushes its value up. Many water resources systems processes contain negative feedback loops which provide useful stability, but which can also resist needed changes. In the face of an external environment that dictates that a system needs to change, it continues on with similar behaviour.

4.4 SYSTEM FORMULATION EXAMPLES

General systems theory offers a new way for formulating water resources management problems. A number of simple examples are developed in this section to illustrate the problem formulation phase, which calls for a transfer of the conceptual system definition into a practical problem domain. I have found from my experience that this is the most difficult phase for novices in the field.

4.4.1 City water supply

A city is growing in population and the water works department projects the need for an increased water supply. The current water supply is drawn from a groundwater aquifer and is of good quality. However, alternative sources for future water supply have various problems. Water from a nearby stream is available, but its hardness level is too high unless it is mixed with a lower-hardness water from another source. Many domestic water users are concerned about the hardness of their water. Hard water requires more soap and synthetic detergents for home laundry and washing, and contributes to scaling in boilers and other appliances. (Hardness is caused by compounds of calcium and magnesium, and by a variety of other metals. Water is an excellent solvent and readily dissolves minerals it comes in contact with. As water moves through soil and rock, it dissolves very small amounts of minerals and holds them in solution. Calcium and magnesium dissolved in water are the two most common minerals that make water 'hard'.) The acceptable water hardness is limited to 150 kg per million litres (kg/ml). Water from a distant lake is of sufficient quality, but the cost to pump the water to the distribution plant is quite high and a pipeline would have to be built.

The city is conducting its planning in stages. The first stage is to plan for ten years from the present. The three sources are source A, the current supply, source B, the nearby stream, and source C, the distant lake. The costs to obtain water in dollars per million litres, the supply limits in millions of litres per day (mld), and the hardness in milligrams per litre, are given in Table 4.1. For example, if the present water supply is developed further, up to 75 million litres per day could be made available from source A at a cost of $150 per million litres, and each additional million litres from this source would contribute 100 kg of hardness to the total water supply. A total of 450 additional million litres per day is needed after ten years. The city council members are interested in a least-cost strategy for expanding the water supply, while ensuring that it remains of sufficient quality.

Table 4.1 *Cost, supply and quality of available water resources*

	Source A	Source B	Source C
Cost ($/mld)	150	330	650
Supply limit (mld)	75	360	300
Hardness (kg/ml)	100	1,100	380

Decision variables

Let us consider the following decision variables:

x_1 millions of litres per day to be drawn from source A;
x_2 millions of litres per day to be drawn from source B;
x_3 millions of litres par day to be drawn from source C.

Objective function

The city council would like to minimize the cost of water supply. Therefore the objective function can be expressed as:

$$\text{Maximize}_{\{x_i\}} Z = 150x_1 + 330x_2 + 650x_3 \tag{4.28}$$

Constraints

The problem is subject to water demand constraint:

$$x_1 + x_2 + x_3 \geq 450 \tag{4.29}$$

a water quality constraint:

$$100x_1 + 1100x_2 + 380x_3 \leq 150 \times 450 \tag{4.30}$$

and water availability constraints:

$$x_1 \leq 75 \tag{4.31}$$

$$x_2 \leq 360 \tag{4.32}$$

$$x_3 \leq 300 \tag{4.33}$$

The logical requirement is that all decision variables are non-negative:

$$x_1, x_2, x_3 \geq 0. \tag{4.34}$$

The objective (4.28) is a minimum-cost choice of water resources. The first constraint (4.29) indicates the requirement for 450 million litres per day, somehow divided among three sources. The second constraint (4.30) limits the kilograms of hardness in the blended water. If the hardness concentration is limited to 150 kg per million litres and 450 million litres are required, the total hardness is limited to 150 × 450 = 67,500 kg.

If we look at the mathematical structure of this problem it is evident that both the objective function (4.28) and all constraints (4.29–4.34) are linear functions of decision variables. Therefore, this mathematical model of the city water supply can be classified as a linear optimization problem. Since all the decision variables and parameters are known in advance, the problem is of a deterministic nature. Both decision variables and parameters do not change with time and space, and therefore the problem is static with lumped parameters. An appropriate technique for solving this problem is LP.

4.4.2 Operation of a multi-purpose reservoir

A regional water agency is responsible for the operation of a multi-purpose reservoir used for (a) municipal water supply, (b) groundwater recharge, and (c) the control of water quality in the river downstream from the dam. Allocating the water to the first two purposes is, unfortunately, in conflict with the third purpose. The agency would like to minimize the negative effect on the water quality in the river, and at the same time maximize the benefits from the municipal water supply and groundwater recharge.

The available data are listed in Table 4.2. The following assumptions are made:

- One time period is involved; $t = 0, 1$.
- Allocation is limited to two restrictions: (a) pump capacity is 8 hours per period; and (b) labour capacity is 4 person-hours per period.
- The total amount of water in the reservoir available for allocation is 72 units.
- The pollution in the river increases by three units per unit of water used for water supply and two units per unit of water used for groundwater recharge.

Table 4.2 Available data for an illustrative example

	Water supply	Groundwater recharge
Number of units of water delivered	x_1	x_2
Number of units of water required	1.00	5.00
Pump time required (hr)	0.50	0.25
Labour time required (person-hour)	0.20	0.20
Direct water costs ($)	0.25	0.75
Direct labour costs ($)	2.75	1.25
Sales price of water per unit ($)	4.00	5.00

Decision variables

Let us consider the following decision variables:

x_1 number of units of water delivered for water supply;
x_2 number of units of water delivered for groundwater recharge.

Objective function

From the problem description we note that there are two objectives: minimization of the increase in river pollution, and maximization of benefits. Trade-offs between these two objectives are sought to assist the water agency in the decision-making process.

Based on the available information, the objective functions of the problem can be formulated. The contribution margin (selling price/unit less variable cost/unit) of each allocation can be calculated:

Municipal water supply
$4.00 – $0.25– $2.75 = $1.00 per unit of water delivered
Sales   Direct water   Direct
price   cost   labour

Groundwater recharge
$5.00 – $0.75 – $1.25 = $3.00 per unit of water delivered
Sales   Direct water   Direct
price   cost   labour

and the objective function for profit becomes:

$$\text{Maximize} Z_1 = x_1 + 3x_2 \tag{4.35}$$

The objective function for pollution is:

$$\text{Minimize} Z'_2 = 3x_1 + 2x_2 \tag{4.36}$$

This function can be modified to $Z_2 = -3x_1 - 2x_2$ so that the maximization criterion is appropriate for both of the objective functions.

Constraints

The technical constraints are due to pump capacity:

$$0.5x_1 + 0.25x_2 \leq 8 \tag{4.37}$$

labour capacity:

$$0.2x_1 + 0.2x_2 \leq 4 \tag{4.38}$$

and water availability:

$$x_1 + 5x_2 \leq 72 \tag{4.39}$$

All decision variables are non-negative:

$$x_1, x_2 \geq 0. \tag{4.40}$$

The mathematical structure of the above problem shows that both of the objective functions (4.35 and 4.36) and all constraints (4.37–4.40) are linear functions of decision variables. Therefore, this mathematical model of the multi-purpose reservoir can be classified as a linear multi-objective analysis problem. Since all the decision variables and parameters are known in advance, the problem is of a deterministic nature. Both decision variables and parameters do not change with time and space, and therefore the problem is static with lumped parameters. Various techniques of multi-objective analysis are available for solving this problem.

4.4.3 Wastewater treatment

Let us consider a wastewater treatment problem outlined by Drobny et al (1971). A rolling mill (Figure 4.8) generates two distinct types of liquid wastes. One is pickling waste and the other, process water. These wastes can either be discharged directly into a river, subject to an effluent tax, or else may be treated in the existing plant, which removes 90 per cent of the pollutants. The pickling wastes require pretreatment before being passed through the treatment plant. It is required to determine the level of production that maximizes expected profits, which in this case equal income minus waste disposal costs. The waste costs comprise both waste treatment costs at the factory and effluent tax charges. The steel mill already has pretreatment and treatment facilities with known operating costs.

The nearby municipality controls the effluent tax on untreated wastes entering the city sewers. The higher the effluent tax, the more wastes the plant will treat itself. However, this could cause a decline in plant production which would directly affect the economy of the community. On the other hand, the lower the effluent tax, the less wastes the company will treat and the higher production will be. This means the community will be subsidizing the treatment of the plant's waste at the city sewage plant.

The following data are available:

- income on steel manufactured = $25/ton
- pickle wastes generated = 0.1 thousand litres/ton of steel
- process water generated = 1 thousand litres/ton of steel
- cost of pretreating pickle waste = $0.2/thousand litres
- cost of treating process water and pretreated pickle waste = $0.1/thousand litres
- pretreatment plant capacity = 2 million litres/day
- treatment plant capacity = 50 million litres/day
- treatment plant efficiency = 0.9 (both wastes)
- effluent tax on untreated pickle waste = $0.15/thousand litres
- effluent tax on untreated process water = $0.05/thousand litres.

The city has also imposed restrictions on how much untreated waste can be put into the sewer. No more than 500,000 litres/day of pickle waste or 10 million litres/day of process water may be discharged without treatment.

Decision variables

Let us consider the following decision variables:

x_1 tons of steel to be made per day;
x_2 thousand litres of pickling waste to be treated per day;
x_3 thousand litres of process water to be treated per day.

Objective function

The objective function can be divided into (a) income, (b) waste treatment cost and (c) effluent charges:

(a) The income is from the tons of steel x_1 made at $25 per ton.
(b) The waste treatment costs comprise the pretreatment ($0.20x_2$) and the treatment ($0.10[x_2 + x_3]$). Therefore

$$\text{Waste Treatment Costs} = 0.20x_2 + 0.10(x_x + x_3) \tag{4.41}$$

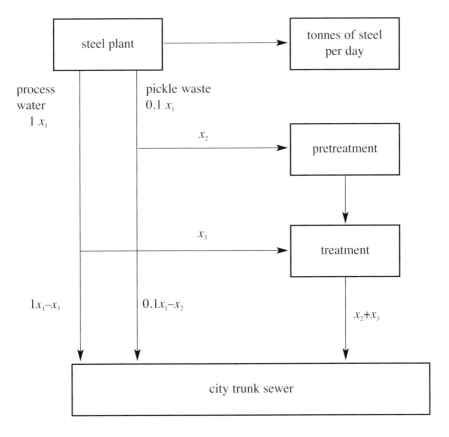

Figure 4.8 *Layout of the waste facilities*

(c) Tax on the untreated process water $(0.05[1.0\ x_1 - x_3])$ and charges for the untreated pickle waste, $(0.15[0.1\ x_1 - x_2])$, are levied against the company for discharging wastes directly into the sewer. However, since the treatment processes are only 90 per cent efficient, the city charges tax on 10 per cent of all treated sewage, $0.15(0.10)x_2 + 0.05(0.10)x_3$. Therefore:

$$\text{Effluent Charges} = (0.05\ [1.0x_1 - x_3]) + (0.15\ [0.1x_1 - x_2]) + \\ (0.15\ (0.10)x_2 + 0.05\ (0.10)x_3). \qquad (4.42)$$

It should be noted that income and waste treatment costs are directly controlled by the steel company. However, the city can indirectly affect income and waste treatment costs since the city sets the effluent charges.

The objective of the steel company is to maximize profits P:

$$P = \text{income} - \text{waste treatment costs} - \text{effluent charges}$$

or

$$P = 25\ x_1 - [0.20\ x_2 + 0.10\ (x_2 + x_3)] - (0.05 \times [1.0x_1 - x_3]) + \\ (0.15 \times [0.1x_1 - x_2]) + (0.15 \times (0.10)x_2 + 0.05 \times (0.10)x_3) \qquad (4.43)$$

Constraints

The capacities of the pretreatment facilities:

$$x_2 \leq 2{,}000 \qquad (4.44)$$

and treatment facilities:

$$x_2 + x_3 \leq 50{,}000 \qquad (4.45)$$

Also the city has imposed a restriction on how much untreated waste can be put into the sewer. No more than 500,000 litres per day of pickle waste or 10 million litres per day of process water may be discharged without treatment.

$$0.1x_1 - x_2 \leq 500 \qquad (4.46)$$

$$1.0x_1 - x_3 \leq 10{,}000 \qquad (4.47)$$

In order that the flow of waste be from the steel plant to the sewer then:

$$1.0x_1 - x_3 \geq 0 \tag{4.48}$$

$$0.1x_1 - x_2 \geq 0 \tag{4.49}$$

For non-negativity it is necessary that

$$x_1, x_2, x_3 \geq 0. \tag{4.50}$$

If we look at the mathematical structure of this problem it is evident that both the objective function (4.43) and all constraints (4.44–4.50) are linear functions of decision variables. Therefore, this mathematical model of wastewater treatment can be classified as a linear optimization problem. Since all the decision variables and parameters are known in advance, the problem is deterministic. Both decision variables and parameters do not change with time and space, and therefore the problem is static with lumped parameters. An appropriate technique for solving this problem is LP.

4.4.4 Irrigation flow control

Let us consider the irrigation canal intake shown in Figure 4.9. As the float drops it turns on the switch that opens the weir to admit water. The rising water volume causes the float to rise, and this in turn gradually shuts off the weir.

(a) What variables represent the system state, action and decision?
(b) Using feedback loop notation let us develop flow diagrams of the intake, (i) without the float and (ii) with the float.
(c) Develop a causal influence diagram of the irrigation canal intake.

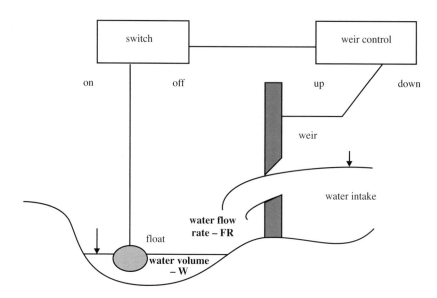

Figure 4.9 *Irrigation canal intake*

(a) The *water volume* corresponds to the system state. If the *water volume* is known, everything else about the system can be deduced. The *water flow rate* is the action variable. It is the *water flow* that causes the system to change. Here the float is coupled to the weir control through the switch. The time necessary for the float to adjust the weir opening is very short compared with the filling time of the canal. Therefore we can make the simplification of using the true system state – *water volume* – as a direct input to the decisions.

(b) A flow diagram of the irrigation system intake without the float is in Figure 4.10a, and with the float in Figure 4.10b.

The feedback loop in Figure 4.10a is broken, and this is an example of an open system. The *water level* (volume) no longer controls the *weir position*. On the other hand the system in Figure 4.10b is a closed or feedback system. The *water level* (volume) directly controls the *weir position*.

(c) The irrigation canal intake system in Figure 4.9 shows the canal–weir feedback system. The *flow rate* is determined by the mechanical design of the system and by the *water volume*. Figure 4.11 is a causal influence diagram of the irrigation canal intake system.

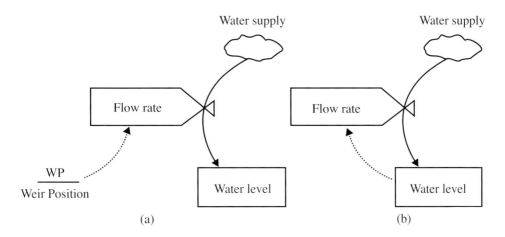

Figure 4.10 *Flow diagram of the irrigation system (a) without the float and (b) with the float*

Let us start from the element *weir position* at the bottom of Figure 4.11. If the *weir position* is increased (that is, the weir is opened further), the *water flow* increases. Similarly, if the *water flow* increases, then the *water volume* in the canal will increase. The next element along the chain of causal influences is the *difference*, which is the difference between the *desired canal water volume* and the (actual) *water volume*. (That is, *Difference = Desired Canal Water Volume – Water Volume*.) From this definition, it follows that an increase in *water volume* decreases the difference. Finally, to close the causal loop back to the *weir position*, a greater value for *difference* presumably leads to an increase in *weir position* (as an attempt is made to fill the canal). There is one additional link in this diagram, from *desired canal water volume* to *difference*.

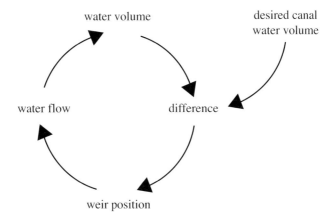

Figure 4.11 *A causal influence diagram of the irrigation canal intake system*

4.4.5 Eutrophication of a lake

Many lakes are becoming more and more oxygen-depleted because of increasing concentrations of nutrients. They have higher and higher algae populations and accumulate thicker and thicker layers of algal detritus, eventually becoming bogs, then dry land (Anderson, 1973). While this is a natural process, it is greatly accelerated by a variety of human activities. Eutrophication is of considerable concern to many people, especially those who happen to gain economic benefit from a lake in its present state. Ironically, those who stand to lose most from the eutrophication of a lake are frequently the largest contributors to that process. The most obvious example is the lake-front property owner who fertilizes his or her lawn, producing runoff, which greatly adds to the nutrient load of the lake.

This is a statement of the problem. A model of the system to be used for making plausible arguments about the system's behaviour is needed next. The first model is a causal influence diagram developed to explain the behaviour of the system. For the generic lake we shall compose a diagram of the situation with a number of feedback loops among ten elements: nutrient, oxygen in the air and epilimnion, decay, solution, respiration, growth, detritus, oxygen in hypolimnion, biomass and death. Definitions of these elements are needed to link them properly.

- Nutrient refers to the levels of phosphorus and nitrogen, which primarily serve as nutrients for the plant population of the lake.
- Epilimnion is the upper layer of the lake.
- Solution is the process of oxygen going into the hypolimnion.
- Respiration is the consumption of oxygen and the release of carbon dioxide by the plants. (Note: The materials given off and absorbed are just the opposite of those in photosynthesis, which the model ignores – appropriately enough for a polluted lake where light penetrates only a thin layer at the top.) Respiration also includes other metabolic processes here, so that some nutrients are returned to the lake in respiration.
- Growth and Death refer to rates of change of plants in the lake.
- Detritus is the dead plant material.
- Hypolimnion is the remainder of the lake. The bulk of the water is in this layer.
- Biomass can be taken to be the mass of all the plants in the lake.

(a) Using the definitions as given, and our own logic or knowledge, we can connect the following pairs of elements with causal arrows. Any pair may be connected in either or both directions.

Detritus   *Decay*

Nutrient	*Growth*
Growth	*Biomass*
Biomass	*Death*
Biomass	*Respiration*
Respiration	*Oxygen in Hypolimnion*
Solution	*Oxygen in Hypolimnion*
Decay	*Oxygen in Hypolimnion*
Growth	*Oxygen in Air and Epilimnion*
Solution	*Oxygen in Air and Epilimnion*

(b) Let us search for larger loops. How many loops are there?
(c) We can add to the causal influence diagram an external source of nutrients added to the lake. Connect this element to the internal element *Nutrient*.
(d) Figure 4.12 shows a causal influence diagram for the lake eutrophication problem.
(e) Careful observation of the diagram in Figure 4.12 should enable you to identify 11 feedback loops.

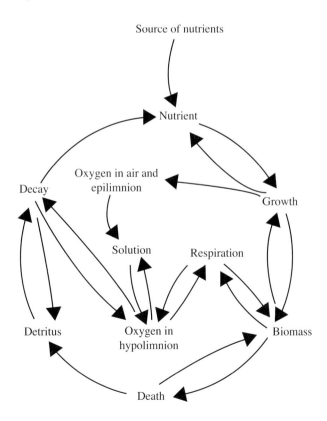

Figure 4.12 *Lake eutrophication loop diagram*

Policy choice

Policy choice is the identification of a particular set of actions as a possible means of altering the problem behaviour. Proposed solutions will undoubtedly change as the system behaviour is better understood.

While it is not necessary to state a policy when defining a problem, having a policy in mind shapes the construction of the loop diagram. Subsequent policy choices may require extensive redrawing of the diagram.

A great deal of patience and care are needed to keep the problem definition clearly in mind as the diagram is built up, torn down and rebuilt several times. The process of writing down the different dimensions for a problem provides a convenient description of the problem definition to use as a guide in building the causal influence diagram.

4.5 REFERENCES

Anderson, J.M. (1973), 'The Eutrophication of Lakes', in *Toward Global Equilibrium: Collected Papers*, Wright-Allen Press, Cambridge, MA

Blanchard, B.S. and W.J. Fabrycky (1990), *Systems Engineering and Analysis*, Prentice-Hall, Englewood Cliffs, NJ

de Neufville, R. (1990), *Applied Systems Analysis: Engineering Planning and Technology Management*, McGraw Hill, New York

Drobny, N.L., H.E. Hull and R.F. Testin (1971), *Recovery and Utilization of Municipal Solid Waste: A Summary of Available Cost and Performance Characteristics of Unit Processes and Systems*, Publication SW-10c. US Environmental Protection Agency, Washington, DC

Forrester, J.W. (1990), *Principles of Systems*, Productivity Press, Portland, OR, first publication in 1968

Hillier, F.S. and G.J. Lieberman (1990), *Introduction to Operations Research*, McGraw Hill, New York

Richmond, B. (1993), 'Systems Thinking: Critical Thinking Skills for the 1990s and Beyond', *System Dynamics Review*, 9(2): 113–133

Senge, P.M. (1990), The *Fifth Discipline: The Art and Practice of the Learning Organization*, Doubleday, New York

Shannon, C.E. (1949), 'A Mathematical Theory of Communications', *Bell System Technical Journal*, 27: 379–423, 623–656.

von Bertalanffy, L. (1969), *General System Theory: Foundation, Development and Applications*, George Braziller, New York

Wiener, N. (1948), *Cybernetics: or Control and Communication in the Animal and the Machine*, MIT Press, Cambridge, MA

4.6 EXERCISES

1. What is a system?
2. Identify and contrast a physical and a conceptual system.
3. Identify and contrast a closed and an open system.
4. Define in your own words and give examples of:
   a. Uncontrolled system input.
   b. Neutral system output.
   c. Feedback.
5. What is a mathematical model? Why do we develop mathematical models? List what are, in your opinion, the main purposes of mathematical models.
6. Describe one water resources system consisting of various interdependent components. What are the inputs to the system and what are its outputs? How did you decide what to include in the system and what not to include? How did you decide on the level of spatial and temporal detail to be included?
7. For the following problems specify in words possible objectives, the unknown decision variables whose values need to be determined, and the constraints that must be met by any solution of the problem.
   a. Exploring and drilling for water.
   b. Locating and deciding the capacity of a water treatment plant.
   c. Determining the size of a reservoir for a water supply of a small community.
   d. Locating new sites for irrigation water intakes.
   e. Allocating funds to urban renewal programmes.
   f. Selecting a most efficient flood control alternative.
   g. Determining the number and location of pumping stations for drainage of a large agricultural region.
8. A city has just approved the use of its municipal budget for building a new water treatment facility. You have been hired as a consultant to assist in the selection of a location for this new facility. List all the objectives to be optimized in selecting the location. Prioritize these objectives from the perspective of:
   a. the Mayor of a city;
   b. the residents of a city;
   c. the manufacturing industrial facility located in a city.
9. According to the mathematical structure of each problem in Exercise 7, classify the models using classification presented in Section 4.1.8.
10. The water supply system that serves the food factory in a city is nearing capacity. The factory is starting to explore alternatives to avoid disruption in its water supply, which is essential for the food manufacturing process. Options that have been proposed include: (i) expanding the current water supply system; (ii) developing a new water supply system; (iii) buying water from another water supply

system; and (iv) modifying the manufacturing process in order to conserve the water and, at the same time, expanding the existing recycling programme. Develop an optimization model to help evaluate these alternatives.

11 Give an example of a watershed process that exhibits exponential growth.
12 Give an example of a watershed process that exhibits exponential decay.
13 More inhabitants of a region are causing an increase in the municipal water demand. This increased demand prompts the expansion of the water supply system capacity. Develop a simple feedback loop representation of the problem. Identify the type of feedback relationship.
14 Develop a small water resources problem that includes a minimum of two feedback loops. Describe the problem and develop a causal influence diagram for it.

PART III

Water Resources Systems Management

5

Introduction to Methods of Water Resources Systems Management

Water resources systems management, as defined in this text, is an iterative process of integrated decision-making regarding the uses and modifications of waters and related lands within a geographic region. It relies on the application of a systems approach to formulating water resources management problems, and the use of systems analysis in finding their solutions. To use a systems approach calls for a change in our basic categories of thought about the physical reality under consideration. In contemporary water resources management we are forced to deal with complexities: with wholes or systems. This implies a basic reorientation in thinking.

System analysis is the use of rigorous methods to help determine preferred plans, design and operations strategies for complex, often large-scale, systems. Its methods depend on the use of the computer for practical application. This part of the book provides a basic introduction to some of the techniques used: simulation, optimization and multi-objective analysis.

5.1 SIMULATION

Simulation models describe how a system operates, and are used to predict what changes will result from a specific course of action. Such models are sometimes referred to as cause-and-effect models. They describe the state of the system in response to various inputs, but give no direct measure of what decisions should be taken to improve the performance of the system. Therefore, simulation is a problem-solving technique. It contains the following phases:

1. Development of a model of the system.
2. Operation of the model (i.e. generation of outputs resulting from the application of inputs).
3. Observation and interpretation of the resulting outputs.

The essence of simulation is modelling and experimentation. Simulation does not directly produce the answer to a given problem.

Simulation includes a wide variety of procedures. In order to choose among them, and use them effectively, the potential user must know how they operate, how they can be expected to perform, and how this performance relates to the problem under investigation. The generic simulation procedure involves decomposition of the problem in order to aid in the system description. When the main elements of the system are identified, the proper mathematical description is provided for each of them. The procedure continues with computer coding of the mathematical description of the model. Each model parameter is then calibrated, and the model performance is verified using data that has not been seen during the calibration process. The completed model is then operated using a set of input data. Detailed analysis of the resulting output is the final step in the simulation procedure.

A completed model can be reused many times with alternative input data. If there is a need for modification of the system description or model structure, the whole process starts again and the model has to be recoded, calibrated and verified again before its use.

The major components of a simulation model are:

- **Input**: quantities that 'drive' the model. (In water resources systems management models, for example, a principal input is the set of streamflows, rainfall sequences, pollution loads, water and power demands, etc.)
- **Physical relationships**: mathematical expressions of the relationships among the physical variables of the system being modelled (continuity, energy conservation, reservoir volume and elevation, outflow relations, routing equations, etc.).
- **Non-physical relationships**: those that define economic variables, political conflicts, public awareness, etc.
- **Operation rules**: the rules that govern operational control.
- **Outputs**: the final product of operations on inputs by the physical and non-physical relations in accordance with operating rules.

Simulation models play an important role in water resources systems management. They are widely accepted within the water resources community and are usually designed to predict the response of a system under a particular set of conditions. Early simulation models were constructed by a relatively small number of highly trained individuals. Many generalized, well-known simulation models were developed primarily in the FORTRAN language. These models include, among many others, SSARR (streamflow synthesis and reservoir regulation – US Army Corps of Engineers, North Pacific Division), RAS (river analysis system – Hydrologic Engineering Center), QUAL (stream water quality model – Environmental Protection Agency), HEC-5 (simulation of flood control and conservation systems – Hydrologic Engineering Center), SUTRA (saturated-unsaturated transport model – US Geological Survey), and KYPIPE (pipe network analysis – University of Kentucky). These models are quite complex, however, and their main characteristics are not readily understood by non-specialists. Also, they are inflexible and difficult to modify to accommodate site-specific conditions or planning objectives that were not included in the original model.

The most restrictive factor in the use of simulation tools is that there is often a large number of feasible solutions to investigate. Even when combined with efficient techniques for selecting the values of each variable, quite substantial computational effort may lead to a solution that is still far from the best possible.

5.2 SYSTEM DYNAMICS SIMULATION

System dynamics is an academic discipline introduced in the 1960s by researchers at the Massachusetts Institute of Technology. System dynamics was originally rooted in the management and engineering sciences but has gradually developed into a tool useful in the analysis of social, economic, physical, chemical, biological and ecological systems (Forrester, 1990; Sterman, 2000). In the field of system dynamics, as in the context of this book, a system is defined as a collection of elements which continually interacts over time to form a unified whole. The underlying pattern of interactions between the elements of a system is called the *structure* of the system. One familiar water resources example of a system is a reservoir. The structure of a reservoir is defined by the interactions between inflow, storage, outflow and other variables specific to a particular reservoir location (storage curve, evaporation, infiltration, etc.). The structure of the reservoir includes the variables important in influencing the system.

The term *dynamics* refers to change over time. If something is dynamic, it is constantly changing in response to the stimuli influencing it. A dynamic system is thus a system in which the variables interact to stimulate changes over time. System dynamics is a methodology used to understand how systems change over time. The way in which the elements or variables comprising a system vary over time is referred to as the *behaviour* of the system. In the reservoir example, the behaviour is described by the dynamics of reservoir storage growth and decline. This behaviour is due to the influences of inflow, outflow, losses and environment, which are elements of the system.

One feature that is common to all systems is that a system's structure determines its behaviour. System dynamics links the behaviour of a system to its underlying structure. It can be used to analyse how the structure of a physical, biological or any other system can lead to the behaviour that the system exhibits. By defining the structure of a reservoir, it is possible to use system dynamics simulation to trace out the behaviour over time of the reservoir.

The system dynamics simulation approach relies on understanding complex interrelationships among different elements within a system. This is achieved by developing a model that can simulate and quantify the behaviour of the system. The simulation of the model over time is considered essential to understanding the dynamics of the system. The major steps that are carried out in the development of a system dynamics simulation model include:

- understanding the system and its boundaries;
- identifying the key variables;
- describing the physical processes that affect variables through mathematical relationships;
- mapping the structure of the model;
- simulating the model for understanding its behaviour.

Advances made during the last decade in computer software provide considerable simplification in the development of system dynamics simulation models. Software tools like STELLA (High Performance Systems, 1992), DYNAMO (Lyneis et al, 1994), VENSIM (Ventana Systems, 1996) and POWERSIM (Powersim Corp., 1996) use the principles of object-oriented programming for the development of system dynamics simulation programs. They provide a set of graphical objects with their mathematical functions for easy representation of the system structure and the development of computer code. Simulation models can be easily and quickly developed using these software tools. The resulting models are easy to modify, easy to understand, and present results clearly to a wide audience of users. They are able to address water management problems with highly nonlinear relationships and constraints.

So what are the advantages of system dynamics simulation over the classical simulation discussed earlier?

- The power and simplicity of use of system dynamics simulation applications is not comparable with those developed in functional algorithmic languages. In a very short period of time, the users of the system dynamics simulation models can experience the main advantages of this approach. The power of simulation lies in the ease of constructing what-if scenarios and tackling big, messy, real-world problems.
- The general principles upon which the system dynamics simulation tools are developed apply equally to social, natural and physical systems. Using these tools in water resources systems management allows for the enhancement of models by adding social, economic and ecological sectors to the model structure.
- The structure–behaviour link of system dynamics models allows analysis of how structural changes in one part of a system might affect the behaviour of the system as a whole. Perturbing a system allows one to test how the system will respond under varying sets of conditions. To return to the example of a reservoir, someone can test the impact of a drought on the reservoir, or analyse the impact of the elimination of a particular user on the behaviour of the entire system. The manipulation of graphical objects in the system dynamics model that describes the structure of a system is as easy as a click of the computer mouse button.

> **Box 5.1** River life
>
> Chao Praya River in Bangkok is the life artery of the city. Our boat was moving fast along the banks, entering small canals and getting us around the Chao Praya world. People live on the river. Houses are built on stilts, leaving some space between the river and the ground floor.
>
> At one spot two elderly people were taking a bath in the river, hiding their naked bodies in the water when the boat passed; at another spot a woman was washing dishes in the river. Across the canal, kids were playing in the water, and two houses down a mother was bathing a baby.
>
> Life at the river, river of life.
>
> (A memory from 1996)

- For well-defined systems with sufficient and good data, the system dynamics simulation offers predictive functionality, determining the behaviour of a system under particular input conditions. However, the ability to use system dynamics simulation models and extend water resources simulation models to include social, ecological, economic and other non-physical system components offers learning functionality – the discovery of unexpected system behaviour under

particular input conditions. This is one of the main advantages of system dynamics over traditional simulation.
- In addition to relating system structure to system behaviour and providing users with a tool for testing the sensitivity of a system to structural changes, system dynamics requires a person to take an active part in the rigorous process of modelling system structure. Since the use of system dynamics software is very simple, the modelling process can be done directly by most experienced stakeholders. Modelling a system structure forces a user to consider details typically glossed over within a mental model. System dynamics simulation can easily become a group exercise, providing for the active involvement of all stakeholders and an interactive platform for the resolution of conflicts among them. In the literature this process has been called *shared vision modelling* (Palmer et al, 1999).

The rest of this book focuses on system dynamics simulation as one of the methods for water resources systems management. A detailed description of system dynamics simulation modelling and its application to water resources systems management follows in Part IV.

5.3 OPTIMIZATION

The procedure of selecting the set of decision variables that maximizes/minimizes the objective function, subject to the systems constraints, is called the *optimization procedure*. Numerous optimization techniques are used in water resources systems management.

A general mathematical form of an optimization problem, as given earlier in Part II, is:

$$\text{Min or Max } f(x) \tag{5.1}$$

subject to:

$$g_j(x) \leq b_j \quad j = 1, 2, ..., m$$
$$x \geq 0 \quad i = 1, 2 ... n$$

where:

x = a vector of decision variables
n = total number of decision variables

g = constraints
b = known right-hand-side values
j = constraint number
m = total number of constraints.

Most water resources allocation problems are addressed using linear programming (LP) solvers. LP is applied to problems that are formulated in terms of separable objective functions and linear constraints, as in:

$$Max(or Min) \quad x_0 = \sum_{j=1}^{n} c_j x_j \tag{5.2}$$

subject to:

$$\sum_{j=1}^{n} a_{ij} x_j = b_i \text{ for } i = 1,2,...,m$$

$$x_j \geq 0 \quad \text{for } j = 1,2,...,n$$

where:

c_j = objective function coefficient
a_{ij} = technological coefficient
b_i = right-hand side coefficient
x_o = objective function
x_j = decision variable
m = total number of constraints
n = total number of decision variables

The objective is usually to find the best possible water allocation (for water supply, hydropower generation, irrigation, etc.) within a given time period in complex water systems. For most practical water management applications, the nonlinearity of the objective function and/or constraints mean that many modifications have been used to convert nonlinear problems for the use of LP solvers.

Nonlinear programming is an optimization approach used to solve problems when the objective function and the constraints are not all in the linear form. In general, the solution to a nonlinear problem is a vector of decision variables that optimizes a nonlinear objective function subject to a set of nonlinear constraints. Successful applications are available for some special classes of nonlinear programming problems such as unconstrained problems, linearly constrained problems, quadratic problems, convex problems, separable problems, non-convex problems and geometric problems. The main limitation in applying nonlinear programming to water management problems is in the fact that it is generally unable to distinguish between a local optimum and a global optimum (except by finding another better local optimum).

Dynamic programming (DP) offers advantages over other optimization tools since the shape of the objective function and constraints do not affect it, and as such, it has been frequently used in water resources systems management. DP requires discretization of the problem into a finite set of stages. At every stage a number of possible conditions of the system (states) are identified, and an optimal solution is identified at each individual stage, given that the optimal solution for the next stage is available. An increase in the number of discretizations and/or state variables would increase the number of evaluations of the objective function and core memory requirement per stage.

However, the complexity of real water resources management problems today exceeds the capacity of traditional optimization algorithms (Simonovic, 2000). I have selected two problems to illustrate the point: a real-time hydropower operation problem, and a water resources network flow problem.

In an attempt to solve problems involving multiple periods and multiple-reservoir operations for hydropower generation, we need to keep in mind the enormity of the problem formulation and the computational power required to solve the problem in real time. Traditional optimization algorithms still suffer from one or more of three limitations: computational intractability, the requirement to calculate derivatives, or a need for too many assumptions for the problem to fit into a standard form of the optimization technique (ranging from linearization of the objective function and constraints, to incorporating problem-specific information into the formulation).

In the case of the real-time operation of multiple-reservoir systems for hydropower generation, the computational time required to solve the problem can be more than the actual time within which a decision is required for implementation. For example, an hourly scheduling problem for a weekly time horizon and four reservoirs with a total of 168 time intervals (hours), formulated as a mixed-integer nonlinear program (MINLP) and solved using the GAMS optimization solver (Brooke et al, 1996), failed to produce the results after four hours of real-time computation on the powerful SUN workstation (Teegavarapu and Simonovic, 2000). This can be explained by the limited capabilities of the optimization algorithm and the complexity associated with the problem formulation. The MINLP formulation suffers from a combinatorial explosion problem. The use of DP algorithm has limitations because of the curse of dimensionality. From experimentation with both MINLP and DP formulations, it was concluded that this hourly scheduling problem for a time horizon of a week cannot be solved within the practicable time limit of one hour.

The most common approach to the optimization of water resources networks is based on using the cost-capacitated network representation of a river basin, and solving the corresponding linear minimum cost flow problem. Virtually all models still rely on LP, mainly since the objective function related to water licensing priorities and other allocation objectives is linear. Typically, most of the models use the Out-of-Kilter algorithm (Barr et al, 1974). The main deficiencies in using LP for network flow optimization are:

- Water allocation models based on LP solvers are unable to incorporate a non-linear change of flow along a river or canal reach without an iterative procedure.
- The Out-of-Kilter and other LP solvers assume the instantaneous availability of water from any potential source (inflow or reservoir) to any existing user in the network.
- Non-linearities associated with the flow bounds are handled by applying successive iterations within a time step if necessary. It should be noted that each time an iteration is performed, a slightly different problem is submitted to the optimizer, resulting in a new solution which becomes the starting point for the next iteration. There is no guarantee that this process will result in convergence to the global optimum even when the objective function is convex. The problem being solved is nonlinear in terms of its flow bounds, and the guessing process solves successive linear approximations of a nonlinear problem.

In the recent past, most researchers have been looking for new approaches that combine efficiency and an ability to find the global optimum for complex water resources systems management problems. One group of techniques, known as *evolutionary algorithms*, seems to have a high potential since it holds a promise to achieve both these objectives (Simonovic, 2000; Ranjithan, 2005). Evolutionary techniques are based on similarities with the biological evolutionary process. In this concept, a population of individuals, each representing a search point in the space of feasible solutions, is exposed to a collective learning process, which proceeds from generation to generation. The population is arbitrarily initialized and subjected to the process of selection, recombination and mutation through stages known as *generations*, such that newly created generations evolve towards more favourable regions of the search space. In short, the progress in the search is achieved by evaluating the fitness of all individuals in the population, selecting the individuals with the highest fitness value, and combining them to create new individuals with increased likelihood of improved fitness. The entire process resembles the Darwinian rule known as the *survival of the fittest*.

Evolutionary algorithms are becoming more prominent in the water resources systems management field. Significant advantages of evolutionary algorithms include:

- no need for an initial solution;
- ease of application to nonlinear problems and to complex systems;
- production of acceptable results over longer time horizons;
- generation of several solutions that are very close to the optimum (and that give added flexibility to a water manager).

This book focuses on only two optimization methods: LP and evolutionary optimization. The first one is presented for its academic and practical significance. The second is discussed as one of the methods for current use in water resources systems optimization. A detailed description of both methods, together with their practical implementation, follows in Part IV.

5.4 MULTI-OBJECTIVE ANALYSIS

The management of complex water resources systems rarely involves a single objective. A multi-objective programming problem is characterized by a *p*-dimensional vector of objective functions:

$$\mathbf{Z}(\mathbf{x}) = [Z_1(\mathbf{x}), Z_2(\mathbf{x}), ..., Z_p(\mathbf{x})] \qquad (5.3)$$

subject to:

$$\mathbf{x} \in \mathbf{X}$$

where **X** is a feasible region:

$$\mathbf{X} = \{\mathbf{x}: \mathbf{x} \in \mathbf{R}^n, g_i(\mathbf{x}) \leq 0, x_j \geq 0 \; \forall \; i, j\} \qquad (5.4)$$

where:

R = set of real numbers
$g_i(\mathbf{x})$ = set of constraints
x = set of decision variables.

The word 'optimization' has been purposefully kept out of the definition of a multi-objective programming problem since one cannot, in general, optimize a priori a vector of objective functions. The first step of the multi-objective problem consists of identifying the set of non-dominated solutions within the feasible region **X**. So instead of seeking a single optimal solution, a set of non-dominated solutions is sought.

The essential difficulty with multi-objective analysis is that the meaning of the optimum is not defined as long as we deal with multiple objectives that are truly different. For example, suppose we are trying to determine the best design of a system of dams on a river, with the objectives of (a) promoting national income, (b) reducing deaths by flooding, and (c) increasing employment. Some designs will be more profitable, but less effective at reducing deaths. How can we state which is better when the objectives are so different, and measured in such different terms? How can we state with any accuracy what the relative value of a life is in terms of national income? If we resolve that question, then how would we determine the relative value of new jobs and other objectives? The answer is, with extreme difficulty. The attempts to set values of these objectives are, in fact, most controversial.

To obtain a single global optimum over all objectives requires that we either establish or impose some means of specifying the value of each of the different objectives. If all objectives can be valued on a common basis, the optimization can be stated in terms of that single value. The multi-objective problem disappears and the optimization proceeds relatively smoothly in terms of a single objective.

In practice it is frequently awkward if not indefensible to give every objective a relative value. The relative worth of profits, lives lost, the environment, and other such objectives is unlikely to be established easily by anyone, or to be accepted by all concerned. We cannot hope, then, to be able to determine an acceptable optimum analytically.

The focus of multi-objective analysis in practice is to sort out the mass of clearly dominated solutions, rather than determine the single best design. The result is the identification of a small subset of feasible solutions that is worthy of further consideration. Formally, this result is known as the *set of non-dominated solutions*.

Multiple-objective decisions do not have an optimal solution, unless one solution completely dominates every other solution for every objective. This does not usually happen in water resources management. As a result, methods are developed for assessing trade-offs between alternatives based on using more than one objective. In the last three decades of multi-objective research, efforts have been made in objective quantification, the generation of alternatives, and selection of the preferred alternative.

Early work focused on alternative generation, providing decision-makers with a complete spectrum of non-dominated solutions. Contemporary research into multi-objective analysis has shifted away from continuous theoretical models, and explored issues in evaluating a discrete set of alternatives (e.g. Srdjevic et al, 2007). There are plenty of options when it comes to choosing a multi-objective method. Cohon and Marks (1975) provided an early comparison of models, and suggested the surrogate worth trade-off (SWT) method for water resources problems because of its interactive nature (Haimes and Hall, 1974). Hobbs et al (1992) compared multi-objective methods for their appropriateness, ease of use and validity for water resources management decisions. They suggested that simpler transparent methods, or no formal method at all, were preferred by experienced water resources managers.

The shortcoming of most multi-objective methods is that they rely on an a priori articulation of preferences – an expression of the importance of each objective to a decision-maker. The difficulty for group decision-making is that conflicts arise, and complicate the evaluation process by tying decision-makers to their articulation of preference. Prior articulation methods are typified by an effort to aggregate the objectives of decision-makers and reduce the problem to a multiple-participant multiple-objective problem. Exceptions to prior articulation are methods that employ a progressive articulation of preferences. These are the true interactive conflict-capable multi-objective methods.

This book concentrates on the introduction of an efficient discrete multi-objective method with a progressive articulation of preferences, known as *Compromise Programming*, and originally introduced by Zeleny (1973). Various forms of the Compromise programming method and their practical implementation for water resources systems management are presented in Part IV.

5.5 REFERENCES

Barr, R.S., F. Glover and D. Klingman (1974), 'An Improved Version of the Out-Of-Kilter Method and Comparative Study of Computer Codes', *Mathematical Programming*, 7(1): 60.

Brooke, A., D. Kendrik and A. Meeraus (1996), *GAMS: A User's Guide*, The Scientific Press, Redwood City, CA, 286 pp.

Cohon, J. and D. Marks (1975), 'A Review and Evaluation of Multiobjective Programming Techniques', *Water Resources Research*, 11(2): 208219.

Forrester, J.W. (1990), *Principles of Systems*, Productivity Press, Portland, OR, first publication in 1968.

Haimes, Y. and W. Hall (1974), 'Multiobjectives in Water Resource Systems Analysis: The Surrogate Worth Trade-off Method', *Water Resources Research*, 10(4): 615–624

High Performance Systems (1992), *Stella II: An Introduction to Systems Thinking*, High Performance Systems, Inc., Nahover, NH

Hobbs, B., V. Chankong, W. Hamadeh, and E. Stakhiv (1992), 'Does Choice of Multicriteria Method Matter? An Experiment in Water Resources Planning', *Water Resources Research*, 28(7): 1767–1779

Lyneis, J., R. Kimberly and S. Todd (1994), 'Professional Dynamo: Simulation Software to Facilitate Management Learning and Decision Making', in *Modelling for Learning Organizations*, J. Morecroft and J. Sterman, eds., Pegasus Communications, Waltham, MA

Palmer, R. N., W. J. Werick, A. MacEwan and A. W. Woods (1999), 'Modelling Water Resources Opportunities, Challenges and Trade-Offs: The Use of Shared Vision Modelling for Negotiation and Conflict Resolution', in *Proceedings ASCE Water Resources Planning and Management Conference*, Erin M. Wilson, ed., 6–9 June 1999, Tempe, AZ

Powersim Corporation (1996), *Powersim 2.5 Reference Manual*, Powersim Corporation Inc., Herndon, VI

Ranjithan, S.R. (2005), 'Role of Evolutionary Computation in Environmental and Water Resources Systems Analysis', *ASCE Journal of Water Resources Planning and Management*, 131(1): 1–2

Simonovic, S.P. (2000), 'Last Resort Algorithms for Optimization of Water Resources Systems', *CORS – SCRO (Canadian Operational Research Society) Bulletin*, 34(1): 9–19

Srdjevic, B., Y.D.P. Medeiros and A.S. Faria (2007), 'An Objective Multicriteria Evaluation of Water Management Scenarios', *Water Resources Management*, 18(1): 35-54

Sterman, J.D. (2000), *Business Dynamics: Systems Thinking and Modelling for a Complex World*, McGraw Hill, New York

Teegavarapu, R.S.V. and S.P. Simonovic (2000), 'Short-Term Operation Model for Coupled Hydropower Reservoirs', *ASCE Journal of Water Resources Planning and Management*, 126(2): 98–107

Ventana Systems (1996), *Vensim User's Guide*, Ventana Systems Inc., Belmont, MA

Zeleny, M., (1973), 'Compromise Programming', in J. Cochrane and M. Zeleny, eds, *Multiple Criteria Decision Making*, University of South Carolina Press, DC

5.6 EXERCISES

1 What is the difference between simulation and optimization?
2 What is the difference between system dynamics simulation and classical simulation?
3 Define system structure. What does the term 'dynamics' refer to?
4 What is the outcome of optimization analysis? Are there multiple optimal solutions to a single optimization problem?
5 What is a feasible solution? Is an optimal solution always feasible?
6 Is there an optimal solution for a problem with multiple objectives? Why?
7 What is a non-dominated solution?
8 Describe, using words and a flow diagram, how you might simulate the operation of a water intake over time. List all assumptions. To simulate a water intake, what data do you need to have or know? Identify a feedback relationship/s in your model.

9 You are hired to determine the allocation of water X_j to four different users j – municipality, irrigation district, car factory and hydropower plant. Each of the users obtains benefits $R_j(X_j)$. The total water available is Q. Produce a flow chart showing how you can find the allocation to each user that results in the maximum total benefits.

10 Consider the following seven alternatives for the use of a storage reservoir to produce energy (10^3 kwh/day) and reduce average annual flood damage (10^6 $/year):

Alternative	Energy production	Flood damage reduction
1	25	78
2	32	56
3	18	100
4	7	112
5	27	71
6	20	92
7	11	105

a. Which alternative would be the best in your opinion, and why?
b. Which alternative would be the worst in your opinion, and why?
c. Provide an argument for selecting alternative 6, even though other alternatives exist that can give more hydropower energy and higher levels of flood protection.
d. What relative weight would you assign to these two objectives, and why?

6

Water Resources Systems Management Under Uncertainty – a Fuzzy Set Approach

Uncertainty is defined in plain language as lack of certainty. It has important implications for what can be achieved by water resources systems management. All water management decisions should take uncertainty into account. Sometimes the implications of uncertainty involve *risk*, in the sense of significant potential unwelcome effects of water resources system performance. Then managers need to understand the nature of the underlying threats in order to identify, assess and manage the risk. Failure to do so is likely to result in adverse impacts on performance, and in extreme cases, major performance failures.

Sometimes the implications of uncertainty involve an opposite form of risk, significant potential welcome effects. Then managers need to understand the nature of the underlying opportunities in order to identify and manage the associated decrease in risk. Failure to do so can result in a failure to capture good luck, which can increase the risk. For example, a development of a regional water supply system which generates unexpectedly rapid urbanization of the area may prove a disaster if the increasing demand cannot be met in the future; a construction project activity which finishes early may not result in a following activity starting early, and later delays will not be avoided by this good luck if it is wasted; a structural flood protection measure which generates new opportunities for the development of a floodplain may increase the future damage in the case of a more severe flood event.

In any given water resources management situation both threats and opportunities are usually involved, and both should be managed. A focus on one should never be allowed to eliminate concern for the other. Opportunities and threats (benefits and risks) can sometimes be treated separately, but they are seldom independent, just as two sides of the same coin can be examined separately, but they are not independent when it comes to tossing the coin.

There are often management options available which reduce or neutralize potential threats and simultaneously offer opportunities for positive improvements in system performance. It is rarely advisable to concentrate on reducing risk without considering the benefits resulting from associated opportunities, just as it is inadvisable to pursue opportunities without regard for the associated risk. Because resources used on risk management may mean reduced effort on the pursuit of opportunities, and vice versa, the effort spent on each needs to be balanced, in addition to balancing the total effort in relation to the benefits (Brugnach et al, 2007; Refsgaard et al, 2007).

To emphasize the desirability of a balanced approach to opportunity and threat management, It is proposed here to use the term *uncertainty management* in preference to the more established 'risk management'. However, uncertainty management is not just about managing perceived threats and opportunities, and their risk implications. It is also about managing the various sources of uncertainty that give rise to and shape risk, threat and opportunity (Antunes and Dias, 2007). Understanding the nature and significance of this uncertainty is an essential prerequisite for its efficient and effective management.

6.1 SOURCES OF UNCERTAINTY IN WATER RESOURCES SYSTEMS MANAGEMENT

Uncertainty is in part about variability in relation to the physical characteristics of water resources systems. But uncertainty is also about ambiguity (Ling, 1993; Simonovic, 1997). Both variability and ambiguity are associated with a lack of clarity because of the behaviour of all system components, a lack of data, a lack of detail, a lack of structure to consider water resources management problems, working and framing assumptions being used to consider the problems, known and unknown sources of bias, and ignorance about how much effort it is worth expending to clarify the management situation.

Uncertainty caused by variability is a result of inherent fluctuations in the quantity of interest (hydrologic variables). The three major sources of variability are temporal, spatial and individual heterogeneity. *Temporal variability* occurs when values fluctuate with time. Other values are affected by *spatial variability*: that is, they are dependent on the location of an area. The third category of *individual heterogeneity* effectively covers all other sources of variability. In water resources management variability is mainly associated with the spatial and temporal variation of hydrological variables (precipitation, river flow, water quality parameters, etc.).

The more elusive type of uncertainty is *ambiguity*, which is caused by a fundamental lack of knowledge. It occurs when the particular values that are of interest cannot be presented with complete confidence because of a lack of understanding or limitation of knowledge. The main sources of uncertainty because of a lack of knowledge are depicted in Figure 6.1.

Model and structural uncertainties refer to the knowledge of a process. Models are simplified representations of real-world processes, and model uncertainties can arise from oversimplification or from the failure to capture important characteristics of the process under investigation. Addressing this type of uncertainty is the coarse-tuning function of the analysis. This type of uncertainty is best understood by studying its major sources.

In water resources systems management, the modelling process includes *surrogate variables* (that is, the substitution of variables for quantities that are difficult to assess). They are an approximation of the real value. The second source of model uncertainty stems from *excluded variables* (variables deemed insignificant in a model). The removal of certain variables or factors introduces large uncertainties into the model. For example, many water resources risk assessment methods do not consider how hazardous chemicals can be propagated through vegetation. Attempting to address excluded variables raises a paradox: we do not know when we have forgotten something until is too late. The *impact of abnormal situations* on models is the third source of uncertainty. The very nature of a water resources model requires model calibration and verification using a set of broad circumstances. The problem occurs when a model is used for a situation that lies outside the set of situations used in the process of model calibration and verification. *Approximation uncertainty* is the fourth source of model uncertainty. This source covers the remaining types of uncertainty as a result of model generalizations. Examples of approximation uncertainty in water resources management can be found in the use of discrete probability distributions to represent a continuous process. The final type of model uncertainty, *incorrect form* (the correctness of the model being used to represent the real world) is initially the most obvious. To properly address this source, we must remember that all results are directly dependent on the validity of the model being used as a representation of the true process.

The next general category of uncertainty is *parameter uncertainty*. It involves the fine-tuning of a model, and cannot cause the large variations found in model uncertainty. The most common uncertainty in this category is caused by random error in direct measurements. It is also referred to as metric error, measurement error, random error or statistical variation. This error occurs because no measurement of water resources can be exact. Imperfections in the measuring instruments and observation techniques lead to imprecision and inaccuracies of measurements.

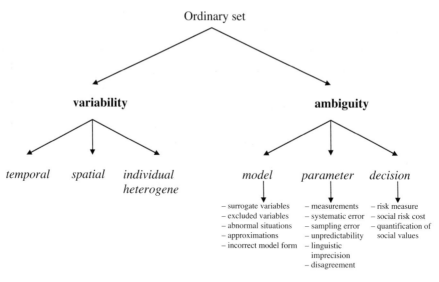

Sources: Ling (1993) and Simonovic (1997)

Figure 6.1 *Sources of uncertainty*

The second, and largest, source of parameter uncertainty is systematic error (error as a result of subjective judgement). Measurements involve both random and systematic error. The latter is defined as the difference between the true value and the mean of the value to which the measurements converge. The third type of error is sampling error (error in drawing inferences from a limited number of observations). Sampling causes uncertainty in the degree to which the sample represents the whole. Well-developed statistical techniques such as confidence intervals, coefficient of variation and sample size are used in water resources management to quantify this type of uncertainty. The fourth type of parameter uncertainty is caused by the unpredictability of an event. Limitations in knowledge and the presence of inherent unpredictability in the process make it impossible to predict, for example, wind direction and velocity at a future date. The fifth source of uncertainty is linguistic imprecision. Everyday language and communication is rather imprecise. It is possible to reduce linguistic uncertainty through clear specifications of events and values. The final source of uncertainty is derived from disagreement (conflicting expert opinion).

The third category of uncertainty is *decision uncertainty*, which arises when there is controversy or ambiguity over how to compare and weigh social objectives. It influences water resources decision-making after parameter and model uncertainties have been considered. The first decision uncertainty includes uncertainty in the selection of an index to measure risk. The measure must be as technically correct as possible while still being measurable and meaningful. The second source of decision uncertainty lies in deciding the social cost of risk (transforming risk measures into comparable quantities). The difficulties in this process are clearly illustrated in the concept of developing a monetary equivalent for the value of life in flood management. The quantification of social values is the third source of uncertainty. Once a risk measure and the cost of risk are generated, controversy still remains over what level of risk is acceptable. This level is dependent upon society's attitude to risk.

Uncertainty management as addressed in this book is about effective and efficient water resources decision-making, given this comprehensive and holistic view of uncertainty.

6.2 THE SCOPE FOR UNCERTAINTY IN THE WATER RESOURCES SYSTEMS MANAGEMENT PROCESS

The scope for uncertainty in any water resources decision situation is considerable. We can see part of this scope by considering a generic water resources systems decision-making framework, which we defined as a sequence of stages (in Section 3.3 of the book), each of which involves associated sources of uncertainty, as shown in Table 6.1.

The first stage in the decision process involves continuous monitoring of the system and its current operations. At some point problem recognition occurs, when decision-makers realize there is a need to make one or more decisions to address an emergent problem. However, uncertainty associated with ambiguity about the completeness, reality and accuracy of the information received, the meaning of the information, and its implications, may make ambiguity associated with the emergence of the problem important. Further, defining problems may not be straightforward. Different decision-makers may have different views about the significance or implications of an existing situation, and differing views about the need for action. Problems may be recognized as either threats or opportunities, which need to be addressed either reactively or proactively.

Alternatively, problems may be expressed in terms of weaknesses in system performance which need to be remedied, or particular benefits which could be exploited more extensively. Problems may involve relatively simple concerns within a given ongoing operation, but they may involve the possible emergence of a major system change or a revision to a key aspect of system management strategy. The decisions involved may be first-order decisions, or they may be higher-order decisions: deciding how to decide. Ambiguity about the way problems are identified and defined implies massive scope for uncertainty.

The definition of short- and long-term goals will depend on how a problem is defined. It involves determining which activities are relevant to addressing the problem, who is already involved with the problem, who should be involved and, importantly, the extent to which other areas of decision-making need to be linked with this decision process.

Determining the objectives involves identifying the performance criteria of concern, deciding how these will be measured, and determining appropriate priorities and trade-offs between the criteria. As with problem recognition, the tasks comprising these two stages can present significant difficulties, particularly if multiple decision-makers with differing criteria or priorities are involved. The determination of alternative courses of action may involve considerable effort to search for, or design, a set of feasible alternatives.

Table 6.1 *Sources of uncertainty in the water management process structure*

Stage in the decision process	Uncertainty about
Assess the nature and status of the water system	Completeness, reality and accuracy of information received, meaning of information, interpretation of implications
Define short- and long-term goals for the system	Significance of issue, urgency, need for action, appropriate frame of reference, scope of relevant activities, who is involved, who should be involved, extent of separation from other decision issues
Determine objectives (criteria) and actions to achieve goals	Relevant performance criteria, whose criteria, appropriate metrics, appropriate priorities and trade-offs between different criteria Nature of actions available (scope, timing and logistics involved), what is possible, level of detail required, time available to identify alternative actions
Assess benefits and costs of each action	Consequences, nature of influencing factors, size of influencing factors, effects and interactions between influencing factors (variability and timing), nature and significance of assumptions made How to weigh and compare predicted outcomes
Implement desired actions	How will alternatives work in practice?
Evaluate the effects of actions	What to monitor, how often to monitor?
Re-evaluate goals and objectives	When to take further action, in what direction?

The assessment of benefits and costs of each action builds on this to identify the factors that are likely to influence the performance of each identified alternative action, estimating their size and combined effect. The final step of selecting the desired action involves comparing the evaluations obtained, often by comparing relative performance on more than one performance criterion. Ambiguity about how best to manage all these steps and the quality of their output is a further massive source of uncertainty.

The final three stages, implementation of the desired alternative, evaluation of the effects of actions, and consequent re-evaluation of goals and objectives, might be regarded as outside the decision process. However, if the concern is problem resolution, it is important to recognize these three steps and consider them in iterative implementation of all earlier steps of the decision process.

Experience, as well as this brief overview of sources of uncertainty in a water resources decision process structure, tells us that the scope for making poor-quality decisions is considerable. Difficulties arise at every stage. The uncertainties listed in Table 6.1 indicate the nature of what is involved. Have we correctly interpreted information about the system environment? Have we correctly identified problems in a timely manner? Have we adopted the most appropriate scope for our decision? Are we clear about the performance criteria and their relative importance to us? Have we undertaken a sufficiently thorough search for alternative solutions? Have we evaluated alternatives adequately in a way that recognizes all relevant sources of uncertainty? And so on.

In order to manage all this uncertainty, water resources decision-makers seek to simplify the decision process by making assumptions about the level of uncertainty that exists, and by considering a model of the decision components. The value of this approach is a starting position for this book. A key aim of this section is to demonstrate that the quality of water resources systems management can be greatly improved by the use of formal decision support processes to manage associated uncertainty.

6.3 CONCEPTUAL RISK DEFINITIONS

An attempt by risk analysis experts in the late 1970s to come up with a standardized definition of risk concluded that a common definition is perhaps unachievable, and that authors should continue to define risk in their own way. As a result, numerous definitions can be found in recent literature, ranging from the vague and conceptual to the rigid and quantitative. At a conceptual level, we defined risk above as a significant potential unwelcome effect of water resources system performance, or the predicted or expected likelihood that a set of circumstances over some timeframe

> **Box 6.1** Water quality
>
> - The quality of natural water in rivers, lakes and reservoirs and below the ground surface depends on a number of interrelated factors. These factors include geology, climate, topography, biological processes and land use.
> - The most frequent sources of pollution are human waste (with 2 million tons a day disposed of in watercourses), industrial waste and chemicals, and agricultural pesticides and fertilizers. Key forms of pollution include faecal coliforms, industrial organic substances, acidifying substances from mining aquifers and atmospheric emissions, heavy metals from industry, ammonia, nitrate and phosphate pollution and pesticide residues from agriculture, and sediments from human-induced erosion to rivers, lakes and reservoirs.
> - One litre of oil can contaminate up to 2 million litres of water.
> - The United Nations Educational, Scientific and Cultural Organization (UNESCO) (2004) provides estimates of the volume of wastewater produced by each continent, giving a global total in excess of 1500 cubic kilometres (km^3) for 1995. Then there is the contention that each litre of wastewater pollutes at least 8 litres of freshwater so, based on this figure, some 12,000 km^3 of the globe's water resources are not available for use.
> - Levels of suspended solids in rivers in Asia have risen by a factor of four over the last three decades.
> - Bangladesh is grappling with the largest mass 'poisoning' (concentrations of arsenic in drinking water) in history, potentially affecting between 35 and 77 million of the country's 130 million inhabitants.
> - Excessive amounts of fluoride in drinking water can also be toxic. Discoloration of teeth occurs worldwide, but crippling skeletal effects caused by long-term ingestion of large amounts are prominent in at least eight countries, including China, where 30 million people suffer from chronic fluorosis.
>
> Source: UNESCO (2004)

will produce some harm that matters. More pragmatic treatments view risk as one side of an equation, where risk is equated with the probability of failure or the probability of load exceeding resistance. Other symbolic expressions equate risk with the sum of uncertainty and damage, or the quotient of hazards divided by safeguards (Lowrance, 1976).

Here we shall start with a risk definition based on the concept of load and resistance, terms borrowed from structural engineering. In the field of water resources systems management these two variables have a more general meaning, as shown in Table 6.2 (modified after Ganoulis, 1994). Load **l** is a variable reflecting the behaviours of the system under certain external conditions of stress or loading. Resistance **r** is a characteristic variable which describes the capacity of the system to overcome an external load (Plate and Duckstein, 1988). When the load exceeds the resistance (**l** > **r**) there should be a failure or an incident. A safety or reliability state is obtained if the resistance exceeds or is equal to the load (**l** ≤ **r**). From Table 6.2 it can be seen that load and resistance may take different meanings, depending on the specific problem domain.

Perhaps the most expressive definition of risk is the one that conveys its multidimensional character by framing risk as the set of answers to three questions: What can happen? How likely is it to happen? If it does happen, what are the consequences? (Kaplan and Garrick, 1981). The answers to these questions emphasize the notion that risk is a prediction or expectation which involves a hazard (the source of danger), uncertainty of occurrence and outcomes (the chance of occurrence), adverse consequences (the possible outcomes), a timeframe for evaluation, and the perspectives of those affected about what is important to them. The answers to these questions also form the basis of conventional quantitative risk analysis methodologies.

Three cautions surrounding risk must be taken into consideration: risk cannot be represented objectively by a single number alone, risks cannot be ranked on strictly objective grounds, and risk should not be labelled as real. Regarding the caution of viewing risk as a single number, the multidimensional character of risk can only be aggregated into a single number by assigning implicit or explicit weighting factors to various numerical measures of risk. Since these weighting factors must rely on value judgements, the resulting single metric for risk cannot be objective. Since risk cannot objectively be expressed by a single number, it is not possible to rank risks on strictly objective grounds. Finally, since risk estimates are evidence-based, risks cannot be strictly labelled as real. Rather, they should be labelled *inferred* at best.

A major part of the risk management confusion relates to an inadequate distinction between three fundamental types of risk:

- Objective risk (real, physical), R_o, and objective probability, p_o, which is the property of real physical systems.
- Subjective risk, R_s, and subjective probability, p_s. Probability is here defined as the degree of belief in a statement. R_s and p_s are not properties of the physical systems under consideration (but may be some function of R_o and p_o).
- Perceived risk, R_p, which is related to an individual's feeling of fear in the face of an undesirable possible event, is not a property of the physical systems but is related to fear of the unknown. It may be a function of R_o, p_o, R_s and p_s.

Table 6.2 *Examples of loads and resistance*

Physical system	Scientific discipline	Load	Resistance	Type of failure
Hydraulic structure (dam, levee, gate, etc.)	Civil engineering	force wind load flood rate stress	resisting stress dam height levee height	structural failure
Water system (lake, aquifer, river, etc.)	Water resources management	water demand pollutant load energy demand	water supply reservoir capacity receiving capacity	water shortage water pollution energy shortage
Hydrologic system (watershed, reservoir, etc.)	Hydrology	flow rate flood rainfall evaporation	threshold flow rate flood rainfall	exceedance floods
Ecosystem	Biological sciences	exposure	ecosystem capacity	ecosystem damage
Human organism	Health sciences	exposure	human capacity	health damage
Economic system	Economics	investment needs capital interest rate	money supply threshold interest rate	fiscal failure lack of capital
Social system	Social sciences	change of system perception acceptance	acceptance level flexibility resistance capacity	change of population culture change war

Because of the confusion between the concepts of objective and subjective risk, many characteristics of subjective risk (Kreps, 1988) are believed to be valid also for objective risk (Slovic, 2000). Therefore, it is almost universally assumed that the imprecision of human judgement is equally prominent and destructive for all water resources risk evaluations and all risk assessments. This is perhaps the most important misconception that blocks the way to more effective societal risk management (Palenchar and Heath, 2007). The ways society manages risks appear to be dominated by considerations of perceived and subjective risks, while it is objective risks that kill people, damage the environment and create property loss.

6.4 CHANGING THE PARADIGM, FROM PROBABILITY TO FUZZINESS

6.4.1 A brief discussion of probability

Probability is a concept widely accepted and practised in water resources systems management. To perform operations associated with probability, it is necessary to use sets – collections of elements, each with some specific characteristics (Modarres et al, 1999). Boolean algebra provides a means for evaluating sets. In probability theory, the elements that comprise a set are outcomes of an experiment. Thus, the universal set Ω represents the mutually exclusive listing of all possible outcomes of the experiment, and is referred to as the *sample space* of the experiment. In examining the outcomes of rolling a dice, the sample space is $S = (1,2,3,4,5,6)$. This sample space consists of six items (elements) or sample points. In probability concepts, a combination of several sample points is called an *event*. An event is, therefore, a subset of the sample space. For example, the event of 'an odd outcome when rolling a dice' represents a subset containing sample points 1, 3 and 5.

Associated with any event E of a sample space S is a probability, shown by $\Pr(E)$ and obtained from the following equation:

$$\Pr(E) = \frac{m(E)}{m(S)} \tag{6.1}$$

where $m(.)$ denotes the number of elements in the set $(.)$.

The probability of getting an odd number when tossing a dice is determined by using

$m(odd\ outcomes) = 3$

and

$m(sample\ space) = 6$

In this case,

$\Pr(odd\ outcomes) = 3/6 = 0.5$

Note that equation (6.1) represents a comparison of the relative size of the subset represented by the event E and the sample space S. This is true when all sample points are equally likely to be the outcome. When all sample points are not equally likely to be the outcome, the sample points may be weighted according to their relative frequency of occurrence over many trials or according to expert judgement.

In water resources management practice we use three major conceptual interpretations of probability.

Classical interpretation of probability (equally likely concept)

In this interpretation, the probability of an event E can be obtained from equation (6.1), provided that the sample space contains N equally likely and different outcomes, i.e. $m(S) = N$, n of which have an outcome (event) E, i.e. $m(E) = n$. Thus $Pr(E) = n/N$. This definition is often inadequate for engineering applications. For example, if failures of a pump to start in a water supply plant are observed, it is unknown whether all failures are equally likely to occur. Nor is it clear whether the whole spectrum of possible events is observed. That case is not similar to rolling a perfect die, with each side having an equal probability of 1/6 at any time in the future.

Frequency interpretation of probability

In this interpretation, the limitation on knowledge about the overall sample space is remedied by defining the probability as the limit of n/N as N becomes large. Therefore, $Pr(E) = \lim_{N \to \infty} (n/N)$. Thus if we have observed 2000 starts of a pump in which 20 failed, and if we assume that 2000 is a large number, then the probability of the pump failure to start is $20/2000 = 0.01$.

The frequency interpretation is the most widely used classical definition in water resources management today. However, some argue that because it does not cover cases in which little or no experience (or evidence) is available, or cases where estimates concerning the observations are intuitive, a broader definition is required. This has led to the third interpretation of probability.

Subjective interpretation of probability

In this interpretation, $Pr(E)$ is a measure of the degree of belief one holds in a specified event E. To better understand this interpretation, consider the probability of improving a system by making a design change. The designer believes that such a change will result in a performance improvement in one out of three missions in which the system is used. It would be difficult to describe this problem through the first two interpretations. That is, the classical interpretation is inadequate since there is no reason to believe that performance is as likely to improve as to not improve. The frequency interpretation is not applicable because no historical data exist to show how often a design change resulted in improving the system. Thus, the subjective interpretation provides a broad definition of the probability concept.

6.4.2 Problems with probability

One of the main goals of water resources management is to ensure that a system performs satisfactorily under a wide range of possible future conditions. This premise is particularly true of large and complex water resources systems. Water resources systems usually include conveyance facilities such as canals, pipes and pumps, treatment facilities such as sedimentation tanks and filters, and storage facilities such as reservoirs and tanks. These elements are interconnected in complicated networks serving broad geographical regions. Each element is vulnerable to temporary disruption in service because of natural hazards or human error, whether unintentional as in the case of operational errors and mistakes, or from intentional causes, such as a terrorist act.

Natural phenomena such as storm surges, excessive precipitation, floods and earthquakes can cause serious damage or the total failure of water resources systems. Terrorism is a new source of potential hazard to water resources systems. Although only a few terrorist threats against water resources systems are documented, the repair of serious damage caused by these acts has proved to be very costly (Haimes et al, 1998). Human error can also affect functioning of water resources systems. For example, in May 2000 the City of Walkerton (Canada) experienced *E-coli* bacteria contamination of its drinking water supply. The Walkerton Inquiry Report (O'Connor, 2002) concluded that the Walkerton Public Utility Commission (PUC) operators engaged in a host of improper operating practices which led to the contamination and the crisis that followed.

The sources of uncertainty are many and diverse, as was discussed earlier, and as a result they provide a great challenge to water resources systems management. The goal to ensure failsafe system performance may be unattainable. Adopting high-safety factors is one way to avoid the uncertainty of potential failures. However, making safety the first priority may render the system solution infeasible. Therefore, known uncertainty sources must be quantified.

The problem of engineering system reliability has received considerable attention from statisticians and probability scientists. Probabilistic (stochastic) reliability analysis has been used extensively to deal with the problem of uncertainty in water resources systems management. A prior knowledge of the probability density functions of both resistance and load, and their joint probability distribution function, is a prerequisite of the probabilistic approach. In practice, data on previous failure experience is usually insufficient to provide such information. Even if data are available to estimate these distributions, approximations are almost always necessary to calculate system reliability (Ang and Tang, 1984). Subjective judgement of the water resources decision-maker in estimating the probability distribution of a random event – the subjective probability approach of Vick (2002) – is another approach to deal with a lack of data. The third approach is Bayes's theory, where engineering judgement is integrated with observed information.

The choice of a Bayesian approach or any subjective probability distribution presents real challenges. For instance, it is difficult to translate prior knowledge into a meaningful probability distribution, especially in the case of multi-parameter problems (Press, 2003). In both subjective probability and Bayesian approaches, the degree of accuracy is strongly dependent on a realistic estimation of the decision-maker's judgement.

Until recently the probabilistic approach was the only approach for water resources systems reliability analysis. However, it fails to address the problem of uncertainty which goes along with human input, subjectivity, a lack of history and records. There is a real need to convert to new approaches that can compensate for the ambiguity or uncertainty of human perception.

Fuzzy set theory was intentionally developed to try to capture judgemental belief, or the uncertainty that is caused by the lack of knowledge. Relative to probability theory, it has some degree of freedom with respect to aggregation operators, types of fuzzy sets (membership functions) and so on, which enables it to be adapted to different contexts. During the last 40 years, fuzzy set theory and fuzzy logic have contributed successfully to technological development in different application areas such as mathematics, algorithms, standard models and real-world problems of different kinds (Zadeh, 1965; Zimmermann, 1996).

This book explores the utility of fuzzy set theory in addressing various uncertainties in water resources systems management. Since there is no previous book that applies fuzzy set theory to water resources management, this part of the book introduces in detail the basic concepts of fuzzy sets and fuzzy arithmetic. In Part IV, fuzzy theory is discussed directly in relation to the simulation, optimization and multi-objective analysis of water resources systems under uncertainty.

6.4.3 Fuzziness and probability

Shortly after fuzzy set theory was first developed in the late 1960s, there were a number of claims that fuzziness was nothing but probability in disguise. Probability and fuzziness are related, but they are different concepts. Fuzziness is a type of deterministic uncertainty. It describes the *event class ambiguity*. Fuzziness measures the *degree to which* an event occurs, not whether it occurs. At issue is whether the event class can be unambiguously distinguished from its opposite. Probability, in contrast, arises from the question of *whether or not* an event occurs. Moreover, it assumes that the event class is crisply defined and that the law of non-contradiction holds – that is, that for any property and for any definite subject, it is not the case both that the subject possesses that property and that the subject does not possess that property. Fuzziness occurs when the law of non-contradiction (and equivalently the law of excluded middle – for any property and for any individual, either that individual possesses that property or that individual does not possess that property) is violated. However, it seems more appropriate to investigate fuzzy probability for the latter case, than to completely dismiss probability as a special case of fuzziness. In essence, whenever there is an experiment for which we are not capable of 'computing' the outcome, a probabilistic approach may be used to estimate the likelihood of a possible outcome belonging to an event class. A fuzzy theory extends the traditional notion of a probability when there are outcomes that belong to several event classes at the same time, but to different degrees. Fuzziness and probability are orthogonal concepts which characterize different aspects of human experience. Hence, it is important to note that neither fuzziness nor probability governs physical processes in nature. These concepts were introduced by humans to compensate for our own limitations.

Let us review two examples that show a difference between fuzziness and probability.

Russell's paradox

That the laws of non-contradiction and excluded middle can be violated was pointed out by Bertrand Russell with the tale of the barber. Russell's barber is a bewhiskered man who lives in a town and shaves a man if and only if the man does not shave himself. The question is, who shaves the barber? If he shaves himself, then by definition he does not. But if he does not shave himself, then by definition he does. So he does and he does not. This is a contradiction or *paradox*. It has been shown that this paradoxical situation can be numerically resolved as follows. Let S be the proposition that the barber shaves himself and not-S the proposition that he does not. Since S implies not-S and vice versa, the two propositions are logically equivalent, i.e. S = not-S. Fuzzy set theory allows for an event class to coexist with its opposite at the same time, but to different degrees, or in the case of paradox to the same degree, which is different from zero or one.

Misleading similarities

There are many similarities between fuzziness and probability. The largest, but superficial and misleading, similarity is that both systems quantify uncertainty using numbers in the unit interval [0,1]. This means that both systems describe and quantify the uncertainty numerically. The structural similarity arising from lattice theory is that both systems algebraically manipulate sets and propositions associatively, commutatively and distributively. These similarities are misleading because a key distinction comes from what the two systems are trying to model. Another distinction is in the idea of observation. Clearly, the two models possess different kinds of information: fuzzy memberships, which quantify similarities of objects to imprecisely defined properties; and probabilities, which provide information on expectations over a large number of experiments.

6.5 INTRODUCTION TO FUZZY SET THEORY

The material presented in this section of the text is based on Kaufmann and Gupta (1985), Zimmermann (1996) and Pedrycz and Gomide (1998). First I shall present basic definitions of fuzzy sets, followed by algebraic operations, which will then form the basis of further consideration in Part IV.

6.5.1 Basic definitions

The basic concept of set theory is a collection of objects that has similar properties or general features. Humans tend to organize objects into sets so as to generalize knowledge about objects through the classification of information. The ordinary set classification imposes a dual logic. An object either belongs to a set or does not belong to it, as set boundaries are well defined. For example, if we consider a set A in a universe X, as shown in Figure 6.2, it is obvious that object x_1 belongs to set A, while x_2 does not. We denote the acceptance of belonging to a set by 1 and rejection of belonging by 0. The classification is expressed through a characteristic membership function $\mu_{\tilde{A}}$, for $x \in X$:

$$\mu_{\tilde{A}}(x) = \begin{cases} 1, & \text{if } x \in A \\ 0, & \text{if } x \notin A \end{cases} \qquad (6.2)$$

where $A(x)$ is the characteristic function denoting the membership of x in set A.

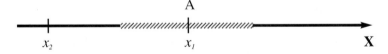

Source: Pedrycz and Gomide (1998)

Figure 6.2 *Ordinary set classification*

The basic notion of fuzzy sets is to relax this definition, and admit intermediate membership classes to sets. Therefore, the characteristic function can accept values between 1 and 0, expressing the grade of membership of an object in a certain set. According to this notion, the fuzzy set will be represented as a set of ordered pairs of elements, each of which presents the element together with its membership value to the fuzzy set:

$$\tilde{A} = \{(x, \mu_{\tilde{A}}(x)) | x \in X\} \tag{6.3}$$

Example 1

Assume the existence of an ordinary set B with three values, 1, 2 and 3, belonging to it. The set is mathematically represented as:

$$B(x) = \{1,2,3\} \tag{6.4}$$

where $B(x)$ is the ordinary set, and $1, 2, 3 \in X$ are elements of the universe belonging to set $B(x)$.

Now let us take $\tilde{B}(y)$, which is a fuzzy set, with three objects belonging to it, 4, 5 and 6, with membership values 0.6, 0.2 and 1.0 respectively. This set can be represented as follows:

$$\tilde{B}(y) = (4, 0.6), (5, 0.2), (6, 0.1) \tag{6.5}$$

where $\tilde{B}(y)$ is the fuzzy set; and $4, 5, 6 \in X$.

In both representations, the other elements in the universe X that do not belong to the ordinary set $B(x)$, and the elements that have membership values of 0, are not listed. Figure 6.3 depicts the difference in representation between the ordinary set and the fuzzy set. The horizontal axis represents the elements of the universe and the vertical axis represents the grade of membership of elements.

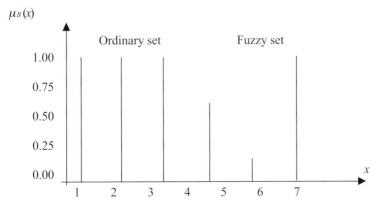

Figure 6.3 *Ordinary and fuzzy set representation*

The membership function is the crucial component of a fuzzy set, therefore all operations with fuzzy sets are defined through their membership functions. The basic definition of a fuzzy set is that it is characterized by a membership function mapping the elements of a domain, space or universe of discourse X to the unit interval $[0,1]$:

$$\tilde{A} : X \rightarrow [0,1] \qquad (6.6)$$

where $\tilde{A}$ is the fuzzy set in the universe of discourse X, and X is the domain, or the universe of discourse.

The function in equation (6.6) describes the membership function associated with a fuzzy set $\tilde{A}$. A fuzzy set is said to be a *normal fuzzy set* if at least one of its elements has a membership value of 1.

The crisp set of elements that belong to the fuzzy set $\tilde{A}$ at least to the degree α is called the α-*level set*:

$$A_\alpha = \{x \in X \mid \mu_{\tilde{A}}(x) \geq \alpha\} \qquad (6.7)$$

and

$$A'_\alpha = \{x \in X \mid \mu_{\tilde{A}}(x) > \alpha\} \qquad (6.8)$$

is called a *strong α-level set* or *strong α-level cut*. α is also known as the *credibility level*.

Example 2

An engineer wants to present possible flood protection levee options to a client. One indicator of protection is the levee height in metres. Let $X = \{1, 1.1, 1.2, 1.3, 1.4, 1.5, 1.6, 1.7, 1.8\}$ be the set of possible levee heights. Then the fuzzy set 'safe levee' may be described as:

$$\tilde{A} = \{(1.3, .2), (1.4, .4), (1.5, .6), (1.6, .8), (1.7, 1), (1.8, 1)\}$$

Figure 6.4 shows the fuzzy set of the 'safe levee'.

A fuzzy set is convex if all α-level sets are convex, or:

$$\mu_{\tilde{A}}(\lambda x_1 + (1-\lambda)x_2) \geq \min(\mu_{\tilde{A}}(x_1), \mu_{\tilde{A}}(x_2)) \qquad (6.9)$$

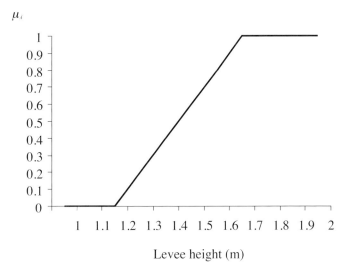

Figure 6.4 *'Safe levee' fuzzy set*

where:

$x_1, x_2 \in X$

and

$\lambda \in [0,1]$

Example 3

Let us refer again to Example 2, and list possible α-level sets:

$A_{.2} = \{1.3, 1.4, 1.5, 1.6, 1.7, 1.8\}$

$A_{.5} = \{1.5, 1.6, 1.7, 1.8\}$

$A_{.8} = \{1.6, 1.7, 1.8\}$

$A_{1} = \{1.7, 1.8\}$

The membership function may have different shapes and may be continuous or discrete, depending on the context in which it is used. Figure 6.5 shows the four most common types of continuous membership function. Families of parameterized functions, such as the following triangular membership function, can represent most of the common membership functions explicitly:

$$\mu_{\tilde{A}}(x) = \begin{cases} 0, & \text{if } x \leq a \\ \dfrac{x-a}{m-a}, & \text{if } x \in [a, m] \\ \dfrac{b-x}{b-m}, & \text{if } x \in [m, b] \\ 0, & \text{if } x \geq b \end{cases} \qquad (6.10)$$

where m is the modal value, and a, b are the lower and upper bounds of the non-zero values of membership.

A *fuzzy number* is a special case of a fuzzy set, having the following properties:

- it is defined in the set of real numbers;
- its membership function reaches the maximum value, 1.0, i.e. it is a normal fuzzy set;
- its membership function is unimodal (it is a convex fuzzy set).

A fuzzy number can be defined as follows:

$$\tilde{X} = \{(x, \mu_{\tilde{X}}(x)) : x \in R;\, \mu_{\tilde{X}}(x) \in [0,1]\} \qquad (6.11)$$

where:
$\tilde{X}$ = the fuzzy number
$\mu_{\tilde{X}}(x)$ = the membership value of element x to the fuzzy number
R = the set of real numbers.

A *support of a fuzzy number* is the ordinary set, which is defined as follows

$$S(\tilde{X}) = \tilde{X}(0) = \{x : \mu_{\tilde{X}}(x) > 0\} \qquad (6.12)$$

The fuzzy number support is the 0-level set and includes all the elements with the credibility level α higher than 0. Figure 6.6 illustrates these definitions.

6.5.2 Set-theoretic operations for fuzzy sets

Operations with fuzzy sets are defined using their membership functions. I present the concepts here as originally suggested by Zadeh (1965), since this provides a

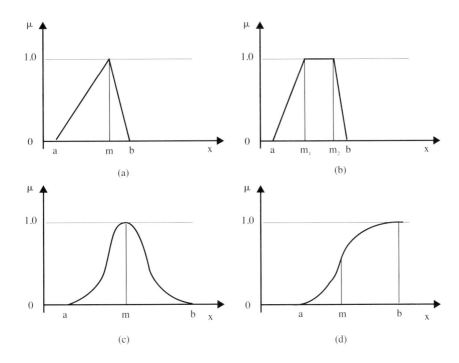

Figure 6.5 *(a) Triangular, (b) trapezoid, (c) Gaussian and (d) sigmoid membership functions*

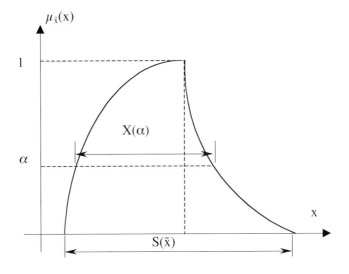

Figure 6.6 *Credibility level (α-cut) and support of a fuzzy set*

consistent framework for the theory of fuzzy sets. However, this is not the only possible way to extend classical set theory to the fuzzy domain.

Intersection

The membership function $\mu_{\tilde{C}}$ of the intersection $\tilde{C} = \tilde{A} \cap \tilde{B}$ is defined by:

$$\mu_{\tilde{C}}(x) = \min \{\mu_{\tilde{A}}(x), \mu_{\tilde{B}}(x)\}, \quad x \in X \tag{6.13}$$

where:
$\mu_{\tilde{C}}(x)$ = the membership of the fuzzy intersection of $\tilde{A}$ and $\tilde{B}$
min ( ) = the ordinary minimum operator
$\mu_{\tilde{A}}(x)$ = the membership of fuzzy set $\tilde{A}$
$\mu_{\tilde{B}}(x)$ = the membership of fuzzy set $\tilde{B}$.

Union

The membership function $\mu_{\tilde{C}}(x)$ of the union $\tilde{C} = \tilde{A} \cup \tilde{B}$ is defined by:

$$\mu_{\tilde{C}}(x) = \max \{\mu_{\tilde{A}}(x), \mu_{\tilde{B}}(x)\}, \quad x \in X \tag{6.14}$$

where:
$\mu_{\tilde{C}}(x)$ = the membership of the fuzzy union of $\tilde{A}$ and $\tilde{B}$
max ( ) = the ordinary maximum operator
$\mu_{\tilde{A}}(x)$ = the membership of fuzzy set $\tilde{A}$
$\mu_{\tilde{B}}(x)$ = the membership of fuzzy set $\tilde{B}$.

Complement

The membership function $\mu_{\tilde{C}}(x)$ of the complement of fuzzy set $\tilde{C}$ is defined by:

$$\mu_{\tilde{C}}(x) = 1 - \mu_{\tilde{C}}(x) \quad x \in X \tag{6.15}$$

where:
$\mu_{\tilde{C}}(x)$ = the membership of the complement of fuzzy set $\tilde{C}$, and
$\mu_{\tilde{C}}(x)$ = the membership of fuzzy set $\tilde{C}$.

Figures 6.7a and 6.7b show the fuzzy intersection and union operators on fuzzy sets.

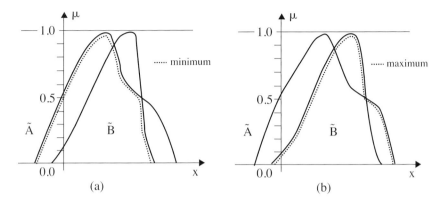

Figure 6.7 *Fuzzy intersection (a) and union (b) of two fuzzy sets*

Example 4

Let $\tilde{A}$ be the fuzzy set of 'safe levee' from Example 2 and $\tilde{B}$ the fuzzy set 'available budget', defined as:

$$\tilde{B} = \{(1.1,.2),(1.2,.5)(1.3,.8),(1.4,1),(1.5,.7),(1.6,.3)\}$$

Figure 6.8 shows both fuzzy sets.

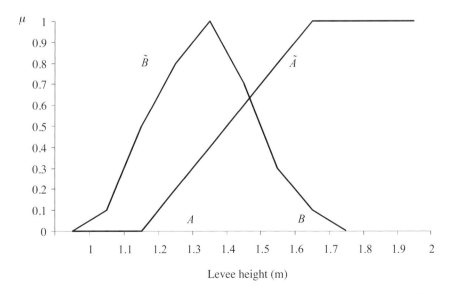

Figure 6.8 *Fuzzy sets of the 'safe levee' $\tilde{A}$ and 'available budget' $\tilde{B}$*

Now we are in the position to find the intersection of the fuzzy sets:

$\tilde{C} = \tilde{A} \cap \tilde{B}$

$\tilde{C} = \{(1.3,.2), (1.4,.4), (1.5,.6), (1.6,.8), (1.7,1), (1.8,1)\}$

and the union of the fuzzy sets:

$\tilde{D} = \tilde{A} \cup \tilde{B}$

$\tilde{D} = \{(1.1,.2), (1.2,.5), (1.3,.8), (1.4,1), (1.5,.7), (1.6,.8), (1.7,1), (1.8,1)\}$

It was mentioned earlier that *min* and *max* are not the only operators that can be used to model the intersection and union of fuzzy sets respectively. Zimmermann (1996) presents an interpretation of intersection as 'logical and' and the union as 'logical or'.

AND–OR operators

Assuming that ∧ denotes the fuzzy AND operation and ∨ denotes the fuzzy OR operation, the definitions for both operators are:

$$\mu_{\tilde{A} \wedge \tilde{B}}(x) = \min \{\mu_{\tilde{A}}(x), \mu_{\tilde{B}}(x)\}, \quad x \in X \tag{6.16}$$

and

$$\mu_{\tilde{A} \vee \tilde{B}}(x) = \max \{\mu_{\tilde{A}}(x), \mu_{\tilde{B}}(x)\}, \quad x \in X \tag{6.17}$$

6.5.3 Fuzzy arithmetic operations on fuzzy numbers

One of the most basic concepts of fuzzy set theory, which can be used to transfer crisp mathematical concepts to fuzzy sets, is the *extension principle*. It is defined as follows. Let X be a cartesian product of domains $X = X_1,...,X_r$. Function f is mapping from X to a domain Y, $y = f(x_1,...,x_r)$. Then the extension principle allows us to define a fuzzy set in Y by:

$$\tilde{B} = \{(y, \mu_{\tilde{B}}(y)) | y = f(x_1,...,x_r), (x_1,...,x_r) \in X\} \tag{6.18}$$

where:

$$\mu_{\tilde{B}(y)} = \begin{cases} \sup_{\{(x_1,\ldots,x_r) \subset f^{-1}(y)\}} \min\{\mu_{\tilde{A}_1}(x_1),\ldots,\mu_{\tilde{A}_r}(x_r)\} & \text{if } f^{-1}(y) \neq 0 \\ 0 & \text{otherwise} \end{cases} \quad (6.19)$$

where:
f^{-1} = the inverse of f.

At any α-level, the fuzzy number $\tilde{A}$ can be represented in the interval form as follows:

$$\tilde{A}(\alpha) = [a_1(\alpha), a_2(\alpha)] \quad (6.20)$$

where:
$\tilde{A}(\alpha)$ = the fuzzy number at α-level
$a_1(\alpha)$ = the lower bound of the α-level interval
$a_2(\alpha)$ = the upper bound of the α-level interval.

From this definition and the extension principle, the arithmetic operations on intervals of real numbers (crisp sets) can be extended to the four main arithmetic operations for fuzzy numbers, i.e. addition, subtraction, multiplication and division. The fuzzy operations of two fuzzy numbers $\tilde{A}$ and $\tilde{B}$ are defined at any α-level cut as follows:

$$\tilde{A}(\alpha) \, (+) \, \tilde{B}(\alpha) = [a_1(\alpha) + b_1(\alpha), a_2(\alpha) + b_2(\alpha)] \quad (6.21)$$

$$\tilde{A}(\alpha) \, (-) \, \tilde{B}(\alpha) = [a_1(\alpha) + b_2(\alpha), a_2(\alpha) - b_1(\alpha)] \quad (6.22)$$

$$\tilde{A}(\alpha) \, (\cdot) \, \tilde{B}(\alpha) = [a_1(\alpha) \cdot b_1(\alpha), a_2(\alpha) \cdot b_2(\alpha)] \quad (6.23)$$

$$\tilde{A}(\alpha) \, (/) \, \tilde{B}(\alpha) = [a_1(\alpha) / b_2(\alpha), a_2(\alpha) / b_1(\alpha)] \quad (6.24)$$

Note that for multiplication and division:

$$(\tilde{A}(/)\tilde{B})(\cdot) \, \tilde{B} \neq \tilde{A}.$$

This is also true for addition and subtraction:

$$(\tilde{A}(-)\tilde{B})(+) \, \tilde{A} \neq \tilde{A}.$$

Example 5

Assume the two triangular fuzzy numbers shown in Figure 6.9 and compute their sum, where:

$$\mu_{\tilde{A}} = \begin{cases} 0, & x \leq -5, \\ x/3+5/3, & -5 \leq x \leq -2, \\ -x/3+1/3, & -2 \leq x \leq 1, \\ 0, & x \geq 1. \end{cases} \quad \forall x \in R \quad (6.25)$$

and

$$\mu_{\tilde{B}} = \begin{cases} 0, & x \leq -3, \\ x/7+3/7, & -3 \leq x \leq 4, \\ -x/8+12/8, & 4 \leq x \leq 12, \\ 0, & x \geq 12. \end{cases} \quad \forall x \in R \quad (6.26)$$

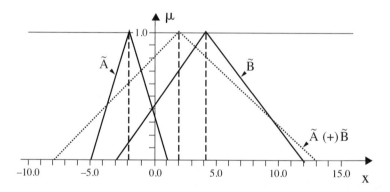

Figure 6.9 *Addition of two triangular fuzzy numbers (Example 5)*

To compute the intervals of confidence for each level α, the triangular shapes will be described by functions of α in the following manner.
From (6.25),

$$\alpha = a_1^\alpha /3+5/3 \text{ and } \alpha = -a_2^\alpha /3+1/3.$$

Hence, the interval of confidence at the level α is given by:

$$A_\alpha = [a_1^\alpha, a_2^\alpha] \tag{6.27}$$
$$A_\alpha = [3\alpha - 5, -3\alpha + 1].$$

From (6.26):

$$\alpha = b_1^\alpha / 7 + 3/7 \text{ and } \alpha = -b_2^\alpha / 8 + 12/8.$$

Therefore:

$$B_\alpha = [b_1^\alpha, b_2^\alpha] \tag{6.28}$$
$$B_\alpha = [7\alpha - 3, -8\alpha + 12].$$

Adding (6.26) and (6.27) gives

$$A_\alpha(+)B_\alpha = [a_1^\alpha + b_1^\alpha, a_2^\alpha + b_2^\alpha] \tag{6.29}$$
$$A_\alpha(+)B_\alpha = [10\alpha - 8, -11\alpha + 13].$$

So at the end we obtain the fuzzy addition as

$$\mu_{\tilde{A}+\tilde{B}}(x) = \begin{cases} 0, & x \leq -8, \\ x/10 + 8/10, & -8 \leq x \leq 2, \\ -x/11 + 13/11, & 2 \leq x \leq 13, \\ 0, & x \geq 13. \end{cases} \tag{6.30}$$

Example 6

Let us consider two fuzzy numbers, $\tilde{A}$ and $\tilde{B}$, with a triangular shape (Figure 6.10), and find the difference $(\tilde{A} - \tilde{B})$ if:

$$\mu_{\tilde{A}} = \begin{cases} 0, & x \leq 7, \\ x/7 + 1, & 7 \leq x \leq 14, \\ -x/5 + 19/5, & 14 \leq x \leq 19, \\ 0, & x \geq 19. \end{cases} \quad \forall x \in R \tag{6.31}$$

and

$$\mu_{\tilde{B}} = \begin{cases} 0, & x \le 3, \\ x/2-3/2, & 3 \le x \le 5, \\ -x/5+10/5, & 5 \le x \le 10, \\ 0, & x \ge 10. \end{cases} \quad \forall x \in R \qquad (6.32)$$

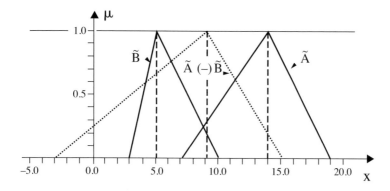

Figure 6.10 *Subtraction of two fuzzy numbers (Example 6)*

Now using (6.31), let

$$\alpha = a_1^\alpha /7-1 \text{ and } \alpha = -a_2^\alpha /5+19/5.$$

from which

$$A_\alpha = [a_1^\alpha, a_2^\alpha] \qquad (6.33)$$
$$A_\alpha = [7\alpha - 7, -5\alpha + 19].$$

Now using (6.32), we obtain

$$\alpha = b_1^\alpha /2-3/2 \text{ and } \alpha = -b_2^\alpha /5+10/5.$$

and

$$B_\alpha = [b_1^\alpha, b_2^\alpha] \qquad (6.34)$$
$$B_\alpha = [2\alpha + 3, -5\alpha + 10].$$

Subtracting (6.34) from (6.33) gives

$$A_\alpha(-)B_\alpha = [a_1^\alpha - b_1^\alpha, a_2^\alpha - b_2^\alpha] \tag{6.35}$$
$$A_\alpha(-)B_\alpha = [12\alpha - 3, -7\alpha + 16].$$

and we obtain fuzzy subtraction

$$\mu_{\tilde{A}-\tilde{B}}(x) = \begin{cases} 0, & x \leq -3, \\ x/12 + 3/12, & -3 \leq x \leq 9, \\ -x/7 + 16/7, & 9 \leq x \leq 16, \\ 0, & x \geq 16. \end{cases} \tag{6.36}$$

Example 7

For this example we shall again use triangular fuzzy numbers because they are easy to work with. Let us find the product of two fuzzy numbers $\tilde{A}(\cdot)\tilde{B}$ in Figure 6.11 if their membership functions are:

$$\mu_{\tilde{A}} = \begin{cases} 0, & x \leq 2, \\ x - 2, & 2 \leq x \leq 3, \\ -x/2 + 5/2, & 3 \leq x \leq 5, \\ 0, & x \geq 5. \end{cases} \quad \forall x \in R \tag{6.37}$$

$$\mu_{\tilde{B}} = \begin{cases} 0, & x \leq 3, \\ x/2 - 3/2, & 3 \leq x \leq 5, \\ -x + 6, & 5 \leq x \leq 6, \\ 0, & x \geq 6. \end{cases} \quad \forall x \in R \tag{6.38}$$

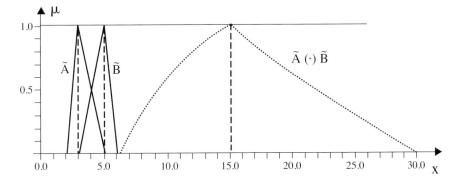

Figure 6.11 *Multiplication of two fuzzy numbers (Example 7)*

For the level α in the Figure 6.11 and using (6.37) we have

$$\alpha = a_1^\alpha - 2 \quad \text{and} \quad \alpha = -a_2^\alpha /2 + 5/2.$$

Hence,

$$A_\alpha = [a_1^\alpha, a_2^\alpha]$$
$$A_\alpha = [\alpha + 2, -2\alpha + 5]. \tag{6.39}$$

Using (6.38) we also have

$$\alpha = b_1^\alpha /2 - 3/2 \quad \text{and} \quad \alpha = -b_2^\alpha + 6.$$

Hence,

$$B_\alpha = [b_1^\alpha, b_2^\alpha]$$
$$B_\alpha = [2\alpha + 3, -\alpha + 6]. \tag{6.40}$$

Thus we obtain the multiplication

$$A_\alpha(\cdot)B_\alpha = [a_1^\alpha \cdot b_1^\alpha, a_2^\alpha \cdot b_2^\alpha]$$
$$A_\alpha(\cdot)B_\alpha = [2\alpha^2 - 7\alpha + 6, 2\alpha^2 - 17\alpha + 30]. \tag{6.41}$$

Now we have two equations to solve:

$$2\alpha^2 - 7\alpha + 6 - x = 0 \tag{6.42}$$

and

$$2\alpha^2 - 17\alpha + 30 - x = 0. \tag{6.43}$$

We shall retain only two roots in [0,1]. For (6.42)

$$\alpha = (-7 + \sqrt{1 + 8x}) / 4,$$

and for (6.41)

$$\alpha = (17 - \sqrt{49 + 8x}) / 4.$$

Finally,

$$\mu_{\tilde{A}(\cdot)\tilde{B}} = \begin{cases} 0, & x \le 6, \\ (-7 + \sqrt{1 + 8x})/4, & 6 \le x \le 15, \\ (17 - \sqrt{49 + 8x})/4, & 15 \le x \le 30, \\ 0, & x \ge 30. \end{cases} \quad \forall x \in R^+ \quad (6.44)$$

The resulting multiplication curve is shown in Figure 6.11. Note that $\tilde{A}(.)\tilde{B}$ does not result in a triangular shape.

Example 8

Let us find the division of two triangular fuzzy numbers shown in Figure 6.12 and given by:

$$\mu_{\tilde{A}} = \begin{cases} 0, & x \le 18, \\ x/4 - 18/4, & 18 \le x \le 22, \\ -x/11 + 3, & 22 \le x \le 33, \\ 0, & x \ge 33. \end{cases} \quad \forall x \in R^+ \quad (6.45)$$

and

$$\mu_{\tilde{B}} = \begin{cases} 0, & x \le 5, \\ x - 5, & 5 \le x \le 6, \\ -x/2 + 4, & 6 \le x \le 8, \\ 0, & x \ge 8. \end{cases} \quad \forall x \in R^+ \quad (6.46)$$

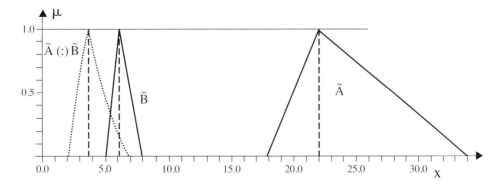

Figure 6.12 *Division of two fuzzy numbers (Example 8)*

In equation (6.45), let

$$\alpha = a_1^\alpha / 4 - 18/4 \quad \text{and} \quad \alpha = -a_2^\alpha / 11 + 3$$

from which

$$A_\alpha = [a_1^\alpha, a_2^\alpha] \tag{6.47}$$
$$A_\alpha = [4\alpha + 18, -1\alpha + 33].$$

In equation (6.46), let

$$\alpha = b_1^\alpha - 5 \quad \text{and} \quad \alpha = -b_2^\alpha / 2 + 4$$

from which

$$B_\alpha = [b_1^\alpha, b_2^\alpha] \tag{6.48}$$
$$B_\alpha = [\alpha + 5, -2\alpha + 8].$$

Thus,

$$A_\alpha(/)B_\alpha = [a_1^\alpha / b_2^\alpha, a_2^\alpha / b_1^\alpha] \tag{6.49}$$
$$A_\alpha(/)B_\alpha = [4\alpha - 18, -11\alpha + 33](/)[\alpha + 5, -2\alpha + 8]$$
$$A_\alpha(/)B_\alpha = \left(\frac{4\alpha + 18}{-2\alpha + 8}, \frac{-11\alpha + 33}{\alpha + 5} \right)$$

Finally we find

$$\mu_{\tilde{A}(/)\tilde{B}} = \begin{cases} 0, & x \leq 9/4, \\ \dfrac{8x - 18}{2x + 4} & 9/4 \leq x \leq 11/3, \\ \dfrac{-5x + 33}{x + 11} & 11/3 \leq x \leq 33/5, \\ 0, & x \geq 33/5. \end{cases} \quad \forall x \in R^+$$

6.5.4 Comparison operations on fuzzy sets

An important application of fuzzy set theory to water resources systems management involves the ranking of fuzzy sets. The relative performance of different water management alternatives may be visually intuitive when looking at its fuzzy representation. However, in cases where many alternatives display similar characteristics, it may be impractical or even undesirable to make a visual selection. A method for ranking alternatives can automate many of the visual interpretations, and create reproducible results. A ranking measure may also be useful in supplying additional insight into decision-maker preferences, such as distinguishing relative risk tolerance levels.

The selection of a ranking method is subjective, and specific to the form of problem and the fuzzy set characteristics that are desirable. There exists an assortment of methods, ranging from horizontal and vertical evaluation of fuzzy sets to comparative methods. Some of these methods may independently evaluate fuzzy sets, while others use competition to choose among a selection list. Horizontal methods are related to the practice of defuzzifying a fuzzy set by testing for a range of validity at a threshold membership value, and are not dealt with in this text. Vertical methods tend to use the area under a membership function as the basis for evaluation, such as the centre of gravity. The comparative methods introduce other artificial criteria for judging the performance of a fuzzy set, such as the compatibility of fuzzy sets.

There follows a discussion of the properties of some selected methods, which are vertical and comparative.

Weighted centre of gravity (WCoG) measure

Given the desirable properties of a ranking method for many fuzzy applications to water resources systems management, one technique that may qualify as a candidate is the centroid method, as discussed by Yager (1981) in terms of its ability to rank fuzzy sets in the range [0,1]. The centroid method appears to be consistent in its ability to distinguish between most fuzzy sets. One weakness, however, is that the centroid method is unable to distinguish between fuzzy sets that may have the same centroid, but differ greatly in their degree of fuzziness. This weakness can be somewhat alleviated by the use of weighting. If high membership values are weighted higher than low membership values, there is some indication of degree of fuzziness when comparing rankings from different weighting schemes. However, in the case of symmetrical fuzzy sets, weighting schemes will not distinguish relative fuzziness.

A weighted centroid ranking measure can be defined as follows:

$$WCoG = \int g(x)\mu(x)^q \, dx \tag{6.50}$$

where:
$g(x)$ = the horizontal component of the area under scrutiny and
$\mu(x)$ = the membership function values.

In practice, WCoG can be calculated in discrete intervals across the valid domain. It allows parametric control in the form of the exponent q. This control mechanism allows ranking for cases ranging from the modal value ($q = \infty$) – which is analogous to an expected case or most likely scenario – to the centre of gravity ($q = 1$) – which signifies some concern over extreme cases. In this way, there exists a family of valid ranking values (which may or may not change too significantly). The final selection of appropriate rankings is dependent on the level of risk tolerance from the decision-maker.

The ranking of fuzzy sets with WCoG is by ordering from the smallest to the largest value. The smaller the WCoG measure, the closer the centre of gravity of the fuzzy set to the origin. As a vertical method of ranking, WCoG values act on the set of positive real numbers.

Fuzzy acceptability measure

Another ranking method which shows promise is a fuzzy acceptability measure, *Acc*, based on Kim and Park (1990). Kim and Park derive a comparative ranking measure, which builds on possibility to signify an optimistic perspective, and supplements it with a pessimistic view similar to necessity. Therefore their measure relies on the concept of *compatibility*.

The compliance of two fuzzy membership functions can be quantified using the fuzzy compatibility measure. The basic concepts of *possibility* and *necessity* lead to the quantification of the compatibility of two fuzzy sets. The possibility measure quantifies the overlap between two fuzzy sets, while the necessity measure describes the degree of inclusion of one fuzzy set in another fuzzy set.

The *possibility* is then defined as:

$$Poss(\tilde{A},\tilde{B}) = \sup[\min\{\mu_{\tilde{A}}(x), \mu_{\tilde{B}}(x)\}], \quad x \in X \tag{6.51}$$

where:
$Poss(\tilde{A},\tilde{B})$ = the possibility measure of fuzzy numbers and
$\sup[\ ]$ = the least upper bound value, i.e. *supremum*
$\mu_{\tilde{A}}(x), \mu_{\tilde{B}}(x)$ = the membership functions of the fuzzy numbers $\tilde{A}$ and $\tilde{B}$ respectively.

The possibility measure is a symmetrical measure, that is:

$$Poss(\tilde{A},\tilde{B}) = Poss(\tilde{B},\tilde{A}) \tag{6.52}$$

The *necessity* measure is defined as:

$$Nec(\tilde{A},\tilde{B}) = \inf[\max\{\mu_{\tilde{A}}(x), \mu_{\tilde{B}}(x)\}], \quad x \in X \tag{6.53}$$

where:
$Nec(\tilde{A},\tilde{B})$ = the necessity measure of fuzzy numbers and
$\inf[\ ]$ = the greatest lower bound value, i.e. *infimum*
$\mu_{\tilde{A}}(x), \mu_{\tilde{B}}(x)$ = the membership functions of the fuzzy numbers $\tilde{A}$ and $\tilde{B}$ respectively.

The necessity measure is an asymmetrical measure, that is:

$$Nec(\tilde{A},\tilde{B}) \neq Nec(\tilde{B},\tilde{A}) \tag{6.54}$$

Both measures hold the following relation:

$$Nec(\tilde{A},\tilde{B}) + Poss(\overline{\tilde{A}},\tilde{B}) = 1 \tag{6.55}$$

where:
$Nec(\tilde{A},\tilde{B})$ = the necessity measure of fuzzy numbers $\tilde{A}$ and $\tilde{B}$
$Poss(\overline{\tilde{A}},\tilde{B})$ = the possibility measure of fuzzy numbers $\overline{\tilde{A}}$ and $\tilde{B}$
$\overline{\tilde{A}}$ = the fuzzy complement of fuzzy number $\tilde{A}$.

These two measures, *Poss* and *Nec*, can be combined to form an acceptability measure *(Acc)* as follows:

$$Acc = \omega Poss(G,L) + (1-\omega) Nec(G,L) \tag{6.56}$$

Parametric control with the acceptability measure *(Acc)* is accomplished with the ω weight and the choice of a fuzzy desirable state such as a goal, G. The ω weight controls the degree of optimism and degree of pessimism, and indicates (an overall) level of risk tolerance. The choice of a fuzzy goal is not so intuitive. It should normally include the entire range of alternative options L, but it can be adjusted to a smaller range either for the purpose of exploring the shape characteristics of L, or to provide an indication of necessary stringency. By decreasing the range of G, the decision-maker becomes more stringent in that the method rewards higher membership values closer to the desired value. At the extreme degree of stringency, G becomes a non-fuzzy number that demands the alternatives be ideal. As a function, G may be linear, but can also be adapted to place more emphasis or less emphasis near the best value.

The ranking of fuzzy sets using *Acc* is accomplished by ordering values from the largest to the smallest. That is, the fuzzy set with the greatest *Acc* is most acceptable. *Acc* values are restricted in the range [0,1] since both the *Poss* and *Nec* measures act on [0,1].

The method of Chang and Lee

Chang and Lee (1994) have simplified their Overall Existence Ranking Index (OERI) for use with convex fuzzy sets. Equation (6.57) corresponds to their ranking index.

$$OERI(j) = \int_0^1 \omega(\alpha)[\chi_1(\alpha)\mu_{jL}^{-1}(\alpha) + \chi_2(\alpha)\mu_{jR}^{-1}(\alpha)]d\alpha \qquad (6.57)$$

where:
- j = alternative set j
- α = the degree of membership
- $\chi_1(\alpha)$ and $\chi_2(\alpha)$ = the subjective weightings indicating neutral, optimistic and pessimistic preferences of the decision-maker, with the restriction that $\chi_1(\alpha) + \chi_2(\alpha) = 1$
- $\omega(\alpha)$ = weighting parameter for a particular level of membership
- $\mu_{jL}^{-1}(\alpha)$ = the inverse of the left part, and
- $\mu_{jR}^{-1}(\alpha)$ = the inverse of the right part of the membership function.

It should be noted that both linear and nonlinear functions for the subjective type weighting are possible, thus giving the user more control in the ranking. For χ_1 values greater than 0.5, the left side of the membership function is weighted more than the right side, which in turn makes the decision-maker more optimistic. Of course, if the right side is weighted more, the decision-maker is considered more of a pessimist. In summary, the risk preferences are: if χ_1 <0.5, the user is a pessimist (risk averse); if χ_1 = 0.5, the user is neutral; and if χ_1 >0.5, the user is an optimist (risk taker). Simply stated, Chang and Lee's (1994) OERI is a sum of the weighted areas between the membership axis and the left and right inverses of a fuzzy set.

The method of Chen

Chen (1985) based his method on maximizing and minimizing fuzzy sets according to the following equations:

$$\mu_M(x) = \begin{cases} [(x - x_{min})/(x_{max} - x_{min})]^r w, & x_{min} \le x \le x_{max} \\ 0, & \text{otherwise} \end{cases} \qquad (6.58)$$

$$\mu_m(x) = \begin{cases} [(x - x_{max}) / (x_{min} - x_{max})]^r w, & x_{min} \leq x \leq x_{max} \\ 0, & otherwise \end{cases} \quad (6.59)$$

where:
j = alternative set j
$w_j = sup(\mu_j(x))$
$w = inf(w_j)$
$x_{min} = inf(x)$
$x_{max} = sup(x)$

The operator *sup* represents the supremum or global maximum, and the operator *inf* stands for the infimum or global minimum. The participation of the decision-maker is controlled by the constant r. If $r = 1$ the decision-maker is conservative or neutral; if $r = 0.5$ the decision-maker is a risk taker or an optimist; and if $r = 2$ the decision-maker is risk averse or a pessimist. Of course, values of r below 0.5 represent extreme optimism, while r values greater than 2 represent extreme pessimism. The maximization and minimization sets are presented in Figure 6.13 for a neutral ($r = 1$) decision-maker.

To rank the alternatives, right ($U_M(j)$) and left ($U_m(j)$) utility values are calculated as follows:

$$U_M(j) = sup(\mu_j(x) \cap \mu_M(x)) \quad (6.60)$$

$$U_m(j) = sup(\mu_j(x) \cap \mu_m(x)) \quad (6.61)$$

In plain language, $U_M(j)$ represents the larger intersection of the maximizing set with the right portion of the set being ranked (the alternative under consideration), and $U_m(j)$ is the larger intersection of the minimizing set with the left part of the membership function of the fuzzy set of alternative j.

At the end, the total utility values, $U_t(j)$, are computed for every alternative fuzzy set of the problem as follows:

$$U_T(j) = (U_M(j) + w - U_m(j))/2 \quad (6.62)$$

The total utility values are then ordered from the smallest to the largest, the smallest being the highest-ranked alternative.

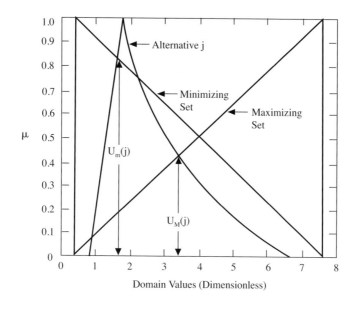

Figure 6.13 *Graphical illustration of Chen's method for r = 1*

Two concerns about this method must be noted. First, the presence of an alternative with a membership function that is far to the left (or far to the right) of other alternatives influences the way maximizing and minimizing sets are obtained. If just one alternative is far away from the rest, it increases (or decreases) the value of the parameter x_{max} (or x_{min}), which in turns shapes the maximizing and the minimizing sets. If all alternatives are relatively close together, Chen's method gives reasonable results. Second, since this method uses only two degrees of membership (the degrees of membership associated with the left and the right utility values), an objection can be raised that not enough fuzzy information is used in the ranking.

6.6 DERIVATION OF A FUZZY MEMBERSHIP FUNCTION FOR WATER RESOURCES SYSTEMS MANAGEMENT UNDER UNCERTAINTY

Two basic forms of uncertainties detailed in Section 6.1 are uncertainty caused by inherent stochastic variability and uncertainty because of a fundamental lack of knowledge. Intuitively, the second form appears to be readily modelled by fuzzy sets. However, it is not only the type of uncertainty that determines the appropriate way of modelling, but also data sufficiency and availability. If sufficient data are available to fit a probability density distribution, the probabilistic approach will be the best way to quantify the uncertain values. On the other hand, if the requirements of water resources systems sustainability are to be addressed, such as the needs of future generations, expanded spatial and temporal scales and long-term consequences, then the information available is scarce. In this case the fuzzy set approach can successfully utilize the information that is available.

The quantification of complex qualitative criteria, a process often encountered in water resources management, is a typical example where fuzzy systems modelling is a good technique. Water quality, flood control, recreation and many other qualitative criteria are still far from precise analytical descriptions. Intuitive linguistic formulations are worth considering since fuzzy set theory provides a successful way to operate them (using *linguistic calculus*). Linguistic formulations can be represented using fuzzy membership functions. The main problem here is with the word 'intuitive'. Many qualitative criteria in water resources systems management are far too complex to be intuitively understood. This complexity is a natural characteristic of the system, but in addition it is a result of the technical, environmental, societal, institutional, political and economic aspects of water resources management. One way to deal with this complexity is to carefully derive the decomposition model for each complex criterion. The model would then provide an easy evaluation of criteria through the evaluation of their components, which are generally less complex. Furthermore, decomposition models can be used repeatedly in various situations, once they have been developed.

From the definition of a membership function given in Section 6.5.1, we can see that constructing a fuzzy set $\tilde{A}$ is equivalent to the construction of its membership function $\mu_{\tilde{A}}$. When only the universal set X and property x_1 are known, the actual construction process usually involves subjective assessment. In most cases, membership functions are designed by experts whose knowledge of the system under consideration is crucial. Depending on what kind of initial assessments are made, the construction methods used when dealing with intrinsic fuzziness can be classified into *direct* and *indirect* methods for membership function estimation. Direct methods refer to those methods where the assessments are made directly on the relations between the elements of X and the values from the interval $[0,1]$. Indirect methods, on the other hand, make initial assessments of certain relations among the elements within the universal set X. These results are then used for the construction of the membership function $\mu_{\tilde{A}}$.

In the course of membership function evaluation, it is quite natural to pay attention to the comparison process among the elements within the universal set X. That is, we can first perform the comparison evaluation among the elements in X and then mathematically transform the obtained data into function $\mu_{\tilde{A}}$ from X to $[0,1]$. Our everyday life is full of examples when we make some kind of comparison, either consciously or unconsciously. However, it is always done in one of two contextually different ways. Things are either compared with one or more reference points (called *prototypes*) or they are compared pairwise without any reference point. Both ways have advantages and disadvantages when applied to membership evaluation.

A brief overview of all the methods (direct and indirect) for membership construction can be found in Pedrycz (1995), while an extensive survey of methods for membership evaluation with a comparative analysis is given in Blishun (1989).

When fuzzy sets are employed to model less complex parameters, some of the direct and indirect methods for membership construction can be applied directly. Table 6.3 lists some sources of uncertainty in water resources management, along with the corresponding methods for membership evaluation whenever a fuzzy set approach appears to be an appropriate way of modelling.

The classification and the choice of the method for membership evaluation for different types of uncertainties are important considerations before applying any kind of operations based on fuzzy set theory. Still, regardless of what method is used, it is even more important to decide how to ask the questions, what questions to ask, and who is to be asked in order to build the appropriate and reliable membership functions.

This section of the book presents an approach for membership construction based on reducing the complexity of a system by implementing a hierarchical analysis of its less complex constituents. The whole process is carried in three steps:

1. Decomposition of a system.
2. Evaluation of the components found at the lowest hierarchical level (the least complex components).
3. Aggregation of the obtained evaluations from step 2 backwards into the unique membership value.

This way of dealing with informational fuzziness is not new in the literature. A similar approach was suggested by Zimmermann (1987) for the evaluation of creditworthiness. The three-step process introduced above can be used to construct an automated evaluation tool. The automation process requires simulation of the reasoning used in step 3 by a suitable aggregation operator, and construction of fuzzy sets that would truly represent the components at the lowest hierarchical level. Although both of these activities might be context-dependent, this is far more likely to be the case with fuzzy set construction than with the reasoning used in an aggregation process. This presentation will be illustrated with the construction of fuzzy sets for flood management. However, the focus of the discussion is on finding an appropriate model for the aggregation process. The recommended methodology has a much higher generic value in water resources systems management when informational fuzziness is present.

Table 6.3 *Interpretation of fuzziness for various problems in water resources*

Areas in water resources	Interpretation of uncertainty	Membership evaluation method
Degree to which model truly represents the actual system Human subjectivity in system analysis	Likelihood view	Neural-fuzzy techniques Horizontal methods
Errors in parameter estimation Qualitative criteria in decision-making	Random set view	Interval estimation Vertical methods
Weight estimation for criteria of unequal importance	Utility view	Pairwise comparison
Reliability and sensitivity levels of different operations	Measurement view	Clustering methods Pairwise comparison Direct rating method

6.6.1 Theoretical discussion of system decomposition techniques and aggregation operators

We have defined a set S as the set of all elements in X having the property p. If the property p is such that it clearly separates the elements of X into two classes (those that have p and those that do not have p) we say that p defines a *crisp* subset S of X. When there is no such clear separation, we say that the property p defines an *ill-defined* subset S of X. A fuzzy membership function, discussed in Section 6.5.1, is a proper mathematical definition of an ill-defined set.

Let us suppose that we have an ill-defined set A that is too complex for the straightforward construction of its corresponding fuzzy set $\tilde{A}$. If we want to design an automated tool for the construction of such a fuzzy set using hierarchical analysis, the procedure in Table 6.4 may be applied.

Each of the three steps in Table 6.4 is important and must be worked out carefully. For Step 1, some input from experts in the subject area of the ill-defined set A is usually necessary. In Step 2, the choice of a suitable method for membership evaluation has to be made.

Table 6.4 *Hierarchical approach for construction of a fuzzy set*

Step 1	Decomposition of the complex fuzzy information (an ill-defined set A) into less complex components (ill-defined sets B_i). If any of the ill-defined sets B_i is still complex, continue with decomposition until the ill-defined sets obtained at the lowest hierarchical level are such that either construction of the corresponding fuzzy sets is not complex any more, or they cannot be decomposed any further
Step 2	Construction of the fuzzy sets corresponding to the ill-defined sets at the bottom of concept hierarchy using the existing methods for membership evaluation
Step 3	Development of an aggregation model that will be able to perform reliable aggregation of the fuzzy sets constructed in Step 2 into a fuzzy set $\tilde{A}$ corresponding to the initial ill-defined set A

Let's concentrate now on some theoretical considerations of the decomposition process (Step 1), while the practical considerations of this process are left as a part of the case study presented in the next section.

The main condition that a proper decomposition procedure should satisfy is that the components B_i must provide complete information (or at least substantially complete information) about set A. In order to test the completeness of the decomposition of A into B_i, we should ask the question, can the membership value $\mu_{\tilde{A}}(x)$ vary (significantly) if the values $\mu_{\tilde{B}}(x)$ are all fixed? If the answer is yes, then the decomposition is not complete, and we must find an additional component that causes the membership value $\mu_{\tilde{A}}(x)$ to change under such circumstances. It may not be always possible (or practical) to form a perfectly complete decomposition. The *degree of incompleteness* can be measured as the maximum change $\Delta\mu_{\tilde{A}}(x)$ of the membership value $\mu_{\tilde{A}}(x)$ with all of the values $\mu_{\tilde{B}}(x)$ fixed. Analytically,

$$\text{degree of incompleteness} = \max \{\Delta\mu_{\tilde{A}}(x) : \Delta\mu_{\tilde{B}}(x) = 0, i = 1, 2, ...\} \quad (6.63)$$

Thus Step 1 of the decomposition/aggregation process will be completed successfully only if the degree of incompleteness is sufficiently small. On the other hand, if we are satisfied with the degree of incompleteness, it is still possible for some of the components B_i to be redundant. In other words, it is possible that if we removed one or more components B_i from the decomposition, the degree of incompleteness would remain unchanged. Any such redundant components that are detected should be removed. This may be particularly important for the aggregation process (Step 3) since it becomes simpler and more accurate when the number of components B_i is smaller.

When dealing with complex environments, such as water resources systems, it is good practice to decompose the ill-defined set A into more levels with fewer components as opposed to fewer levels with many components. This will simplify the aggregation process in Step 3. Of course, the overall decomposition should be intuitively clear in order to provide an effective evaluation of membership values in Step 2. In summary, if enough attention is given to the decomposition process in Step 1, than the complexity of Step 2 and Step 3 can be reduced significantly.

The rules by which membership functions are aggregated are discussed next. I shall concentrate only on a single decomposition of a set A into n components B_i ($i = 1, 2, ... , n$). In the case of more levels the same process can be replicated at each level.

Zimmermann's γ-family of operators

The model for membership aggregation in Zimmermann (1987) is developed in the first place to provide a good empirical fit, while at the same time it satisfies the basic mathematical requirements for easy computation. This family subscribes to the class of commutative operators. It emphasizes the importance of selecting an appropriate operator by which two or more membership functions are to be aggregated. In this respect, Zimmermann also gives a list of rules for selecting the 'best' operator. From experimental observations, it has been concluded that for different phenomena we need different operators in order to build an adequate model.

There is an infinite number of operators that can be used to describe any particular model. However, even within one decomposition, depending on particular context, we might have to change operator again and again. Instead of formulating a new operator every time, Zimmermann suggests using the family of operators determined by a single parameter γ. When applied to membership aggregation, this family of operators is defined in the following way:

$$\mu_{\tilde{A}}(x) = \left(\prod_{i=1}^{n} \mu_{\tilde{B}}(x)\right)^{1-\gamma} \cdot \left(1 - \prod_{i=1}^{n}(1 - \mu_{\tilde{B}}(x))\right)^{\gamma}, \; 0 \leq \gamma \leq 1 \qquad (6.64)$$

Changing the value of γ from 0 to 1, the membership value ranges from the value that would be obtained by the product operator (γ = 0), and the value that would be obtained by algebraic sum (γ = 1). If the product operator and algebraic sum are considered as two extreme operators representing non-compensatory 'and' and full compensatory 'or' respectively, then parameter γ can be interpreted as a *grade of compensation*. Each value of γ taken from the interval [0,1] represents a different operator, and according to Zimmermann can successfully model the empirical results of membership aggregation.

Further modification of equation (6.64) is necessary if the components B_i are not of equal importance with respect to the decomposed set A. This is achieved by assigning weights for each set B_i. Then (6.64) becomes

$$\mu_{\tilde{A}}(x) = \left(\prod_{i=1}^{n}(\mu_{\tilde{B}_i}(x))^{w_i}\right)^{1-\gamma} \cdot \left(1-\prod_{i=1}^{n}(1-\mu_{\tilde{B}_i}(x))^{w_i}\right)^{\gamma}, \quad 0 \leq \gamma \leq 1, \sum w_i = n \quad (6.65)$$

To preserve the structure of the model, the weights w_i should add up to the total number of components n in a single decomposition.

Unlike the weights w_i, the parameter γ is much less intuitively clear for direct estimation unless the decomposition of A contains only two sets, B_1 and B_2. Even in this case, the best way to obtain γ is through experimentally obtained membership values for some selected elements x. For each selected object x_j, we can find γ_j by solving equation (6.64) and then using the average value over all js.

One of the limitations for using this model is that the components B_i in a single decomposition B should have approximately equal 'compensation strength' compared with the other components from B. Only in this case can we have satisfying results for any combination of membership values $\mu_{\tilde{B}_i}(x)$. Another important limitation comes from the fact that if some x from B_i happen to have the membership value $\mu_{\tilde{B}_i}(x) = 0$, then for any grade of compensation $\gamma \neq 1$, the resulting membership value is always $\mu_{\tilde{A}_i}(x) = 0$. In other words, the γ-operator does not give any compensation if any of the components in the decomposition is estimated to be zero. Thus, before we decide to use this method for aggregation, we must make sure that the zero membership value of any element B_i logically leads to a zero membership value of the set A, regardless of the membership values of the other elements. Only after this is approved can we start assigning relative weights and calculating the parameter γ from the empirical results.

Ordered weighted averaging (OWA) aggregation operators

Another family of commutative operators for membership aggregation is the family of OWA operators introduced by Yager (1988). An OWA operator combines membership values $\mu_{\tilde{B}_i}(x)$ into the resulting membership value $\mu_{\tilde{A}}(x)$ by the simple formula:

$$\mu_{\tilde{A}}(x) = w_1 a_1 + w_2 a_2 + \ldots + w_n a_n = \sum w_i a_i \quad (6.66)$$

where:
a_i = the i-th largest element in the collection $\mu_{\tilde{B}_i}(x)$ for $i = 1, 2, \ldots, n$, and
w_i = the parameters that have to satisfy the conditions $w_i \in [0,1]$ and $\sum w_i = 1$.

This is an elegant mathematical description of the intuitive process often used in various judgements. Equation (6.66) can be also written in the following form:

$$\mu_{\tilde{A}}(x) = W^T A \tag{6.67}$$

where:
$W^T = [w_1 \ w_2 \ ... \ w_n]$ and
$A = [a_1 \ a_2 \ ... \ a_n]^T$.

Vector W, also called the *weighting vector*, is the one we want to determine in order to perform aggregation of the membership values $\mu_{\tilde{B}}(x)$. When the ill-defined sets B_i are of equal importance relative to the ill-defined set A, then the weights w_i can be directly estimated by the decision-maker.

It is easy to show that *min* and *max* operators are the lower and the upper limit for any OWA operator. Like the product operator and algebraic sum in the family of γ-operators, the *min* and *max* in the family of OWA operators may represent the logical connectives *and* and *or*. Therefore, Yager (1988) introduces the *degree of orness* associated with each weighted vector W as:

$$orness(W) = \frac{1}{n-1} \sum_{i=1}^{n}(n-i)w_i \tag{6.68}$$

The degree of orness has a similar meaning as the degree of compensation γ discussed earlier. However, the important distinction is that the same degree of orness may correspond to many different OWA operators (i.e. different weighting vectors W), while each degree of compensation γ corresponds to strictly one γ-operator. This comes from the fact that only one parameter is included in the formulation of the γ-operator regardless of the number of elements to be aggregated, while the OWA operator is defined by $n - 1$ independent parameters, with n being the number of set elements.

The degree of orness is an important parameter when aggregating elements of unequal importance. Yager (1988) employs fuzzy systems modelling to find the *importance transform function* $G(u_{B_i}, \mu_{\tilde{B}_i}(x))$, where $u_{B_i} \in [0, 1]$ is the importance of the element B_i. For a simple fuzzy system model, formulated on two possible states of the degree of orness (high and low), this function is:

$$G(u_{B_i}, \mu_{\tilde{B}_i}(x)) = (1 - u_{B_i})(1-\theta) + u_{B_i} \cdot \mu_{\tilde{B}_i}(x) \tag{6.69}$$

where θ represents the degree of orness.

The function G is applied to the membership values first, and then the standard procedure for OWA aggregation is used for such modified membership values. Aggregation performed in this way will successfully achieve the appropriate ordering of different combinations of membership values $\mu_{\tilde{B}}(x)$ using the corresponding aggregated values $\mu_{\tilde{A}}(x)$. However, if our goal is also to obtain the right aggregated values, then for the elements of unequal importance, the application of transform function G is much more limited.

Direct estimation of the weights w_i does not appear to be very attractive when the elements B_i are of unequal importance. When direct estimation of the weights is not a viable option, we can use a learning mechanism to find the weights from the set of experimental data. An even more acute problem comes to mind here, namely the problem of the uncertainty within the estimated values $\mu_{\tilde{A}}(x)$. This problem is particularly important when the values for $\mu_{\tilde{A}}(x)$ are obtained directly from experts.

Composition under pseudomeasures (CUP)

Orlovski (1994) considers a set of properties $B=\{B_1, B_2, B_3,...\}$ representing one decomposition of $\tilde{A}$, and introduces the following definitions.

Definition 1
The set U is called a *complete class of subsets of B* if:

(a) $B \in U \Rightarrow B_i \backslash B \in U$, and
(b) $\varnothing \in U$.

Definition 2
A function $\eta : U \to [0,1]$ is called a *pseudomeasure* (or *fuzzy measure*) on a space (B,U) if:

(a) $\eta(\varnothing) = 0$;
(b) $\forall B_1, B_2 \in U,\ B_1 \subseteq B_2 \Rightarrow \eta(B_1) \le \eta(B_2)$, and
(c) $\eta(B) = 1$.

For any set of properties $B \in U$, the value $\eta(B)$ is understood as the *degree to which an object shows the property $\tilde{A}$ under the condition that it has all the properties from the set B*. Using the above definitions, and taking care that the class U should be sufficiently rich to allow for the evaluation of all meaningful collections of properties possessed by objects from the set X, the membership function $\mu_{\tilde{A}}(x)$ is aggregated in the following manner:

$$\mu_{\tilde{A}}(x) = \sup_{B \in U} \min \{\mu(B), \min_{B_i \in B} \mu_{\tilde{B}_i}(x)\} \tag{6.70}$$

To obtain the membership function $\mu_{\tilde{A}}(x)$ using equation (6.70), we first have to evaluate a pseudomeasure η on U. This can be done through empirical evaluation using the intuitive definition of a pseudomeasure η given above.

When this method is compared with the aggregation methods that use commutative operators, like the families of γ and OWA operators, we can observe two important distinctions. First, the evaluation of a pseudomeasure η is rather cumbersome if the number of components $B_i \in B$ is large. If the complete class U is the set of all subsets of $B = \{B_1, B_2, ..., B_n\}$ then we might have to perform evaluation for as much as $2^n - 2$ different subsets $B \in U$ (unless we use a special class of pseudomeasures, e.g. the class of additive pseudomeasures). On the other hand, the number of parameters to be determined within the class of commutative operators is significantly smaller. Considering the case of aggregation of n components of unequal importance, we must determine $n+1$ parameters for the γ-operator, and $2n - 1$ parameters for the OWA operator. Therefore, the composition under pseudomeasure is much more limited by the number of components within a decomposition than the γ- and OWA operators are. However, with an increasing number of components (especially if the components are of unequal importance), the difficulty and uncertainty of estimating the appropriate parameters for γ or OWA aggregation also increases.

Second, the way the importance of each component is included in the class of commutative operators is rather rigid, and does not allow for a very flexible decomposition structure. In other words, if we want to use the commutative class of operators for aggregation, we must provide that the components in a decomposition are not strongly dependent on each other. Conversely, CUP is much more flexible in this respect, and can also handle local variation of degrees of importance, which makes it possible to successfully aggregate even very interrelated components within a decomposition. This quality should not be confused with compensation strength, which is a considerably weaker point of this composition and, together with the limited number of components, represents the main constraint on the use of this method.

Polynomial composition under pseudomeasure (P-CUP)

In the previous section, we saw that CUP allows a rather flexible decomposition structure. This is an important feature for dealing with a number of complex systems found in the real-world environment that are characterized by the variety of relationships among the constituents of a system. CUP is based on the *min* operator, and therefore contains many intervals where the increase in membership value of one constituent within a system does not affect the resulting membership value of the whole system. In order to provide more smooth composition, while at the same time preserving the flexibility of CUP, an original method has been developed for application in water resources systems management (Despic and Simonovic, 2000), using the polynomial function which satisfies both qualities. Polynomial composition under pseudomeasure (P-CUP) has the following form:

$$\mu_{\tilde{A}}(x) = a_1\mu_{\tilde{B}_1}(x) + a_2\mu_{\tilde{B}_2}(x) + \ldots + a_n\mu_{\tilde{B}_n}(x) +$$

$$+ a_{12}\mu_{\tilde{B}_1}(x)\mu_{\tilde{B}_2}(x) + a_{13}\mu_{\tilde{B}_1}(x)\mu_{\tilde{B}_3}(x) + \ldots + a_{n-1,n}\mu_{\tilde{B}_{n-1}}(x)\mu_{\tilde{B}_n}(x) + \quad (6.71)$$

$$\ldots + a_{12\ldots n}\mu_{\tilde{B}_1}(x)\mu_{\tilde{B}_2}(x)\ldots\mu_{\tilde{B}_n}(x)$$

Since the practical applicability of this function is limited by the number of components B_i in the same way as CUP is, it will be sufficient to explain the meaning of the coefficients $a_{i\ldots j}$ by using a decomposition B which contains only three elements, $B = \{B_1, B_2, B_3\}$. Thus, for the function:

$$\mu_{\tilde{A}}(x) = a_1\mu_{\tilde{B}_1}(x) + a_2\mu_{\tilde{B}_2}(x) + a_3\mu_{\tilde{B}_3}(x) +$$

$$+ a_{12}\mu_{\tilde{B}_1}(x)\mu_{\tilde{B}_2}(x) + a_{13}\mu_{\tilde{B}_1}(x)\mu_{\tilde{B}_3}(x) + a_{23}\mu_{\tilde{B}_2}(x)\mu_{\tilde{B}_3}(x) + \quad (6.72)$$

$$+ a_{123}\mu_{\tilde{B}_1}(x)\mu_{\tilde{B}_2}(x)\mu_{\tilde{B}_3}(x).$$

the parameters $a_{i\ldots j}$ are obtained in the following way:

$$a_1 = \eta\{B_1\}$$
$$a_2 = \eta\{B_2\}$$
$$a_3 = \eta\{B_3\}$$
$$a_{12} = \eta\{B_1, B_2\} - \eta\{B_1\} - \eta\{B_2\}$$
$$a_{13} = \eta\{B_1, B_3\} - \eta\{B_1\} - \eta\{B_3\}$$
$$a_{23} = \eta\{B_2, B_3\} - \eta\{B_2\} - \eta\{B_3\}$$
$$a_{123} = \eta\{B_1, B_2, B_3\} + \eta\{B_1\} + \eta\{B_2\} + \eta\{B_3\} - \eta\{B_1, B_2\} - \eta\{B_1, B_3\} - \eta\{B_2, B_3\}$$

where η is a pseudomeasure as defined earlier in this section.

It is easy to show that for any combination of membership values $\mu_{\tilde{B}_1}(x)$, $\mu_{\tilde{B}_2}(x)$ and $\mu_{\tilde{B}_3}(x)$ taken from the set $[0,1]$, the resulting membership value $\mu_{\tilde{A}}(x)$ using equation (6.72) is the same as $\mu_{\tilde{A}}(x)$ in (6.71). Thus, in three-dimensional reasoning, we can notice that the pillars at the corners of the domain (points $(0, 0)$, $(0,1)$, $(1, 0)$, $(1, 1)$) stay at the same height as defined by a pseudomeasure η, while the surface carried by these pillars has changed from a rather stiff shape into a relaxed tent spanned by these four pillars (see Figures 6.14 and 6.15, Plates 7 and 8).

P-CUP has all the advantages and disadvantages of CUP discussed earlier. However, the polynomial composition has a greater flexibility in its structure since it is not limited by the *min* operator, and can reflect various natures of empirically obtained data. It is self-explanatory that in situations where aggregation based on the *min* operator is expected, CUP will be the preferred choice.

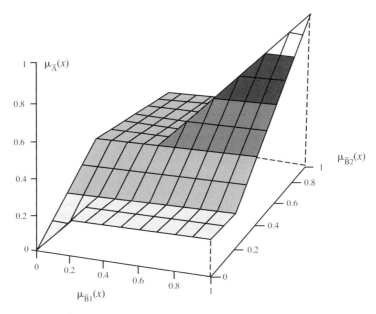

Note: See Plate 7 for a colour version.

Figure 6.14 *Composition under pseudomeasure for $\eta(B_1) = 0.2$ and $\eta(B_2) = 0.4$*

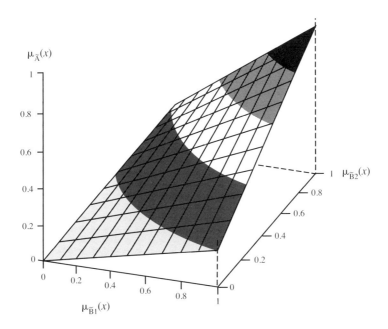

Note: See Plate 8 for a colour version.

Figure 6.15 *Polynomial composition under pseudomeasure for $\eta(B_1) = 0.2$ and $\eta(B_2) = 0.4$*

The method described in this section is the result of an attempt to find an appropriate aggregation method for the water resources management domain. A case study – qualitative evaluation of flood control, with all the empirical data obtained from three flood control experts – is presented in the next section to illustrate the application of the method.

6.7 DERIVATION OF A FUZZY MEMBERSHIP FUNCTION FOR FLOOD CONTROL

The main purpose of this example is to illustrate the application of the four aggregation methods presented in Section 6.6 and to suggest the one that is most appropriate for water resources systems management under uncertainty. Three experts for flood control participated in the evaluation process, but only one of them (Expert 1) participated throughout the whole experiment. The membership functions for the sets from the lowest hierarchical level in the flood control concept hierarchy are not evaluated because the primary focus is on the aggregation reasoning, and evaluation of these membership functions should be done within a specific setting. Nevertheless, the shape of these functions may be a subject for a priori evaluation. The experiment had four distinctive stages, which are presented below in separate subsections.

6.7.1 Definition of 'flood control' and its decomposition

Flood control is difficult to represent via a fixed number of assessable parameters, and therefore I shall stay here on a lower analytical level and define it through its purpose. Good flood control should provide adequate protection against flooding and effective emergency management measures if flooding occurs. Both flood protection and emergency management measures are designed to prevent serious harm to people, and reduce material and environmental damage. Flood protection is highly associated with the flood return period, and subject to numerous uncertainties associated with eventual flooding.

The three major areas of concern can be inferred from this introduction. The first concern is about the safety level, below which flooding will not occur. The second one includes all the actions aimed at decreasing the negative impact of a flood should it occur. The third concern is with the total potential 'bill' after a flood has occurred. At this point, it seems appropriate to start developing a concept hierarchy for flood control, since this will significantly reduce the complexity of further analytical thinking.

We can start by dividing Flood Control into the three major components as described above (see also Figure 6.16). Further decomposition can then be performed

Box 6.2 Flooding

- Flooding, including flash and riverine floods, coastal floods, snowmelt floods and floods related to ice jams and mud flows, is the most taxing water-related natural hazard to humans and material assets, as well as to cultural and ecological resources.
- Each year flooding affects about 520 million people and their livelihoods, claiming about 25,000 lives worldwide.
- The annual cost to the world economy of flooding and other water-related disasters, is between US$50 billion and US$60 billion.
- When flooding occurs in developing countries, it can result in thousands of deaths and lead to epidemics, as well as effectively wiping out decades of investment in infrastructure and seriously crippling economic prosperity.
- Agriculture-centred developing economies largely depend on fertile floodplains for food security and poverty alleviation efforts.
- The wetlands in floodplains contribute to biodiversity and provide employment opportunities. It is estimated that 1 billion people, one-sixth of the global population – the majority of them among the world's poorest inhabitants – live on floodplains today.
- In Asia, the continent with the greatest potential flood hazard, floods claimed an average of 22,800 lives per annum between 1987 and 1997 and caused an estimated US$136 billion in economic damage.
- The 2002 floods in Europe claimed 100 lives and caused US$20 billion in damage.
- With the frequency and variability of extreme flood events changing because of urbanization, coupled with population growth in flood-prone areas, deforestation, potential climate change and a rise in sea levels, the number of people vulnerable to devastating floods worldwide is expected to rise.

Source: UNESCO (2004)

for each of these components. In the case of the risk of flooding (component B_1 in Figure 6.16), we can identify two basic pieces of information needed in order to evaluate the risk. These are the risk level – calculated from the available hydrological data – and the reliability of the data. We also need to distinguish between the level of risk that will be steady over time and the level of risk that may significantly change in the near future, for example because of changes in land use upstream. The arrangement of these influential factors (one of several possible) that was adopted as our decomposition model for the component B_1 is shown in Figure 6.16.

In similar fashion, the other two components, B_2 and B_3, are analysed and further decomposed. Components C_{31} and C_{32} are decomposed in the same way. The only difference is in the flood volume for which the total damage has to be determined. This classification can extend to as many different flood volumes as are appropriate in a particular flooding region. The total material damage corresponding to each flood volume can be obtained from a stage-damage curve. The evaluation of overall flood damage is achieved by taking the weighted average of membership values for flood volume I, flood volume II and so on. The weights should correspond to the probability of each event.

Material damage, component D_{311}, can be expressed using a crisp number, and the appropriate membership values can easily be assigned. The components E_{3111}, E_{3112}, E_{3113} and E_{3114} are included only to indicate that the importance of different categories of capital concentration may vary.

However, we should disconnect the issue of importance from the potential impact on people and the environment, since this impact is considered separately through the component D_{312}. Therefore, the aggregation of components C_{31} and C_{32} into B_3, as well as the aggregation of components E_{3111}, E_{3112}, E_{3113} and E_{3114} into D_{311}, was excluded from the experiment.

6.7.2 Questionnaire preparation and definition of the extremes

Every single component within the flood control concept hierarchy must be carefully formulated and prepared for evaluation by defining its extremes, i.e. the worst and the best conditions, which will correspond to the membership values 0 and 1. Experts not only need to agree on these boundary conditions, they also need to agree on their meaning. In this case study, these requirements were not met and the experts were exposed individually to these definitions through query forms. (Samples of two query forms, one corresponding to the bottom and the other one to the top level of the concept hierarchy, are shown in Tables 6.5 and 6.6.) The preparation of a good questionnaire is an important step of the process, and some useful guidelines are available in Bates (1992).

In this experiment, the chosen scale for evaluation was from 1 to 5, according to the preference of Expert 1. This was later converted into a scale from 0 to 1, which is more appropriate for most fuzzy aggregation methods.

6.7.3 Evaluation

All three experts were provided with two sets of query forms. The first set was prepared with input containing all the combinations of 5s and 1s for each decompo-

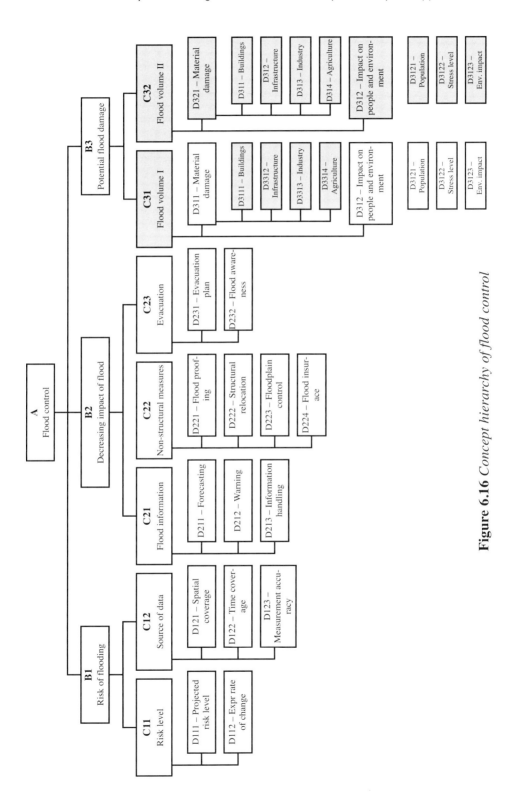

Figure 6.16 *Concept hierarchy of flood control*

sition. The evaluation of these combinations yielded the pseudomeasures necessary for the use of aggregation methods. The second set of query forms had input randomly generated from the set $\{1, 2, 3, 4, 5\}$, where 3 was understood to symbolize the midpoint between the conditions defined by 1 and 5. Likewise, 2 and 4 were taken to represent the midpoints between 1 and 3, and 3 and 5. For both sets of queries the experts were allowed to chose any number from the whole interval [1,5] for their evaluation of a decomposed criterion. In addition to this, we also wanted to obtain a direct evaluation of the relative weights of importance of each set within each decomposition. When all these results had been collected, it was possible to test the appropriateness of each of the aggregation methods discussed in Section 6.6, for the simulation of the experts' reasoning.

Although this was the original intention, the evaluation procedure turned out to be somewhat different. The data required were not obtained from all the experts, because neither Expert 2 nor Expert 3 agreed to provide ratings. They argued that the complexity of flood control makes it very difficult to simultaneously process all the factors involved and assign this kind of satisfaction level to them. Expert 3 argued that the grades he would assign for the second set of sample data would be much less credible than the grades he assigned for the first set (the pseudomeasures). Expert 2 made much the same point but from a slightly different perspective: 'I must admit that they [the query forms from the first set of the sample data] are tough enough, so I would imagine that the sheets with 2s, 3s and 4s would be much tougher.'

Table 6.5 *Example of the query form given to experts for evaluation – bottom level*

C_{22} **Non-structural measures**															
D_{221} Flood proofing 1 = not existing 5 = ready, easy to do, reliable protection	1	5	1	1	1	5	5	5	1	1	1	5	5	5	1	5
D_{222} Structural relocation 1 = no plan 5 = relocation can be performed effectively	1	1	5	1	1	5	1	1	5	5	1	5	5	1	5	5
D_{223} Floodplain control 1 = no control 5 = effective control over the whole area	1	1	1	5	1	1	5	1	5	1	5	5	1	5	5	5
D_{224} Flood insurance 1 = no good flood insurance plan 5 = flood insurance covers most of the potential damage	1	1	1	1	5	1	1	5	1	5	5	1	5	5	5	5
Non-structural measures? 1 = not included or poorly developed 5 = maximum possible, actions are well planned and organized																

Table 6.6 *Example of the query form given to experts for evaluation – top level*

A. Flood control								
B_1 Risk of flooding (risk level, potential changes of the risk level in the future, spatial and time coverage of the data used for calculation and the accuracy of these data) 1 = high risk, no data at all, frequency of flooding is expected to increase in the future 5 = very low risk of flooding, reliable hydrological data, no major increase expected	1	5	1	1	5	5	1	5
B_2 Decreasing impact of flood (forecasting techniques, warning system, efficiency of distributing this information, what types of non-structural measures are considered for protection and how easy it is to implement them, whether there is an evacuation plan and the local population is aware and prepared for the eventual flooding) 1 = no forecasting or warning system, no extra measures for flood protection planned, no evacuation plan 5 = up-to-date forecasting and warning systems, an excellent communication system for information handling, plan for evacuation is very effective and people are flood educated	1	1	5	1	5	1	5	5
B_3 Potential flood damage (the total material value in the flood zone, how many people live there, their age structure and their overall mobility, whether there is any possibility of major environmental impairment in the case of flooding) 1 = level of material damage is very high, flood zone is highly populated, flood can cause major problems to the whole environment 5 = flood practically does not endanger anything	1	1	1	5	1	5	5	5
Flood control? 1 = very unsatisfactory 5 = excellent; no need for improvement								

All the experts were allowed to use any number from the interval [1,5] for the evaluation. Although Experts 2 and 3 were not participating in the problem formulation, they were not necessarily expected to assign the value 1 to a criterion when all its components were given the value 1. In the same way, combinations with all individual scores of 5 were allowed to be given a different score in the unconstrained evaluation. These 'loose ends' led to few results different from 1 and 5 for the samples with all 1s and all 5s. However, neither expert had strong objections on the way the criteria were decomposed: these disparities were mainly due to their different understanding of extremes.

The experts also noted that the decomposition was limited to some extent in its generality. For example, Expert 2 wrote:

> I realized that the numbers that I was assigning to the combinations of 1s and 5s are reflective of my experience with flooding in my region, and that I would probably be assigning different numbers in some cases if my experience was in different settings. For example, flooding in my region usually occurs with ample warning (many days or weeks), so I do not really consider the situation where large floods can develop with only few hours of warning. As well, I assume that it is quite easy to communicate with local authorities/governments about flooding, even if no pre-set, pre-planned communication strategy/plan was developed.

The results of experts' evaluations for the first set of sample data (combinations of 1s and 5s) are given in Table 6.7, while the evaluation for the second set of data, performed only by Expert 1, is given in Table 6.8.

Table 6.7 *The results of experts' evaluations for the first set of sample data*

(a) Component D_{312} – Impact on people and environment

Sample data			Evaluation for D_{312}		
E_{3121}	E_{3122}	E_{3123}	Expert 1	Expert 2	Expert 3
1	1	1	1.0	1.0	2
5	1	1	2.5	2.5	4
1	5	1	2.0	1.7	3
1	1	5	2.0	1.6	1
5	5	1	3.5	3.5	5
5	1	5	3.0	1.7	4
1	5	5	3.0	3.5	5
5	5	5	5.0	5.0	5

(b) Component B_3 – Potential flood damage

Sample data		Evaluation for B_3		
D_{311}	D_{312}	Expert 1	Expert 2	Expert 3
1	1	1.0	1.0	1
5	1	3.5	2.5	4
1	5	3.0	2.0	3
5	5	5.0	5.0	5

(c) Component C_{23} – Evacuation

Sample data		Evaluation for C_{23}		
D_{231}	D_{231}	Expert 1	Expert 2	Expert 3
1	1	1	1	1
5	1	3	2	1
1	5	3	3	2
5	5	5	5	4

(d) Component C_{22} – Other non-structural measures

Sample data				Evaluation for C_{22}		
D_{221}	D_{222}	D_{223}	D_{224}	Expert 1	Expert 2	Expert 3
1	1	1	1	1.0	1.0	1
5	1	1	1	3.0	1.5	2
1	5	1	1	2.5	1.8	2
1	1	5	1	2.5	2.0	3
1	1	1	5	2.5	1.4	1
5	5	1	1	3.5	2.0	2
5	1	5	1	3.5	2.5	4
5	1	1	5	3.0	1.9	3
1	5	5	1	3.0	2.7	4
1	5	1	5	3.0	1.9	3
1	1	5	5	2.5	2.3	4
5	5	5	1	4.0	3.5	5
5	5	1	5	4.0	3.0	3
5	1	5	5	4.0	2.8	4
1	5	5	5	3.5	2.7	5
5	5	5	5	5.0	5.0	5

(e) Component C_{21} – Flood information

Sample data			Evaluation for C_{21}		
D_{211}	D_{212}	D_{213}	Expert 1	Expert 2	Expert 3
1	1	1	1.0	1.0	0.5
5	1	1	3.0	1.0	1.0
1	5	1	2.5	1.5	1.0
1	1	5	2.0	1.5	1.0
5	5	1	3.5	1.5	2.0
5	1	5	3.0	2.0	3.0
1	5	5	3.0	1.5	2.5
5	5	5	5.0	4.5	4.5

(f) Component B_2 – Decreasing impact of flood

Sample data			Evaluation for B_2		
C_{21}	C_{22}	C_{23}	Expert 1	Expert 2	Expert 3
1	1	1	1.0	1.0	1
5	1	1	3.5	2.0	2
1	5	1	3.0	2.0	3
1	1	5	3.0	1.0	2
5	5	1	4.0	4.7	4
5	1	5	4.0	2.5	4
1	5	5	3.5	2.3	3
5	5	5	5.0	5.0	5

(g) Component C_{12} – Source of data

Sample data			Evaluation for C_{12}		
D_{121}	D_{122}	D_{123}	Expert 1	Expert 2	Expert 3
1	1	1	1.0	1.0	1.0
5	1	1	3.0	1.3	2.0
1	5	1	2.5	1.5	1.0
1	1	5	2.5	1.2	2.0
5	5	1	4.0	1.8	2.5
5	1	5	3.5	1.5	3.0
1	5	5	3.0	1.5	3.5
5	5	5	5.0	4.5	5.0

(h) Component C_{11} – Risk level

Sample data		Evaluation for C_{11}		
D_{111}	D_{112}	Expert 1	Expert 2	Expert 3
1	1	1	1	1
5	1	3	4	2.5
1	5	3	1	2
5	5	5	5	4

(i) Component B_1 – Risk of flooding

Sample data		Evaluation for B_1		
C_{11}	C_{12}	Expert 1	Expert 2	Expert 3
1	1	1	1.0	1
5	1	3	1.5	2
1	5	3	1.0	1
5	5	5	4.5	4

(j) Component A – Flood control

Sample data			Evaluation for A		
B_1	B_2	B_3	Expert 1	Expert 2	Expert 3
1	1	1	1.0	1.0	1.0
5	1	1	3.0	1.2	2.5
1	5	1	2.5	1.0	2.0
1	1	5	2.5	2.0	2.0
5	5	1	4.0	1.4	2.5
5	1	5	3.5	3.7	3.0
1	5	5	3.0	1.2	3.0
5	5	5	5.0	4.0	4.5

Table 6.8 *The results of Expert 1's evaluation for the second set of sample data*

(a) Component D_{312} – Impact on people and environment

Sample data			Expert 1
E_{3121}	E_{3122}	E_{3123}	D_{312}
2	2	3	2.0
4	3	4	3.5
4	1	4	2.5
3	1	3	2.0
3	3	5	3.0
4	4	4	4.0
4	2	4	3.0
1	3	3	2.0

(b) Component B_3 – Potential flood damage

Sample data		Expert 1
D_{311}	D_{312}	B_3
3	4	4.0
3	2	2.5
5	4	5.0
2	3	2.5
4	4	4.5
3	3	3.5
3	5	4.0
1	2	1.5

(c) Component C_{23} – Evacuation

Sample data		Expert 1
D_{231}	D_{232}	C_{23}
2	2	2.0
5	4	5.0
3	3	2.5
2	4	2.5
2	1	1.5
1	2	1.5
5	2	3.0
4	2	2.5

(d) Component C_{22} – Other non-structural measures

Sample data				Expert 1
D_{221}	D_{222}	D_{223}	D_{224}	C_{22}
3	4	1	2	3.0
2	2	3	5	3.0
1	3	4	4	3.0
5	5	4	1	4.0
4	2	3	3	3.5
5	2	3	3	3.5
1	3	2	1	2.0
1	2	1	2	1.5

(e) Component C_{21} – Flood information

Sample data			Expert 1
D_{211}	D_{212}	D_{213}	C_{21}
3	3	4	3.0
1	2	5	2.0
2	5	3	3.0
2	4	2	3.0
3	2	2	2.5
5	5	4	5.0
1	2	2	2.0
5	4	5	4.5

(f) Component B_2 – Decreasing impact of flood

Sample data			Expert 1
C_{21}	C_{22}	C_{23}	B_2
1	4	4	3.5
4	2	2	3.0
1	5	1	3.0
3	2	3	2.5
4	2	4	4.0
2	2	4	2.5
5	2	4	3.5
5	5	3	4.0

(g) Component C_{12} – Source of data

Sample data			Expert 1
D_{121}	D_{122}	D_{123}	C_{12}
5	4	1	3.5
5	3	1	3.0
3	3	4	3.0
3	5	2	3.5
1	2	4	2.0
2	2	5	3.0
4	4	3	3.5
1	5	4	3.0

(h) Component C_{11} – Risk level

Sample data		Expert 1
D_{111}	D_{112}	C_{11}
2	4	3.0
2	1	1.5
4	5	4.5
3	3	3.0
2	2	2.0
2	5	3.0
4	2	3.0
1	3	2.0

(i) Component B_1 – Risk of flooding

Sample data		Expert 1
C_{11}	C_{12}	B_1
3	4	3.5
4	2	3.0
1	3	2.0
5	4	5.0
3	1	2.0
2	2	2.0
4	4	4.0
5	2	3.0

(j) Component A – Flood control

Sample data			Expert 1
B_1	B_2	B_3	A
4	4	2	3.5
4	2	3	3.0
3	4	4	3.5
3	3	5	3.5
4	1	3	3.0
1	3	1	2.0
4	1	4	3.0
4	2	3	3.0

In the next and the last step, the choice was made of an appropriate aggregation operator that could reasonably simulate the reasoning of Expert 1.

6.7.4 Results

Of the four aggregation methods presented in Section 6.6, Zimmermann's γ-operator is not applicable in this experiment because none of the empirically obtained aggregation showed the quality that the aggregated value equalled 0 if at least one of its components was equal to 0 (the scale 1 to 5 was converted to the scale 0 to 1). An adequate application of the OWA operator was also not possible, since there were elements of unequal importance, while at the same time aggregation had to be continued from one level to the next higher level, carrying up the values obtained on the previous level. The OWA operator model for aggregation of elements of unequal importance does not achieve this goal. However, the experiment employed the model for learning the OWA operator from data as a form of linear regression, and used it as an indicator for the effectiveness of the remaining two methods. The data used to illustrate these methods are only the data obtained from Expert 1, as presented in Table 6.8.

Starting with the set D_{312} and its corresponding decomposition, the OWA weights w_i using the best fit solution of equation (6.66) are shown in Table 6.9. The resulting membership functions obtained using OWA operator are shown in the comparison graphs, Figure 6.17.

Table 6.9 *Results of the application of the OWA operator*

Component (see Figure 6.16)	Weighting factors of OWA operator (see equation (6.66))
D_{312}	$w_1 = 0,264 \ w_2 = 0,265 \ w_3 = 0,471$
B_3	$w_1 = 0,573 \ w_2 = 0,427$
C_{23}	$w_1 = 0,442 \ w_2 = 0,558$
C_{22}	$w_1 = 0,406 \ w_2 = 0,118 \ w_3 = 0,216 \ w_4 = 0,260$
C_{21}	$w_1 = 0,340 \ w_2 = 0,203 \ w_3 = 0,457$
B_2	$w_1 = 0,508 \ w_2 = 0,206 \ w_3 = 0,286$
C_{12}	$w_1 = 0,387 \ w_2 = 0,231 \ w_3 = 0,382$
C_{11}	$w_1 = 0,473 \ w_2 = 0,526$
B_1	$w_1 = 0,482 \ w_2 = 0,518$
A	$w_1 = 0,414 \ w_2 = 0,231 \ w_3 = 0,355$

The same set of data, when used to perform aggregation under pseudomeasures (equation 6.70, Section 6.6), gives the membership functions shown in Figure 6.17.

At the end, the aggregation of the same set of data is performed using the general expression of the P-CUP method (equation 6.71, Section 6.6). An equation is generated for each element from Figure 6.16. P-CUP equations for the example are shown in Table 6.10.

Table 6.10 *Results of the application of P-CUP aggregation*

Component (see Figure 6.16)	P-CUP equations (see equation (6.71))
D_{312}	$\mu_{D312}(X) = 0.375 \times \mu_{E3121}(X) + 0.25 \times \mu_{E3122}(X) + 0.25 \times \mu_{E3123}(X)$ $-0.125 \times \mu_{E3121}(X) \times \mu_{E3123}(X) + 0.25 \times \mu_{E3121}(X) \times \mu_{E3122}(X) \times \mu_{E3123}(X)$
B_3	$\mu_{B3}(X) = 0.625 \times \mu_{D311}(X) + 0.5 \times \mu_{D312}(X) - 0.125 \times \mu_{D311}(X) \times \mu_{D312}(X)$
C_{23}	$\mu_{C23}(X) = 0.5 \times \mu_{D231}(X) + 0.5 \times \mu_{D232}(X)$
C_{22}	$\mu_{C22}(X) = 0.5\mu_{D221}(X) + 0.375\mu_{D222}(X) + 0.375\mu_{D223}(X) + 0.375\mu_{D224}(X) - 0.25\mu_{D221}(X)\mu_{D222}(X) - 0.25\mu_{D221}(X)\mu_{D223}(X) - 0.375\mu_{D221}(X)\mu_{D224}(X) - 0.25\mu_{D222}(X)\mu_{D223}(X) - 0.25\mu_{D222}(X)\mu_{D224}(X) - 0.375\mu_{D223}(X)\mu_{D224}(X) + 0.25\mu_{D221}(X)\mu_{D222}(X)\mu_{D223}(X) + 0.375\mu_{D221}(X)\mu_{D222}(X)\mu_{D224}(X) + 0.5\mu_{D221}(X)\mu_{D223}(X)\mu_{D224}(X) + 0.375\mu_{D222}(X)\mu_{D223}(X)\mu_{D224}(X) - 0.375\mu_{D221}(X)\mu_{D222}(X)\mu_{D223}(X)\mu_{D224}(X)$
C_{21}	$\mu_{C21}(X) = 0.5\mu_{D211}(X) + 0.375\mu_{D212}(X) + 0.25\mu_{D213}(X) - 0.25\mu_{D211}(X)\mu_{D212}(X) - 0.25\mu_{D211}(X)\mu_{D213}(X) - 0.125\mu_{D212}(X)\mu_{D213}(X) + 0.5\mu_{D211}(X)\mu_{D212}(X)\mu_{D213}(X)$
B_2	$\mu_{B2}(X) = 0.625\mu_{C21}(X) + 0.5\mu_{C22}(X) + 0.5\mu_{C23}(X) - 0.375\mu_{C21}(X)\mu_{C22}(X) - 0.375\mu_{C21}(X)\mu_{C23}(X) - 0.375\mu_{C22}(X)\mu_{C23}(X) + 0.5\mu_{C21}(X)\mu_{C22}(X)\mu_{C23}(X)$
C_{12}	$\mu_{C12}(X) = 0.5\mu_{D121}(X) + 0.375\mu_{D122}(X) + 0.375\mu_{D123}(X) - 0.125\mu_{D121}(X)\mu_{D122}(X) - 0.25\mu_{D121}(X)\mu_{D123}(X) - 0.25\mu_{D122}(X)\mu_{D123}(X) + 0.375\mu_{D121}(X)\mu_{D122}(X)\mu_{D123}(X)$
C_{11}	$\mu_{C11}(X) = 0.5\mu_{D111}(X) + 0.5\mu_{D112}(X)$
B_1	$\mu_{B1}(X) = 0.5\mu_{C11}(X) + 0.5\mu_{C12}(X)$
A	$\mu_A(X) = 0.5\mu_{B1}(X) + 0.375\mu_{B2}(X) + 0.375\mu_{B3}(X) - 0.125\mu_{B1}(X)\mu_{B2}(X) - 0.25\mu_{B1}(X)\mu_{B3}(X) - 0.25\mu_{B2}(X)\mu_{B3}(X) + 0.375\mu_{B1}(X)\mu_{B2}(X)\mu_{B3}(X)$

Figure 6.17 shows a portion of the results of the exercise. Membership functions obtained using all three aggregation techniques are presented for the elements $A, B_1, B_2, B_3, C_{11}, C_{12}, D_{111}$ and D_{112} of the original concept hierarchy of flood control from Figure 6.16. Despic and Simonovic (2000) calculated the total error (as the sum of squared errors) and the maximum error for each aggregation method to show that P-CUP for this set of data gives a much better fit than the other two methods.

6.7.5 Additional observations

Different experts have different experiences, and this was the keystone for the judgements used in the experiment. Every expert can be expected to be consistent in his or her own judgement, but not necessarily consistent with other experts. Lumping all the data together to find the best aggregation operator would result in modelling no one's reasoning. This fits with the observation of Norwich and

Water Resources Systems Management Under Uncertainty – a Fuzzy Set Approach

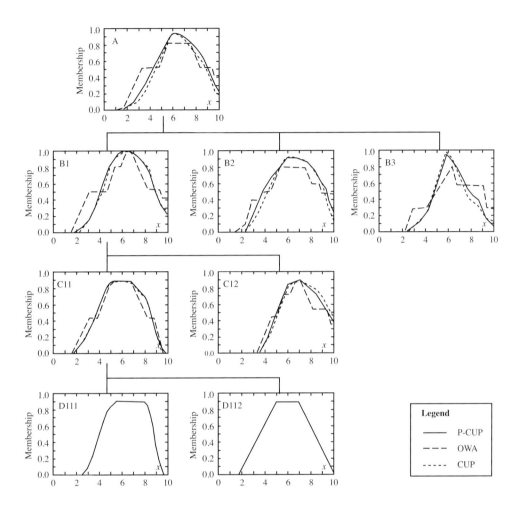

Figure 6.17 *Comparison of membership functions obtained using different aggregation methods*

Turksen (1981) that fuzziness should permit individual interpretability. It is not only that fuzzy relationships are context-dependent: the meaning of a relationship depends on who is expressing it at the linguistic level. In the case study, the best we can hope for is a reliable fuzzy set construction by one person in one specific setting. Based on this exercise it can be concluded that the P-CUP aggregation method has some advantages over other methods for application in water resources management. The primary benefit of the P-CUP methodology is that it provides a reliable and flexible evaluation algorithm which can be used in diverse settings, and if necessary it can easily be modified. The reliability of the algorithm comes from the fact that just making hierarchical decomposition gives a higher precision in the evaluation process. Further advantages of this approach are as follows.

- Complexity is reduced by analysing the system in layers, and the evaluation of the components from the bottom level is easier.
- Partitioning the system allows different individuals to work on different parts of it.
- Aggregation rules can be determined from the least complex combinations of the input data and still give an excellent prediction for some other more complex combinations.
- There is no restriction on using different aggregation methods for different decompositions within a single concept hierarchy.

6.8 FUZZY MEASURES FOR THE ASSESSMENT OF WATER RESOURCES SYSTEMS PERFORMANCE

Let us now explore the utility of fuzzy set theory in the field of water resources systems reliability analysis. Three new fuzzy performance measures are proposed here:

1. a combined reliability–vulnerability index;
2. a robustness index;
3. a resiliency index, based on the work of El-Baroudy and Simonovic (2004).

These measures are illustrated using two simple hypothetical case studies of water supply systems. The indices suggested are able to handle different fuzzy representations and different system conditions. The next section describes the application of these measures to the regional water supply system for London, Ontario, Canada.

The sources of uncertainty discussed in Section 6.1 are many and diverse, and as a result they pose a great challenge for water resources system design, planning and management. The goal of ensuring failsafe system performance may be unattainable. Adopting high-safety factors is one means to avoid the uncertainty of potential failures. However, making safety the first priority may render the system solution infeasible. Therefore, known uncertainty sources must be quantified.

Engineering risk and reliability analysis is a general methodology for the quantification of uncertainty in water resources management and the evaluation of its consequences for the safety of water infrastructure systems (Ganoulis, 1994). The first step in any risk analysis is to identify the risk, clearly detailing all sources of uncertainty that may contribute to the risk of failure. The quantification of risk is the second step, where the effects of the uncertainties are measured using different system performance indices and figures of merit. In this book, three fuzzy measures are developed to evaluate the operational performance of water resources systems. These measures could be useful decision-making aids in a fuzzy environment where subjectivity, human input and lack of previous records impede the decision-making process.

The majority of engineering reliability analyses rely on the use of a probabilistic approach. Both resistance and load as used in the conceptual definition of risk (Section 6.3) are considered to be random variables. However, the characteristics of resistance and/or load cannot always be measured precisely or treated as random variables. Therefore, the fuzzy representation of them is examined. The first use of both fuzzy resistance and fuzzy load can be found in Shrestha and Duckstein (1998).

6.8.1 Key definitions

Partial failure

The calculation of performance indices depends on the exact definition of unsatisfactory system performance. It is difficult to arrive at a precise definition of failure because of the uncertainty in determining system resistance, load and the accepted threshold below which performance is taken to be unsatisfactory. Figure 6.18 depicts a typical system performance (resistance time series), with a constant load during the operation horizon. According to the classical definition, the failure state is the state when resistance falls below the load, margin of safety (difference between the resistance and load) M < 0.0 or safety factor Θ < 1.0, which is represented by the ratio between the system's resistance and load, shown in Figure 6.18 by the dashed horizontal line.

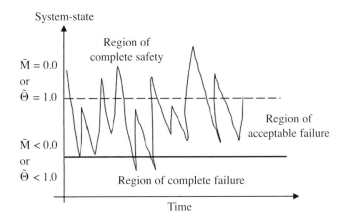

Figure 6.18 *Variable system performance*

Sometimes water resources systems fail to perform their intended function. For example, the available resistance from different sources is highly variable in the case of water resources systems. The actual load may also fluctuate significantly. Consequently, *partial failure* may be acceptable in the design of water resources systems. The precise identification of failure is neither realistic nor practical. It is more realistic to build in the inevitability of partial failure. A degree of acceptable system failure is introduced using the solid horizontal line, as shown in Figure 6.18. The region between the dashed and the full line in Figure 6.18 is the region of partial failure, which will be referred to as *acceptable failure* in the rest of this section.

The boundary of the acceptable or partial failure region is ambiguous, and varies from one decision-maker to another depending on their personal perception of risk. Boundaries cannot be determined precisely. Fuzzy sets are capable of representing the notion of imprecision better than ordinary sets, and therefore the acceptable level of performance can be represented as a fuzzy membership function in the following form:

$$\tilde{M}(m) = \begin{cases} 0, & \text{if } m \leq m_1 \\ \varphi(m), & \text{if } m \in [m_1, m_2] \\ 1, & \text{if } m \geq m_2 \end{cases}$$

or (6.73)

$$\tilde{\Theta}(\theta) = \begin{cases} 0, & \text{if } \theta \leq \theta_1 \\ \varphi(\theta), & \text{if } m \in [\theta_1, \theta_2] \\ 1, & \text{if } \theta \geq \theta_2 \end{cases}$$

where:
$\tilde{M}$ = the fuzzy membership function of margin of safety
$\varphi(m)$ and $\varphi(\theta)$ = functional relationships representing the subjective view of the acceptable risk
m_1, m_2 = the lower and upper bounds respectively of the acceptable failure region (that is, the margin of safety)
$\tilde{\Theta}$ = the fuzzy membership function of factor of safety
θ_1, θ_2 = the lower and upper bounds of the acceptable failure region (that is, the factor of safety), respectively.

Figure 6.19 is a graphical representation of the definition presented in equation (6.73). The lower and upper bounds of the acceptable failure region are introduced in equation (6.73) by m_1 (or θ_1) and m_2 (or θ_2). The value of the margin of safety (or factor of safety) below (or θ_1) is definitely unacceptable. Therefore, the membership function value is zero. The value of the margin of safety (or factor of safety) above m_2 (or θ_2) is definitely acceptable and therefore belongs in the acceptable failure region. Consequently, the membership value is 1. The membership of the in between values varies with the subjective assessment of a decision-maker. Different functional forms may be used for $\varphi(m)$ (or $\varphi(\theta)$) to reflect the subjectivity of different decision-makers' assessments.

High system reliability is reflected through the use of high values for the margin of safety (or factor of safety), i.e. high values for both m_1 and m_2 (or θ_1 and θ_2). The difference between m_1 and m_2 (or θ_1 and θ_2) inversely affects the system reliability, i.e. the higher the difference, the lower the reliability. Therefore, the reliability reflected by the definition of an acceptable level of performance can be quantified in the following way:

$$LR = \frac{m_1 \times m_2}{m_2 - m_1}$$

or (6.74)

$$LR = \frac{\theta_1 \times \theta_2}{\theta_2 - \theta_1}$$

where LR is the reliability measure of the acceptable level of performance.

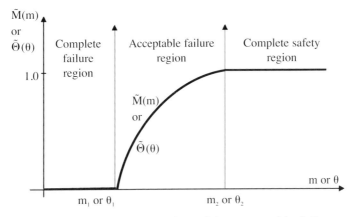

Figure 6.19 *Fuzzy representation of the acceptable failure region*

The subjectivity of decision-makers will always result in a degree of ambiguity of risk perception. This alternative definition of failure allows for a choice between the lower bound, the upper bound, and the function $\varphi(m)$ (or $\varphi(\theta)$). This approach also provides an easy and comprehensive tool for risk communication.

The fuzzy system state

System resistance and load can be represented in a fuzzy form to capture the uncertainty inherent in the system performance. The fuzzy form allows for the determination of the membership function of the resistance and load in a straightforward way even when there is limited available data. Fuzzy arithmetic can be used to calculate the resulting margin of safety (or factor of safety) membership function as a representation of the system state at any time:

$$\tilde{M} = \tilde{X} \; (-) \; \tilde{Y}$$

and (6.75)

$$\tilde{\Theta} = \tilde{X} \; (/) \; \tilde{Y}$$

where:
$\tilde{M}$ = fuzzy margin of safety
$\tilde{X}$ = fuzzy resistance capacity
$\tilde{Y}$ = fuzzy load requirement
$(-)$ = fuzzy subtraction operator
$(/)$ = fuzzy division operator
$\tilde{\Theta}$ = fuzzy factor of safety.

Dynamic systems can be introduced using time-dependent membership functions for the resistance and load. Fuzzy arithmetic can be used to calculate the resulting margin of safety (or factor of safety) membership function as a representation of the system state at any time. In addition, a change in qualitative input dynamics (such as public perception of risk and regulatory trends) can also be expressed through the use of time-dependent membership functions of the acceptable level of performance. As a result, dynamic system performance can be evaluated at any time.

Compatibility

The purpose of comparing two fuzzy membership functions, as introduced in Section 6.5.4, is to illustrate the extent to which the two fuzzy sets match. The reliability assessment, discussed here, involves a comparative analysis of the system-state membership function and the predefined acceptable level of the performance membership function. Therefore, the compliance of two fuzzy membership functions can be quantified using the fuzzy compatibility measure (Balopoulos et al, 2007).

Possibility (equation 6.51) and necessity (equation 6.53) lead to the quantification of the compatibility of two fuzzy sets. The possibility measure quantifies the overlap between two fuzzy sets, while the necessity measure describes the degree of inclusion of one fuzzy set in another fuzzy set. However, in some cases (see Figure 6.20) high possibility and necessity values do not reflect clearly the compliance between the system-state membership function and the acceptable level of performance membership function. Figure 6.20 shows two system-state functions, A and B, which have the same possibility and necessity values. However, system-state A has a larger overlap with the performance membership function than the system-state B (the shaded area in Figure 6.20).

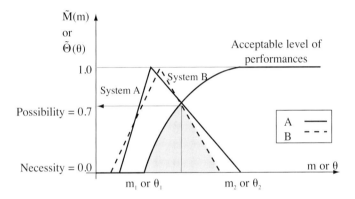

Figure 6.20 *Compliance between the system state and the acceptable level of performance*

The overlap area between the two membership functions, as a fraction of the total area of the system state, illustrates compliance more clearly than the possibility and necessity measures in equations (6.51) and (6.53) respectively:

$$C_{S,L} = \frac{OA_{S,L}}{A_S} \tag{6.76}$$

where:
$C_{S,L}$ = the compliance between the system-state membership function (S) and the acceptable level of performance membership function (L)
$OA_{S,L}$ = the overlap area between the system-state membership function (S) and the acceptable level of performance membership function (L)
A_S = the area of the system-state membership function (S).

An overlap in a high-significance area (that is, an area with high membership values) is preferable to an overlap in a low significance area, as shown in Figure 6.21. Therefore, the compliance measure should take into account the weighted area approach (Verma and Knezevic, 1996).

The compatibility measure can be calculated using:

$$C_{S,L} = \frac{WOA_{S,L}}{WA_S} \tag{6.77}$$

where:
$C_{S,L}$ = compatibility measure between the system-state membership function (S) and the acceptable level of performance membership function (L)
$WOA_{S,L}$ = weighted overlap area between the system-state membership function (S) and the acceptable level of performance membership function (L)
WA_S = weighted area of the system-state membership function (S).

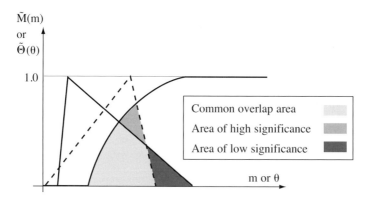

Figure 6.21 *Overlap analysis*

6.8.2 Combined fuzzy reliability – vulnerability measure

Reliability and vulnerability are used to provide a complete description of system performance in the case of failure, and to determine the magnitude of the failure event. Once an acceptable level of performance is determined in a fuzzy form, the anticipated performance in the event of failure as well as the expected severity of failure can be determined.

When certain values are specified for the lower and upper bounds (m_1 and m_2 (or θ_1 and θ_2) in equation (6.73)), thus establishing a predefined acceptable level of performance, the anticipated system failure is limited to a specified range. Systems that are highly compatible with the predefined acceptable level of performance will yield a similar performance, i.e. the expected system failure will be within the specified range ($[m_1, m_2]$ or $[\theta_1, \theta_2]$). In order to calculate system reliability, several acceptable levels of performance must be defined to reflect the different perceptions of decision-makers.

A comparison between the fuzzy system-state membership function and the predefined fuzzy acceptable level of performance membership function provides information about both system reliability and system vulnerability at the same time (see Figure 6.22). The system reliability is based on the proximity of the system state to the predefined acceptable level of performance. The measure of proximity is expressed by the compatibility measure suggested in equation (6.77). The combined fuzzy reliability–vulnerability index is formulated as follows:

$$RE_f = \frac{\max_{i \in K} \{CM_1, CM_2, \ldots CM_i\} \times LR_{max}}{\max_{i \in K} \{LR_1, LR_2, \ldots LR_i\}} \quad (6.78)$$

where:
RE_f = the combined fuzzy reliability–vulnerability index
LR_{max} = the reliability measure of acceptable level of performance corresponding to the system state with maximum compatibility value
LR_i = the reliability measure of the i-th acceptable level of performance
CM_i = the compatibility measure for system state with the *i*-th acceptable level of performance
K = the total number of the defined acceptable levels of performance.

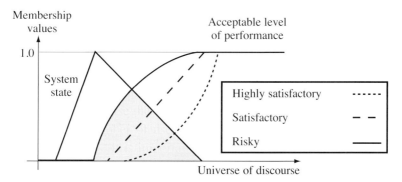

Figure 6.22 *Compatibility of the system state with different levels of the performance membership function*

The reliability–vulnerability index is normalized to attain a maximum value of 1.0, by the introduction of the $\max_{i \in k} \{LR_1, LR_2 \ldots LR_i\}$ value as the maximum achievable reliability.

6.8.3 Fuzzy robustness measure

Robustness measures the system's ability to adapt to a wide range of possible future load conditions, at little additional cost (Hashimoto et al, 1982). The fuzzy form of change in future conditions can be obtained through a redefinition of the acceptable level of performance and a change in the system-state membership function. As a result, the system's robustness is defined as the change in the compatibility measure:

$$RO_f = \frac{1}{CM_1 - CM_2} \tag{6.79}$$

where:
RO_f = fuzzy robustness index
CM_1 = compatibility measure before the change in conditions
CM_2 = compatibility after the change in conditions.

Equation (6.79) reveals that the higher the change in compatibility, the lower the value of fuzzy robustness. Therefore, high robustness values allow the system to adapt better to new conditions.

6.8.4 Fuzzy resilience measure

Resilience measures how fast the system recovers from a failure state. The time required to recover from the failure state can be represented as a fuzzy set. Because the reasons for a failure may be different, system recovery times will vary depending on the type of failure, as shown in Figure 6.23.

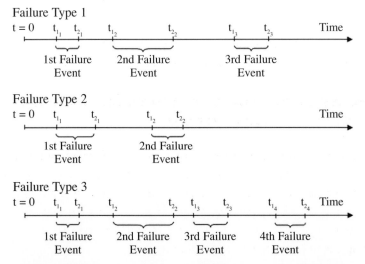

Figure 6.23 *Recovery times for different types of failure*

A series of fuzzy membership functions can be developed to allow for various types of failure. The maximum recovery time is used to represent the system recovery time (Kaufmann and Gupta, 1985):

$$\tilde{T}(\alpha) = \left(\max_{j \in J} [t_{1_1}(\alpha), t_{1_2}(\alpha), \ldots, t_{1_J}(\alpha)], \max_{j \in J} [t_{2_1}(\alpha), t_{2_2}(\alpha), \ldots, t_{2_J}(\alpha)] \right) \quad (6.80)$$

where:
$\tilde{T}$ = the system fuzzy maximum recovery time at the α-level
$t_{1_j}(\alpha)$ = the lower bound of the j-th recovery time at the α-level
$t_{2_j}(\alpha)$ = the upper bound of the j-th recovery time at the α-level
J = total number of failure events.

The centre of gravity of the maximum fuzzy recovery time can be used as a real number representation of the system recovery time. Therefore, system resilience is determined to be the inverse value of the centre of gravity (Klir et al, 1997):

$$RS_f = \left[\frac{\int_{t_1}^{t_2} t\, \tilde{T}(t)\, dt}{\int_{t_1}^{t_2} \tilde{T}(t)\, dt} \right]^{-1} \quad (6.81)$$

where:
RS_f = fuzzy resiliency index;
$\tilde{T}(t)$ = system fuzzy maximum recovery time
t_1 = lower bound of the support of the system recovery time
t_2 = upper bound of the support of the system recovery time.

The inverse operation can be used to illustrate the relationship between the value of the recovery time and the resilience. The longer the recovery time, the lower the system's ability to recover from the failure, and the lower the resilience.

6.8.5 Fuzzy performance measures for multi-component water resources systems

Water resources systems are made up of a variety of interconnected subsystems. Each subsystem has multiple components, where the interconnection configuration affects the overall system performance. Multi-component systems have several system-state membership functions representing the system state of each component. Aggregation of these membership functions results in a system-state membership function for the whole system.

Aggregation of system-state functions

There are three main types of configuration of multi-component systems: serial, parallel and combined. For each component, a fuzzy membership function, representing the component's state, can be calculated based on the component's load and resistance. The overall system state is calculated depending on the system configuration.

Assuming that a serial configuration system is composed of N components, as shown in Figure 6.24a, the n-th component has a state membership function $\tilde{S}_n(u)$, defined on the universe of discourse U. The weakest component, in terms of system-state, controls the whole system state. Therefore, the system state can be calculated as follows:

$$\tilde{S}(u) = \min_N (\tilde{S}_1, \tilde{S}_2, \ldots, \tilde{S}_N) \qquad (6.82)$$

where:
$\tilde{S}(u)$ = the system state, and
$(\tilde{S}_1, \tilde{S}_2, \ldots, \tilde{S}_N)$ are component system states.

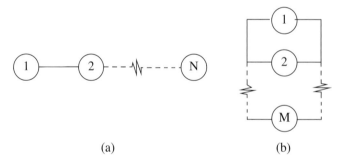

Figure 6.24 *Serial and parallel system configurations: (a) a serial system configuration of N components; (b) a parallel system configuration of M components*

A parallel system configuration is composed of M components, as shown in Figure 6.24b. The m-th component has a state membership function $\tilde{S}_m(u)$, defined on the universe of discourse U. All states of the components contribute to the system state. If all components of the system fail, system failure occurs. Hence, the system state can be calculated as follows:

$$\tilde{S}(u) = \sum_{m=1}^{M} \tilde{S}_m(u) \qquad (6.83)$$

where:
$\tilde{S}_m(u)$ = the m-th component system state, and
M = the total number of parallel components.

Combined systems are systems with both parallel and serial subsystems. The system state in this case can be arrived at by calculating the subsystem states according to equations (6.82) and (6.83).

Aggregation of recovery time membership functions

The aggregation of recovery time membership functions is different from the aggregation of system-state membership functions. The system-state membership function determines the performance (or state) of the system, which can be either satisfactory or unsatisfactory. Therefore, aggregation is based on the contribution of each component to the system state. The recovery time function, on the other hand, represents the system in the failure state.

For serial configuration systems composed of N components, the n-th component has a maximum recovery time membership function $\tilde{T}_n(t)$, defined on the universe of discourse T. The component having the longest recovery time controls the system recovery time. Therefore, the system recovery time can be calculated as follows:

$$\tilde{T}(t) = \tilde{T}_c(t) \qquad (6.84)$$

given

$$S(\tilde{T}_c) = \max_N \left(S(\tilde{T}_1), S(\tilde{T}_2), \ldots, S(\tilde{T}_N) \right)$$

and $\qquad (6.85)$

$$\tilde{T}_c(1) = \max_N \left(\tilde{T}_1(1), \tilde{T}_2(1), \ldots, \tilde{T}_N(1) \right)$$

where:
$\tilde{T}(t)$ = system recovery time
$\tilde{T}_c(1)$ = controlling recovery time
$S(\tilde{T}_c)$ = support of the controlling recovery time fuzzy membership functions
$\left(S(\tilde{T}_1), S(\tilde{T}_2), \ldots, S(\tilde{T}_N) \right)$ = support sets of N components
$\tilde{T}_c(1)$ = controlling recovery time set at the credibility level = 1
$\left(\tilde{T}_1(1), \tilde{T}_2(1), \ldots, \tilde{T}_N(1) \right)$ = recovery time sets at credibility level = 1 of the N components.

In a parallel system configuration, composed of *M* number of components, the *m*-th component has a maximum recovery time membership function $\tilde{T}_m(t)$, defined on the universe of discourse *T*. The total failure event equals the failure of every component in the system. As a result, the membership function of system recovery time can be calculated as follows:

$$\tilde{T}(t) = \max_M (\tilde{T}_1, \tilde{T}_2, \ldots, \tilde{T}_M) \qquad (6.86)$$

where:
$\tilde{T}(t)$ = system recovery time, and
$(\tilde{T}_1, \tilde{T}_2, \ldots, \tilde{T}_M)$ = component recovery time.

The combined system recovery time membership function can be arrived at by calculating subsystems recovery time membership functions according to either equation (6.84) or (6.86).

6.8.6 An illustrative example

The implementation of these three new fuzzy performance measures can be illustrated using two simple hypothetical cases. As shown in Figure 6.25, system A consists of a pump, single pipeline and a reservoir, while system B consists of a pump, two parallel pipelines and a reservoir. The introduction of the two parallel pipelines in system B increases system redundancy, resulting in higher system reliability. Therefore, the reliability value should reflect the difference between the two systems. Note that both systems are exposed to the same load (demand) requirement and have the same resistance (supply capacity).

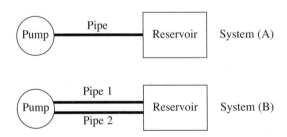

Figure 6.25 *Schematic representation of the hypothetical case studies*

With this small example we can:

- test the hypothesis that system B has a higher reliability measure than system A, introduced through higher redundancy;

- illustrate computational aspects related to the fuzzy performance indices introduced above;
- evaluate the sensitivity of these indices to the use of two different shapes of membership function, i.e. triangular or trapezoidal.

The choice of the membership function, shape and value is a knowledge acquisition problem discussed in Section 6.7. In this example triangular and trapezoidal membership functions were subjectively selected and used to test the sensitivity of the fuzzy reliability measures to the shape. In real applications each problem domain will result in a different shape of membership function (Despic and Simonovic, 2000). In the case of multiple decision-makers, one of the available techniques for integration of the inputs into a single membership function will be applied. An effective method is the fuzzy expected value evaluation tested in the field of flood management (Akter and Simonovic, 2005).

Different elements of each system, i.e. pump, pipes and reservoir, are serially connected. Therefore, the overall system reliability depends on the reliability of the weakest element. Assuming that the reliability of the pipes in both systems controls the overall system reliability, two different scenarios can be applied to system B: (i) both pipes have the same supply capacity, and (ii) one of the pipes has a supply capacity twice as large as that of the other pipe. The sum of the supply capacities of the two pipes, in both scenarios, is equal to the supply capacity of the pipe in system A.

Table 6.11 summarizes the four different input scenarios analysed in this example. In case I, triangular fuzzy membership represents fuzzy supply and fuzzy demand for both systems. The system's supply and demand were distributed between the two pipes in system B with the ratio 1:1 (equal distribution). In case II, triangular fuzzy membership again represents fuzzy supply and fuzzy demand for both systems. System supply and demand were distributed between the two pipes in system B in the ratio 1:2 (unequal distribution). In case III, trapezoidal fuzzy membership represents fuzzy supply and fuzzy demand for both systems. System supply and demand were equally distributed between the two pipes in system B. In case IV, trapezoidal fuzzy membership again represents fuzzy supply and fuzzy demand for both systems. System supply and demand were unequally distributed between the two pipes in system B.

The utility of the fuzzy reliability measure suggested by El-Baroudy and Simonovic (2004) and discussed in this section can be evaluated by comparing the results of different experiments. The sensitivity of these measures to the type of the membership function was also examined using the triangular and trapezoidal membership functions. Recovery time data for several types of failures were used to conduct the calculation of resiliency. The results indicated that simple case studies cannot be used to illustrate the utility of the resiliency measure.

Table 6.11 *Summary of test cases*

Case	Case description	System	Supply capacity (m³/s)	Demand requirement (m³/s)
I	Triangular fuzzy membership with equal distribution between pipes in system B	A	(0.0,3.0,6.0)	(1.0,2.0,4.0)
		B	(0.0,1.5,3.0) (0.5,1.0,2.0)	(0.0,1.5,3.0) (0.5,1.0,2.0)
II	Triangular fuzzy membership with unequal distribution between pipes in system B	A	(0.0,3.0,6.0)	(1.0,2.0,4.0)
		B	(0.0,1.0,2.0) (0.3,0.7,1.3)	(0.0,2.0,4.0) (0.7,1.3,2.7)
III	Trapezoidal fuzzy membership with equal distribution between pipes in system B	A	(0.0,1.0,5.0,6.0)	(1.0,2.0,3.0,4.0)
		B	(0.0,0.5,2.5,3.0) (0.0,0.5,2.5,3.0)	(0.5,1.0,1.5,2.0) (0.5,1.0,1.5,2.0)
IV	Trapezoidal fuzzy membership with unequal distribution between pipes in system B	A	(0.0,1.0,5.0,6.0)	(1.0,2.0,3.0,4.0)
		B	(0.0,0.3,1.7,2.0) (0.0,0.7,3.3,4.0)	(0.3,0.7,1.0,1.3) (0.7,1.3,2.0,2.7)

The utility of the new measures was evaluated using three acceptable levels of performance of the safety factor. These levels are (1) high-safety level, (2) safe level, and (3) low-safety level. These levels are represented by three trapezoidal fuzzy numbers, (0.8, 1.2, 15, 15), (0.7, 1.0, 15, 15), and (0.5, 0.8, 15, 15) respectively. Figures 6.26a and 6.26b illustrate the system state memberships used for cases I and III together with the memberships of the predefined acceptable levels of performance.

The effect of the shape of the membership function is shown in case II, where the system B reliability value was 1.25 higher than the reliability of system A (0.803 and 0.644 respectively). In case IV the system B reliability value was 1.27 higher than system A (0.726 and 0.571 respectively).

Table 6.12 shows that the reliability of system B is higher than the reliability of system A (0.53 and 0.55 respectively for cases I and II; 0.47 to 0.53 respectively for cases III and IV). These results support the main hypothesis that higher redundancy results in greater system reliability. The shape of the membership function does not affect the main conclusion concerning system reliability. The reliability value is not affected by the capacity of system components, 0.55 in cases I and II and 0.53 in cases III and IV.

In order to calculate the system's robustness, the level of acceptable performance membership was changed from the low-safety level to the safe level. Using two parallel pipes (as in Table 6.12) increased the system robustness, as the value of the fuzzy robustness index increased from 16.9 to 114.5 in case I.

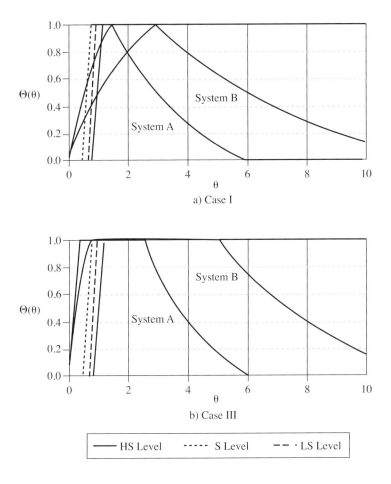

Figure 6.26 *System states for cases I and III with the predefined acceptable levels of performance*

The increase in the case of the triangular membership function is three times the increase in the case of the trapezoidal function. Therefore, the value of system robustness depends on the shape of the supply and demand membership functions and their position in the universe of discourse. The load distribution between the parallel pipes affects the robustness of the system. It increases from 114.5 to 120.8 between cases I and II. No significant change was observed between cases III and IV.

Table 6.12 *Summary results of suggested reliability and robustness measures*

Case	Case description	System	Reliability–vulnerability measure	Robustness measure	Result
I	Triangular fuzzy membership with equal distribution between pipes in system B	A	0.53	16.9	The measure indicates a difference in reliability
		B	0.55	114.5	
II	Triangular fuzzy membership with unequal distribution between pipes in system B	A	0.53	16.9	The measure indicates a difference in reliability
		B	0.55	120.8	
III	Trapezoidal fuzzy membership with equal distribution between pipes in system B	A	0.47	15.4	The measure indicates a difference in reliability
		B	0.53	30.9	
IV	Trapezoidal fuzzy membership with unequal distribution between pipes in system B	A	0.47	15.4	The measure indicates a difference in reliability
		B	0.53	30.7	

6.9 THE APPLICATION OF FUZZY PERFORMANCE MEASURES TO A REGIONAL WATER SUPPLY SYSTEM

This section describes the application of fuzzy performance measures to the City of London (Ontario, Canada) regional water supply system. First the Lake Huron Primary Water Supply System (LHPWSS) and the Elgin Area Primary Water Supply System (EAPWSS) are introduced, then the methodology used for the analysis of both systems is presented. This presentation starts with a description of the procedure for system representation. The method used to construct membership functions for different system components follows. The calculation process for the fuzzy performance measures is presented in detail. Finally the fuzzy performance measures for the LHPWSS and EAPWSS system are presented. The utility of these measures in identifying critical system components is demonstrated in the results.

6.9.1 System description

The City of London regional water supply system (El-Baroudy and Simonovic, 2006) consists of two main components, the LHPWSS and the EAPWSS. The LHPWSS system obtains raw water from Lake Huron. Water is treated and pumped from the lake to a terminal reservoir in Arva, as illustrated in Figure 6.27 (Plate 9).

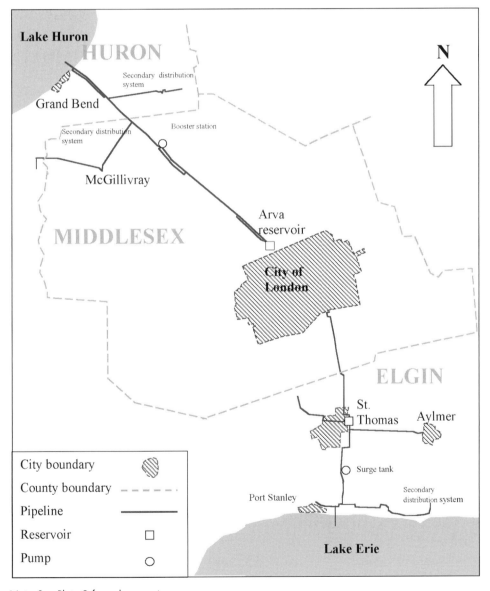

Note: See Plate 9 for colour version.

Figure 6.27 *The City of London, Ontario, regional water supply system*

Water from the Arva reservoir is pumped to the north of the City of London, where it enters the municipal distribution system. The system provides water for the City of London as well as a number of smaller neighbouring municipalities (through a secondary system).

The EAPWSS system treats raw water from Lake Erie and pumps the treated water to a terminal reservoir located in St Thomas. Water from the reservoir is pumped to the south of the City of London where it enters the municipal distribution system, as shown in Figure 6.27. In case of emergency, the city can obtain additional water from a number of wells located inside the city and in the surrounding areas. Table 6.13 lists the different system components for both systems.

6.9.2 Lake Huron Primary Water Supply System (LHPWSS)

The Lake Huron treatment facility has a treatment capacity of 336,400 cubic metres a day (m^3/day). The plant's individual components are designed with a 35 per cent overload capacity, resulting in a maximum capacity of 454,600 m^3/day (Earth Tech Canada, 2001). The water treatment system employs conventional and chemically assisted flocculation and sedimentation systems, dual-media filtration, and chlorination as the primary disinfection. Both the treatment system and the water quality are continuously monitored using a computerized Supervisor Control and Data Acquisition (SCADA) system. A schematic representation of the treatment process is shown in Figure 6.28.

Raw water flows by gravity from Lake Huron through a reinforced concrete intake pipe to the low-lift pumping station. The intake pipe discharges raw water through mechanically cleaned screens into the pump-well of the low-lift pumping station. Chlorine is injected into the intake crib through the screens, or to the low-lift pumping station for zebra mussel control (pre-chlorination). The low-lift pumping station, which consists of six pumps, is located on the shore of Lake Huron at the treatment plant site.

Water is discharged from the low-lift pumping station into the treatment plant, where it bifurcates into two parallel streams. It flows by gravity from the flash mix chambers to the flocculation tanks. The first treatment step takes place in the flash mix chambers, where alum is added (for coagulation) together with powdered activated carbon (PAC) (seasonally added for taste and odour control) and a polymer (as an aid to coagulation). Chlorine is also used for disinfection upstream of the flash mixers. A mechanical flocculation process takes place in both treatment lines. Each flocculation tank is divided into two zones, primary and secondary, where paddle mixers perform the mixing. Water flows through the two zones to the clarifiers/settlers. Twelve high-rate gravity filters remove particulate matter from water flowing from the clarifiers. Water flows to any of the twelve filters from both

Table 6.13 *Summary list of LHPWSS and EAPWSS system components*

LHPWSS system				EAPWSS system			
No	Component	No	Component	No	Component	No	Component
1	Intake crib	34	(2) Flash mix cells	1	Intake crib	32	Polymer mix tank
5	(4) chlorinators	35	PAC transfer pump	2	Sodium hypochlorite	33	Polymer day tank
6	RC intake pipe	43	(8) Flocculation cells	3	RC intake pipe	37	(4) Polymer metering pump
7	Travelling screens	47	(4) Settling tanks	4	Drain intake pipe	38	Alum transfer pump
8	Pumping wells	59	(12) Filters	5	Travelling screens	40	(2) Alum metering Pumps
12	(4) Single-speed pumps	62	(3) Clear-wells	7	(2) Pumping well	44	(4) Flocculation cells
14	(2) Variable-speed pump	67	(5) High-lifting pumps	9	(2) High-discharge pump	46	(2) Sedimentation tanks
16	(2) PAC storage tanks	69	(2) Mains	11	(2) Low-discharge pump	48	(2) Chlorinators
18	(2) PAC transfer pump	73	(4) Boosting pumps	12	RC intake pipe	50	(2) Fluoride storage tanks
20	(2) Alum storage tanks	74	Intermediate storage	15	(3) PAC storage tanks	52	(2) Fluoride feed pumps
22	(2) Alum transfer pump	78	(4) Terminal storage cells	17	(2) PAC transfer pump	56	(4) Filters
24	(2) Flash mix cells			21	(4) PAC metering pumps	57	Clear-well
26	(2) Polymer storage tanks			24	(3) Alum storage tanks	58	On-site reservoir
28	(2) Polymer transfer pump			25	Alum transfer pump	62	(4) High-lifting pumps
30	(2) Polymer mix tanks			29	(4) Alum metering pumps	63	Main
32	(2) Polymer feed pump			31	(2) Flash mix cells	65	(2) Terminal storage cells

treatment lines. Filtered water is then discharged into three clear-wells where chlorine is added for post-chlorination. The treated water is then pumped from the clear-wells through the transmission main to the terminal reservoir at Arva by high-lift pumps. There are an intermediate reservoir and booster station in McGillivary Township. The intermediate reservoir serves the users in this area. Water from the

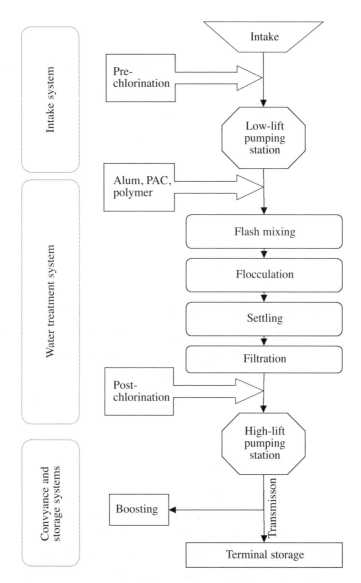

Figure 6.28 *Schematic representation of the LHPWSS treatment process*

reservoir can be withdrawn back into the primary transmission main during high demand periods, by four high-lift pumps at the booster station.

6.9.3 Elgin Area Primary Water Supply System (EAPWSS)

The Elgin water treatment facility was constructed in 1969 to supply water from Lake Erie to the City of London, St Thomas and a number of smaller municipalities.

In 1994 the facility was expanded to double its throughput to its current 91,000 m^3/day capacity. A series of upgrades took place from 1994 to 2003 to add surge protection and introduce fluoridation treatment. The design capacity of the treatment facility is 91,000 m^3/day, with an average daily flow of 52,350 m^3/day, which serves a population of about 94,400 (Earth Tech Canada, 2000).

The water treatment in EAPWSS employs almost the same conventional treatment methods used in LHPWSS. The only exception is that the facility uses the fluoridation treatment system, adding fluoride with the aim of preventing dental decay among users. As in LHPWSS, the treatment system and water quality are continuously monitored using a computerized SCADA system. The finished treated water is pumped to a terminal reservoir located in St Thomas. A schematic of the treatment process is shown in Figure 6.29.

Raw water, drawn from Lake Erie, is pumped through the intake conduit to the low-lift pumping station at the shore of the lake. In case of an emergency, the plant drain serves as an alternative intake. The low-lift pumping station houses two clear-wells. Each well has two independent vertical turbine pumps which discharge into a transmission main to the water treatment plant. The raw water discharged from the low-lift pumping station is metered and split evenly into two parallel streams, as in the LHPWSS. The split continues from the head-works to the filtration process. The first treatment process is flash mixing, where alum is added as a coagulation agent, together with PAC. Water flows by gravity from the flash mix chamber to the flocculation tanks. The flocculation system consists of two banks of flocculation tanks.

Polymer can be added at any point in the series of flocculation tanks. Water flows directly from the flocculation tanks into the sedimentation system. There is a gravity sedimentation tank in each process stream. Pre-chlorination takes place after the sedimentation process and before filtration.

Finally, particulate matter is removed using four gravity filters during the filtration process. The treatment is no longer split into two parallel streams as the water can be directed to any of the four filters. The filtered water is collected in the filtered water conduit underlying the filters, and flows into a clear-well and the on-site reservoir. Post-chlorination takes place in the conduit leading from the on-site reservoir to the high-lift pumping station. The high-lift pumping station delivers finished water through the transmission main to the terminal reservoir in St Thomas. It also delivers water to the secondary distribution system. A surge facility was constructed in 1994 to protect the transmission main from damage as a result of system transit pressure conditions during cycling of the high-lift pumps.

Through the valve chamber, upstream of the terminal reservoir, water from the transmission main is directed to one or both reservoirs at the Elgin-Middlesex facility. Water can bypass the reservoirs and flow directly to each of the secondary pumping stations.

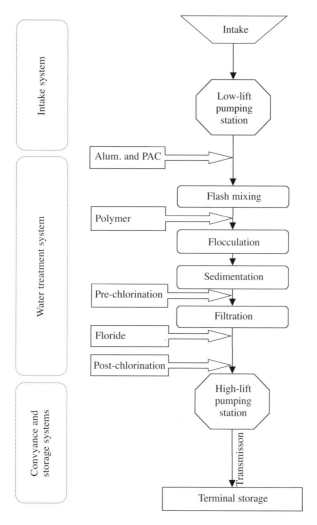

Figure 6.29 *Schematic representation of the EAPWSS treatment process*

6.9.4 Methodology for the application of fuzzy performance measures

A detailed introduction to the methodology used in this case study is presented in Section 6.8.

Acceptable level of performance

The calculation of system performance measures depends on the exact identification of the unsatisfactory system state, as was demonstrated in Section 6.8. However, it is difficult to arrive at a precise definition of a failure event because of the uncertainty in determining system supply, demand, and the accepted unsatisfactory performance threshold. The fuzzy reliability analysis quantifies this uncertainty through the use of the appropriate fuzzy membership function to describe the fuzzy system's state of safety and the fuzzy failure events.

The fuzzy failure event is represented by a fuzzy membership function which includes different failure events. This membership function is a formal mathematical representation of the acceptable level of system performance. The acceptable level of performance also reflects the decision-maker's ambiguous and imprecise perception of risk, as was shown in Section 6.8. The acceptable level of performance is quantified by equation (6.74) and illustrated in Figure 6.19.

System-state membership function

System reliability analysis uses load and resistance as the fundamental concepts to define the risk of system failure (Simonovic, 1997). In water supply systems, load and resistance are replaced by demand and supply respectively, to reflect the specific domain variables. Therefore, the system demand is defined as the variable that reflects different water requirements that may be imposed over the useful life of the system. System supply is defined as the system characteristic variable that describes the capacity of the system to meet the demand.

The fuzzy reliability analysis uses membership functions to express the uncertainty in both supply and demand of each system component. Construction of the membership function is based on the system design data and choice of a suitable shape. Triangular and trapezoidal shapes are the simplest membership function shapes that meet this requirement. In this application, all component-state membership functions are formulated in terms of a fuzzy margin of safety using the fuzzy subtraction operator (equation (6.75)).

Fuzzy performance measures

The compatibility between the system state and the acceptable level of performance is the basis for the calculation of the fuzzy combined reliability–vulnerability performance measure. It is illustrated in equation (6.78) and Figure 6.22.

Fuzzy robustness measures the system's ability to adapt to a wide range of possible future load conditions. The fuzzy form of change in future conditions is obtained through a redefinition of the acceptable level of performance and a change in the system-state membership function. As a result, the system robustness is defined as the change in the compatibility measure (equation (6.79)).

The time required to recover from the failure state is represented as a fuzzy set. System recovery times vary depending on the type of failure. Therefore, a series of fuzzy membership functions is developed to represent recovery from different types of failure. The centre of gravity of the maximum fuzzy recovery time is used as a real number representation of the system recovery time. Thus, a system fuzzy resiliency measure is determined to be the inverse value of the centre of gravity (equation (6.81)).

Multi-component system representation

A water supply system is a multi-component system which includes a collection of conveyance, treatment and storage components. The key step in the evaluation of system performance is the appropriate representation of different relationships between system components. This representation should reflect the effect of the performance of each component on the overall system performance. For example, the chemical treatment of raw water in a water supply system depends on adding different chemicals at certain locations in the treatment process. This process requires the availability of chemicals in the storage facility and the ability to transfer them to the required location on time. Storage and conveyance facilities, responsible for delivering those chemicals to the mixing chambers, are not part of the raw water path. The failure of those process components directly affects the water treatment process and might cause a total failure of the water treatment system. Therefore, they must be considered when performing a system reliability analysis.

Figure 6.30 shows the layout of one part of a typical water treatment plant, where the stored chemicals are conveyed to the mixing location via the feed pump. It is evident that taking those components into consideration in the system reliability analysis is difficult because of the need to identify the functional relationships between them and the other system components. Similar relationships are required for all non-carrying water components. If those components are not taken into consideration, the chance of an improper estimation of system reliability will increase.

Representing a multi-component system as an integrated system of components, each having different failure relationships, can be used as an effective means to integrate water-carrying and non-water-carrying components into one system. For example, any two components are considered serially connected if the failure of one component leads to the failure of the other. Two components are considered to have a parallel connection if the failure of one component does not lead to the failure of the other. A clear identification of the failure relationship between different components facilitates the calculation of system fuzzy performance measures. Figure 6.31

Water Resources Systems Management Under Uncertainty – a Fuzzy Set Approach 221

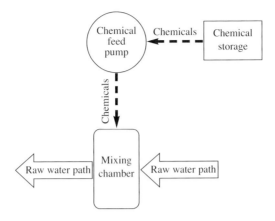

Figure 6.30 *Typical water supply system layout*

shows the integrated layout for the previous example. In this figure, the system representation integrates components carrying chemicals into the path of raw water.

Calculation of the system fuzzy performance measures based on the integrated layout will be fairly easy, as there is a clear link between the failure of the components carrying chemicals and the components carrying raw water.

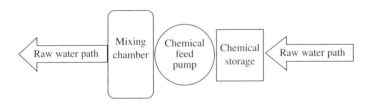

Figure 6.31 *System integrated layout for the fuzzy reliability analysis*

System-state membership function

Membership functions of system components are aggregated to calculate fuzzy system performance measures using fuzzy operators. Therefore, all membership functions must be expressed in the same units. This is achieved through standardization of the membership functions (i.e. division by the unit maximum capacity value). The membership function of each system component has a maximum value of 1. For example, a triangular membership function representing a reservoir capacity (m^3) is defined as follows:

$$\mu_{\tilde{A}}(x) = \begin{cases} 0, & \text{if } x \leq a \\ \dfrac{x-a}{m-a}, & \text{if } x \in [a, m] \\ \dfrac{b-x}{b-m}, & \text{if } x \in [m, b] \\ 0, & \text{if } x \leq b \end{cases} \qquad (6.87)$$

where:
m = the modal value and
a and b = the lower and upper bounds of the non-zero values of the membership.

This membership function can be standardized to the following (dimensionless) membership function:

$$\mu_{\tilde{A}}(x) = \begin{cases} 0, & \text{if } x \leq (a/b) \\ \dfrac{x - (a/b)}{(m/b) - (a/b)}, & \text{if } x \in [(a/b)], (m/b)] \\ \dfrac{1-x}{1-(m/b)}, & \text{if } x \in [(m/b), 1] \\ 0, & \text{if } x \leq 1 \end{cases} \qquad (6.88)$$

where
(m/b) = the modal value and
$(a/b), 1$ = the lower and upper bounds of the non-zero values of the membership.

Aggregation of these membership functions results in the system-state membership function of the whole system. Parallel and redundant components are aggregated into a number of serially connected components. For a group of M parallel (or redundant) components, the m-th component has a component-state membership function defined $\tilde{S}_m(u)$ on the universe of discourse U. All the component-states contribute to the system-state membership function. Failure of the group occurs if all components fail. Therefore, the system state is calculated using equation (6.82).

For a system of N serially connected groups, where the n-th group has a state-membership function $\tilde{S}_n(u)$, the weakest component controls the system state or causes the failure of the system. Therefore, the system state is calculated using equation (6.81).

System-failure membership function

The system-failure membership function is used to calculate the fuzzy resiliency index. This membership function represents the system's time of recovery from the failure state. Multi-component systems have several system-failure membership functions representing the system failure for each component. Aggregation of these membership functions results in a system-failure membership function of the system.

Parallel and redundant components are aggregated into serial groups using the fuzzy maximum operator. For a parallel system configuration consisting of M components, the m-th component has a maximum recovery time membership function $\tilde{T}_m(t)$, defined on the universe of discourse T. Therefore, the system-failure membership function (i.e. the membership function that represents the system recovery time) is calculated using equation (6.85). The system-failure membership function for the N serially connected components is calculated using equation (6.83).

6.9.5 Results of the fuzzy reliability analyses

LHPWSS contains 78 components combined in an integrated layout. Component-state and component-failure membership functions are constructed based on data extracted from Earth Tech Canada (2000, 2001), American Water Services Canada (AWSC) (2003a, 2003b), and DeSousa and Simonovic (2003) for the LHPWSS. Three acceptable levels of performance are defined on the universe of the margin of safety. They are expressed as trapezoidal membership functions with values of (0.6, 0.7, 5.0, 5.0), (0.6, 1.2, 5.0, 5.0) and (0.6, 5.0, 5.0, 5.0). They are selected to reflect three different views of decision-makers. Their reliability measures are 4.20, 1.20 and 0.68 respectively. They are referred to as reliable level (level 1), neutral level (level 2) and unreliable level (level 3).

As shown in Table 6.14, the combined reliability–vulnerability measure for LHPWSS is 0.699. Therefore, the reliability of the system is relatively high, taking into account that the system is almost 70 per cent compatible with the highest level of performance (level 1). LHPWSS is highly robust as it has a fuzzy robustness measure of –2.12. This is supported by the fact that increasing the system redundancy, by adding parallel and standby components, increases the capacity of the overall system to meet the demand. A negative robustness value indicates that the compatibility measure after changing the system conditions is larger than before the change. The LHPWSS system has more than 20 parallel groups and 7 redundant components (about 70 components out of the total 78 system components). It must be noted that the plant's individual components are designed with a 35 per cent overload capacity, which results in an increase in the overall system supply and consequently its reliability and robustness.

The combined reliability–vulnerability measure for EAPWSS is 0.042, as shown in Table 6.14. This value is very low, taking into account that the system is only 4 per cent compatible with the highest level of performance (level 1). The fuzzy robustness measure for the EAPWSS is –1.347. This value is the inverse of change in the overlap area, as defined in equation (6.78). EAPWSS has low robustness as

Table 6.14 *The LHPWSS and EAPWSS systems' fuzzy performance measures for different membership function shapes*

Fuzzy performance index	LHPWSS		EAPWSS	
	Triangular	Trapezoidal	Triangular	Trapezoidal
Combined reliability–vulnerability	0.699	0.642	0.042	0.017
Robustness (level 2–level 1)**	NA*	NA*	1.347	3.314
Robustness (level 3–level 1)**	−2.120	−2.473	NA*	NA*
Robustness (level 3–level 2)**	−2.120	−2.473	−1.347	−3.314
Resiliency	0.017	0.017	0.054	0.054

Notes: * Value not available as there is no change in the overlap area.
  ** Robustness based on the change in conditions from the first level, in brackets, to the second level, in brackets.

the overlap area is reduced by more than 74 per cent. Low reliability and robustness are the result of the small number of parallel and redundant components in the system. The EAPWSS system has 16 parallel groups and 4 redundant components (about 54 components out of the total of 66 system components). Unlike LHPWSS, EAPWSS components are not designed with an overload capacity, which significantly reduces its reliability and robustness.

6.9.6 Utility of the fuzzy performance measures

The reliability of a system depends on the reliability of its components. However, not all components are of equal importance. For example, serial components have a more significant effect on the overall system reliability than parallel components, because failure of any serial component leads to a failure of the whole system. Therefore, a system's performance can only be enhanced by improving the performance of critical components. A critical component is a component that significantly affects the system performance, reflected by the area of the system-state membership function, and accordingly the fuzzy performance measures of the system.

The developed computational procedure is used to identify the critical components of the system. The calculation procedure transforms the multi-component system into a system of serially connected components. The fuzzy summation operator is used to aggregate parallel and redundant components into single entities with equivalent component-state membership functions. Then the fuzzy minimum operator is used to sum all serial components and entities into the system-state membership function. The change in the system-state membership function is used to identify critical system components.

In the case of the LHPWSS, for example, the change in the system-state membership function is shown in Figure 6.32. The system-state membership function changes significantly with the addition of the PAC transfer pump. This is the point where the flash mix components are introduced into the system. The improvement of flash mix system components will lead to the improvement of the overall system performance. Looking into the components of the flash mix system, it is found that the PAC transfer pump has the smallest component-state membership function relative to the other flash mix components.

If the supply capacity of the PAC transfer pump is increased, the area of the component-state membership function will increase. This will lead to a direct improvement in the overall system performance. Table 6.15 summarizes the fuzzy performance measures for both cases (before and after changing the PAC transfer pump's supply capacity by 300 per cent). The combined reliability–vulnerability measure increased from 0.699 to 0.988, an increase of 41 per cent. The fuzzy robustness index also increased, from -2.120 to -1.127, indicating a corresponding improvement in the system robustness. Negative robustness values indicate that the compatibility measure, after changing the surrounding conditions, is larger than before the change, as shown in equation (6.78). In other words, the system is more compatible after the change, expressed as a change in the acceptable level of performance, than before it.

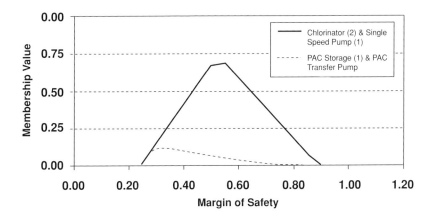

Figure 6.32 *The LHPWSS system-state membership function change for different system components*

This procedure can be extended to identify the optimum improvement of the critical components. Table 6.16 shows three different changes in the supply capacity of the PAC transfer pump and their impact on the system fuzzy performance measures. A

20 per cent increase in the maximum capacity of the PAC transfer pump results in a 30 per cent increase in the combined reliability–vulnerability measure and an increase of 24 per cent in the robustness measure.

Table 6.15 *Change in system fuzzy performance measures due to an improvement in the PAC transfer pump capacity*

Fuzzy performance index	Before supply capacity increase (300%)	After supply capacity increase (300%)
Combined reliability–vulnerability	0.699	0.988
Robustness (level 2–level 1)**	NA*	NA*
Robustness (level 3–level 1)**	−2.120	−1.127
Robustness (level 3–level 2)**	−2.120	−1.127

Notes: * Value not available as there is no change in overlap area.
       ** Robustness based on the change in conditions from the first level, in brackets, to the second level, in brackets.

Table 6.16 *Change in system fuzzy performance measures due to a change in the supply capacity of the PAC transfer pump*

Fuzzy performance index	Percentage change of the supply capacity		
	300%	20%	5%
Combined reliability–vulnerability	0.988	0.921	0.749
Robustness (level 2–level 1)**	NA*	NA*	NA*
Robustness (level 3–level 1)**	−1.127	−1.607	−2.100
Robustness (level 3–level 2)**	−1.127	−1.607	−2.100

Notes: * Value not available as there is no change in the overlap area.
       ** Robustness based on the change in conditions from the first level, in brackets, to the second level, in brackets.

6.10 REFERENCES

Akter, T. and S.P. Simonovic (2005), 'Aggregation of Fuzzy Views of a Large Number of Stakeholders for Multi-objective Flood Management Decision Making', *Journal of Environmental Management*, 77(2): 133–143

American Water Services Canada-AWSC (2003a), *Elgin Area Water Treatment Plant 2003 Compliance Report*, Technical Report, Joint Board of Management for the Elgin Area Primary Water Supply System, London, Ontario, Canada, http://www.watersupply.london.ca/Compliance_Reports/Elgin_Area2003_Compliance_Report.pdf (accessed November 2004)

American Water Services Canada-AWSC (2003b), *Lake Huron Water Treatment Plant 2003 Compliance Report*, Technical Report, Joint Board of Management for the Lake Huron Primary Water Supply System, London, Ontario, Canada, http://www.watersupply.london.ca/Compliance_Reports/Huron003_Compliance_Report.pdf (accessed November 2004)

Ang, H.-S. and H. Tang (1984), *Probability Concepts in Engineering Planning and Design*, John Wiley, Hoboken, NJ

Antunes, C.H. and L.C. Dias (2007), 'Managing Uncertainty in Decision Support Models', *European Journal of Operations Research*, 181(1): 1425–1426

Balopoulos, V., A.G. Hatzimichailidis and B.K. Papadopulos (2007), 'Distance and Similarity for Fuzzy Operators', *Information Sciences*, 177(11): 2336–2348

Bates, G.D. (1992), 'Learning How to Ask Questions', *Journal of Management in Engineering*, 8(1): editorial.

Blishun, F. (1989), 'Comparative Analysis of Methods of Measuring Fuzziness', *Soviet Journal of Computer and System Sciences*, 27(1): 110–126

Brugnach, M., A. Tahh, F. Keil and W.J. deLange (2007), 'Uncertainty Matters: Computer Models at the Science–Policy Interface', *Water Resources Management*, 21(7): 1075–1090

Chang, P.T. and E.S. Lee (1994), 'Ranking of Fuzzy Sets Based on the Concept of Existence', *Computers and Mathematics with Applications*, 27(9): 1–21

Chen, S.H. (1985), 'Ranking Fuzzy Numbers with Maximizing Set and Minimizing Set', *Fuzzy Sets and Systems*, 17: 113–129

DeSousa, L. and S. Simonovic (2003), *Risk Assessment Study: Lake Huron and Elgin Primary Water Supply Systems*, Technical Report, University of Western Ontario, Facility for Intelligent Decision Support, London, Ontario, Canada

Despic, O. and S.P. Simonovic (2000), 'Aggregation Operators for Soft Decision Making in Water Resources', *Fuzzy Sets and Systems*, 115(1): 11–33

Earth Tech Canada Inc. (2000), *Engineers' Report: Elgin Area Primary Water Supply System*, Earth Tech Canada, Inc., London, Ontario, Canada

Earth Tech Canada Inc. (2001), *Engineers' Report: Lake Huron Primary Water Supply System*, Earth Tech Canada, Inc., London, Ontario, Canada

El-Baroudy, I. and S.P. Simonovic (2004), 'Fuzzy Criteria for the Evaluation of Water Resources Systems Performance', *Water Resources Research*, 40(10): W10503

El-Baroudy, I. and S.P. Simonovic (2006), 'Application of the Fuzzy Performance Measures to the City of London Water Supply System', *Canadian Journal of Civil Engineering*,

33(3): 255-266

Ganoulis, J.G. (1994), *Engineering Risk Analysis of Water Pollution*, VCH, New York

Haimes, Y. Y., C. Nicholas, J. Lambert, B. Jackson and J. Fellows (1998), 'Reducing Vulnerability of Water Supply Systems to Attack', *Journal of Infrastructure Systems*, 4(4): 164–177

Hashimoto, T., J.R. Stedinger and D.P. Loucks (1982), 'Reliability, Resiliency, and Vulnerability Criteria for Water Resources System Performance Evaluation', *Water Resources Research*, 18(1): 14–20

Kaplan, S. and B.J. Garrick (1981), 'On the Quantitative Definition of Risk', *Risk Analysis*, 1(1): 165–188

Kaufmann, A. and M.M. Gupta (1985), *Introduction to Fuzzy Arithmetic: Theory and Applications*, Van Nostrand Reinhold Company, New York

Kim, K. and K. Park (1990), 'Ranking Fuzzy Numbers with Index of Optimism', *Fuzzy Sets and Systems*, 35: 143–150

Klir, G.J., U.H. St.Clair and B. Yuan (1997), *Fuzzy Set Theory: Foundations and Applications*, Prentice-Hall, Englewood Cliffs, NJ

Kreps, D.M. (1988), *Notes on the Theory of Choice*, Westview Press Inc., Boulder, CO

Ling, C.W. (1993), *Characterising Uncertainty: A Taxonomy and an Analysis of Extreme Events*, MSc Thesis, School of Engineering and Applied Science, University of Virginia, VA

Lowrance, W.W. (1976), *Of Acceptable Risk*, William Kaufman, Inc., Los Altos, CA

Modarres, M., M. Kaminskiy and V. Krivtsov (1999), *Reliability Engineering and Risk Analysis: A Practical Guide*, Marcel Dekker, Inc., New York

Norwich, M. and I. B. Turksen (1981), 'Measurement and Scaling of Membership Functions', in G.E. Lasker ed., *Applied Systems and Cybernetics*, 6: 2851–2858, Pergamon Press, Oxford

O'Connor, D.R. (2002), *Report of the Walkerton Inquiry: The events of May 2000 and Related Issues*, Queen's Printer for Ontario, Ottawa, Canada

Orlovski, S.A. (1994), *Calculus of Decomposable Properties*, *Fuzzy Sets and Decisions*, Academic Press, New York

Palenchar, M.J. and R.J. Heath (2007), 'Strategic Risk Communication: Adding Value to Society', *Public Relations Review*, 33(2): 120–129

Pedrycz, W. (1995), *Fuzzy Sets Engineering*, CRC Press, Boca Raton, FL

Pedrycz, W. and F. Gomide (1998), *An Introduction to Fuzzy Sets*, MIT Press, Cambridge, MA

Plate, E.J. and L. Duckstein (1988), 'Reliability Based Design Concepts in Hydraulic Engineering', *Water Resources Bulletin*, 24: 234–245

Press, S.J. (2003), *Subjective and Objective Bayesian Statistics: Principles, Models, and Applications*, John Wiley, Hoboken, NJ

Refsgaard, J.C., J.P. van der Sluijs, A.L. Hojberg and P.A. Vanrolleghem (2007),

'Uncertainty in the Environmental Modelling Process – A Framework and Guidance', *Environmental Modelling and Software*, 22(11): 1543–1556

Shrestha, B. and L. Duckstein (1998), 'A Fuzzy Reliability Measure for Engineering Applications', *Uncertainty Modelling and Analysis in Civil Engineering*, CRC Press, Boca Raton, 120–135

Simonovic, S.P. (1997), 'Risk in Sustainable Water Resources', in *Sustainability of Water Resources Under Increasing Uncertainty*, D. Rosbjerg et al, eds, IAHS Publication, 240: 3–17

Slovic, P. (2000), *The Perception of Risk*, Earthscan, London

UNESCO (UN Educational, Scientific and Cultural Organization) (2004) UNESCO water portal, http://www.unesco.org/water (accessed December 2005)

Verma, D. and J. Knezevic (1996), 'A Fuzzy Weighted Wedge Mechanism for Feasibility Assessment of System Reliability During Conceptual Design', *Fuzzy Sets and Systems*, 38: 179–187

Vick, S. G. (2002), *Degrees of Belief: Subjective Probability and Engineering Judgment*, ASCE Press, New York

Yager, R. (1981), 'A Procedure for Ordering Fuzzy Subsets of the Unit Interval', *Information Science*, 24: 143–161

Yager, R. (1988), 'On Ordered Weighted Averaging Aggregation Operators in Multicriteria Decision Making', *IEEE Transactions of Systems, Man, Cybernetics*, 18(1): 183–190

Zadeh, L.A. (1965), 'Fuzzy Sets', *Information Control*, 8: 338–353

Zimmermann, H.J (1987), *Fuzzy Sets, Decision Making and Expert Systems*, Kluwer Academic Publishers, Boston, MA

Zimmermann, H.J. (1996), *Fuzzy Set Theory – and its Applications*, 2nd revised edn, Kluwer Academic Publishers, Boston, MA

6.11 EXERCISES

1   What are the main differences between the probabilistic and the fuzzy set approaches for addressing water resources management uncertainties?
2   Model the following expressions as fuzzy sets:
    a. Large flow.
    b. Minimum flow.
    c. Reservoir release between 10 and 20 cubic metres per second (m^3/s).
3   Assume that X = {1, 1.5, 2, 2.5, 3, 3.5, 4, 4.5, 5} is the set of possible Simon River flows in m^3/s. Using a fuzzy set describe 'available flow' according to the following assumptions:
    a. The available flow membership function is unimodal and convex.
    b. The available flow membership function reaches a maximum value of 1 for

a flow value of 3 m³/s.
c. The available flow membership function reaches a minimum value of 0 for a flow value of 1 m³/sec, as well as for a flow value of 5 m³/s.
d. Show the available flow membership function in a graphical form.

4   What is the support of an available flow fuzzy set?
5   What is the $\alpha = 0.5$ level set of an available flow fuzzy set?
6   Assume that a small municipality of Trident will use water for meeting its domestic water demand from the Simon River. Let the fuzzy set = {(2.5, 0),(6, 1)} represent the Trident municipal water demand. Show the Trident water demand membership function in graphical form.
7   Find the intersection of the Simon River available flow and the Trident city domestic water demand. What is the linguistic interpretation of the obtained fuzzy set?
8   Find the union of the Simon River available flow and the Trident city domestic water demand. What is the linguistic interpretation of the fuzzy set obtained?
9   Find the complement of the Simon River available flow and the Trident city domestic water demand. What is the linguistic interpretation of the fuzzy set obtained?
10 Let $\tilde{A} = \{(1,0.3), (2,1), (3,0.4)\}$ and $\tilde{B} = \{(2,0.7), (3,1), (4,0.2)\}$.
    a. Compute $\tilde{A}(+)\tilde{B}$.
    b. Compute $\tilde{A}(-)\tilde{B}$.
    c. Compute $\tilde{A}(\times)\tilde{B}$.
    d. Compute $\tilde{A}(/)\tilde{B}$.
11 Assess the compatibility of the two fuzzy membership functions from Exercise 10.
    a. Find the possibility $Poss\ (\tilde{A},\tilde{B})$.
    b. Find the necessity $Nec(\tilde{A},\tilde{B})$.
12 Use the flood control example from Section 6.7 and derive a fuzzy membership function for 'water supply'.
    a. Identify experts for water supply to participate in the derivation process.
    b. Develop a hierarchy of 'water supply' similar to the one in Figure 6.16 developed for 'flood control'.
    c. Prepare a questionnaire for the evaluation of 'water supply' by experts.
    d. Start by evaluating membership functions for the sets from the lowest hierarchical level in the 'water supply' concept hierarchy. This evaluation can be done in a separate set of sessions with experts.
    e. Evaluate the components of the 'water supply' concept hierarchy following the procedure presented in Section 6.7.3.
    f. Show the results of your evaluation in tabular form. Use the conversion process (from the scale you selected in c. to the scale from 0 to 1) and show

your results using the 0–1 scale.
13. Aggregate the results of your evaluation from Exercise 12 according to four aggregation methods (from Section 6.6) to develop a membership function for 'water supply'.
   a. Check the empirically obtained aggregations (from Exercise 12 f.) and find out whether any of them show the quality that the aggregated values equals 0 if at least one of the components is equal to 0 (after the conversion of your scale to the scale 0 to 1). If yes proceed with b. If not skip to c.
   b. Perform the aggregation using Zimmermann's γ-operator. Show all membership functions in the concept hierarchy in graphical form.
   c. Perform the aggregation using the OWA operator. Show all membership functions in the concept hierarchy in graphical form.
   d. Perform the aggregation under the pseudomeasures – (CUP). Show all membership functions in the concept hierarchy in graphical form.
   e. Perform the aggregation using polynomial composition under the pseudomeasures – (P-CUP). Show all membership functions in the concept hierarchy in graphical form.
   f. Which aggregation method, in your opinion, provides the best fit for your set of data?
14. Using your own words define a partial failure.
   a. Give an example in hydropower generation.
   b. Give an example in irrigation water supply.
15. Let us consider further modification of the example from Section 6.8.6 by adding a System C as shown below:

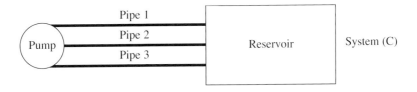

   a. Calculate the combined fuzzy reliability and vulnerability and fuzzy robustness measures for systems A, B and C. Perform your calculation for the two cases listed in the table below, and use three acceptable levels of performance of the safety factor: (i) high-safety level represented by the trapezoidal fuzzy number (0.8, 1.2, 15, 15); (ii) safe level represented by the trapezoidal fuzzy number (0.7, 1.0, 15, 15); and (iii) low-safety level represented by the trapezoidal fuzzy number (0.5, 0.8, 15, 15).

Case	Description	System	Supply capacity (m³/s)	Demand (m³/s)
1	Triangular fuzzy membership with equal distribution between pipes in systems B and C	A	(0.0,3.0,6.0)	(1.5,1.8,4.5)
		B	(0.0,1.5,3.0)	(0.75,0.9,2.25)
			(0.0,1.5,3.0)	(0.75,0.9,2.25)
		C	(0.0,1.0,2.0)	(0.5,0.6,1.5)
			(0.0,1.0,2.0)	(0.5,0.6,1.5)
			(0.0,1.0,2.0)	(0.5,0.6,1.5)
2	Trapezoidal fuzzy membership with equal distribution between pipes in systems B and C	A	(0.0,1.5,4.5,6.0)	(1.5,1.8,3.0,4.5)
		B	(0.0,0.75,2.25,3.0)	(0.75,0.9,1.5,2.25)
			(0.0,0.75,2.25,3.0)	(0.75,0.9,1.5,2.25)
		C	(0.0,0.5,1.5,2.0)	(0.5,0.6,1.0,1.5)
			(0.0,0.5,1.5,2.0)	(0.5,0.6,1.0,1.5)
			(0.0,0.5,1.5,2.0)	(0.5,0.6,1.0,1.5)

b. Compare your results for cases 1 and 2 from the aspects of selected membership function shape and system structure.

7

Water Resources Systems Management for Sustainable Development

Decision-making in water resources systems management has been influenced in the last two decades by the introduction of a sustainability paradigm. It can safely be assumed that sustainability is the major unifying concept promoted, accepted and discussed by governments throughout most of the world. However, as yet there is no consensus on its precise meaning or on how to measure it. The Brundtland Commission's report *Our Common Future* (WCED, 1987) introduced the concept of sustainable development as 'The ability to meet the needs of the present, without compromising the needs of future generations'. This vision of sustainable development may never be realized, but it is clearly a goal worthy of serious consideration. In this chapter we shall focus on ways of measuring relative sustainability – the sustainability of one development or management option in comparison with another (Simonovic et al, 1997).

7.1 PRINCIPLES OF SUSTAINABLE WATER RESOURCES DECISION-MAKING

Planning, design and operational water resources management decisions today are made under complex circumstances of multiple objectives, conflicting interests and the participation of multiple stakeholders. Many of the decision-making problems are subjective, or non-quantifiable, and deal with the dynamic interactions between the human population and natural processes, both internal and external. A common point of intersection among principal decision-makers from science, industry and government is their goal of sustainable development. The ethos of sustainable development not only reinforces but also extends the main principles of decision-making.

There are many definitions of sustainable development. One of the strengths regarding sustainable development discussions is the diversity of views expressed. Without going into a major debate about the definition of sustainable development, we can observe that there is a pressing need to evaluate developmental decisions with respect to sustainability. In spite of that need, it has been extremely difficult to define just what sustainability is in terms more specific than those suggested by the Brundtland Commission.

Applying principles of sustainability to water resources decision-making requires major *changes in the objectives* on which decisions are based and an understanding of the complicated interrelationships between existing ecological, economic and social factors. The broadest objectives for achieving sustainability are environmental integrity, economic efficiency and equity (Young, 1992). Another important aspect of sustainable decision-making is *the challenge of time* (i.e. identifying and accounting for long-term consequences). We are failing to meet the basic water needs of more than 2 billion people today (UNESCO – WWAP, 2003), and therefore are not at the starting point in terms of dealing with the needs of future generations. For some developments, the prediction of long-term consequences is difficult.

The third aspect of the sustainable decision-making context is the *change in procedural policies* (implementation). Pursuing sustainable water project development and selection will require major changes in both substantive and procedural policies. The diverse policy questions raised include: How should the decision-making methods and processes be used? What should be the reliance on market as opposed to regulatory mechanisms? And what should be the role of public and interest groups in decision-making?

Box 7.1 Water and religion

- Water plays a central role in many religions and belief systems around the world. As the source of life, it represents (re)birth. Water cleans the body, and by extension purifies it.
- Water is often perceived as a god, goddess or divine agency in religions.
- Religious water is never neutral and passive. It is considered to have powers and capacities to transform this world, annihilate sins and create holiness. Water carries away pollution and purifies in both a physical and symbolic sense. Water is living and spiritual matter, working as a mediator between humans and gods. It often represents the border between this world and the other.

- **Buddhism**: water is used in Buddhist funerals. It is poured and overflows into a bowl placed before the monks and the dead body. As it fills and pours over the edge, the monks recite, 'As the rains fill the rivers and overflow into the ocean, so likewise may what is given here reach the departed.'
- **Christianity**: water is intrinsically linked to baptism, a public declaration of faith and a sign of welcome into the Christian church. When baptised, one is fully or partially immersed in water. In baptism, water symbolizes purification, the rejection of original sin.
- In the New Testament, 'living water' or 'water of life' represents the spirit of God: that is, eternal life.
- **Hinduism**: water is imbued with powers of spiritual purification for Hindus, for whom morning cleansing with water is an everyday obligation. All temples are located near a water source, and followers must bathe before entering the temple. Many pilgrimage sites are found on river banks; sites where two, or even three, rivers converge are considered particularly sacred.
- **Islam**: for Muslims, water serves above and beyond all for purification. There are three sorts of ablutions. The first and most important involves washing the whole body; it is obligatory after sex, and recommended before the Friday prayers and before touching the Koran. Second, before each of the five daily prayers, Muslims must bathe their heads, wash their hands, forearms and feet. Finally, when water is scarce, followers of Islam use sand to cleanse themselves.
- **Judaism**: Jews use water for ritual cleansing to restore or maintain a state of purity. Hand-washing before and after meals is obligatory. Although ritual baths, or *mikveh*, were once extremely important in Jewish communities, they are less so now. Men attend *mikveh* on Fridays and before large celebrations, women before their wedding, after giving birth and after menstruation.
- **Shinto**: this religion is based on the veneration of the *kami*, innumerable deities believed to inhabit nature. Worship of the *kami* must always begin with a ritual of purification with water. This act restores order and balance between nature, humans and the deities. Waterfalls are considered sacred in Shinto.

Source: UNESCO water portal, http://www.unesco.org/water (last accessed December 2005)

In the implementation of the three principles of sustainable development, one thing is clear: a more specific discussion is needed on how to evaluate development decisions with respect to sustainability. However, in spite of that need, it has been extremely difficult to operationalize sustainability principles. Sustainability is an integrating process. It encompasses technology, ecology and the social and political infrastructure of society. As already indicated, it is not a state that may ever be reached completely. It is, however, one for which water planners and decision-makers strive. While it may never be possible to identify with certainty what is sustainable and what is not, it is possible to develop some measures that permit us to compare the performances of alternative systems with respect to sustainability. Four such measures can be identified here: measures of fairness, risk, reversibility and consensus.

Work on measuring sustainability has been conducted with two main emphases. The first focuses on *sustainability indicators*. Sustainability indicators can be defined as conditions strictly associated with a sustainable development so that their presence is indicative of its existence (Spangenberg, 2004). The second emphasis is on the development of *criteria for measuring sustainability*. In this context a criterion is defined as a standard on which a decision may be based (Sahely et al, 2005). Indicators are quantitative or qualitative variables that can be measured or described and that demonstrate trends. They cannot, however, be used directly in decision-making. Their major role is to fulfil analytical, communication, warning, mobilization and coordination functions. They can be used as a measure of the potential for management considering sustainability. To be useful, the condition shown by various indicators must be compared with a past or a desired future in order to quantify the extent to which projects may contribute to sustainable development.

Four practical sustainability criteria are proposed here for application in water resources management and decision-making:

1 fairness;
2 risk;
3 reversibility;
4 consensus.

Fairness provides a meaningful format for assessing the distribution of benefits. Risk has measurable qualities, provided the proper risk events are identified and the probabilities can be calculated. Reversibility evaluates the degree to which the aggregated set of anticipated or unanticipated impacts of a development project can be mitigated. Consensus as a sustainability metric describes the level in which stakeholders are satisfied with a solution to a problem under consideration.

7.1.1 Enhancement of the decision-making process

Decision-making is not easily defined since it is a process rather than an act. In the field of water resources it involves the following activities (see also Section 3.3 and Goodman, 1984):

1   Establishment of goals and objectives.
2   Problem identification and analysis.
3   Solution identification and impact assessment.
4   Formulation of alternatives and analysis.
5   Recommendations.
6   Decisions.
7   Implementation.
8   Operation and management.

Our focus here is on the analysis phase, which follows the formulation of alternatives. This step includes the selection of criteria and procedures for the comparison of alternatives. An array of systematic procedures and mathematical frameworks have been developed to assist with the evaluation of alternatives. Both optimization and simulation techniques are used in deterministic and stochastic environments to identify the best alternative according to a preselected set of criteria. They are introduced in Chapter 3 and are presented in detail in Chapters 8, 9 and 10.

7.1.2 Multi-objective decision-making

Since sustainability is a function of various economic, environmental, ecological, social and physical goals and criteria (in this text the words 'criteria' and 'objectives' are used interchangeably), analyses must involve trade-offs among multiple objectives in a multidisciplinary and multi-participatory decision-making process. No single discipline, profession or interest group has the knowledge to make these trade-offs. Appropriate trade-offs can only be determined through a political process involving all interested and impacted stakeholders. The participants must at least attempt to take into account the likely preferences of those not present in this decision-making process, namely those who will be living in the future and who will be impacted by current decisions (UNESCO, 1998).

In the last three decades, there has been an increased awareness of the need to identify and consider simultaneously several criteria in the analysis and solution of some problems, in particular those dealing with large-scale systems. Many national and international organizations are providing guidelines for project evaluation that take into account multiple objectives. Project selection and implementation negotiates a best compromise decision where conflicting objectives exist. To combine project impacts into a measure of relative worth so that alternative projects can be ranked clearly involves making compromises among conflicting objectives. Typically, multi-objective decision-making techniques are used to make objectives that were initially non-commensurate comparable with each other, so that a decision can be selected that achieves each of the objectives to some degree.

The performance of the objective and the acceptable trade-off between objectives can be based on the orientation of the decision-makers. This orientation can be expressed as optimism or pessimism with respect to the outcome. Thus the inclusion of multiple objectives, as opposed to a single objective, in the decision-making process broadens and complicates the decision-making framework.

It is possible to conduct a sensitivity analysis on the objectives in the multi-objective decision-making process. To do this, the objectives are given different weights to indicate how significant their effects are on system performance. These weights are then varied and the effect on the overall decision taken is evaluated. The sensitivity analysis may simplify the decision-making process by allowing objectives that do not significantly affect the outcome of the solution to be omitted.

The 'best' solution of a multi-objective decision analysis is the solution that provides the most desirable cross-section of trade-offs between the objectives. Performing a sensitivity analysis on the objectives makes it possible to determine which solutions are most robust. Instead of maximizing or minimizing the expected payoff, a robust solution determines which alternative satisfies most of the objectives, with a favourable outcome most of the time. A robust solution is one that has acceptable impacts for the majority of the objectives. Chapter 10 in Part IV is devoted to the presentation of multi-objective analysis.

Sustainable decision-making requires that the selection process considers the present and the future. The life spans of projects and their long-lasting effects require that the future impacts of decisions be considered. The concept of planning for the future creates a disadvantage for decision-makers by increasing the uncertainty and complexity of the decision-making problem.

7.1.3 Sustainability and other objectives

The decision-making process involves the creation of alternative choices that appear to satisfy developmental objectives, and are financially feasible and institutionally acceptable. Developmental criteria may be categorized as economic and financial, while their implementation must consider social and environmental impact assessments. Decision assessment predicts and evaluates, to the extent possible, all alternative decision impacts, costs and benefits to all affected individuals. Impacts may be seen as direct, indirect or cumulative. *Direct impacts* are changes in system components and processes that result immediately from a decision. *Indirect impacts* are changes in system components and processes that are consequences of direct impacts. *Cumulative impacts* are the aggregates of direct and indirect impacts resulting from two or more decisions in the same area.

Economic objectives

Economic objectives, which are dominant in contemporary water resources decision-making, are derived from cost–benefit analyses. Costs include those necessary to implement a selected alternative, such as investment costs, operation and maintenance expenses, and other direct costs. Benefits are usually divided into direct and indirect benefits. Direct (user, primary or expansion) benefits are defined as the total change in income brought about by a selected alternative. This definition applies to a fully developed economy where no externalities are brought into the analyses. When the major reason for a proposed development is to stimulate economic growth, it is anticipated that the economic benefits will include direct and indirect (induced or secondary) benefits. Economic objectives should also take into account external economies or diseconomies. Credits for the use of unemployed labour may be considered in areas of surplus labour when the prices do not fully reflect the supply–demand relationship of a free market. When the costs and benefits cannot be determined from existing prices because of inadequate or nonexistent markets, it may be appropriate to use shadow prices, as is often done in developing countries. The use of economic objectives ends with the selection of alternatives with positive net economic benefits.

Financial objectives

The results of economic analyses do not provide sufficient information on the financial viability of different development alternatives. Once an alternative is selected based on economic objectives, financial analyses are made to determine the needs for financing the project construction and handling the flows of costs, revenues and subsidies after the project goes into operation. Although these financial criteria are important, they are seldom controlling in the case of federal government agencies. However, for regional or local governments, for private sponsors and others, the financial feasibility and the ranking of alternatives on a financial basis may be more important than the results in economic terms.

Social impact assessments

Societal values and norms are shifting from a position of favouring 'unchecked' economic growth to one of concern for the environment and well-being of people. Social impact assessments have become an integral part of the water resources decision-making process. The success of large developments often depends on an effective balance of local costs and distributed benefits. Social assessments are useful in suggesting appropriate trade-offs between these costs and benefits. Each development is an intervention into a social system. The baseline social conditions are a dynamic system of interactions. Technology can be applied to increase the resilience of the social system or to mitigate the effects of natural events on that system. Social impact assessment assists the decision-making process in developing the profile of this dynamic system, projecting future states with and without the development, and identifying and evaluating the impacts.

Environmental impact assessments

The quantification of the environmental effects of developmental alternatives is already a widely accepted phase of the decision-making process. The main measure used in many countries today is environmental quality. Attributes of environmental quality are the ecological, cultural and aesthetic properties of natural and cultural resources that sustain and enrich human life. *Ecological attributes* are components of the environment and their interactions that directly and indirectly sustain dynamic and diverse ecosystems. *Cultural attributes* are evidence of past and present habitation that can be used to reconstruct or preserve human life. *Aesthetic attributes* are stimuli that provide diverse and pleasant surroundings for human enjoyment and appreciation (Goodman, 1984). In order to effectively utilize environmental objectives in decision-making, a standard, threshold, optimum or other desirable level for an indicator is required that provides the basis for a judgement on whether an effect is beneficial or adverse.

Sustainability criteria

Very limited work has been reported on the development of original objectives capable of capturing the spirit of the definition of sustainable development. The United Nations Educational, Scientific and Cultural Organization (UNESCO) (1998), for example, revisits statistical measures (reliability, resiliency and vulnerability) and recommends their combination as a measure of relative sustainability over time. Duckstein and Parent (1994) suggest measuring sustainability as a combination of high resiliency and low vulnerability.

The sustainability definition emphasizes the integral treatment of three subsystems: economic, social and ecological. The questions raised by the sustainable development perspective of decision-making reveal major gaps in knowledge about the behaviour of diverse natural, economic and human subsystems. In general, developmental decision-making becomes progressively more complex with the growing recognition of the need to consider comprehensive linkages between natural (ecological), economic and human (socio-political–institutional) subsystems. A number of important issues are making sustainable decision-making more challenging. They include:

- expansion of spatial and temporal scale;
- risk and uncertainty;
- multi-objective analysis;
- the participation of multiple stakeholders.

The selection of inter-temporal fairness, risk, reversibility and consensus as sustainability criteria is an attempt to address some of these issues.

With the introduction of sustainable development principles comes the necessity to consider much wider spatial boundaries (*spatial scale*). Systems should be examined as subsystems of biophysical, socio-economic and political systems. The recognition of important interdependencies among subsystems at the international, national and regional levels must be considered. The evaluation of the fair distribution of impacts between different subsystems justifies the choice of fairness criterion.

Expanded spatial boundaries, lengthened timescales and other issues related to sustainable decision-making may contribute to the identification of deficiencies in our knowledge of the behaviour of a wide range of natural and human systems under consideration. If we recognize that many of these deficiencies cannot rapidly be eliminated, this makes it evident that risk and uncertainty are inherent concepts related to sustainable water resources project development and management. Therefore, risk has been proposed as the sustainability criterion to be investigated.

Box 7.2 How much water does it take?

To produce 1 kg of oven dry wheat grain, it takes 715–750 litres of water.
For 1 kg of maize, 540–630 litres.
For 1 kg of soybeans, 1650–2200 litres.
For 1 kg of paddy rice, 1550 litres.
For 1 kg of potatoes, 1000 litres.
For 1 kg of beef, 50,000–100,000 litres.
For 1 kg of clean wool, 170,000 litres.
For 1 litre of gasoline, 10 litres.
For 1 kg of paper, 300 litres.
For 1 metric tonne of steel, 215,000 litres.

Source: Environment Canada, http://www.ec.gc.ca/water/en/e_quickfacts.htm (last accessed December 2005)

Addressing the needs of future generations requires much longer timescales than have traditionally been used. Analysis of long time periods provides the information base for inter-temporal fairness. An extension of the timescale implies the examination of the long-term consequences of proposed developments, and the possibilities for reversing the consequences of past development decisions. A reversibility criterion is therefore an obvious choice for the examination of long-term consequences.

As public involvement in water project development, licensing and operations becomes more vocal, proponents of water resources development will also need to be more aware of public goals. It has become necessary to recognize all significant interests, and consider a full range of options for sound, comprehensive water management to be achieved. This may be accomplished through open collaboration, or within a technical framework for incorporating uncertainty, or a combination of the two. It is proposed here that *consensus* be used as a sustainability metric that describes the level to which stakeholders are satisfied with a solution to a problem. Consensus assumes that an appropriate group of stakeholders is able to collaborate in assessing proposed solutions to environmental problems, or development initiatives. It also assumes that the collective best a group of stakeholders has to offer implicitly provides insight into the needs of future generations.

7.1.4 A methodology for applying sustainability criteria

The addition of four new objectives to be used in decision-making places a much larger weight on replacing single-objective optimization with multi-objective analysis. The importance of considering more than one objective at a time is emphasized by the existence of several conflicting and non-commensurate objectives at every step of the sustainable water resources decision-making process. Multi-objective analysis, discussed in detail in Chapter 10, considers that objectives might change over time as a result of, for example, a change in technology, weather or population. Multi-objective analysis begins with the construction of a *payoff table* (see Table 7.1). In row k of this table, the maximum value of the kth objective is M_k and the associated values of the other objectives are Z^k_i. The set of decision-making objectives is enhanced by the addition of sustainability objectives to the set of criteria Z_p. The last row of the payoff table includes the relative importance of the objectives, W_i, provided by the decision-maker(s).

Sustainable water resources decision-making is subject to four major sources of complexity regarding the application of multi-objective analysis:

1 the addition of sustainability objectives to the payoff table;
2 the quantification of different objectives;

Table 7.1 *Payoff table*

Alternative	Value of *k*th objective/criterion							
	Economic objectives/criteria			Environmental objectives/criteria	Social objectives/criteria			Sustainability objectives/criteria
	Z_1	Z_2	...	Z_k	...	...	...	Z_p
a^1	M_1			Z_k^1				
a^2		M_2		Z_k^2				
...			...	...	...			
a^k	Z_1^k	Z_2^k	...	M_k	...	...	...	Z_p^k
...				...	...	...		
a^p				Z_k^p				M_p
W_i	W_1	W_2		W_k				W_p

3 getting preferences from decision-makers regarding the different objectives;
4 avoiding overlap between different objectives.

7.2 FAIRNESS AS A SUSTAINABILITY CRITERION IN WATER RESOURCES DECISION-MAKING

Fairness, or equity, is an important consideration in the selection of water resources project alternatives. The consideration of the fairness of impacts of a project is important both to ensure the maintenance of social well-being and to secure project acceptance by affected stakeholders. The fairness of a distributive situation can be quantified using a variety of distance-based approaches which result in both intra-temporal and inter-temporal fairness measures. The proposed fairness measure was developed by Matheson (1997) and applied to different examples by Lence et al (1997) and Matheson et al (1997).

There are three possible components or norms of fairness: equality, need and proportionality. *Equality* refers to a uniform distribution of benefits and costs among stakeholders. Such a distribution may be considered fair if there is no basis on which to differentiate between the stakeholder groups. However, an equal distribution is not always an equitable one. A more equitable distribution considers need and proportionality. *Need* addresses the different requirements of each stakeholder group. *Proportionality* requires that an individual user's level of benefit be determined by his or her level of input or contribution towards achieving that benefit. An effective evaluation of fairness considers equality, need and proportionality for distance- and temporally-based distributions of costs and benefits.

7.2.1 Distance-based fairness measures

Distance-based fairness measures have been developed and utilized by a variety of disciplines, from economics and engineering to psychology and other social sciences. However, there is no good method discussed in the literature for choosing the most practical of these for a given situation. Several distance-based measures have been evaluated and recommended as appropriate for the sustainable project evaluation process.

Distance-based measures are grouped according to whether they are essentially measures of proportionality, equality or need. Evaluation of each of the distance-based measures is done using a number of principles, the most important of which are the fundamental principle and the principle of transfers. The *fundamental principle* is summarized as: 'When a group's relative outcome remains constant, the group's outcome should increase monotonically with that group's input.' Basically, a group's outputs should be maintained in proportion to its inputs. The *principle of transfers* specifies that: 'Measures show an improvement in equality when a unit amount of some benefit is transferred from someone better off to someone worse off.'

After evaluation of the proposed fairness measures, two equality-based measures, two proportionality-based measures and four need-based measures were recommended as satisfying most of the principles used to evaluate them. These measures are summarized in Table 7.2.

7.2.2 Temporal distribution-based fairness measures

Temporal considerations, as discussed by the Brundtland Commission and subsequent works, consider inter-generational fairness (equity between generations) and intra-generational fairness (equity within generations). The formulae listed in Table 7.2 were not constructed to account for temporal variation. An expansion of the recommended distance-based measures to incorporate temporal considerations is required. Intra-temporal fairness is evaluated using an appropriate distance-based measure across all groups in a single time period. Inter-generational equity is evaluated by applying one of the distance-based measures over each time period during which the group in question exists. Following the recommendation of Matheson (1997), we shall consider only equality-based measures. Equations (7.1) and (7.2) are the proposed measures used for evaluating intra-generational and inter-generational fairness.

Table 7.2 *Distance-based fairness measures*

Name	Fairness norm	Formula
Walster formula	Proportionality	$\sum_{i=1}^{I}\sum_{j=1}^{I}\left\lvert\dfrac{E(i)-A(i)}{A(i)}\right\rvert$
Equal excess formula	Proportionality, Need	$\sum_{i=1}^{I}\lvert E(i)-A(i)\rvert$
Coefficient of variation	Equality	$\dfrac{\left[\sum_{i=1}^{I}(E(i)-\overline{E})\right]^{1/2}}{\overline{E}}$
Gini coefficient	Equality	$\dfrac{\sum_{i=1}^{I}\sum_{j=1}^{I}\lvert E(i)-E(j)\rvert}{2I^{2}\overline{E}}$
Adams formula	Need	$\sum_{i=1}^{I}\sum_{j=1}^{I}\left\lvert\dfrac{E(i)}{A(i)}-\dfrac{E(j)}{A(j)}\right\rvert$
Coulter method	Need	$\left[\dfrac{1}{I}\sum_{i=1}^{I}\left[\dfrac{E(i)}{\overline{E}}-\dfrac{A(i)}{\overline{A}}\right]^{2}\right]^{1/2}$
Coulter method #1	Need	$\sum_{i=1}^{I}\left[\dfrac{E(i)}{A(i)}-\dfrac{\overline{E}}{\overline{A}}\right]^{2}$
Hoovers concentration index	Need	$\dfrac{1}{I}\sum_{i=1}^{I}\left[\dfrac{E(i)}{\overline{E}}-\dfrac{A(i)}{\overline{A}}\right]$

Notes:
- I = total of all groups evaluated
- i, j = individual group indexes
- $E(i)$ = actual impacts experienced by group I
- $\overline{E}$ = average impact experienced by all groups
- $A(i)$ = impact that group i deserves to receive
- $\overline{A}$ = average impact that all groups deserve to receive.

A lower value for either the intra-generational or inter-generational fairness measure indicates a more equitable alternative. These measures are capable of considering any number of impacts. In the case of only one impact, the impact weighting component (w_g) included in equations (7.1) and (7.2) is not required.

$$B_2(x) = \frac{1}{GI}\sum_{t=1}^{T}\sum_{g=1}^{G}\left[\frac{w_g\sum_{i=1}^{I}\sum_{j=1}^{I}\lvert E(i,g,t,x)-E(j,g,t,x)\rvert}{2I^2\overline{E}_{gtx}}\right] \qquad (7.1)$$

$$B_2^1(x) = \frac{1}{GI} \sum_{g=1}^{G} \left\{ w_g \sum_{i=1}^{I} \left[\frac{\sum_{s=1}^{T} \sum_{t=1}^{T} |E(i, g, s, x) - E(i, g, t, x)|}{2T^2 \overline{E_{gtx}}} \right] \right\}$$ (7.2)

where:
$B_2(x)$ = intra-generational fairness measure which is the weighted sum of deviations from an equal distribution of impacts
$B_2^1(x)$ = inter-generational fairness measure which is the weighted sum of deviations from an equal distribution of impacts
G, I, T, X = number of different impacts, number of groups, number of time steps and number of alternatives
i, g, t, x = indices for group, impact type, time step and alternative
j, s = group and time indexes that are required for pairwise comparisons
w_g = weights on impact types
$E(i, g, t, x)$ = magnitude of impact type g acting on group i during time timestep t that results from alternative x
$\overline{E_{gtx}}$ = average impact over all groups for a given combination of impact type, time step and alternative
$\overline{E_{igx}}$ = average impact over all time steps for a given combination of group, impact type and alternative.

7.2.3 Practical considerations for intra-generational and inter-generational fairness measures

Application of the intra-generational or the inter-generational fairness measures may be complicated by inaccurate or uncertain predictions of future societal values, impact distributions and types, and group demographics (Lence et al, 1997). Due to changing societal values, the viewpoints on which distributive fairness is based may change with time. For example, as an economy develops, the applicable fairness measure may shift from a need to an equality to an equity-based measure. Rather than predicting the values of future generations, and applying a fairness measure that reflects such values, it may be more useful to select projects based on their robustness in terms of meeting all fairness viewpoints. That is, the distributive fairness measures for satisfaction of equality, equity and need can be evaluated and projects can be selected based on the individual values, the sum, the average or the weighted average of measures for these.

The magnitude of the impacts on different groups and the number of impacts that affect each group may change over time. The impacts of a given project alternative may evolve as a result of the cumulative effects of other projects, or of changes in the world in the future; for example, an ecological impact may change to an economic effect. Some of these impacts may be easier to predict than others. Even in the present some impacts may be better known, relative to other impacts, for a given project alternative than for another, or better known under one project alternative than under another. Furthermore, the uncertainty in the impacts may have different degrees of importance depending on the impact type. It may be easier to accept uncertainty in income distribution than in contamination from toxic spills, for example.

Uncertainty in impact distributions and types may be addressed by developing significantly different scenarios of the future and of the impacts of project alternatives, based initially on the projects that are currently being planned. The length of time examined may depend on the impact types being examined. If the impact is the use of a resource that is being depleted, and it is a non-renewable resource, a long time horizon may be necessary. Thus, the choice of time length relates to reversibility, because if a project is irreversible the impacts far in the future need to be estimated, but if it is reversible, this may not be necessary. Using scenarios of the future, we could evaluate the fairness measures under all scenarios and select a project alternative based on these, or we could develop distributions of the expected impacts, based on different scenarios, and use the expected value of the impacts in the evaluation of the fairness measure. Since the relative uncertainty about the future impacts may vary depending on the impact type and time of occurrence – uncertainty may increase with time – the weights (w_g) may be selected to reflect this.

The composition and relative sizes of the groups, and even their existence over long time horizons, are not known with certainty, and changes in these will influence the groups' contributions to projects, their characteristics and their needs. Some of these changes may be easier to predict than others, and some may depend on whether the groups are defined based on the spatial, demographic or physical characteristics of the project region. Again, weights may be used in such cases to indicate the degree of confidence in the definition of group attributes and needs over time. Regardless of this, equity and need-based fairness measures in which impacts are compared with group attributes or needs may have greater prediction error than equality-based measures because of the potential for change in several variables. This gives further justification for including all fairness viewpoints in the selection of projects.

Finally, the set of required and desirable characteristics for intra-generational fairness measures may need to be revised or expanded for analysing appropriate inter-generational measures. Other desirable characteristics that may be introduced for evaluating inter-generational fairness measures might include the ability to incorporate the relative importance of the impacts and the relative confidence in predictions of those impacts.

7.3 RISK AS A SUSTAINABILITY CRITERION IN WATER RESOURCES DECISION-MAKING

Risk exists when there is the possibility of negative social, environmental or economic impacts associated with a project. Risk is estimated as the product of the magnitude of the negative consequence and the probability of occurrence of the consequence. Risk can be estimated using combinations of historical and empirical data, heuristic knowledge and cultural perceptions. A composite measure of risk is influenced by the weighting that is given to the various components of the risk measure. A proposed risk criterion has been developed by Kroeger (1997) and Kroeger and Simonovic (1997). Additional resources on sustainable water resources management under uncertainty can be found in Chang (2005) among others.

The general definition of risk, as introduced in Chapter 6, can be based on the concept of load and resistance. In many other fields these two variables have a more general meaning. *Load* (l) is a variable reflecting the behaviours of the system under certain external conditions of stress or loading. *Resistance* (r) is a characteristic variable that describes the capacity of the system to overcome an external load. When the load exceeds the resistance ($l > r$) there should be a failure or an incident. A safety, or reliability, state is obtained if the resistance exceeds or is equal to the load ($l \leq r$).

7.3.1 Operational risk definition

Risk is traditionally defined as being the product of the magnitude of an event or act and the probability of that event occurring. There is a shortcoming associated with this definition, in that it fails to address the difference between the products of low-probability–high-magnitude events and high-probability–low-magnitude events. An operational definition of risk can be developed based on the principles of sustainable development, existing definitions of risk and how these definitions are applied to decision-making.

> Risk is present when the possibility exists of a negative social, environmental or economic consequence. A negative consequence may take the form of a negative impact, or the possibility of missing a positive impact. The risk can be represented in an estimate by the product of the consequence's value and the probability of its occurrence. The estimate can be determined using historical and empirical data, heuristic knowledge and cultural perceptions. The estimate is as plausible and reliable as its individual components. The emphasis given to the components in the estimate influences the evaluation of risk. The resulting representation, or measure, and acceptability of the risk are relative to the specific circumstances analysed and the perception of those affected by the risk.

This definition of risk acknowledges that risk is a function of societal and environmental values, as well as economic efficiency. In order to rejuvenate damaged social and environmental systems the chance of missing a positive impact, or opportunity, is also perceived as a risk to sustainable development. The risk is estimated using the probability of its occurrence and the value of the consequence. The reliability of a risk estimate is affected by the extent to which an effort has been made to understand the risk. A poor understanding of a risk can be expected to result in a poor estimate of its consequences. Better communication between the public and decision-makers contributes to better understanding of a risk and thus a better estimate of its consequences.

In keeping with the three facets of sustainable development, we shall take into consideration social, economic and environmental risks. Unfortunately, many of the variables are measurable only on a subjective scale. Also, different stakeholders are likely to have different ideas of what is an acceptable risk. Therefore the risk model requires that the different stakeholder groups be identified and consulted as part of the risk assessment process.

The risk measure algorithm (Kroeger and Simonovic, 1997) to be used as one of the sustainable project evaluation criteria consists of the following steps:

Step 1: Identification of the risks that contribute to the analysis.
Step 2: Estimation of the probability of the risks occurring in each alternative.
Step 3: Calculation of the risk value for each risk by each participant, using sustainable development category weights and risk weights.
Step 4: Estimation of risk separately for each alternative and each participant.
Step 5: Comparing the alternatives by combining the participant risk estimates in a joint risk estimate.

The first step in this process requires the identification of all risks (r_c). The second step, estimation of risk probabilities, is subjective. Experts in the relevant field should be consulted for their assessment of the risk probability ($p_{r_c i} = P\{r_c\}_i$). The third step requires the consultation of as many stakeholder groups as can be identified. Each group is given the opportunity to weight both the importance that they attach to each of the sustainable development categories (d_{cj}) and also their willingness to accept each of the identified risks (k_{rcj}). This is a significant improvement over traditional risk assessment methodologies, in that individual stakeholders are actually consulted and their values and preferences are used in the analysis. In Step 4, each alternative receives a ranking from each of the participant groups. The two types of information obtained from the stakeholders are combined using:

$$v_{r_cj} = \frac{d_{cj} \times k_{r_cj}}{\sum d_{cj}} \quad \forall\, j, r, c \tag{7.3}$$

where:
v_{r_cj} = the value for risk r_c for participant j
d_{cj} = the sustainable development category weight for participant j
k_{r_cj} = the risk weight assigned to risk r_c by participant j.

Once the risk values assigned by the participants are calculated, the risk values are multiplied with the probability values.

$$e_{ij} = \sum p_{r_ci} \times v_{r_cj} \quad \forall\, i, j, r, c \tag{7.4}$$

where e_{ij} is the risk estimate of alternative i relative to participant j.

A lower risk value indicates a less risky alternative. In Step 5, the values for different participant groups may be combined according to an average or weighted average, to obtain one aggregate ranking for each alternative.

7.4 REVERSIBILITY AS A SUSTAINABILITY CRITERION IN WATER RESOURCES DECISION-MAKING

Reversibility is viewed as a measure of the degree to which the aggregated set of anticipated and unanticipated impacts of a project can be mitigated. Development projects that are highly reversible should allow the users of the affected system to continue their normal use. A high degree of reversibility requires the imposition of the least amount of disturbance to the natural environment. The reversal of adverse effects is often not technically feasible, but mitigation plans, or the provision of substitute resources, can help to reduce the negative effects. A proposed reversibility criterion has been developed by Fanai (1996) and applied by Fanai and Burn (1997).

7.4.1 Reversibility criterion algorithm

Ideally, the reversibility should be high so that any negative implications that may result from present decisions can be mitigated in the future. An operational concept to be used here is that decisions that are highly reversible result in the stakeholders being able to maintain their traditional uses of the system. The algorithm for calculating the reversibility criterion consists of four stages:

Stage 1: Select impacts and characteristics and classify from general to specific.
Stage 2: Determine the units of measurement for each impact and determine the expected value in each scenario, along with the best and worst possible values.
Stage 3: Apply R-metric.
Stage 4: Perform sensitivity analysis.

The first stage in the process is the most important and the most time-consuming. This stage involves identifying the impacts and classifying them as economic, environmental or social, as well as indicating whether they are true impacts or characteristics of impacts. After all the possible impacts are identified, it becomes important to determine how each of the impacts can be measured (Stage 2).

Some impacts may be easily quantifiable, and at this point in the analysis the units of measurement should be specified. Often, however, the impacts are not easily quantifiable and must be measured on a subjective scale. In these cases, experts may be consulted to provide a qualitative estimate of the impacts. The reversibility criterion requires that the best and worst possible values for each impact be known or estimated. The expected value for each scenario is then either derived or predicted.

The third stage of the reversibility framework involves the application of an analytical formulation to the set of quantified impact values and weights. The impacts or their characteristics are used as metrics to derive a measure of the reversibility of a scenario. Because the impacts are not measured on a common scale, they must be either rendered commensurate or converted into dimensionless numbers. Commensuration is accomplished by employing a simplified version of the distance metric, termed the *R-metric*:

$$R_{cj} = \left(\sum_{i=1}^{N_c} w_{ci}^2 \left| \frac{M_{ci} - f_{cij}}{M_{ci} - m_{ci}} \right|^2 \right)^{1/2} \qquad (7.5)$$

where:
c = index for category ($c = 1$ for environmental, 2 for economic, 3 for social)
j = index for alternative
R_{cj} = reversibility index in category c for alternative j
i = index for impact
N_c = total number of impacts in category c
w_{ci} = weight assigned for impact i in category c (scale from 0–1)
M_{ci} = best value for impact i in category c
m_{ci} = worst value for impact i in category c
f_{cij} = expected value of impact i from implementing alternative j in category c.

The inputs M_{ci}, m_{ci} and f_{cij} are quantified in Stage 2. The impact weights, w_{ci}, allow the researcher to account for the fact that all impacts may not be of equal importance. The weighting values are selected on a scale from 0 to 1 depending on the perceived importance of each impact. After the R-metric is calculated for each scenario, the threshold values can be calculated. The threshold values set the boundaries of the R-metric. The minimum reversibility, Tc_{min}, is simply equal to 0 while the reversibility limit, Tc_{max}, is calculated by assigning all the impacts their worst possible values. These values are useful for comparison purposes as they establish the range of possible values for each scenario. A perfectly reversible alternative would then receive an R-metric value of 0, as all the impacts are at their ideal value. A larger R-metric value represents a more irreversible (less desirable) alternative.

7.4.2 The sensitivity analysis

The final stage in the reversibility framework is the sensitivity analysis. This is important because of the subjective nature of the values assigned to the impacts. This analysis helps to evaluate how changes in the subjective impact values affect the final index. The first sensitivity analysis involves individual manipulation of the qualitative impact values to judge their effect on the outcome of the R-metric. For this analysis, the researcher will replace individual values and recalculate the R-metric, comparing the outcome of the original calculation with the altered version. The second sensitivity analysis addresses the issue of the impact weightings. Recall that the original impact weightings are assigned by the researcher according to the perceived importance of each impact to stakeholders in the region. Admittedly this process leaves considerable room for the researcher's personal bias to enter the analysis. The second sensitivity analysis is designed to address this issue. This analysis involves assigning random weight values to the impacts and recalculating the R-metric. If the random weights significantly alter the results, then the robustness of the original calculation is brought under scrutiny.

7.5 CONSENSUS AS A SUSTAINABILITY CRITERION IN WATER RESOURCES DECISION-MAKING

Consensus, as a concept for promoting sustainability in water resources decision-making, is a criterion quite unlike any of the others previously described. Consensus has no units of measurement. It is measured in a brief moment of time, but may implicitly consider future events and uncertainties. Consensus is a high-level indicator, dependent on value judgements which may in turn depend on lower-level indicators derived from facts concerning problem characteristics. It applies only to decision-making situations where more than one decision-maker is involved. The definition for consensus in *Webster's Dictionary* is 'a general agreement in opinion'. It relies on a qualitative and subjective opinion, and the qualifying condition is a general agreement. How much do decision-makers need to agree? There may not be an adequate answer to this question, but a consensus approach may indicate more than just when to stop an exploration of alternatives. A proposed consensus criterion was developed by Bender (1996) and used by Bender and Simonovic (1997).

7.5.1 Definition of consensus

Sustainability can be defined in many ways, as was pointed out earlier. If consensus leads to sustainability, what is consensus but sustainability in an operational form? Let us start by giving the following definition for consensus as it relates to sustainability:

> Consensus is an equitable compromise which is robust with regard to (a) resource management uncertainties, and (b) stakeholder perspectives.

Water resources management uncertainties include data uncertainty, model uncertainties and technological uncertainties. Stakeholder perspectives are related to the value systems of relevant decision-makers. This definition is not yet operational, but its constituent parts might be manageable. Some assumptions also need to be made. It is assumed that the appropriate stakeholders have been included in the decision-making process. By stakeholder, we refer to interested parties who may be impacted in some way by any decision that is made (a political choice). The second major assumption is that all stakeholders voluntarily cooperate in the decision-making process.

It is implicit in an idea of consensus sustainability that the needs and values of future generations are some equitable combination of the values and needs exhibited by today's generation. The overwhelming set of externalities and uncertainties associated with caring for future generations cannot be so easily minimized by this last assumption. However, an appropriate set of stakeholders could bring all of the important issues to bear on the decision, circumventing economic models which do not handle externalities.

7.5.2 Development of a consensus measure

A consensus measure of sustainability requires a method for ranking alternative choices. One of the common methods based on the distance metric is Compromise programming (Zeleny, 1982), which is presented in detail in Chapter 10. All alternatives are ranked according to their distance metric values. The alternative with the smallest distance metric is typically selected as the 'best compromise solution'. The following mathematical formulation is used to compute the distance metric values (L_j) for a set of n objectives and m alternatives.

$$x_j = \left[\sum_{i=1}^{n} w_i^p \left| \frac{f_i^* - f_{i,j}}{f_i^* - f_{i,w}} \right|^p \right]^{1/p} \tag{7.6}$$

where:
x_j = distance metric
f_i^* = optimal value of the ith objective
$f_{i,j}$ = value of the ith objective for alternative j
$f_{i,w}$ = worst value of the ith objective
w_i = weights indicating decision-maker preferences with respect to different objectives
p = parameter ($1 \leq p \leq \infty$).

In equation (7.6), each objective is to be given a level of importance, or weight w_i, provided by the decision-makers. The parameter p is used to represent the importance of the maximal deviation from the ideal point. If $p = 1$, all deviations are weighted equally; if $p = 2$, the deviations are weighted in proportion to their magnitude.

In order to apply this distance metric technique, decision-makers must choose a weight to describe the importance of each objective. Unfortunately, there are several decision-makers, all with their own priorities. As each decision-maker uses his or her individual set of weights, the rankings may change. The choice of alternative is no longer a straightforward decision which results in a strong ordinal ranking. Each set of weights and each choice of distance measure provide a strong ranking, but there are uncertainties in ranking related to subjective priorities.

In a consensus-based approach for achieving sustainability through decision-making, the decision process becomes iterative, using an extra step to evaluate progress in discussions among decision-makers. The distance metrics used previously can be used to assess the degree of consensus among decision-makers. The following are five measures for the degree of consensus found in the literature (Kuncheva, 1994).

$$\gamma^1 = 1 - \min_{j \neq k} \left| w_j x_j - w_k x_k \right|, j,k = 1,...n \tag{7.7}$$

$$\gamma^2 = 1 - \min \left| w_j x_j - w_k x_k \right|, j,k = 1,...n \tag{7.8}$$

$$\gamma^3 = 1 - \frac{1}{n}\sum_{j=1}^{n-1}\left|w_j x_j - u\right| \tag{7.9}$$

$$\gamma^4 = 1 - \frac{2}{n(n-1)}\sum_{j=1}^{n-1}\sum_{k=j+1}^{n}\left|w_j x_j - w_k x_k\right| \tag{7.10}$$

$$\gamma^5 = 1 - \max\left|w_j x_j - u\right|, j = 1,\ldots n \tag{7.11}$$

and:

$$u = \frac{1}{n}\sum_{j=1}^{n} w_j x_j$$

where:
- n = number of decision-makers
- x_j = distance metric value for decision-maker j
- w_j = provides parametric control and possible weighting of decision-makers
- $\gamma^l \in [0\ 1]$ = degree of consensus measure for an alternative, indexed by $l \in [1, 5]$.

Of course, some care must be taken to preserve a consistent and meaningful mathematical form in γ^l. That is, distance metrics (x) and weights on decision-makers (w) must be set appropriately. To be of use, γ^l must operate on the range [0,1], with due regard to the sensitivity of selecting unequal weights for decision-makers. Weights would normally be set at $w_j = 1$.

The highest coincidence measure (γ^1) checks, for each alternative, whether any decision-makers agree on the rank (the distance metric value in our case). $\gamma^1 = 1$ if at least two decision-makers agree on the rank (actually, the value of the distance metric).

The highest discrepancy measure (γ^2) checks whether any decision-makers disagree on the distance metric value of an alternative. The two decision-makers who disagree most strongly are chosen to represent the consensus measure. $\gamma^2 = 1$ if all decision-makers are in agreement.

The integral mean coincidence measure (γ^3) records the (average) variability of disagreement among decision-makers, using the average distance metric value (u) as the basis for summation. $\gamma^3 = 1$ if all decision-makers are in complete agreement.

The integral pairwise coincidence measure (γ^4) cycles through comparisons of every possible pair of decision-makers, measures any discrepancy, and computes an average value. $\gamma^4 = 1$ if all decision-makers are in complete agreement. γ^4 is very similar to γ^3, but provides slightly different information about the same general aspect of consensus. Instead of expecting an average distance metric value and focusing on decision-makers with extreme views (such as with γ^3), γ^4 gives a better indication of the relative grouping of decision-makers.

The integral highest discrepancy measure (γ^5) focuses on the single most extreme perspective, using an average distance metric value as the basis for judging extremes. $\gamma^5 = 1$ if all decision-makers are in complete agreement.

Each measure for the degree of consensus illuminates or captures a different aspect of consensus. The three coincidence measures focus on identifying common ground. The two discrepancy measures are focused on identifying sources of disagreement. Besides the provision of numerical feedback to the decision process, decision-makers can be identified as supportive or otherwise, including the identification of significant pairs of decision-makers.

The degree of consensus indicates the relative strength of ranking. That is, the worst alternative may have a high degree of consensus because everyone agrees that it is the worst alternative! The result is a weak ordering of alternatives, and complete transitivity may not be achieved.

Using consensus as the measure for sustainability, decision-makers have the opportunity to explore their values with different sets of weights to find a robust solution. Decision-makers also have an opportunity to evaluate the strength of their decisions as well as progress in negotiations. The encouragement of iterative, interactive feedback to a negotiation process is motivated by possible spontaneous creativity in resolving differences of opinion. Other searches may identify clusterings or groupings of individuals on the basis of their ranking. An advanced use of the degree of consensus may even identify aspects of the system as candidates for adaptation, as an attempt to improve the non-dominated frontier of solutions towards more sustainable solutions.

7.5.3 Discussion

Degree of consensus, as a measure for achieving sustainability, calculates the level of agreement between the set of interested or affected stakeholders about the judgement of rank for each alternative. The iterative process that this measure promotes may also provide insight into specific issues on which to focus the planning of water resources use or development.

A consensus sustainability approach may not be capable of calculating the correct answer (is there a 'correct answer'?). Instead, consensus measures provide sources of feedback designed to assist in the following ways:

- whittling down the number of appropriate alternatives;
- identifying sources of disagreement;

- tracking the progress of negotiations;
- adding additional insight to our perceived degree of robustness.

As the concept of sustainability continues to evolve, more and more attempts are being made to provide relevant feedback to decision-makers, as opposed to trying to calculate sustainability itself.

7.6 THE SUSTAINPRO COMPUTER PROGRAM

The SUSTAINPRO computer program enclosed on the accompanying CD-ROM provides support for the computation of four sustainability measures discussed in previous chapters. SUSTAINPRO is a computer package that contains four programs for the computation of four sustainability measures. In the SUSTAINPRO folder on the CD-ROM you will find a read.me file that provides instructions for the installation of the program. In addition, the Examples sub-folder contains all the examples from the text. All programs have been developed using the user-friendly Windows™ environment, with a detailed tutorial provided in the form of an interactive menu system. A first-time user is advised to browse through the menu system before starting to use the program.

7.7 MANAGEMENT OF THE ASSINIBOINE DELTA GROUNDWATER AQUIFER (MANITOBA, CANADA): CASE STUDY

This section presents a case study of groundwater use in the Assiniboine delta aquifer (ADA) region for testing sustainability criteria. This study was completed by McLaren (1998) and reported by McLaren and Simonovic (1999a, 1999b). The ADA is centred on the town of Carberry, Manitoba, approximately 50 km east of Brandon (see Figure 7.1, Plate 10). The aquifer extends over an area of 3885 square kilometres (km^2). The estimated capacity of the aquifer is 12 million acre-ft (14.6 billion cubic metres (m^3) with an annual recharge capacity of 166,000 acre-ft/yr (201.5 million m^3/yr). Of the annual recharge capacity, 106,000 acre-ft/yr (128.7 million m^3/yr) is considered developable and 50 per cent of this, or approximately 58,000 acre-ft/yr (64.4 million m^3/yr), is made available for allocation. This 58,000 acre-ft/yr (64.4 million m^3/yr) represents the maximum allowable allocation under current Manitoba legislation and policy. In 1999 when this study was conducted, water in the ADA was allocated beneath this level (Render, 1988).

Major human water use from the aquifer can be divided into three broad

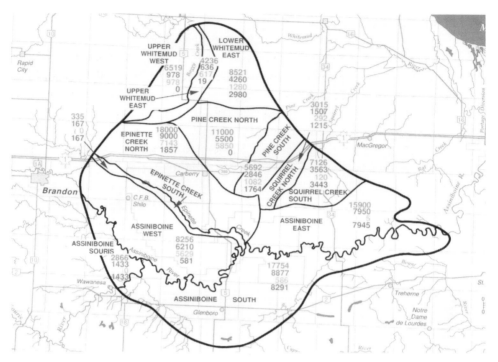

Note: See Plate 10 for a colour version.
Source: courtesy of the Province of Manitoba, 1998

Figure 7.1 *The Assiniboine delta aquifer, Manitoba, Canada*

categories: irrigation, industrial and domestic. Kulshreshtha (1994) estimates that 11 per cent of the water withdrawn for human purposes is used in industrial processes, 20 per cent for domestic uses and 69 per cent for irrigation. It is also noted that the major threats to the aquifer are overuse and point and non-point pollution.

Manitoba Water Stewardship (formerly the Department of Natural Resources) currently monitors water levels in the aquifer. In 1994, the Manitoba Crop Diversification Centre (MCDC) initiated a small-scale monitoring programme to quantify the chemicals present in water from the aquifer. Both programmes provide baseline data for evaluating sustainability.

The ADA provides the region with water which is valuable for a number of purposes. Farmers, industry and the local people all have an interest in using this resource sustainably. Four proposed criteria of sustainability were used to assess the sustainability of current groundwater uses, as well as to measure the success of any future sustainable development initiatives.

7.7.1 Case study objectives

As discussed in the previous sections, the four sustainability criteria can be used as part of a larger multi-objective decision-making process to evaluate the sustainability of different project alternatives. For this case study, no particular project proposal was at hand so instead the alternatives took the form of different policy scenarios. Each policy scenario represented a unique approach to managing the aquifer, and consisted of a set of programmes, policies and legislation designed to reflect a different set of goals and objectives for water resource allocation and use. Care and attention was taken to ensure that the three policy scenarios were as realistic and practical as possible. However, there were some assumptions placed on the development of the policy scenarios which may have affected their accuracy. First, all of the policy scenarios were set under the assumption that they would take place in similar climatic years – climate undoubtedly has a major role in the regulation of aquifer water levels. Second, the policy scenarios were defined only in terms of actions that could be taken directly by humans for irrigation and domestic uses. Environmental requirements like transpiration and wildlife uses were not considered.

7.7.2 Data needs

Assessment of sustainability is data intensive (Bagheri and Hjorth, 2005). Demographic data specific to the ADA region are difficult to obtain. Most provincial data in Canada are collected on a rural municipality (RM) basis. For the purposes of this case study, RM figures were modified based on the land area within the ADA boundaries. Table 7.3 summarizes the land area considered to fall within the ADA boundaries. Data collection for this study fell into three broad categories (McLaren and Simonovic, 1999a):

1   Existing data collected by government or other agencies.
2   Primary data collection of information previously uncollected.
3   Expert opinions and values.

Existing data

This source of data was used wherever possible. All of the data used for establishing the parameters of the policy scenarios and much of the environmental and economic data used for calculating the sustainability criteria were of this type. Data in this category were collected from agencies such as the Manitoba Provincial Government Departments of Agriculture, Health and Natural Resources, the Prairie Farm Rehabilitation Administration, the Manitoba Crop Diversification Centre, P.M. Associates Ltd and Mid-West Foods Ltd.

Table 7.3 *Rural municipality areas included in the ADA region*

Rural municipality	% of land area considered part of the ADA region
North Cypress	100
South Cypress	100
Langford	60
Victoria	50
Cornwallis	40
Lansdowne	40
North Norfolk	40
Elton	20
South Norfolk	20

Primary data collection

Primary data collection was necessary for aspects of the risk consensus criteria. Collection of this information was undertaken during a series of workshops conducted in June 1998 (McLaren, 1998). Three workshops were held and attended by irrigators, environmentalists and government employees (federal and provincial). Eligible participants were identified through organization membership lists and transcripts of previous public meetings on issues concerning the aquifer. Participants were sent an invitation by mail and attendance was voluntary.

Expert opinion

Some data requirements for the risk and reversibility criteria were not easily measurable. In these cases, such as the risk probabilities or impacts measured on a qualitative scale for the reversibility criterion, an expert in the field was consulted to provide an informed estimate for the value. This was perhaps the weakest source of data used in this case study because it allowed personal bias to affect the analysis. However, expert opinion was only used where other estimates or indicators were absent, and expert opinion, as used in the reversibility criterion, was subjected to a sensitivity analysis to determine the impact of changes in the inputs on the outcome of the analysis. The results of the sensitivity analyses are discussed later in this section.

Table 7.4 lists the data types used for this case study and the sources of the data. The second part of the implementation strategy involved establishing the parameters of the alternatives being compared using the criteria. The following section discusses the qualitative and quantitative aspects of each of the policy scenarios.

Table 7.4 *Data requirements and sources for application of sustainability criteria in the ADA region*

Data type	Criteria				Source
	Fairness	Risk	Reversibility	Consensus	
Change in arable land			X	X	Census data
Water quality		X	X	X	Expert estimated and workshop responses
Streamflow and water table levels			X	X	Water Resources Branch, Manitoba Natural Resources*
Erosion impact		X	X	X	Expert estimate and workshop responses
Irrigated potato acreage			X	X	P.M. Associates survey
Livestock populations			X	X	Census data
Processing employment figures			X	X	Mid-West Food
Domestic water availability			X	X	Expert estimate
Landscape aesthetic values		X	X	X	Expert estimate and workshop responses
Stakeholder value of increased agricultural revenue		X		X	Workshop responses
Stakeholder value of increased employment		X X		X X	Workshop responses
Stakeholder value of increased infrastructure costs		X		X	Workshop responses
Stakeholder value of increased recreation opportunity		X		X	Workshop responses
Stakeholder value of wildlife habitat		X		X	Workshop responses
Water licensing figures	X			X	Water Resources Branch, Manitoba Natural Resources*

Note: * currently Water Resources Branch, Manitoba Water Stewardship.

7.7.3 Alternative policy scenarios

The policy scenarios used for the case study were developed by modifying water use data in four broad categories: domestic, irrigation, industrial and livestock. This section provides a qualitative description of each scenario along with a brief outline of the policy tools which might be used to achieve the water budget.

Status quo scenario

The status quo scenario represented the prevailing water use policy of 1996–1997. As such it was a result of policy tools and legislation from 1996 such as the Water Rights Act. Tables 7.5 to 7.7 list the demographic and agricultural statistics by RM. Human water use was based on an estimate of 100 gallons/person/day (378.5 litres/person/day) and irrigation water use was estimated at 6 in (15.4 cm) irrigation coverage. Livestock water use was calculated using average rate values provided by the American National Research Council's 'Nutrient Requirements of Domestic Animals' series (National Research Council, 1985, 1988, 1989, 1994, 1996).

Table 7.5 *Population of ADA region by RM and township*

Municipality	Population
North Cypress	1671
South Cypress	673
Langford	370
Victoria	724
Cornwallis	1280
Lansdowne	379
North Norfolk	1323
Elton	274
South Norfolk	238
Carberry	1544
Glenboro	746
CFB Shilo	1213
Total	**10,435**

Source: Manitoba Health (1997)

Table 7.6 *Irrigated crop areas within the ADA region*

Crop	Irrigated acreage (1 acre = 4,046.8 m²)
Potatoes	6353
Wheat	610
Rye	600
Beets	100
Grass	30
Forage	5
Barley	22
Bent grass	6
Canola	31
Mixed vegetables	108
Linola	64
Total	**8029**

Source: P.M. Associates (1996)

Table 7.7 *Livestock populations by RM*

Rural municipality	Cattle	Poultry	Hogs	Sheep	Horses
North Cypress	22,549	88,184	21,597	498	268
South Cypress	11,202	0	19,209	0	2298
Langford	4552	16	1423	211	201
Victoria	4542	23	5146	47	650
Cornwallis	1848	32,127	2239	335	266
Lansdowne	6238	0	5450	126	315
North Norfolk	8972	34,780	7766	180	529
Elton	1865	32,520	2242	336	266
South Norfolk	2375	0	4578	50	30
Total	**64,143**	**187,650**	**69,650**	**1783**	**4823**

Source: Manitoba Agriculture (1998)

The current water allocation system in Manitoba is based on the western prior appropriation model. Prior appropriation starts from the premise that the first person either to put water flowing in defined channels or to put percolating water to beneficial use acquires an enforceable water right. Subsequent appropriators also obtain rights, but these are subject to prior appropriators receiving their full share. As flow diminishes in periods of drought, appropriators are required to close their intakes in reverse order from the date of the first appropriations. This allocation system originated in the western regions of the United States during the late-19th century gold rush as an alternative to the traditional riparian allocation model (Lucas, 1990).

Water licensing is administered under the Manitoba Water Rights Act, according to the following use hierarchy:

1. domestic;
2. municipal;
3. agricultural (non-irrigation);
4. industrial;
5. irrigation;
6. other uses.

This status quo scenario establishes a baseline condition which is modified through different policy changes in the other two scenarios.

Development scenario

The development scenario represented a policy change which encourages expansion and continued development. Under this scenario, more money is earmarked for expanding agricultural facilities and funding water development infrastructure. Water licensing restrictions are eased and agricultural production, both crop and livestock, increases. The Mid-West processing plant is taken to operate at its economic short-run capacity. The accelerated economic growth of the region attracts more residents, increasing the population. All these developments increase the demand on the water budget of the aquifer, the quantitative description of which is included later in this chapter.

Conservation scenario

The conservation scenario represented a policy change that emphasizes waste minimization and frugality. Under this scenario the population remains stable but domestic use decreases through behavioural change induced by a public education campaign. Livestock populations and associated water use decrease slightly.

Irrigation, particularly of potatoes, decreases resulting in somewhat lower yields. These lower yields impact the processing plant, resulting in lowered production. The single greatest change between the conservation scenario and the status quo scenario is a revision of the water licensing procedure.

For the purposes of this case study, the conservation scenario considered an alternative water allocation system, for irrigation only, that took the form of a tradable water share system. It should be noted that at this stage the water share system was merely a proposal and had received no formal approval from any agency with a water rights jurisdiction. The tradable water share system was a modification of the tradable emissions permit system used for some environmental amenities in the US. The province and the community were taken to enter into a co-management agreement for managing the aquifer. The province retained the title to the resource, but relinquished management and use rights to the community for a period of 25 years. The Assiniboine Delta Aquifer Advisory Board (ADAAB), or a similar entity, was taken to be responsible for the day-to-day management of the aquifer.

It was taken that under the new administration, allocations within sub-basins did not exceed the current limits imposed by the Water Resources Branch (WRB) of Manitoba Water Stewardship. Each water share was valid for the five-year term of each management plan submitted by the ADAAB. Within that five-year term, shareholders were allowed to transfer their shares within their sub-basin. Thus water shares could be bought by landowners who required more than their initial allocation, from landowners who had more shares than they needed. At the end of each five-year management term, landowners had to reapply to the ADAAB for the next period. This allowed new landowners, or previous landowners who wished to begin irrigating, the opportunity to obtain a water share.

7.7.4 Quantitative description of scenarios

Now that we have discussed the qualitative elements of the policy scenarios, we can focus on the quantitative aspects of the water budget under each policy regime. A detailed modelling of the aquifer water budget was not done. It should be noted therefore that these water budgets represented only one possible physical manifestation of the policy structure. Care was taken, however, to ensure that the water budgets did represent a feasible set of circumstances. Table 7.8 shows the population and agricultural statistics projected to exist under each of the policy alternatives, while Table 7.9 presents the human use water budget for the aquifer under each scenario.

The water budgets are a function of both the demographic statistics and the use rates associated with each activity. In the development scenario, human and livestock populations increase by 3 per cent but the use rate associated with each

Table 7.8 *Projected agricultural and population statistics under three policy scenarios*

	Status quo	Development	Conservation
Population	10,931*	11,259	10,931
Poultry	187,650**	197,033	183,897
Cattle	64,143**	67,350	62,860
Hogs	69,650**	73,133	68,257
Sheep	1783**	1872	1747
Horses	4823**	5064	4727
Irrigated acres (1 acre = 4046.8 m²)	8029***	8832	8029

Sources: * Manitoba Health (1997); ** Manitoba Agriculture (1998); *** P.M. Associates (1996)

Table 7.9 *Projected water budgets under three policy scenarios*

Use category	Status quo	Development	Conservation
Domestic	1224.4	1261.2	979.5
Livestock	693.8	728.5	679.9
Irrigation	4014.5	4415.9	2673.7
Industrial	2016.0	2217.0	1814.4
Total	**7948.7**	**8622.6**	**6147.5**

Note: measured in acre-ft/yr. 1 acre-ft/yr = 1214.04 m3/yr.

individual in those populations remains the same. Similarly, the irrigated crop acreage increases by 10 per cent, but the average use rate remains constant at 6 in (15.4 cm). To accommodate the increase in agricultural yield, the processing plant also increases its use by 10 per cent. In the conservation scenario, the human population remains stable, but domestic water use decreases. For the purposes of this case study, a more conservative consumption decrease of 20 per cent is used. The conservation scenario also considers a 2 per cent decrease in the livestock population and a reduction in the average irrigation rate to 4 in (10.2 cm). The reduced yield in the potato crop then causes the processing plant to reduce output, resulting in an associated decrease in its water consumption.

7.8 APPLICATION OF SUSTAINABILITY CRITERIA

Now after considering the qualitative and quantitative implications of each policy scenario, let me discuss the application of the four sustainability criteria.

7.8.1 Fairness

The current groundwater allocation system in Manitoba employs a use hierarchy system and the 'first in time, first in right' principle as the equitable standard between users. However, for the purposes of this case study, fairness was modelled using the availability of groundwater to landowners for irrigation as an indicator. A major assumption in this work is that all landowners who wish to spend the initial investment necessary for irrigation should have equal opportunity to obtain a water right. Conditions were examined for two of the thirteen sub-basins of the ADA, Pine Creek North and Lower Whitemud East. Pine Creek North was chosen because it is currently allocated above the level set by the WRB. To contrast with this situation, the Lower Whitemud East sub-basin was chosen because as of 1997 it had some irrigation development but still had considerable amounts of water available for allocation.

Both inter- and intra-generational fairness were considered by modelling the access opportunity of landowners in different sub-basins both in the first year of each policy scenario and over a 30-year planning horizon. Information for the first year of the model was based on the 1997 licensing data from the WRB. Subsequent users were added by extrapolating the 1997 user application rate (calculated from the previous ten-year average) for the status quo scenario. This user application rate was then modified for the development and conservation scenarios. Allocations were also based on the historical averages for each sub-basin. Users who might exist in the future after a sub-basin resource was fully allocated were still considered in the analysis but were assigned a zero value for their allocation. This allowed the equity analysis to consider those users who would apply for a water licence but could not receive one because the sub-basin resource was allocated to its capacity. Table 7.10 shows the number of users in each sub-basin for each year under the different policy scenarios. It is important to note that the users in the first year are users who held, or had made application for, a water licence as of 1 January 1997. Users in other time periods represent potential users, who might desire a licence in that time step. In some cases, such as the status quo and development scenarios in the Pine Creek North sub-basin, these new users have no access to a water licence because the sub-basin is fully allocated under those policy scenarios.

Table 7.11 summarizes the results of the intra-generational and intergenerational fairness analysis respectively for the two sub-basins. The values in

Table 7.10 *Number of users in each year in two sub-basins under different policy scenarios*

Year	Pine Creek North sub-basin lower			Whitemud East sub-basin		
	Status quo scenario	Development scenario	Conservation scenario	Status quo scenario	Development scenario	Conservation scenario
1	45	45	45	11	11	11
2	48	50	48	13	13	12
3	53	55	52	14	15	13
4	57	61	56	17	16	15
5	59	65	58	18	17	16
6	63	71	61	19	18	19
7	66	74	64	20	20	20
8	70	78	68	20	21	21
9	75	83	70	21	24	23
10	78	87	74	24	25	24
11	81	90	79	25	26	25
12	85	92	83	27	27	26
13	88	97	86	27	29	27
14	92	101	89	28	30	29
15	96	106	93	31	31	29
16	98	111	97	32	34	30
17	103	113	99	33	35	31
18	106	116	103	34	36	33
19	109	121	106	35	38	35
20	113	125	109	36	39	35
21	118	131	112	37	41	36
22	122	136	118	39	42	37
23	126	141	122	40	43	37
24	132	144	124	41	43	39
25	137	150	128	42	45	40
26	140	156	131	44	48	41
27	146	160	134	45	49	41
28	152	163	139	46	52	43
29	157	167	143	47	53	44
30	160	172	145	48	55	45

these tables were generated using equations (7.1) and (7.2) from Section 7.2.2, with the number of impacts (*G*) being 1 (the individual user's access to a water right), the number of groups (*I*) being equal to the number of individuals in each time step (as listed in Table 7.10), the number of time steps (T) being set at 30 and the number of alternatives (X) being 3, representing the status quo, development and conservation scenarios. The test statistics $B_2(x)$ (intra-generational fairness) and $B'_2(x)$ (inter-generation fairness) are impacted by both the number of users in a given time step and the difference in the size of water licences allocated to different users.

The review of inter-generational fairness in Table 7.11 shows that the conservation scenario receives a higher (less equitable) ranking in both sub-basins compared with the development or status quo scenarios. In fact, both the status quo and development scenarios receive a zero value because no user's access to water is changing during the planning horizon. This indicates that under the current legislation and policy directives, users have somewhat more security in knowing that the size of their water allocation will remain the same for several time periods into the future.

The comparison of intra-generational equity summarized in Table 7.11 shows that the conservation scenario receives a much lower (more equitable) rating than either the development or status quo scenarios. In fact the magnitude of the difference between the conservation scenario and the development and status quo scenarios suggests that it is the allocation procedure used in the conservation scenario that is responsible for most of the difference in the fairness rating, and not the number of users in each scenario. Table 7.12 summarizes the intra-generational equity calculations for each time period, policy scenario and sub-basin. At five-year intervals, the conservation scenario receives a zero value as the equalization period sets in under the tradable water share system. The Lower Whitemud East sub-basin receives zero values from the fifth to the twenty-third time period because it is not yet fully allocated during these periods and therefore new users can obtain a water share outside of the regular five-year planning interval. The five-year equalization period, where all interested landowners are given a water right based on the amount of land they own, results in a more equitable intra-generational distribution.

Table 7.11 *Intra- and inter-generational fairness calculations for access to irrigation licensing for two sub-basins of the ADA*

Fairness	Pine Creek North			Lower Whitemud East		
	Status quo scenario	Development scenario	Conservation scenario	Status quo scenario	Development scenario	Conservation scenario
Intra-generational	0.56	0.60	0.09	0.15	0.19	0.02
Inter-generational	0.00	0.00	0.25	0.00	0.00	0.09

Table 7.12 *Intra-generational equity calculations for two sub-basins of the ADA under different policy scenarios*

Year	Pine Creek North			Lower Whitemud East		
	E(SQ)	E(DE)	E(CO)	E(SQ)	E(DE)	E(CO)
1	0.15	0.15	0.15	0.12	0.12	0.12
2	0.20	0.24	0.21	0.11	0.11	0.11
3	0.27	0.31	0.27	0.11	0.11	0.11
4	0.33	0.38	0.32	0.09	0.11	0.11
5	0.36	0.42	0.00	0.10	0.11	0.00
6	0.40	0.46	0.05	0.10	0.11	0.00
7	0.42	0.49	0.09	0.10	0.11	0.00
8	0.46	0.51	0.15	0.10	0.11	0.00
9	0.49	0.55	0.17	0.11	0.11	0.00
10	0.50	0.57	0.00	0.10	0.10	0.00
11	0.53	0.58	0.06	0.10	0.10	0.00
12	0.55	0.59	0.11	0.10	0.10	0.00
13	0.57	0.61	0.14	0.10	0.10	0.00
14	0.59	0.63	0.17	0.10	0.09	0.00
15	0.60	0.65	0.00	0.11	0.09	0.00
16	0.61	0.66	0.04	0.11	0.10	0.00
17	0.63	0.67	0.06	0.11	0.12	0.00
18	0.64	0.68	0.10	0.10	0.15	0.00
19	0.65	0.69	0.12	0.10	0.19	0.00
20	0.66	0.70	0.00	0.10	0.21	0.00
21	0.68	0.71	0.03	0.12	0.25	0.00
22	0.69	0.72	0.08	0.17	0.27	0.00
23	0.70	0.73	0.11	0.19	0.29	0.00
24	0.71	0.74	0.12	0.21	0.29	0.05
25	0.72	0.75	0.00	0.23	0.32	0.00
26	0.73	0.76	0.02	0.26	0.36	0.01
27	0.74	0.76	0.04	0.27	0.37	0.01
28	0.75	0.77	0.08	0.28	0.41	0.09
29	0.76	0.77	0.10	0.30	0.42	0.10
30	0.76	0.78	0.00	0.31	0.44	0.00
Sum	16.86	18.03	2.79	4.42	5.77	0.71
$B_2(x)$	**0.56**	**0.60**	**0.09**	**0.15**	**0.19**	**0.02**

Notes: SQ = status quo scenario; DE = development scenario; CO = conservation scenario.

Whether the equalization period occurs every year, every five years or every ten years is not as important as the fact that an equalization period seems to have a greater impact on the overall value of the fairness calculation than the number of users present in each period.

7.8.2 Risk

The value attached to a particular risk varies with the value systems of the decision-maker, analyst or person subject to the risk. Valuation of risk is therefore a highly sensitive undertaking because of the subjective personal nature of the issue. The risk criterion used in this case study took this into account by considering both traditional risk probability assessments and the values and preferences of stakeholder groups. The list of impacts was generated through brainstorming, a review of previous studies of the region and informal conversations with landowners and government officials. It is likely that it does not represent a complete list of all conceivable impacts associated with each policy scenario, but it does summarize the major concerns. A list of possible risks associated with the different management regimes for the ADA is shown in Table 7.13.

As is apparent from the table, it is important to consider not only the likelihood and magnitude of negative impacts, but also positive impacts which may be forgone or decreased under different scenarios. For the purposes of this study, impacts that might generally be considered negative are marked (–) while generally positive impacts are marked (+).

Table 7.13 *Possible risks associated with the three policy scenarios*

Risk name	Risk type
Erosion (–)	Environmental
Loss of wildlife habitat (–)	Environmental
Point-source water pollution (–)	Environmental
Streamflow variability (–)	Environmental
Increased agricultural revenue (+)	Economic
Increased employment (+)	Economic
Increased infrastructure costs (–)	Economic
Increased recreation opportunity (+)	Social
Aesthetic changes (–)	Social

Erosion

This impact considers the risk of increased wind or water erosion. Increased erosion leads to loss of topsoil and can ultimately lead to losses in crop yields. Erosion is an important issue to farmers, and since farming is the primary economic activity in the ADA region, it concerns all residents. Increased erosion could be associated with development, land clearing or increasing traffic.

Loss of wildlife habitat

Wildlife habitat could be lost to woodlot clearing or wetland draining for development purposes (municipal, agricultural, infrastructure or other reasons). This impact could also reflect increasing disturbance or degraded quality of habitat as a result of exotic species, noise or contaminants. This risk impact is included because the ADA region contains some of the last examples of tall grass prairie habitat in Manitoba. Several environmentalists and area residents were concerned about the impact that increasing development would have on unique species and habitat.

Point-source water pollution

This impact reflects the risk of increased water pollution associated with fuel, industrial or agricultural spills. Such pollution could affect surface or groundwater. Currently, the water quality in the ADA is excellent and most area residents rely on water from the aquifer for all their domestic uses. Replacing the water supply from the aquifer used for domestic purposes would be extremely costly.

Streamflow variability

Some residents expressed concern that surface water bodies, particularly smaller ones like the Squirrel or Pine Creek, could be impacted negatively by increased use of groundwater. Some of the small creeks in the ADA region have run dry in drought years. Several area residents are concerned about the stability of surface water flows being compromised by increased groundwater usage.

Increased agricultural revenue

This impact considers the risk of forgoing possible increases in farm gate receipts for crops or livestock. Increased development would tend to increase the likelihood of increased agricultural production while some of the measures associated with a water conservation policy would be likely to decrease agricultural production. Agriculture is the primary source of economic revenue for the region. Policies or programmes that could impact agricultural revenue, either positively or negatively, are of considerable interest to many area residents.

Increased employment

Increased agricultural revenue would be likely to lead to increased employment. Such increases could be directly attributable to agriculture in the form of processing or farm labour jobs, or indirectly attributable, in the service, health and education sectors. This positive impact would be lessened in a scenario that emphasized conservation measures. Agriculture is the economic foundation of the region, and policies that impact agricultural jobs will have a ripple effect through the rest of the local economy.

Increased infrastructure costs

Increasing development comes at the price of greater wear and tear on both private and public capital. Roads and machinery would have to be repaired or replaced more frequently, and taxes or operating costs could increase as a result. Several residents expressed concern that with increased development, certain individuals or sectors would make greater use of public facilities. There was also concern that the cost of this increased demand on public services would be borne by all residents, in the form of higher municipal and provincial taxes.

Increased recreation opportunities

Development could also lead to the expansion or improvement of recreation opportunities. Higher farm incomes would also mean more time and money for recreation. Slowing development would be likely to result in fewer recreation opportunities.

Aesthetic changes

Development generally necessitates changes to the landscape. Clearing areas for building or agriculture changes the character of the area. The species composition of some areas may change. Some residents stated that they felt the character of their region had been negatively impacted by development. Quantification of such concern is difficult, but this impact was included to give respondents the opportunity to express their concern over such aesthetic changes.

As presented in Section 7.4, the risk criterion requires three types of data: the risk probabilities, sustainable development category weights and the risk value preferences of different stakeholder groups. Risk probabilities under each scenario were estimated by a professional in the water resources management field who was familiar with the case study.

Personal risk preferences and sustainable development category weights were obtained from three identified stakeholder groups: government managers, irrigators and environmentalists. Government managers were identified as specific employees of the Manitoba Departments of Agriculture, Natural Resources, Environment and Rural Development as well as employees of the Prairie Farm Rehabilitation Administration and the Manitoba Crop Diversification Centre. Irrigators were identified as members of the Assiniboine Delta Aquifer Irrigators Association. 'Environmentalists' is a broad stakeholder category that included farmers from the region who do not currently irrigate, non-farming landowners and members of environmental non-government organizations (NGOs). McLaren (1998) pointed out that these three categories of stakeholders provide a representative cross-section of the different opinions and value systems held with respect to the management of the aquifer.

Individuals identified as belonging to a certain stakeholder group were invited by letter to one of a series of workshops. One workshop was held for each stakeholder group. The workshop format was considered to be the most appropriate way to obtain the risk preference and sustainable development category weight data.

The workshops began with a brief introduction to the study, including a discussion of the purpose and objectives of the study. A general discussion on aquifer management issues important to the participants accounted for most of the workshop time. The quality of the discussion was high. Issues discussed at the workshops ranged from the feasibility of alternative water distribution systems and the effectiveness of local political structures to the economic and social benefits of irrigation and environmental integrity. At the end of each workshop, participants were asked to assign a value to each of the three sustainable development categories (environmental, economic and social), such that the total value for all three was 100. Participants were also asked to respond to a brief survey which asked them to rank the value they attached to a particular risk on a scale from 0 to 10. For negative risks, marked (−), 0 indicated the respondent had no particular aversion to experiencing the risk impact, while 10 meant the respondent would prefer to avoid the risk impact at all costs. For positive risks, marked (+), 0 indicated the respondent was not concerned about achieving the benefit associated with the impact, while 10 indicated that the respondent wished to achieve the benefit at all costs. It is important to note that respondents were asked to fill out the surveys according to their individual views and preferences. The responses should therefore not be interpreted as representing the official viewpoint of any organization or agency to which an individual respondent may belong.

Tables 7.14 and 7.15 list the participant risk weights and sustainable development category weights obtained from the surveys administered during the workshops. It should be noted that the values presented here are the average values of the respondents in each stakeholder category.

The values in these two tables were converted to the participant risk values by means of equations (7.3) and (7.4). The individual risk values were first divided by the total for that risk class and stakeholder group. This modified risk value was then multiplied by the sustainable development category weight assigned by the stakeholder group to that particular risk class. The resulting number was then divided by 100. Table 7.16 shows the results of these calculations. The probabilities of each risk occurring were estimated using the scale in Table 7.17 by a professional from the Prairie Farm Rehabilitation Administration in the field of water management, after reading the scenario descriptions.

For negative impacts, marked (−), the probability estimate reflects the likelihood of the risk impact occurring. For positive impacts, marked (+), the probability estimate indicates the probability that the impact will be forgone. Table 7.18 summarizes the probability estimates.

Table 7.14 *Average risk preferences for three stakeholder groups in the ADA*

Risk	Risk class	Government employees	Irrigators	Environ-mentalists
Loss of wildlife habitat (−)	Environmental	5.2	6.8	7.7
Increased erosion (−)	Environmental	6.0	7.8	7.3
Increased water pollution (−)	Environmental	8.0	7.8	9.0
Streamflow variability (−)	Environmental	5.8	6.0	7.7
Total		**25.0**	**28.4**	**31.7**
Increased agricul. revenue (+)	Economic	5.8	7.8	5.0
Increased employment (+)	Economic	6.0	6.0	7.7
Increased infrastruct. costs (−)	Economic	4.8	5.0	6.0
Total		**16.6**	**18.8**	**18.7**
Aesthetic change (−)	Social	4.4	5.8	7.3
Increased recreation (+)	Social	3.6	7.5	4.3
Total		**8.0**	**13.3**	**11.6**

Table 7.15 *Average sustainable development categories weights as indicated by three stakeholder groups in the ADA*

Sustainable development category	Government employees	Irrigators	Environmentalists
Environment	39.8	21.3	48.3
Economy	32.6	65.0	30.0
Social	27.6	13.8	21.7

Table 7.16 *Scaled risk preferences for three stakeholder groups in the ADA*

Risk	Government employees	Irrigators	Environmentalists
Loss of wildlife habitat (–)	0.08	0.05	0.12
Increased erosion (–)	0.10	0.06	0.11
Increased water pollution (–)	0.13	0.06	0.14
Streamflow variability (–)	<0.01	<0.01	0.12
Increased agricultural revenue (+)	0.11	0.27	0.08
Increased employment (+)	0.12	0.21	0.12
Increased infrastructure costs (–)	0.09	0.17	0.10
Aesthetic change (–)	0.15	0.06	0.14
Increased recreation (+)	0.12	0.08	0.08

Table 7.17 *Qualitative scale and quantitative equivalent used for estimating risk probabilities in the ADA*

Qualitative	Quantitative	Qualitative	Quantitative
None	0.00	Moderate	0.50
Negligible	0.05	Moderately high	0.65
Very low	0.10	High	0.80
Low	0.20	Very high	0.95
Moderately low	0.35	Certain	1.00

Table 7.18 *Summary of risk probability estimates for three policy scenarios in the ADA*

Risk	Status quo	Development	Conservation
Loss of wildlife habitat (–)	0.35	0.65	0.20
Increased erosion (–)	0.20	0.35	0.10
Increased water pollution (–)	0.35	0.50	0.35
Streamflow variability (–)	0.05	0.05	0.05
Increased agricultural revenue (+)	0.65	0.20	0.80
Increased employment (+)	0.80	0.50	0.95
Increased infrastructure costs (–)	0.20	0.35	0.10
Aesthetic change (–)	0.20	0.20	0.20
Increased recreation (+)	0.80	0.65	0.80

The risk probabilities were multiplied by the participant risk values developed earlier. Tables 7.19 to 7.21 summarize the risk criteria analysis for the three policy scenarios, while Table 7.22 provides an average of the participant weights for each of the three scenarios.

Table 7.19 *Risk estimates for status quo scenario in the ADA*

Risk	Government employees	Irrigators	Environmentalists
Loss of wildlife habitat (–)	0.03	0.02	0.04
Increased erosion (–)	0.02	0.01	0.02
Increased water pollution (–)	0.04	0.02	0.05
Streamflow variability (–)	0.00	0.00	0.01
Increased agricultural revenue (+)	0.07	0.18	0.05
Increased employment (+)	0.09	0.17	0.10
Increased infrastructure costs (–)	0.02	0.03	0.02
Aesthetic change (–)	0.03	0.01	0.03
Increased recreation (+)	0.10	0.06	0.06
Totals	**0.40**	**0.50**	**0.38**

Table 7.20 *Risk estimates for development scenario in the ADA*

Risk	Government employees	Irrigators	Environmentalists
Loss of wildlife habitat (–)	0.05	0.03	0.08
Increased erosion (–)	0.03	0.02	0.04
Increased water pollution (–)	0.06	0.03	0.07
Streamflow variability (–)	0.00	0.00	0.01
Increased agricultural revenue (+)	0.02	0.05	0.02
Increased employment (+)	0.06	0.10	0.06
Increased infrastructure costs (–)	0.03	0.06	0.03
Aesthetic change (–)	0.03	0.01	0.03
Increased recreation (+)	0.08	0.05	0.05
Totals	**0.36**	**0.35**	**0.39**

Table 7.21 *Risk estimates for the conservation scenario in the ADA*

Risk	Government employees	Irrigators	Environmentalists
Loss of wildlife habitat (–)	0.02	0.01	0.02
Increased erosion (–)	0.01	0.01	0.01
Increased water pollution (–)	0.04	0.02	0.05
Streamflow variability (–)	0.00	0.00	0.01
Increased agricultural revenue (+)	0.09	0.22	0.06
Increased employment (+)	0.11	0.20	0.12
Increased infrastructure costs (–)	0.01	0.02	0.01
Aesthetic change (–)	0.03	0.01	0.03
Increased recreation (+)	0.10	0.06	0.06
Totals	**0.41**	**0.55**	**0.37**

Table 7.22 *Average risk preferences for three stakeholder groups in the ADA*

Participant group	Status quo scenario	Development scenario	Conservation scenario
Government employees	0.40	0.36	0.41
Irrigators	0.50	0.35	0.55
Environmentalists	0.38	0.39	0.37
Participant average	**0.43**	**0.37**	**0.44**

A lower value for the risk criterion indicates an option that is preferred as less risky. Comparison of the three tables indicates that on average, the three participant groups showed a preference for the development scenario as the least risky alternative. The status quo scenario was generally considered somewhat more risky than the development scenario. Government employees and irrigators considered the conservation scenario the most risky, while environmentalists found it marginally less risky than the development scenario. The preference for the development scenario could indicate a general desire among those surveyed for more support for economic expansion and diversification. Table 7.23 separates the average risk values for positive and negative impacts under each policy scenario for each participant group.

Table 7.23 *Breakdown of positive and negative risk estimates*

Scenario	Participant group	Positive	Negative
Status quo	Government employees	0.26	0.14
	Irrigators	0.41	0.09
	Environmentalists	0.21	0.17
Development	Government employees	0.16	0.20
	Irrigators	0.20	0.15
	Environmentalists	0.13	0.26
Conservation	Government employees	0.30	0.11
	Irrigators	0.48	0.07
	Environmentalists	0.24	0.13

Table 7.23 indicates that the risk of forgoing positive impacts (generally associated with development) contributes the most to the risk criterion. Although the conservation scenario receives the lowest risk value (least risky) associated with negative impacts, the greater likelihood of forgoing positive impacts results in a higher overall risk rating.

Data for the computation of the risk measure for the ADA case study is located on the accompanying CD-ROM, in the directory SUSTAINPRO, subdirectory Examples, file ADA.sdrisk.

7.8.3 Reversibility

The steps involved in calculating the reversibility criterion are laid out in Section 7.4. The first step in the process is the identification of possible impacts associated with development. As noted earlier, some impacts are directly measurable on a quantitative scale while others have to be measured indirectly, on a qualitative scale. Table 7.24 summarizes the identified impacts and the associated units of measurement for this case study. The list of impacts was generated through brainstorming, a review of previous studies of the region and informal conversations with landowners and government officials. It is likely not to represent a complete list of all conceivable impacts associated with each policy scenario, but it does summarize the major concerns of the stakeholders.

Table 7.24 *Reversibility criteria: possible impacts*

Impact name	Impact type	Units
Loss of arable land	Ecological	Acres/year
Water quality	Ecological	Qualitative scale
Riparian water needs	Ecological	Qualitative scale
Erosion	Ecological	Qualitative scale
Irrigated crop production	Economic	cwt/acre
Livestock population	Economic	No. animals
Direct employment – processing	Economic	No. employees
Water availability – domestic	Social	Qualitative scale
Landscape aesthetics	Social	Qualitative scale

Note: acre = 4046.8 m²; cwt = potato production in hundredweight = 0.04536 ton.

The weights necessary for the application of the R-metric, on a scale from 0 to 1, were selected through consultations with the stakeholders based on the perceived importance of each impact. A sensitivity analysis on the impact that the assigned weights have on the R-metric was carried out as part of the analysis. Many of the impacts are not easily quantifiable. Rather than exclude these impacts from the

Note: A barrage is a hydraulic structure, a gated dam. Barrages were typically built on the River Nile or its branches with the purpose of elevating upstream water levels so that all the intakes into the branching irrigation canals were fed gravitationally. This old barrage on the Damietta branch is very close to the delta apex. It was built in 1863 to guarantee perennial irrigation of the delta without pumping.

Source: photo courtesy of Dr Hussam Fahmy

Plate 1 *An old barrage on the Damietta branch of the River Nile*

Source: photo courtesy of Dr Dejiang Long

Plate 2 *Main drainage canal in the Sihu basin*

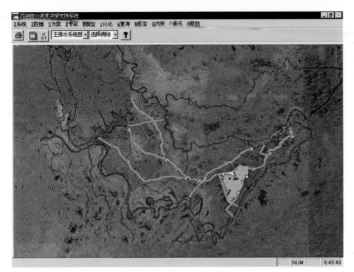

Plate 3 *Main interface of the Sihu flood management decision support system*

Upstream level: 235.1 m – near peak. Downstream level: 232 m. Floodway flow: 1841 m^3 per second. Flow in the Red River (above Floodway) 3905 m^3 per second. Flow in the Red River (below Floodway) 2066 m^3 per second.

Source: photo courtesy of the Province of Manitoba, 1997

Plate 4 *Red River Floodway inlet, 4 May 1997*

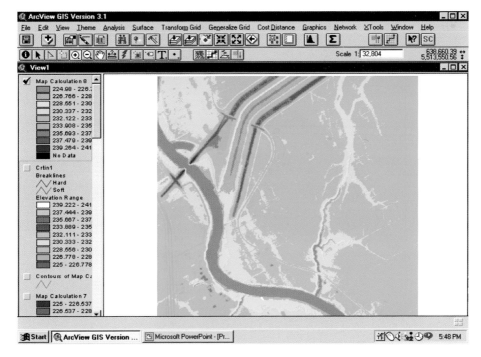

Plate 5 *LIDAR data showing the Winnipeg Floodway inlet on the Red River*

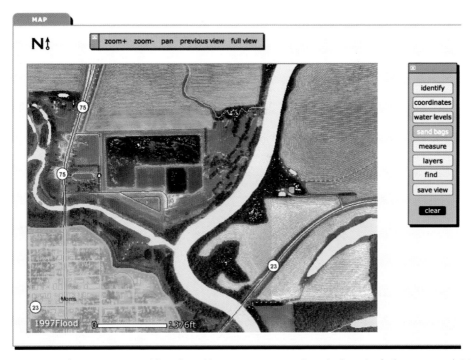

Source: Manitoba water stewardship, http://www.geoapp.gov.mb.ca/website/rrvft (last accessed 11 June 2006)

Plate 6 *A prototype of the Red River basin flood decision support system*

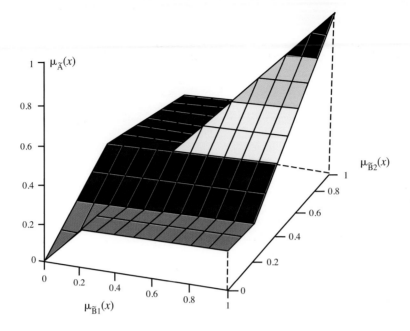

Plate 7 *Composition under pseudomeasure for $\eta(B_1) = 0.2$ and $\eta(B_2) = 0.4$*

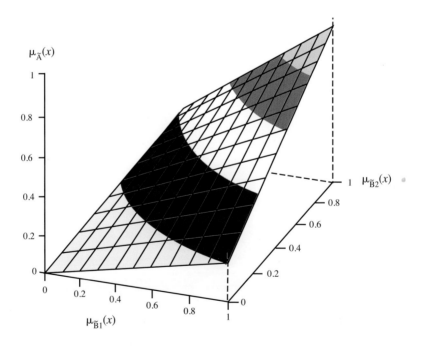

Plate 8 *Polynomial composition under pseudomeasure for $\eta(B_1) = 0.2$ and $\eta(B_2) = 0.4$*

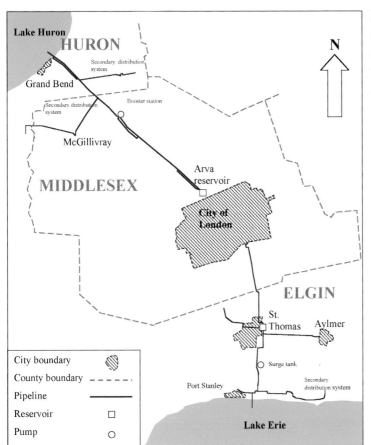

Plate 9 *The City of London, Ontario, regional water supply system*

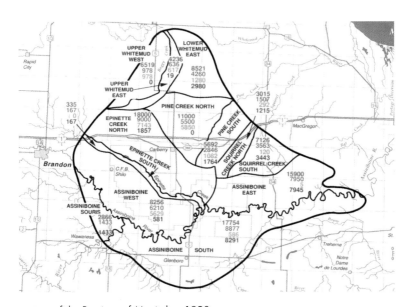

Source: courtesy of the Province of Manitoba, 1998

Plate 10 *The Assiniboine delta aquifer, Manitoba, Canada*

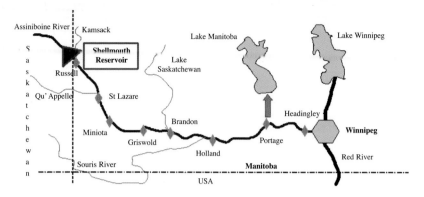

Plate 11 *Shellmouth Reservoir area*

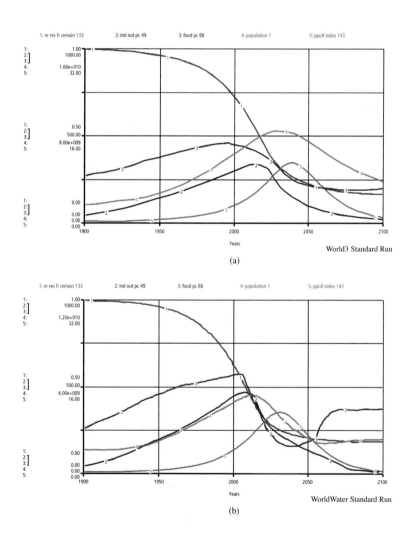

Plate 12 *State of the World: 'Standard run' results of World3 (a) and WorldWater (b) models*

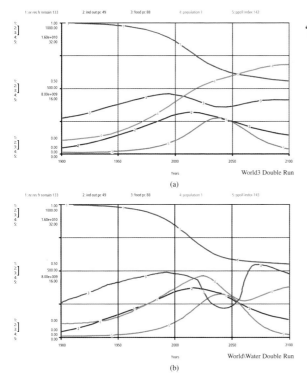

Plate 13 *State of the World: 'Double run' results of World3 (a) and WorldWater (b) models*

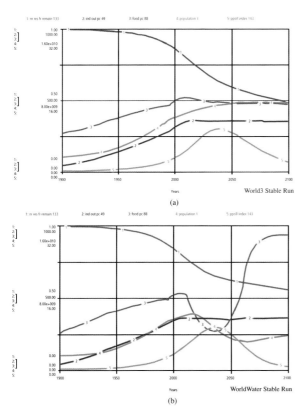

Plate 14 *State of the World: 'Stable run' results of World3 (a) and WorldWater (b) models*

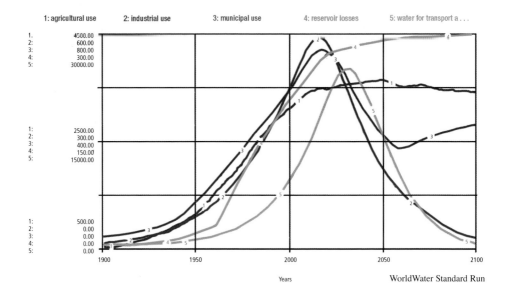

Plate 15 *Use of water: 'Standard run' results of WorldWater*

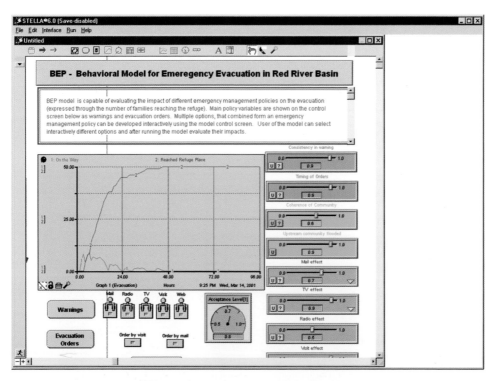

Plate 16 *Evacuation model interface*

analysis, an estimate along a qualitative scale was included. All of the qualitative scale values were estimated along a scale from 0 to 10, with 0 being the worst possible value and 10 being the best possible value. Except where otherwise noted, the impact values were provided by a government professional. The sources of other impact values and the rationale behind the choice of the impact weights are discussed for each selected impact.

Loss of arable land

This is a measure of the amount of arable, or agricultural, land that is removed from production for municipal, infrastructure or other purposes. This impact increases with development, as much of the available land in the region has already been cleared or designated as protected. This impact is classed as an ecological risk as opposed to an economic one because it is indicative of the competing uses for a relatively limited land base. The impact is measured in the total number of acres lost per year, and estimated from 1991 and 1996 Canadian census data (courtesy of Manitoba Agriculture). This impact was given a moderately low weight of 0.40 as it did not seem to be of great concern to those stakeholders surveyed.

Water quality

The quality of the water in the ADA is extremely high, and maintaining this high quality is important to all users of the aquifer. The MCDC has been conducting a water quality monitoring programme since 1994. Results of that study may prove to be useful for future evaluations. However, for this case study, a qualitative scale was used to estimate conditions under the different policy programmes. This impact seemed to be of great concern to residents of the region, and received a high rating of 0.85.

Riparian water needs

The association between groundwater levels and surface water flows is not entirely understood. Conversations with area residents revealed concern that increased groundwater usage would result in destabilization of streamflows. A regression analysis between the water table level at Carberry and the flow rates of the Epinette creek near Carberry, provided by the WRB of the Manitoba Water Stewardship, was used as the basis to estimate the probability of streamflow fluctuations as groundwater usage increases. This impact was assigned a moderately high weight of 0.60.

Erosion

Erosion by wind and water is a concern to both farmers and conservationists in the ADA region. For this case study, a qualitative estimate of the amount of erosion taking place under each scenario was used in the analysis. This impact was mentioned by several stakeholders as being important, and as such received a high weight of 0.70.

Irrigated crop production

Agriculture is the economic life blood of the region. The economic fortunes of the town of Carberry are directly tied to it. Since potato is the major irrigated crop, it provides a good indicator of the economic fortune of the region. No attempt to model the different potato market conditions under the three policy scenarios was made, so instead the estimator for this impact was the amount of potato produced in hundredweight (cwt = 1 hundredweight = 0.04536 ton). It is a function of both the number of acres in production and the productivity. A figure of 250 cwt/acre was used for the status quo and development scenarios, while a more conservative estimate of 200 cwt/acre was used for the conservation scenario. This impact was of considerable concern to many stakeholders, and as such received a very high weight of 0.90.

Livestock

Crop farming is not the only agricultural activity in the region. There is also a significant amount of livestock farming. Livestock production in the region is therefore an important indicator of the health of the local economy. Once again, no attempt was made to model the livestock markets under the three policy scenarios, and as a result the total livestock population was used as the indicator in the analysis. This impact was assigned a high weight of 0.70.

Direct employment – processing

The Mid-West processing plant is also important to the economic life of the community. It adds value to the local potato crop and also directly employs a number of local people. The number of workers directly employed by the plant is another important indicator of the well-being of the local economy, and as such it received a very high weight of 0.85.

Water availability – domestic

Most of the residents of the ADA region live in a rural setting and depend on groundwater as their domestic water source. A major concern of several area residents was that increased development would lead to the water table dropping, thereby causing their household wells to fail. The security of domestic water supplies was estimated on a qualitative scale. This impact was assigned a moderately high weight of 0.60.

Landscape aesthetics

Any policy change that affects development rates have an impact on the way the landscape looks. This indicator, estimated on a qualitative scale, denotes the aesthetic quality of the landscape. This impact was deemed to be relatively unimportant compared with other impacts, and received a moderately low weight of 0.30.

Table 7.25 lists the weights and impact values for each indicator under the three policy scenarios. Recall equation (7.5), the R-metric, from Section 7.4. The information from Table 7.25 was entered into the R-metric, yielding a reversibility index for each alternative and each class of impacts (R_{cj}). Table 7.26 displays the results of the reversibility analysis. The Tc_{max} values were obtained by entering the worst possible value (m_{ci}) into the R-metric. The scaled R_c values were then obtained by dividing the R_c values for each alternative by the appropriate category Tc_{max} value.

As Table 7.26 indicates, the scenarios are ranked differently in the three sustainability categories. The conservation scenario receives the lowest value (least irreversible) in the social and environmental categories, but the highest value (most irreversible) in the economic category. The development scenario receives the best value in the economic category but the lowest in the social and environmental categories. The status quo scenario ranks in the middle on all three criteria.

Data for the computation of reversibility measure for the ADA case study is located on the CD-ROM, in the directory SUSTAINPRO, subdirectory Examples, file ADA.sdrvrs.

Two sensitivity analyses were performed to determine the impact on the analysis of the different impact weightings and the qualitative scale values. A summary of the results of the impact weight sensitivity analysis appears in Table 7.27. This sensitivity analysis was performed by randomly assigning the impact weights a value between 0 and 1. This process was carried out iteratively ten times. Table 7.27 displays the category indices calculated using the R-metric with the randomly assigned weights for each alternative and each of the ten iterations. The number of times an alternative is ranked first (i.e. receives the lowest R-metric value) is listed in the final column of the table. The results of this sensitivity analysis indicate that the reversibility results for this case study are not particularly sensitive to the weighting values.

Table 7.25 *Reversibility impacts and weights for three policy scenarios in the ADA*

Impact	Units	M_{ci}	m_{ci}	w_{ci}	fSQ	fDE	fCO
Loss of arable land	Acres/yr	0	2818	0.40	2562	2818	2306
Water quality	Scale	10	0	0.85	7	7	7
Riparian water needs	Scale	10	0	0.60	6	6	6
Erosion	Scale	10	0	0.70	7	6	8
Irrigated crop production	cwt/acre	1,746,508	0	0.90	1,587,735	1,746,508	1,270,188
Livestock population	No. animals	344,451	0	0.70	328,049	344,451	321,488
Direct employment – processing	No. employees	451	0	0.85	410	451	369
Water availability – domestic	Scale	10	0	0.60	7	6	8
Landscape aesthetics	Scale	10	0	0.30	7	7	7

Notes: SQ = status quo scenario; DE = development scenario; CO = conservation scenario.

Table 7.26 *Reversibility R-metric results*

Index	Category		
	c = 1 Environmental	c = 2 Economic	c = 3 Social
Tc_{min}	0.000	0.000	0.000
Tc_{max}	1.316	1.422	0.671
$R_{c1}(SQ)$	0.547	0.117	0.201
$R_{c2}(DE)$	0.601	0.000	0.256
$R_{c3}(CO)$	0.499	0.294	0.150
Scaled $R_{c1}(SQ)$	0.415	0.083	0.300
Scaled $R_{c2}(DE)$	0.456	0.000	0.382
Scaled $R_{c3}(CO)$	0.379	0.207	0.224

Notes: SQ = status quo scenario, DE = development scenario, CO = conservation scenario.

Table 7.28 displays the results of the impact value sensitivity analysis. In this analysis, individual impact values were modified to demonstrate their impact on the overall analysis. Only impacts ranked on a qualitative scale were modified. The percentage change in the original value (measured by dividing the difference between the two impact values by the original impact value) is listed in the fourth column of Table 7.28. The eighth column of Table 7.28 lists the percentage change in the index value that results from manipulating the impact value. As Table 7.28 illustrates, the ecological component of the reversibility analysis seems relatively stable, despite large changes in individual impact values. The margin of difference between the three alternatives changes, but the recommended alternative in all cases is still the conservation scenario. The social component, however, is more vulnerable to changes in the qualitative value of impacts. A possible explanation for this is the smaller number of impacts considered in this category (two, versus four in the ecological category) and the fact that both impacts are estimated on a qualitative scale. Individual views or biases could therefore impact the criteria value in this category.

Table 7.27 *Results of R-metric weight sensitivity analysis for the ADA case study*

Category	R_{cj}	Random weight sets										# rank 1st
		1	2	3	4	5	6	7	8	9	10	
Social	Rsq	0.48	0.32	0.34	0.24	0.18	0.39	0.21	0.26	0.39	0.20	0
	Rde	0.58	0.37	0.43	0.29	0.22	0.46	0.24	0.29	0.48	0.23	0
	Rco	0.40	0.27	0.28	0.20	0.14	0.33	0.18	0.23	0.31	0.18	10
Ecological	Rsq	0.01	0.01	0.01	0.01	0.01	0.01	0.01	0.01	0.00	0.01	0
	Rde	0.00	0.00	0.00	0.00	0.00	0.00	0.00	0.00	0.00	0.00	10
	Rco	0.05	0.05	0.03	0.03	0.04	0.02	0.06	0.04	0.02	0.02	0
Economic	Rsq	0.09	0.09	0.09	0.09	0.09	0.09	0.09	0.09	0.09	0.09	0
	Rde	0.14	0.14	0.14	0.15	0.12	0.14	0.14	0.13	0.14	0.10	0
	Rco	0.05	0.06	0.06	0.05	0.07	0.05	0.05	0.06	0.06	0.08	10

Table 7.28 *Results of the impact sensitivity analysis for the ADA case study*

	Param. tested	Original impact value	Chang. value	% change	Index affect.	Original index value	Chang. index value	% change	Pref. altern.
Ecological	f112	7	10	+30	R11	0.22	0.19	−14	CO
	f122	7	10	+30	R12	0.27	0.24	−11	CO
	f123	6	10	+66	R12	0.27	0.23	−17	CO
	f113	6	10	+66	R11	0.22	0.18	−18	CO
	f114	7	9	+29	R11	0.22	0.20	−9	CO
	f124	6	9	+50	R12	0.27	0.23	−15	CO
	f134	8	6	−25	R13	0.18	0.22	+22	CO
Social	f312	7	10	+43	R31	0.09	0.06	−33	CO
	f321	6	9	+50	R32	0.14	0.04	−71	DE
	f322	7	10	+43	R32	0.14	0.11	−21	CO
	f331	8	6	−25	R33	0.06	0.14	+57	SQ

A good exercise for using SUSTAINPRO is to do the computation presented in the sensitivity analysis by making appropriate modifications to the data input file ADA.sdrvrs located in the directory SUSTAINPRO, subdirectory Examples.

7.8.4 Consensus

The five consensus measures that are used here are presented in Section 7.5, equations (7.7)–(7.11). Each measure for degree of consensus illuminates or captures a different aspect of consensus. The three coincidence measures, γ^1, γ^3 and γ^4, focus on identifying common ground. The two discrepancy measures, γ^2 and γ^5, are focused on identifying sources of disagreement. The degree of consensus indicates the relative strength of ranking. That is, the worst alternative may have a high degree of consensus because everyone agrees that it is the worst alternative! The result is a weak ordering of alternatives, and complete transitivity may not be achieved.

Based on the data in Tables 7.8, 7.9 and 7.15, the three alternatives, status quo, development and conservation were evaluated according to six criteria. The first criterion (column 1 in Table 7.29) measured the population in the ADA region, while the second captured the use of water and the third the benefits from agriculture measured by a proxy variable, irrigated area. The remaining three criteria were aggregated expressions of fairness (derived from Table 7.11), risk (derived from Table 7.22) and reversibility (derived from Table 7.26). The problem size was as follows:

three alternatives, six criteria, three decision-makers. Weights corresponding to the different decision-makers are listed in the last three rows of Table 7.29.

Table 7.29 ADA groundwater management problem

Criteria Scenario	Population (number of people) (max)	Water use acre-feet/ (year) (min)	Irrigated area (acre = 4046.8 m²) (max)	Fairness (max)	Risk (min)	Reversibility (max)
Status quo	10,931	7950	8029	0.17	0.43	0.225
Development	11,259	8622	8832	0.20	0.37	0.236
Conservation	10,931	6147	8.029	0.10	0.44	0.223
			Weights			
Government employees	1	1	1	3	2	2
Irrigators	2	1	4	1	1	1
Environmentalists	1	3	2	1	2	1

An ordinal ranking of the alternatives was achieved using the distance metrics of Compromise programming, equation (7.6), which provides a strong ranking of the alternatives. One distance measure was used to evaluate the alternatives, defined by the exponent $p = 2$. The decision-makers' weights described the importance of each criterion. Unfortunately there were three decision-makers, each with their own priorities. As each decision-maker used an individual set of weights, the rankings changed. Table 7.30 shows the value of distance metric, x_j, and the corresponding rank for each decision-maker.

Table 7.30 Ranking of alternative scenarios

Decision-maker		Scenario		
		Status quo	Development	Conservation
Government employees	Distance metric Rank	0.302 2	0.100 1	0.436 3
Irrigators	Distance metric Rank	0.470 2	0.100 1	0.480 3
Environmentalists	Distance metric Rank	0.368 3	0.300 1	0.332 2

Table 7.31 shows the degree of consensus for each alternative scenario in the case study, for all five consensus measures (equations (7.7) – (7.11)). Looking at the numbers in Table 7.31, we can see that two decision-makers are in relative agreement on the performance of every alternative (γ^1). However, γ^2 indicates that two decision-makers are in relative disagreement for alternatives {SQ, CO} and that there is relatively little disagreement about the performance of alternative {DE}. The overall level of agreement about the performance of each alternative is quite good, when considering a comparison of each decision-maker with every other decision-maker (γ^4). This result was generally expected, based on the inspection of ordinal ranks supplied by the Compromise programming. Note that the consensus measures are calculated based on distance metric value (x_j), not the ordinal ranks presented in Table 7.30.

Table 7.31 *Degree of consensus measures for ADA case study*

Alternative scenario	Consensus measure				
	γ^1	γ^2	γ^3	γ^4	γ^5
Status quo	0.935	0.833	0.94	0.888	0.91
Development	1	0.8	0.911	0.867	0.867
Conservation	0.956	0.852	0.944	0.901	0.916

Data for the computation of the consensus measure for the ADA case study is located on the CD-ROM, in the directory SUSTAINPRO, subdirectory Examples, file ADA.sdcons.

Using consensus as the measure for sustainability, decision-makers have the opportunity to explore their values with different sets of weights to find a robust solution. Decision-makers also have an opportunity to evaluate the strength of their decisions as well as progress in negotiations. Encouragement of iterative, interactive feedback to a negotiation process is motivated by possible spontaneous creativity in resolving differences of opinion. Other searches may identify clusterings or groupings of individuals in terms of their ranking. Advanced use of the degree of consensus may even identify aspects of the system as candidates for adaptation, as an attempt to improve the non-dominated frontier of solutions towards more sustainable solutions.

7.8.5 Discussion

As noted throughout this section, the four sustainability criteria are not meant to stand alone as decision-making tools. They can add a valuable dimension to the multi-objective decision-making process, which will be discussed later in the book. From the application of sustainability criteria to the case study, we can see that the sustainability issues brought to light in this application can make a valuable contribution to the design and implementation of groundwater management and policy in the ADA region.

On the basis of the case study, some recommendations can tentatively be made for future management and research directions within the ADA region:

- further investigation by Manitoba Water Stewardship into the possibility of a tradable water entitlement system for the ADA and review of the existing Manitoba Water Rights Act;
- continued consultation and involvement of residents from a variety of stakeholder groups in the region;
- serious consideration of including the sustainability criteria discussed in this work in the evaluation of future management and policy decisions concerning the ADA.

7.9 REFERENCES

Bagheri, A. and P. Hjorth (2005), 'Monitoring for Sustainable Development: A Systemic Framework', *International Journal of Sustainable Development*, 8(4):280–301

Bender, M. (1996), 'A Framework for Collaborative Planning and Investigations of Decision Support Tools for Hydro Development'. Ph.D. dissertation, Civil and Geological Engineering, University of Manitoba, Winnipeg, Manitoba, Canada

Bender, M. and S.P. Simonovic (1997), 'Consensus as the Measure of Sustainability', *Hydrological Sciences Journal*, 42(4): 493–500

Chang, N.B. (2005), 'Sustainable Water Resources Management Under Uncertainty', *Stochastic Environmental Research and Risk Assessment*, 19(2): 97–98

Duckstein, L. and E. Parent (1994), 'Systems Engineering of Natural Resources under Changing Physical Conditions: A Framework for Reliability and Risk', in *Natural Resources Management*, L. Duckstein and E. Parent, eds, Kluwer, Dordrecht, The Netherlands

Fanai, N. (1996), 'Reversibility as a Sustainability Criterion for Project Selection', M.Sc. thesis, Civil and Geological Engineering, University of Manitoba, Winnipeg, Manitoba, Canada

Fanai, N. and D.H. Burn (1997), 'Reversibility as a Sustainability Criterion for Project Selection', *International Journal of Sustainable Development and World Ecology*, 4(4):259–272

Goodman, A.S. (1984), *Principles of Water Resources Planning*, Prentice-Hall, Englewood Cliffs, NJ

Kroeger, H.I. (1997), 'Development of a Risk Measure as a Project Selection Criterion for Sustainable Project Development', M.Sc. thesis, Civil and Geological Engineering University of Manitoba, Winnipeg, Manitoba, Canada

Kroeger, H.I. and S.P. Simonovic (1997), 'Development of a Risk Measure as a Sustainable Project Selection Criteria', *International Journal of Sustainable Development and World Ecology*, 4(4):274–285

Kulshreshtha, S.N. (1994), *Economic Value of Groundwater in the Assiniboine Delta Aquifer in Manitoba*, Social Science Series 29, Environmental Conservation Service, Ottawa, Canada, 70pp

Kuncheva, L. (1994), 'Pattern Recognition with a Model of Fuzzy Neuron using Degree of Consensus', *Fuzzy Sets and Systems*, 66: 241–250

Lence, B. J., J. Furst and S.M. Matheson (1997), 'Distributive Fairness as a Criterion for Sustainability: Evaluative Measures and Application to Project Selection', *International Journal of Sustainable Development and World Ecology*, 4(4): 245–257

Lucas, A.R. (1990), *Security of Title in Canadian Water Rights*, Canadian Institute of Resources Law, Calgary, Canada, 102pp

McLaren, R.A. (1998), 'Evaluating Sustainability Criteria: Application in the Assiniboine Delta Aquifer Region', Master in Natural Resources Management thesis, Natural Resource Institute, University of Manitoba, Winnipeg, Manitoba, Canada

McLaren R.A. and S.P. Simonovic (1999a), 'Data Needs for Sustainable Decision Making', *International Journal of Sustainable Development and World Ecology*, 6(2): 103–113

McLaren R.A. and S.P. Simonovic (1999b), 'Evaluating Sustainability Criteria for Water Resources Decision Making: Assiniboine Delta Aquifer Case Study', *Canadian Water Resources Journal*, 24(2): 147–163

Manitoba Agriculture (1998), 'Selected Census Data, 1991 and 1996', Unpublished

Manitoba Health (1997), *Population Reports by Rural Health Authority and Municipality*, Manitoba Health: Health Information Services, Manitbo, Canada

Matheson, A.M. (1997), 'Distributive Fairness Measures for Sustainable Project Selection', M.Sc. thesis, Civil and Geological Engineering, University of Manitoba, Winnipeg, Manitoba, Canada

Matheson, S.M., B.J. Lence and J. Fuerst (1997), 'Distributive Fairness Considerations in Sustainable Project Selection', *Hydrological Sciences Journal*, 42(4): 531–548

National Research Council (1985), *Nutrient Requirements of Sheep*, National Academy Press, Washington, DC, 99pp

National Research Council (1988), *Nutrient Requirements of Swine*, National Academy Press, Washington, DC, 242pp

National Research Council (1989), *Nutrient Requirements of Horses*, National Academy Press, Washington, DC, 100pp

National Research Council (1994), *Nutrient Requirements of Poultry*, National Academy Press, Washington, DC, 155pp

National Research Council (1996), *Nutrient Requirements of Beef Cattle*, National Academy Press, Washington, DC, 242pp

P.M. Associates Ltd. (1996), 'Water Monitoring Survey: 1995', Washington, DC, Unpublished

Render, F.W. (1988), 'Water Supply Capacity of the Assiniboine Delta Aquifer', *Canadian Water Resources Journal*, 13(4): 16–34

Sahely, H.R., C.A. Kennedy and B.J. Adams (2005), 'Developing Sustainable Criteria for Urban Infrastructure', *Canadian Journal of Civil Engineering*, 32(1): 72–85

Simonovic, S.P., D.H. Burn and B.J. Lence (1997), 'Practical Sustainability Criteria for Decision Making', *International Journal of Sustainable Development and World Ecology*, 4(4): 231–244

Spangenberg, J.H. (2004), 'Reconciling Sustainability and Growth: Croteria, Indicators, Policies', *Sustainable Development*, 12(2): 74–86

UNESCO (UN Educational, Scientific and Cultural Organization) (1998), *Sustainability Criteria for Water Resources Systems*, ASCE Press, New York

UNESCO – WWAP (World Water Assessment Program) (2003), *Water for People, Water for Life – UN World Water Development Report (WWDR)*, UNESCO Publishing, Paris, France, co-published with Berghahm Books, UK

WCED (World Commission on Environment and Development) (1987), *Our Common Future*, Oxford University Press, Oxford

Young, M.D. (1992), *Sustainable Investment and Resource Use*, UNESCO, Man and the Biosphere Series, Volume 9, The Parthenon Publishing Group, Casternton Hall, Carnforth, UK

Zeleny, M. (1982), *Multiple Criteria Decision Making*, McGraw-Hill, New York

7.10 EXERCISES

1. Define sustainable development.
2. Define sustainable water resources systems management.
3. What are the main principles of sustainable water resources decision-making?
4. Discuss sustainable water resources systems management with respect to water supply and demand in your region/country.
   a. In your words describe how would you apply the main principles of sustainable water resources systems management (presented in Section 7.1) to water supply and demand management in your region/country.
   b. Indicate what spatial and temporal scales will be applicable to your region/country.
   c. How would you decide whether or not some plan or management policy is sustainable?

For the following exercises use the SUSTAINPRO computer package provided on the enclosed CD-ROM.

5 Familiarize yourself with the SUSTAINPRO computer package. Spend some time reviewing help menus for all four programs. Check the input and output files.

6 Identify the input data for the computation of risk measure for ADA (look in the directory SUSTAINPRO, subdirectory Examples, file ADA.sdrisk).

   a. Calculate the risk estimates if the average sustainable development categories weights are changed from those in Table 7.15 to:

Sustainable development category	Government employees	Irrigators	Environmentalists
Environment	20.3	21.3	58.3
Economy	52.1	65.0	20.0
Social	27.6	13.8	21.7

Compare your results with the results presented in Section 7.8.2. Discuss the difference.

   b. Calculate the risk estimates if the average sustainable development categories weights are changed from those above to:

Sustainable development category	Government employees	Irrigators	Environmentalists
Environment	20.3	11.3	48.3
Economy	52.1	65.0	30.0
Social	27.6	23.8	21.7

Compare your results with the results of Exercise 6a. Discuss the difference.

   c. Change the risk preferences in Table 7.14 and stakeholder preferences in Table 7.15 to those that will capture the preferences (in your opinion) of stakeholders in your region/country. Run SUSTAINPRO and compare the results with those of Section 7.8.2.

7 Identify the input data for the computation of reversibility measure for ADA (look in the directory SUSTAINPRO, subdirectory Examples, file ADA.sdrvrs).

   a. Perform your own sensitivity analysis for the example in Section 7.8.3 by modifying the input data in Table 7.25.
   b What data did you change? Why?
   c What is the difference between your output results and those in Table 7.26?
   d What did you learn from this sensitivity analysis?

8 Identify the input data for the computation of consensus measure for ADA (look

in the directory SUSTAINPRO, subdirectory Examples, file ADA.sdcons).
   a. Change the weights in Table 7.29 in order to capture your understanding of different stakeholders in your region/country.
   b. Why did you decide to use this set of weights?
   c. Run SUSTAINPRO and compare the degree of consensus obtained with those presented in Table 7.31.
   d. What is your interpretation of the differences?
9. Identify one major water resources system in your region/country.
   a. What are the main management issues?
   b. Collect the necessary data to evaluate the sustainability of the system.
   c. Using SUSTAINPRO, calculate the four measures of sustainability presented in this chapter.
   d. How would you use these measures for better water resources systems management?
   e. Present your work to the water authority responsible for the management of selected water resources systems.

PART IV

Implementation of Water Resources Systems Management Tools

8

Simulation

Simulation models *describe* how a system operates, and are used to determine changes resulting from a specific course of action. Such models are referred to as *cause-and-effect* models in the introductory discussion in Section 5.1. They describe the state of the system in response to various inputs but give no direct measure of what decisions should be taken to improve the performance of the system. Therefore, simulation is a problem-solving technique which contains the following phases:

1. Development of a model of the system.
2. Operation of the model (i.e. generation of outputs resulting from the application of inputs).
3. Observation and interpretation of the resulting outputs.

The essence of simulation is an iterative process of modelling and experimentation.

8.1 DEFINITIONS

The *classical simulation* procedure involves decomposition of the problem in order to aid the system description. When the main elements of the system are identified, the proper mathematical description is provided for each of them. The procedure continues with computer coding of the mathematical description of the model. Each model parameter is then calibrated and the model performance is verified using data that has not been seen during the calibration process. The completed model is then operated using a set of input data. Detailed analysis of the resulting output is the final step in the simulation procedure.

A completed model can be reused many times with alternative input data. If there is a need for modification of the system description or model structure, the whole process starts again and the model has to be recoded, calibrated and verified again before its use.

The major components of a simulation model are:

- **input**: quantities that 'drive' the model (in water resources systems management, for example, a principal input is the set of streamflows, rainfall sequences, pollution loads, water and power demands, etc.);
- **physical relationships**: mathematical expressions of the relationships among the physical variables of the system being modelled (continuity, energy conservation reservoir volume and elevation, outflow relations, routing equations, etc.);
- **non-physical relationships**: those that define economic variables, political conflicts, public awareness, etc.;
- **operation rules**: the rules that govern operational control;
- **outputs**: the final product of operations on inputs by the physical and non-physical relations in accordance with operating rules.

System dynamics simulation is introduced in Section 5.2 as a rigorous method of system description, which facilitates feedback analysis via a simulation model of the effects of alternative system structures and control policies on system behaviour. The advantages of system dynamics simulation over classical simulation are presented in Section 5.2. Briefly, they include:

- the simplicity of use of system dynamics simulation applications;
- the applicability of system dynamics general principles to social, natural and physical systems;
- the ability to address how structural changes in one part of a system might affect the behaviour of the system as a whole;
- combined predictive (determining the behaviour of a system under particular input conditions) and learning (discovery of unexpected system behaviour under particular input conditions) functionality;
- active involvement of stakeholders in the modelling process.

The rest of this chapter focuses on system dynamics simulation as a method of water resources systems management.

8.2 SYSTEM DYNAMICS SIMULATION

8.2.1 Introduction

System dynamics is an academic discipline introduced in the 1960s by the researchers at the Massachusetts Institute of Technology. It has gradually developed into a tool useful in the analysis of social, economic, physical, chemical, biological and ecological systems (Forrester, 1990; Sterman, 2000). In the context of this book, in Chapter 4 a *system* was defined as a collection of elements which continually interacts over time to form a unified whole. The underlying pattern of interactions between the elements of a system is called the *structure* of the system. The term *dynamics* refers to change over time. If something is dynamic, it is constantly changing in response to the stimuli influencing it. A dynamic system is thus a system in which the variables interact to stimulate changes over time. The way in which the elements, or variables, composing a system vary over time is referred to as the *behaviour* of the system.

Box 8.1 Water use

- Worldwide water withdrawals from water bodies have risen from 250 cubic metres/person/year in 1900 to over 700 cubic metres today.
- Water consumption usually drops by 18–25 per cent after a water meter is installed.
- Toilets consume nearly one-quarter of the municipal water supply and use over 40 per cent more water than is needed.
- Residential indoor water use in Canada can be broken down into toilet – 30 per cent; bathing and showering – 35 per cent; laundry – 20 per cent; kitchen and drinking – 10 per cent; cleaning – 5 per cent.
- A five-minute shower with a standard shower head uses 100 litres of water. A five-minute shower with a low-flow shower head uses only 35 litres of water
- Among common water uses and consumption are: toilet flush 15–19 litres; five-minute shower 100 litres; tub bath 60 litres; automatic dishwashing 40 litres; dishwashing by hand 35 litres; hand washing (with tap running) 8 litres; brushing teeth (with tap running) 10 litres; outdoor watering 35 litres/min; washing machine cycle 225 litres.
- A single lawn sprinkler spraying 19 litres per minute uses 50 per cent more water in just one hour than a combination of ten toilet flushes, two five-minute showers, two dishwasher loads and a full load of clothes.

Source: Environment Canada, http://www.ec.gc.ca/water/en/e_quickfacts.htm (last accessed December 2005)

One feature that is common to all systems is that a system's structure determines its behaviour. System dynamics links the behaviour of a system to its underlying structure. It can be used to analyse how the structure of a physical, biological or any other system can lead to the behaviour the system exhibits.

The system dynamics simulation approach relies on understanding complex interrelationships existing between different elements within a system, by developing a model that can simulate and quantify the behaviour of the system. The major steps that are carried out in the development of a system dynamics simulation model include:

1 understanding the system and its boundaries;
2 identifying the key variables;
3 describing the physical processes or variables through mathematical relationships;
4 mapping the structure of the model;
5 simulating the model for understanding its behaviour.

Advances made during the last decade in computer software have brought about considerable simplification in the development of system dynamics simulation models. The accompanying CD-ROM includes all the system dynamics models developed in the text, using the state-of-the-art simulation software Vensim PLE (Ventana Systems, 1995), which is available from the Ventana Systems Inc. website (http://www.vensim.com, last accessed December 2005). This software is free for educational use. In the SYSTEMDYNAMICS directory on the CD-ROM there is a read.me file that contains program installation instructions. The Tutorial subdirectory contains a short tutorial for Vensim PLE developed by Professor Craig Kirkwood at Arizona State University. The Examples subdirectory contains all the examples from this chapter.

Vensim PLE is an ideal tool for personal learning of system dynamics. Like similar programs, it uses the principles of object-oriented programming. It provides a set of graphical objects with their mathematical functions for easy representation of the system structure and the development of computer code. Simulation models can be easily and quickly developed using this type of software tool. Such models are easy to modify, easy to understand, and present results clearly to a wide audience of users.

8.2.2 System structure and patterns of behaviour

In starting to consider system structure, we first generalize from the specific events associated with the problem to considering patterns of behaviour that characterize the situation. Usually this requires investigation of how one or more variables of interest change over time (e.g. flow of water, load on a bridge or wind load). That is, we ask, what patterns of behaviour do these variables display? The system dynamics simulation approach gains much of its power as a problem-solving method from the fact that similar patterns of behaviour show up in a variety of different situations, and the underlying system structures that cause these characteristic patterns are known. Thus, once we have identified a pattern of behaviour that is a problem, we can look for the system structure that is known to cause that pattern. If we can find and modify this system structure, there is the possibility of permanently eliminating the problem pattern of behaviour.

Feedback relationships

The difference between the two basic types of feedback is important in understanding dynamic behaviour. Section 4.3 introduced the concept of feedback. A positive, or reinforcing, feedback loop reinforces change with even more change (see Section 4.3.2). This can lead to rapid growth at an ever-increasing rate. This type of growth pattern is often referred to as *exponential growth*. Note that in the early stages of the growth, it seems to be slow, but then it speeds up. Thus the nature of growth in a water resources system that has a positive feedback loop can be deceptive. Examples that fit this category are water demand and population growth. Positive feedback is quite common in managed systems, and may be valuable as an engine of growth. In an engineering system, however, positive feedback is undesirable and should be designed out.

Figure 8.1a shows an example of a generic positive feedback loop, and Figure 8.1b shows the corresponding behaviour.

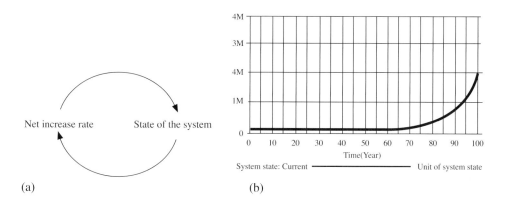

Figure 8.1 *(a) Positive feedback loop; (b) System behaviour*

A negative, or balancing, feedback loop seeks a goal. If the current level of the variable of interest is above the goal, then the loop structure pushes its value down, while if the current level is below the goal, the loop structure pushes its value up. Many water resources management processes contain negative feedback loops which provide useful stability, but which can also resist needed changes (see Section 4.3.3). The essential idea of negative feedback is that, when there is a difference between the desired and actual states of the system, action is generated according to the system's policy, in an attempt to eliminate the difference.

Figure 8.2a shows a negative feedback loop and Figure 8.2b a typical pattern of behaviour.

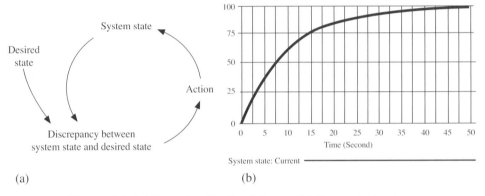

Figure 8.2 *(a) Negative feedback loop; (b) System behaviour*

Example 1

Let us revisit the irrigation flow control example from Section 4.4.4 shown in Figure 8.3. As the float drops, it turns the switch on, which opens the weir to admit the water. The rising water level (volume) causes the float to rise, and this in turn gradually shuts off the weir.

(a) If FR is set at 0.1 cubic metres (m^3/minute, how much water will enter the canal in ten minutes?

$W = 0.1 [m^3 / min] \times 10 [min]$
$W = 1 [m^3]$

(b) The units of measure must always accompany any numerical value to define the quantity. The units of measure of water flow rate are:

$FR\ [m^3/min]$

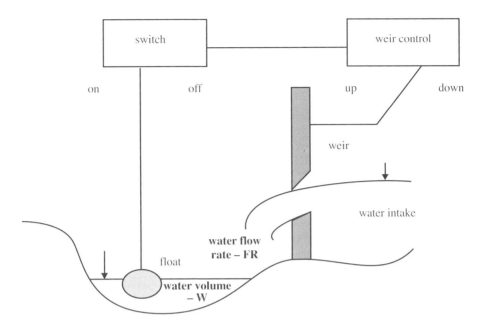

Figure 8.3 *Irrigation water intake*

The water volume is then measured in:

$W[m^3]$

(c) Let us now plot the water flow rate of 0.1 m³/minute. Figure 8.4 shows the plot.

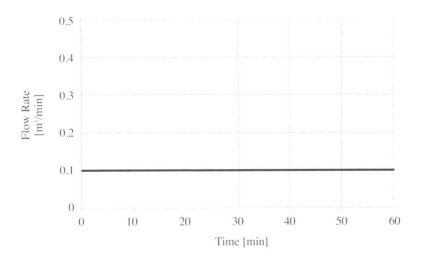

Figure 8.4 *Constant flow rate*

(d) If $FR = 0.04$ m³/minute and if the canal is empty at time 0, calculate the amount of water in the canal after 10 min, 20 min, 40 min and 60 min.

$W_{10} = 0.04 \times 10 = 0.4 [m^3]$
$W_{20} = 0.04 \times 20 = 0.8 [m^3]$
$W_{40} = 0.04 \times 40 = 1.6 [m^3]$
$W_{60} = 0.04 \times 60 = 2.4 [m^3]$

(e) The plot in Figure 8.5 shows the amount of water in the canal for every point in time.

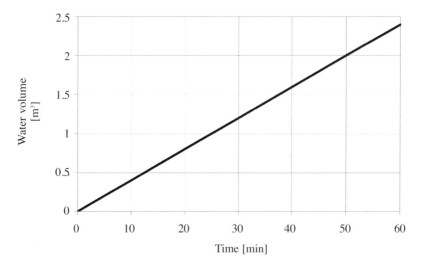

Figure 8.5 *Quantity time graph*

(f) On a similar graph (Figure 8.6), plot the water–time relationship showing the volume of water in the canal for a flow rate of 0.2, 0.1 and 0.02 m³/min.

(g) Using the weir with float control, suppose that the flow rate is 0.2 m³/min when the canal is empty and declines proportionally to zero when the canal contains 4 m³. Show the flow rate versus water volume in the canal on the graph.

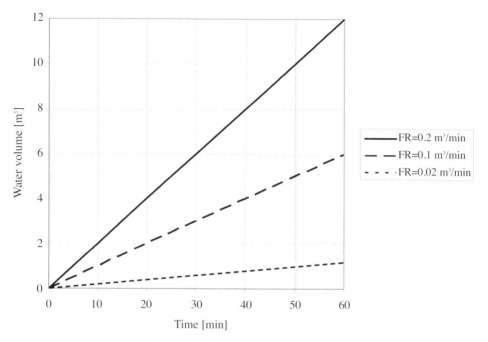

Figure 8.6 *Quantity time graph*

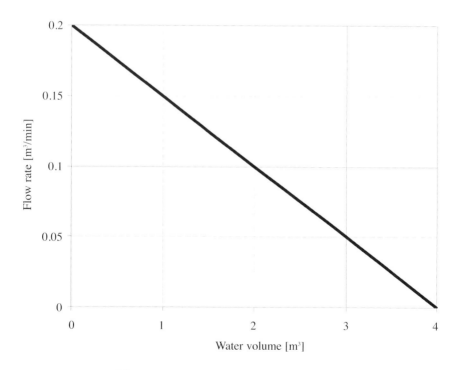

Figure 8.7 *Flow rate water volume graph*

Express FR as an equation.

$$FR = \frac{1}{T}(4 - W) \qquad (8.1)$$

Find T when W = 0.
Replace W = 0 and FR = 0.2 m³/min in equation (8.1) to yield:

$$0.2 = \frac{1}{T}(4 - 0)$$

$$T = \frac{4}{0.2} = 20[\min]$$

Replace T from the previous problem in equation (8.1) and find FR when W = 2.5 m³. Check the value using the graph in Figure 8.7.

$$FR = \frac{1}{20}(4 - 2.5) = 0.075[m^3 / \min]$$

(h) Using a water volume of 0.8 m³, what is the flow rate from the equation $FR = 1/20(4 - W)$?

$$FR = \frac{1}{20}(4 - 0.8) = 0.16[m^3 / \min]$$

(i) How much water will be added to the canal during the next four minutes?

$W = FR \times T$
$W = 0.16 \times 4 = 0.64[m^3]$

(j) Using Table 8.1 calculate the flow rate and water volume every four minutes. Complete the table for one hour and plot the curves of water volume and flow rate (Figure 8.8) as your calculation progresses. (It is important for learning that you actually do these calculations and the plotting. Pay attention to the way in which the variables are changing and why.)

Generic patterns of behaviour

The six patterns of behaviour shown in Figure 8.9 often show up, either individually or in combinations, in water resources systems simulations. In this figure, the vertical axis shows a variable of interest.

With exponential growth, an initial quantity of something starts to grow, and the rate of growth increases. The term exponential growth comes from a mathematical model for this increasing growth process where the growth follows a particular

Table 8.1 *Flow rate and water volume calculation*

Minutes	Change in volume [m³]	Volume [m³]	Flow rate [m³/min]
0.0000		0.0000	0.2000
4.0000	0.8000	0.8000	0.1600
8.0000	0.6400	1.4400	0.1280
12.0000	0.5120	1.9520	0.1024
16.0000	0.4096	2.3616	0.0819
20.0000	0.3277	2.6893	0.0655
24.0000	0.2621	2.9514	0.0524
28.0000	0.2097	3.1611	0.0419
32.0000	0.1678	3.3289	0.0336
36.0000	0.1342	3.4631	0.0268
40.0000	0.1074	3.5705	0.0215
44.0000	0.0859	3.6564	0.0172
48.0000	0.0687	3.7251	0.0137
52.0000	0.0550	3.7801	0.0110
56.0000	0.0440	3.8241	0.0088
60.0000	0.0352	3.8593	0.0070

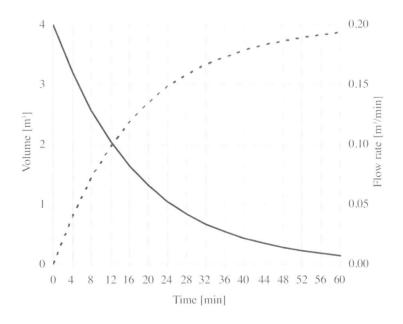

Figure 8.8 *Water volume and flow rate graph*

functional form called the exponential (see Section 4.2.2). Such growth is seen in increases in pollution, water demand, sales of water, etc.

With goal-seeking behaviour, the quantity of interest starts either above or below a goal level and over time moves toward the goal. Figure 8.9 shows one possible case where the initial value of the quantity is above the goal. The curve might represent the way the water is released from a reservoir into an irrigation canal. The change toward the final value is rapid at first and becomes slower as the discrepancy decreases between the present and final value.

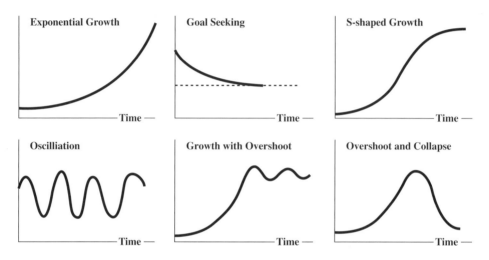

Figure 8.9 *General patterns of system behaviour*

With S-shaped growth, initial exponential growth is followed by goal-seeking behaviour which results in the variable levelling off.

With oscillation, the quantity of interest fluctuates around a level. Note that oscillation initially resembles exponential growth, and then appears to be S-shaped growth before reversing direction.

Common combinations of these four patterns include growth with overshoot. With this behaviour, the quantity of interest will overshoot the goal first on one side and then the other. The amplitude of these overshoots declines until the quantity finally stabilizes at the goal. Such behaviour can result from excessive time delays in the feedback loop or from too violent an effort to correct a discrepancy between the current state of the system and the system goal (as with adjusting the temperature of water in a shower).

Another combination is overshoot and collapse. In this behaviour, when resources are initially ample, the positive growth loop dominates and the state of the system grows exponentially. As it grows, resource adequacy drops. The negative loop gradually gains in strength and, unlike with S-shaped growth, the state of the system starts to decline.

Example 2

Which curve in Figure 8.9 would best describe:

(a) how the temperature of a thermometer changes with time after it is immersed in a hot liquid? Answer: the goal-seeking curve.
(b) the position of a pendulum, which is displaced and allowed to swing? Answer: the oscillation curve (note that these oscillations will gradually reduce with time).
(c) the learning process? Answer: the growth and overshoot curve.
(d) the amount of capital equipment over time in industrialization, where capital equipment is used to produce more capital equipment? Answer: exponential growth curve.

Delays

Delays are a critical source of dynamics in nearly all systems. They are sometimes a source of instability and oscillatory system behaviour. They are omnipresent in water resources management. It takes time to measure and report precipitation or flow. It takes time to make decisions on how to operate a weir or a pump. It also takes time for decisions to affect the state of a system. The simplest definition of a delay is *a process whose output lags behind its input in some fashion* (Sterman, 2000).

There are two types of delays. *Material delay* captures the physical flow of material. Consider, for example, the flow of water from the irrigation intake to an irrigated field. It takes time for water to fill in the canal and then to be transported by the canal from the intake to the field. Other examples of material delay include the construction of buildings and the progression of design tasks. In each there are physical units (cubic metres of water, square metres of space, or engineering drawings) moving through the process.

Other delays represent the gradual adjustment of perceptions or beliefs – these are *information delays*. The delay between a change in the temporary flood protection level and our belief in the flood forecast is an example of information delay. There is a delay between the receipt of new information and the updating of our perception. For example, flood risk perception is directly related to the extent of the flood forecast. If the forecast changes, there will be a delay in the change of our risk perception.

8.2.3 Causal loop diagram

To better understand the system structures that cause the patterns of behaviour discussed in the last section, let us introduce a notation for representing system structures. When an element of a system indirectly influences itself, the portion of the system involved is called a *feedback loop*. A map of the feedback structure – an *annotated causal loop diagram* – of a simple engineering system, such as that shown in Figure 8.10, is a starting point for analysing what is causing a particular pattern of behaviour. This figure considers a simple process, filling a storage tank with water. It includes elements, and arrows (which are called *causal links*) linking them, and also includes a sign (either + or –) on each link. These signs have the following meanings.

- A causal link from one element A to another element B is positive (that is, +) if either (a) A adds to B or (b) a change in A produces a change in B in the same direction.
- A causal link from one element A to another element B is negative (that is, –) if either (a) A subtracts from B or (b) a change in A produces a change in B in the opposite direction.

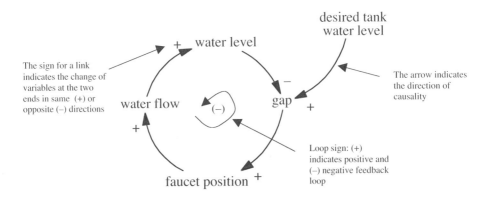

Figure 8.10 *Causal loop diagram for 'filling a storage tank with water'*

Let us start from the element *faucet position* at the bottom of the diagram. If it is increased (that is, the faucet (tap) is opened further) then the water flow increases. Therefore, the sign on the link from *faucet position* to *water flow* is positive. Similarly, if the *water flow* increases, then the *water level* in the tank will increase.

Therefore, the sign on the link between these two elements is positive. The next element along the chain of causal influences is the *gap*, which is the difference between the *desired tank water level* and the (actual) *water level* (i.e. *gap = desired tank water level – water level*). From this definition, it follows that an increase in *water level* decreases *gap*, and therefore the sign on the link between these two elements is negative. Finally, to close the causal loop back to *faucet position*, a greater value for *gap* presumably leads to an increase in *faucet position* (as you attempt to fill the tank), and therefore the sign on the link between these two elements is positive. There is one additional link in this diagram, from *desired tank water level* to *gap*. From the definition of *gap* given above, the influence is in the same direction along this link, and therefore the sign on the link is positive.

In addition to the signs on each link, a complete loop is also given a sign. The sign for a particular loop is determined by counting the number of minus (–) signs on all the links that make up the loop. Specifically, a feedback loop is called positive, indicated by a (+), if it contains an even number of negative causal links, and it is called negative, indicated by a (–), if it contains an odd number of negative causal links.

Thus the sign of a loop is the algebraic product of the signs of its links. Often a small looping arrow is drawn around the feedback loop sign to more clearly indicate that the sign refers to the loop, as is done in Figure 8.10. Note that in this diagram there is a single feedback loop, and that this loop has one negative sign on its links. Since 1 is an odd number, the entire loop is negative.

To start drawing a causal loop diagram, decide which events are of interest in developing a better understanding of system structure. From these events, move to showing (perhaps only qualitatively) the pattern of behaviour over time for the quantities of interest. Finally, once the pattern of behaviour is determined, use the concepts of positive and negative feedback loops, with their associated generic patterns of behaviour, to begin constructing a causal loop diagram which will explain the observed pattern of behaviour.

The following tutorial for drawing causal loop diagrams is based on guidelines by Forrester (1990) and Senge (1990).

Suggestion 1

Think of the elements in a causal loop diagram as variables which can go up or down, but don't worry if you cannot readily think of existing measuring scales for these variables.

- Use nouns or noun phrases to represent the elements, rather than verbs. That is, the actions in a causal loop diagram are represented by the links (arrows), and not by the elements.

- Be sure that the definition of an element makes it clear which direction is positive and which is negative.
- Generally it is clearer if you use an element name for which the positive sense is preferable.
- Causal links should imply a direction of causation, and not simply a time sequence. That is, a positive link from element A to element B does not mean first A occurs and then B occurs. Rather it means, when A increases then B increases.

Suggestion 2
As you construct links in your diagram, think about possible unexpected side-effects that might occur in addition to the influences you are drawing. As you identify these, decide whether links should be added to represent them.

Suggestion 3
For negative feedback loops, there is a goal. It is usually clearer if this goal is explicitly shown along with the gap that is driving the loop towards the goal.

Suggestion 4
A difference between actual and perceived states of a process can often be important in explaining patterns of behaviour. Thus, it may be important to include causal loop elements for both the actual value of a variable and the perceived value. In many cases, there is a lag (delay) before the actual state is perceived. For example, when there is a change in water quality, it usually takes a while before we perceive this change.

Suggestion 5
There are often differences between short-term and long-term consequences of actions, and these may need to be distinguished with different loops.

Suggestion 6
If a link between two elements needs a lot of explaining, you probably need to add intermediate elements between the two existing elements that will more clearly specify what is happening.

Suggestion 7
Keep the diagram as simple as possible, subject to the earlier suggestions. The purpose of the diagram is not to describe every detail of the process, but to show those aspects of the feedback structure that lead to the observed pattern of behaviour.

Example 3

Develop a causal loop diagram of the irrigation canal intake from Example 1 and identify the character of the feedback relationship.

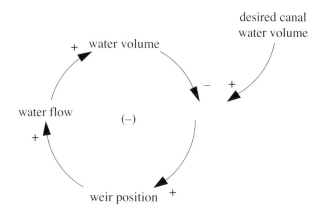

Figure 8.11 *A causal diagram of the irrigation canal intake system*

Figure 8.11 shows a causal diagram for the irrigation intake example. Start from the element *weir position* at the bottom of the diagram. If the *weir position* is increased (that is, the weir is opened further), the *water flow* increases. Therefore, the sign on the link from *weir position* to *water flow* is positive. Similarly, if the *water flow* increases, then the *water volume* in the canal will increase. Therefore, the sign on the link between these two elements is positive.

The next element along the chain of causal influences is the *difference*, which is the difference between the *desired canal water volume* and the (actual) *water volume* (i.e. *difference = desired canal water volume − water volume*). From this definition, it follows that an increase in *water volume* decreases *difference*, and therefore the sign on the link between these two elements is negative. Finally, to close the causal loop back to *weir position*, a greater value for *difference* leads to an increase in *weir position* (as you attempt to fill the canal) and therefore the sign on the link between these two elements is positive. There is one additional link in this diagram, from *desired canal water volume* to *difference*. From the definition of *difference* given above, the influence is in the same direction along this link, and therefore the sign on the link is positive.

In this case we have one negative sign, and therefore this feedback loop is negative or balancing, indicated by a minus sign in parentheses. The causal diagram of the irrigation canal intake is on the CD-ROM, in the directory SYSTEM DYNAMICS, subdirectory Examples, Example 1.

Example 4

Let us revisit the lake eutrophication example from Section 4.4.5. In the loop diagram from Figure 4.12, assign polarities to each of the arrows and each of the feedback loops.

The arrow and loop polarities are shown in Figure 8.12. There are eleven feedback loops in total, three of them positive (ticker lines) and the rest of them negative. The causal diagram is enclosed on the CD-ROM, in the directory SYSTEM DYNAMICS, subdirectory Examples, Example 2.

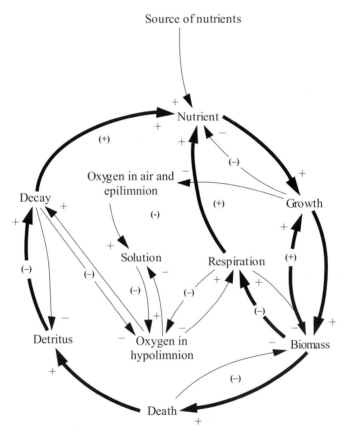

Figure 8.12 *Lake eutrophication causal loop diagram*

8.2.4 Stocks and flows

Causal loop diagrams are very useful in many situations. They are well suited to representing interdependencies and feedback processes. They can be used effectively at the start of any modelling project to capture mental models. However, they suffer from a number of limitations. One of the most important is their inability to capture the stock and flow structure of systems. Stocks and flows, along with feedback, are the two central concepts of system dynamics theory. Stock and flow notation provides a general way to graphically characterize any system process. *Any process?* This ambitious statement by Forrester (1990) is based on the characteristics that are generally shared by all engineering processes and the components that make up these processes. It is a remarkable fact that all such processes can be characterized in terms of variables of two types, stocks (levels, accumulations) and flows (rates).

Stocks are accumulations. They characterize the state of the system and generate the information upon which decisions and actions are based. Stocks give systems inertia and provide them with memory. Stocks create delays by accumulating the difference between the inflow to a process and its outflow. By decoupling rates of flow, stocks are the source of disequilibrium dynamics in systems.

Stocks and flows are familiar to all of us. The amount of water in a reservoir is a stock. The number of people connected to a water supply system is a stock. Stocks are altered by inflows and outflows. Reservoir storage is increased by the inflow of water provided by tributaries and decreased by the release provided to reservoir users and the needs of aquatic systems downstream from the reservoir. Despite everyday experience of stocks and flows, quite often we fail to distinguish clearly between them. Is water shortage a stock or a flow?

Diagramming notation

System dynamics simulation and the computer tools for its implementation use a particular diagramming notation for stocks and flows. The Vensim notation is shown in Figure 8.13.

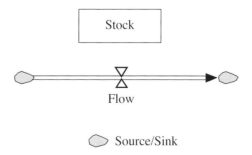

Figure 8.13 *Stock and flow diagramming notation*

Stocks are represented by rectangles (suggesting a container holding the contents of the stock). Stocks can be used to depict both material and non-material accumulations. The magnitudes of stocks within a system persist even if the magnitudes of all the activities fall to zero. When you take a snapshot of a system only the accumulations that the activities had filled and drained would appear in the picture. The picture would show the state of the system at that point in time. Because they accumulate, stocks often act as 'buffers' within a system. In this role, stocks enable inflows and outflows to be out of balance with each other – i.e. out of equilibrium.

Flows are used to depict activities (i.e. things in motion). They are represented by a pipe (arrow) pointing into or out of the stock. Valves (flow regulators) control the flows. Clouds represent the sources and sinks for the flows. A source represents the stock from which a flow originating outside the boundary of the model arises; sinks represent the stocks into which flows leaving the model boundary drain. Sources and sinks are assumed to have infinite capacity and can never constrain the flows they support. If there is an accumulation of something, that accumulation must result from some activity, a flow of something. And if there is a flow of something, there must be an associated build-up or depletion. 'Stuff' flows through the pipe, in the direction indicated by the arrowhead. The flow volume is calculated by the algebraic expression, or number, that you enter into the flow regulator. You can imagine that large volumes cause the spigot to be open wide, while small volumes cause it to be shut down.

Flows can have several attributes. They can be conserved or non-conserved, unidirectional or bidirectional, and not unit-converted or unit converted. A conserved flow draws down one stock as it fills another. The 'stuff' that is flowing is conserved, in the sense that it only changes its location within the system (not its magnitude). In most cases, flows are expressed in the same units of 'stuff' as the stocks to which they are attached. For example, a reservoir holds water. The flow would also be expressed in units of water, with the suffix 'per [time]'. When the units are not converted, the only difference between stock and flow expression is in this suffix.

Stocks and flows are inseparable. Both are necessary for generating change over time, or dynamics. If we want only a static snapshot of reality, stocks alone would be sufficient. But without flows, no change in the magnitude of the stocks could occur. In order to move from snapshots to continuous presentations, we need flows. Figure 8.13 is on the CD-ROM, in the directory SYSTEM DYNAMICS, subdirectory Examples, Example 3.

The structure of all stock and flow diagrams is composed of these elements. Vensim notation offers two more graphical objects that complete the system dynamics syntax. They are auxiliary variables and arrows.

Auxiliary variables often modify the activities within the system. They transform inputs into outputs. They can represent either information or material quantities. They are often used to break out the detail of the logic which otherwise would be buried within a flow regulator. Unlike stocks, auxiliary variables do *not* accumulate. The value for an auxiliary variable is recalculated from scratch in each time step. Auxiliary variables thus have no 'memory'. Auxiliary variables play one of four roles: stock-related, flow-related, stock/flow related and external input-related. In their stock-related role, they can provide an alternative way to measure the magnitude of a stock, and are sometimes used to substitute for a stock. In their flow-related role, they can be used to 'roll up' the net of several flow processes, or to break out the components of the logic of a flow, so as to avoid diagram clutter. Finally, auxiliary variables can be used as external inputs including some time series inputs (often implemented via the graphical function), as well as various built-in functions.

Arrows link stocks to auxiliary variables, stocks to flow regulators, flow regulators to flow regulators, auxiliary variables to flow regulators, and auxiliary variables to other auxiliary variables. Arrows represent inputs and outputs, *not* inflows and outflows! Arrows do not take on numerical values. They only transmit values taken on by other building blocks.

Identifying stocks and flows

The distinction between stocks and flows is very important and sometimes not obvious. In mathematics, system dynamics, control theory and related engineering disciplines, stocks are also known as *integrals* or *state variables*. Flows are also known as *rates* or *derivatives*.

The units of measure can help you distinguish stocks from flows. Stocks are usually a quantity such as amount of water in storage, people employed, or $ in an account. The associated flows must be measured in the same units *per time*: for example, the rate at which water is added per second to the storage, the hiring rate in workers per month, or the rate of expenditure from an account in $/day. Note that the choice of time period is arbitrary. You are free to select any measurement system you like as long as you remain consistent.

System dynamics principles

Forrester (1990) presents a set of principles that are of help in understanding and implementing system dynamics simulations. They are reproduced here.

> *Principle 1. A feedback system is a closed system.* Its dynamic behaviour arises within its internal structure. Any interaction that is essential to the behaviour mode being investigated must be included inside the system boundary.

Principle 2. Every decision is made within a feedback loop. The decision controls action which alters the system state which influences the decision. A decision process can be part of more than one feedback loop.

Principle 3. The feedback loop is the basic structural element of the system. Dynamic behaviour is generated by feedback. The more complex systems are aggregations of interacting feedback loops.

Principle 4. A feedback loop consists of stocks and flows. Except for constants these two are sufficient to represent a feedback loop. Both are necessary.

Principle 5. Stocks are integrations. The stocks integrate the results of action in a system. The stock variables cannot change instantaneously. The stocks create system continuity between points in time.

Principle 6. Stocks are changed only by the flows. A stock variable is computed by the change, due to flow variables, that alters the previous value of the stock. The earlier value of the stock is carried forward from the previous period. It is altered by flows over the intervening time interval. The present value of a stock variable can be computed without the present values of any other stock variables.

Principle 7. Stocks and flows are not distinguished by units of measure. The units of measure of a variable do not distinguish between a stock and a flow. The identification must recognize the difference between a variable created by integration and one that is a policy statement in the system.

Principle 8. No flow can be measured except as an average over a period of time. No flow can control another flow without an intervening stock variable.

Principle 9. Flows depend only on stocks and constants. No flow variable depends directly on any other flow variable. The flow equations of a system are of simple algebraic form. They do not involve time or the solution interval. They are not dependent on their own past values.

Principle 10. Stock and flow variables must alternate. Any path through the structure of a system encounters alternating stock and flow variables.

Principle 11. Stocks completely describe the system condition. Only the values of the stock variables are needed to fully describe the condition of a system. Flow variables are not needed because they can be computed from stocks.

Principle 12. A policy or flow equation recognizes a local goal towards which that decision strives. It compares the goal with the current system condition to detect a discrepancy, and uses the discrepancy to guide the action.

Example 5

Identify at least one accumulation that exists within you.

An example could be 'knowledge'. The input into the 'knowledge' stock is learning, and the output is forgetting.

Example 6

Consider the following situation: I am teaching you what I know. Figure 8.14 provides two possible flow diagrams of that process. For each one, explain what you like and do not like about the map as a representation of the process.

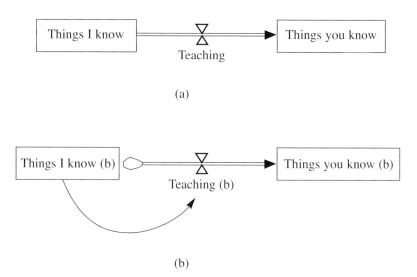

Figure 8.14 *Example flow diagrams*

Figure 8.14 is on the CD-ROM, in the directory SYSTEM DYNAMICS, subdirectory Examples, Example 4. Careful analysis of the two sketches in Figure 8.14 will show a difference in the interpretation of teaching activity. In both graphs 'teaching' is shown as flow. By definition, flow changes the attached stocks. The interpretation of Figure 8.14a is as follows. Through 'teaching', the stock of 'things I know' is being depleted and the stock of 'things you know' is increasing. The last part of this statement is correct but the initial part is not. Teaching as an activity does not drain the stock of the instructor's knowledge. Therefore, Figure 8.14a is not correct. A more appropriate flow diagram of teaching activity is Figure 8.14b, where the stock of 'things I know' provides the source of information that goes into the 'teaching' flow, which helps the stock of 'things you know' to increase.

Example 7

Create an annotated causal diagram and a flow diagram to represent each of the key feedback loop processes described below:

(a) The activity of construction causes a city to grow. As the city grows and more people come, construction also increases since there are more people.
(b) At the height of the Star War (early 3200), the Dreamdisintegration arms race was a cause of great public concern. Since neither the Zonks nor the Grokins had perfect knowledge of the size of the other's actual arsenal, neither side could be quite sure whether it had 'enough' weapons relative to the other. When the Zonks perceived that the Grokins had increased their arsenal, the Zonks stepped up their production of weapons in order to keep up. Similar behaviour was exhibited by the Grokins.
(c) When John performs well in the class, his self-confidence grows. With this increasing self-confidence comes even better performance.

Let us look at each of these in turn.

(a) Figure 8.15 shows an annotated causal diagram (a) and flow diagram (b) for our problem.

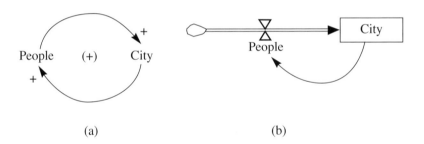

Figure 8.15 *City growth*

Two variables are selected to describe the problem: *City* – a number of people living in the city [people]; and *People* – a number of people moving to the city [people/time interval]. The loop shown in the diagram is a positive feedback loop. Note that construction is represented here as an activity that is implicitly part of the two variables representing stock and flow.

(b) The Dreamdisintegration arms race causal diagram and corresponding flow diagram are shown in Figure 8.16.

The Dreamdisintegration arms build-up is another example of a positive feedback loop with two stocks and two flows. Exponential growth in both stocks is the product of the link that exists between the stocks and flows as shown in the flow diagram (Figure 8.16b).

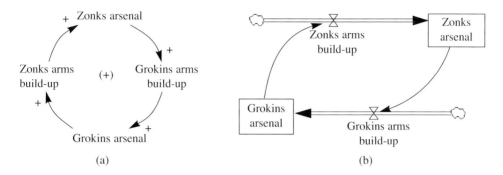

Figure 8.16 *Dreamdisintegration arms race*

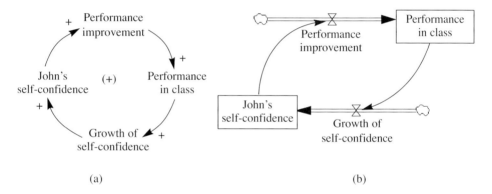

Figure 8.17 *John's performance in class*

(c) John's performance in class is shown in Figure 8.17.

The loop representing John's performance is again a positive feedback loop. In this example both John's performance in class and John's self-confidence are shown as stocks, and the increase in both is shown using flows. Note the reinforcing character of this loop, and consider what units can be used to represent the variables.

Example 8

Develop flow diagrams to represent each of the activities described below:

(a) Generation of noise pollution at the construction site.
(b) Regulation of room temperature with a thermostat.
(c) Generation of solid waste (optional: show the impact of recycling).

(a) Figure 8.18 is a flow diagram for the construction site noise pollution problem.

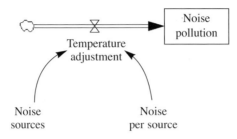

Figure 8.18 *Construction site noise pollution flow diagram*

This diagram shows an assumption that the construction site contains a number of machines that can be considered as *noise sources*. Each of them contributes to the noise accumulation at the location of the receiver. An auxiliary variable *noise per source* is used to assist in finding the cumulative noise at the receiver location.

(b) Figure 8.19 is a flow diagram of thermostat temperature control.

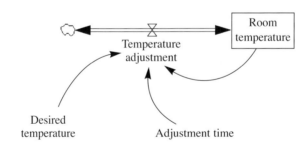

Figure 8.19 *Thermostat temperature control flow diagram*

This flow diagram contains a number of new concepts. *Room temperature* and *desired temperature* are shown as stocks. *Temperature adjustment* is shown as a flow. The existence of arrows on both sides of the flow graphical object is an indication of bidirectional flow. (This has been chosen to make the flow diagram more general.) Flow to the *room temperature* stock corresponds to heating conditions, and flow from the *room temperature* stock describes cooling conditions. Another important concept in this flow diagram is an *adjustment time* auxiliary variable. It is introduced to capture the delay that exists between the observation of a temperature difference and the *room temperature* stock reaching the desired temperature (i.e. the heating or cooling period). This is obviously a function of the furnace or air-conditioning system capacity.

(c) The solid waste generation flow diagram is shown in Figure 8.20.

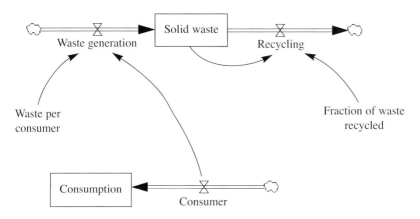

Figure 8.20 *Solid waste generation flow diagram*

The impact of solid waste recycling is introduced through the *recycling* outflow, which reduces the *solid waste* stock taking into calculation the recycling capacity through the auxiliary variable *fraction of waste recycled*.

Figures 8.18, 8.19 and 8.20 are included on the CD-ROM, in the directory SYSTEM DYNAMICS, subdirectory Examples, Example 5.

8.2.5 Introduction to system dynamics simulation

The stock and flow diagramming notation from the previous section is based on a hydraulic metaphor (Sterman, 2000) – the flow of water into and out of a reservoir (Figure 8.21).

Despite the simple metaphor, the stock and flow diagram has a precise mathematical meaning. Stocks accumulate or integrate their flows:

$$Stock(t) = \int_{t_0}^{t} [Inflow(s) - Outflow(s)]ds + Stock(t_0) \qquad (8.2)$$

where *inflow(s)* is the value of the inflow at any time *s* between the initial time t_0 and the current time *t*.

Equivalently, the net rate of change of any stock can be represented with its derivative:

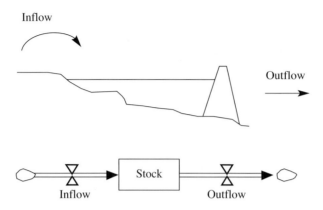

Figure 8.21 *Hydraulic metaphor for stock and flow diagram*

$$\frac{d(Stock)}{dt} = Inflow(t) - Outflow(t) \qquad (8.3)$$

The differential equation (8.3) is the basis of system dynamics simulation. System dynamics simulation software tools like Vensim use the principles of object-oriented programming to help users develop the model structure using objects that represent stocks, flows, auxiliary variables and arrows. The mathematical equations corresponding to a particular structure are written by the tool itself.

System dynamics models are systems of nonlinear ordinary differential equations such as equation (8.3). With simpler notation – replacement of the integral sign with the INTEGRAL () function; replacement of Inflow with I, Outflow with O and Stock with S – we can rewrite equation (8.2) as follows:

$$S_t = INTEGRAL(I_t - O_t, S_{t0}) \qquad (8.4)$$

and express inflows and outflows as:

$$I_t = f(S_t, U_t, C) \qquad (8.5)$$
$$O_t = g(S_t, U_t, C)$$

where U is any exogenous variable and C is a constant.

System dynamics models are systems of nonlinear ordinary differential equations. For any realistic system dynamics model, analytic solutions cannot be found and the behaviour of the model must be computed numerically. Most system dynamics tools like Vensim provide a basic tool for numerical integration known as the

Euler method, and some more sophisticated tools like Runge-Kutta. Details of numerical integration methods can be found, for example, in Atkinson (1985). We shall concentrate here on some basic principles and practical considerations.

Denoting the time interval between periods as *dt*, the assumption of constant flows during the time interval implies:

$$S_{t+dt} = S_t + dt \times (I_t - O_t) \tag{8.6}$$

Equation (8.6) is the most basic technique, known as *Euler integration*. The assumption that the flows remain constant throughout the time interval *dt* is reasonable if the dynamics of the system are slow enough and *dt* is small enough. The definitions of 'reasonable' and 'slow enough' depend on the required accuracy, which depends on the purpose of the model. As the time step gets smaller, the accuracy of Euler's approximation improves. At the limit, when *dt* becomes an infinitesimal moment of time, equation (8.6) reduces to the exact continuous-time differential equation governing the dynamics of the system:

$$\lim_{dt \to 0} \frac{S_{t+dt} - S_t}{dt} = \frac{dS}{dt} = (I_t - O_t) \tag{8.7}$$

Vensim uses Euler integration as its default simulation method. The only difference between the numerical and analytic solution of the underlying differential equation system is the size of *dt*. The differential equation uses an infinitesimal, a true instant. Digital computers use discrete steps and a finite time step. The use of a finite time step and resulting approximations of flows over the interval introduce error, known as integration error or *dt* error. This error depends on how quickly the flows change relative to the time step. The faster the dynamics of the system, or the longer the *dt*, the larger the integration error. That points us to one of the most common questions about using system dynamics simulation: how should we select the time step? Here are some practical recommendations:

- Select a time step for your model that is a power of 2, such as 2, 1, 0.5, 0.25, etc.
- Make sure your time step is evenly divisible into the interval between data points.
- Select a time step one-fourth to one-tenth as large as the smallest time constant in your model.
- Test for integration error by cutting the time step in half and running the model again. If there are no significant differences, then the original value is fine. If the behaviour changes significantly, continue to cut the time step in half until the differences in behaviour no longer matter.
- Note that Euler integration is almost always fine in models where there are large errors in parameters, initial conditions, historical data and especially model

structure. Test the robustness of your results to Euler by running the model with a higher-order method such as fourth-order Runge-Kutta. If there are no significant differences, Euler is fine.

Euler integration is simple and adequate for many water resources management applications. It is advisable to spend more time on improving the model rather than fine-tuning the numerical integration method. However, there are some systems and some model purposes where Euler is not appropriate, because either the errors it generates are too large or the time step required to gain the needed accuracy slows model execution too much.

There are many more advanced techniques for numerical integration of differential equations. The most popular of those available in Vensim are the Runge-Kutta methods. Euler's method assumes the flows at time t remain constant over the entire interval to time $t + dt$, that is, that the average flow over the interval equals the flow at the start of the interval. The Runge-Kutta method finds a better approximation of the average rate between t and $t + dt$. First, provisional estimates of the stocks at $t + dt$ are calculated by Euler's method. Next the flows at time $t + dt$ are calculated from the Euler estimate of the stocks at time $t + dt$. The estimated flows at time t and $t + dt$ are averaged and used to calculate the value of the stocks at $t + dt$. This method is known as second-order Runge-Kutta.

Higher-order Runge-Kutta methods work in essentially the same way but estimate the average flow over subintervals within $[t, t + dt]$ to yield a still better approximation. Vensim offers the fourth-order Runge-Kutta. While Runge-Kutta requires more computation per time step, the accuracy of the approximation is much greater than with Euler's method. Integration errors for a comparable choice of dt are much smaller and propagate at much smaller rates, allowing the modeller to use a larger time step or gain additional accuracy. For the details of Runge-Kutta methods available in Vensim consult the Vensim Reference Manual (Ventana Systems, 2003a) available for download from the Ventana website (http://www.vensim.com, last accessed December 2005).

Example 9

Let us consider a small community that is growing in population. Water supply for the community comes from the groundwater aquifer, which can support the growth up to a certain level. With more *potential customers* in the community, *water production* from the aquifer will increase. With higher *water production*, the more *actual customers* there will be. However, the higher *water production* brings awareness that the aquifer has limited capacity, and that inversely affects *potential customers*. The more water produced reduces the aquifer capacity and makes less water available for new customers.

The annotated causal diagram of this example is in Figure 8.22. We can see that this problem is an example of a negative feedback loop.

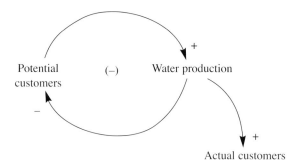

Figure 8.22 *Groundwater aquifer production causal diagram*

The corresponding stock and flow diagram containing two stocks and one flow is shown in Figure 8.23.

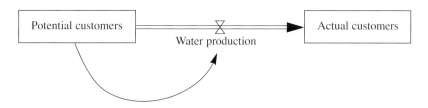

Figure 8.23 *Groundwater aquifer production flow diagram*

The number of *potential customers* at any time *t* is equal to the number of *potential customers* at the starting time minus the number that have left the community because of water production problems. If *water production* is measured in customers supplied per unit time, and there were initially 1 million *potential customers*, then:

$$Potential\ customers(t) = 1{,}000{,}000 - \int_0^t Water\ production(\tau)d\tau \qquad (8.8)$$

where the initial time $t = 0$, and τ is the integration variable.

Similarly, if we assume that there were initially zero *actual customers*, then

$$Actual\ customers = \int_0^t Water\ production(\tau)d\tau \qquad (8.9)$$

The process illustrated by these two equations can be generalized to any stock. In the case of software such as Vensim, once we have drawn a stock and flow diagram like the one shown in Figure 8.23, the program enters the equation for the value of any stock at any time without your having to give any additional information except the initial value for the stock.

However, we must enter the equation for the flows. There are many possible flow equations which are consistent with the stock and flow diagram in Figure 8.23. Let's say that if we provide the accurate information about available water supply to *potential customers*, then 2.5 per cent of the *potential customers* each month may decide to move to the community and become *actual customers* of the water agency. Then the flow equation is:

$$\text{Water production}(t) = 0.025 \times \text{Potential customers}(t) \tag{8.10}$$

If you are familiar with solving differential equations, you can solve equations (8.8) and (8.9) in combination with equation (8.10) to obtain a graph of *potential customers* over time. However, it quickly becomes infeasible to solve such equations by hand as the number of stocks and flows increases, or if the equations for the stocks are more complex than in this case.

The objective here is to illustrate how this is done using Vensim. The stock and flow diagram shown in Figure 8.23 is created using Vensim (it is available on the CD-ROM, in the directory SYSTEM DYNAMICS, subdirectory Examples, Example 6). If we enter the initial values for the two stocks into the model, and also the equation for the flow, we can tell the system to solve the set of equations. This solution process is referred to as system dynamics simulation, and the result is a time history for each of the variables in the model. The time history for any particular variable can be displayed in either graphical or tabular form.

Figure 8.24 shows the Vensim equations for the model using equations (8.8) and (8.9) for the two stocks, and equation (8.10) for the flow. These equations are numbered and listed in alphabetical order. Equation (1) in Figure 8.24 corresponds to equation (8.9), and equation (4) in Figure 8.24 corresponds to equation (8.8). These are the equations for the two stock variables in the model. The notation is straightforward. The function name INTEG stands for 'integration' and has two arguments. The first argument includes the flows into the stock, where flows out are entered with a minus sign. The second argument gives the initial value of the stock. Equation (7) in Figure 8.24 corresponds to equation (8.10). This equation is for the flow variable in the model, and is a straightforward translation of the corresponding mathematical equation. Equation (3) in Figure 8.24 sets the lower limit for the integrals. Thus, the equation INITIAL TIME = 0 corresponds to the lower limits of $t = 0$ in equations (8.8) and (8.9). Equation (2) in the Figure 8.24 sets the last time for which the simulation is to be run. Thus, with FINAL TIME = 100, the values of the various variables will be calculated from the INITIAL TIME (which is zero) until a time of 100 (that is, $t = 100$). Equations (5) and (6) in Figure 8.24 set characteristics of the simulation process.

(1) Actual customers = INTEG(Water production , 0)
(2) FINAL TIME = 100
    The final time for the simulation.
(3) INITIAL TIME = 0
    The initial time for the simulation.
(4) Potential customers – INGEG (– Water production , 1e+006)
(5) SAVEPER = TIME STEP
    The frequency with which output is stored.
(6) TIME STEP = 1
    The time step for the simulation.
(7) Water production = 0.025 * Potential customers

Figure 8.24 *Vensim equations for the groundwater production model*

Figure 8.25 shows the time histories for the *potential customers* and *water production* variables produced using the equations in Figure 8.24. *Water production* decreases in what appears to be an exponential manner from an initial value of 25,000, and similarly *potential customers* also decrease in an exponential manner. The results presented in Figure 8.25 were obtained using Euler's integration (the default method for Vensim).

8.2.6 Formulating and analysing a system dynamics simulation model

Although there is no universally accepted process for developing and using good-quality system dynamics models, there are some basic practices that are quite commonly used (Ford and Flynn, 2005). The following steps are a useful guideline (Ventana Systems, 2003b).

Issue statement

The issue statement is simply a statement of the problem which makes it clear what the purpose of the model will be. Clarity of purpose is essential to effective model development. It is difficult to develop a model of a system or process without specifying how the system needs to be improved or what specific behaviour is problematic. Having a clear problem in mind makes it easier to develop models with good practical applicability.

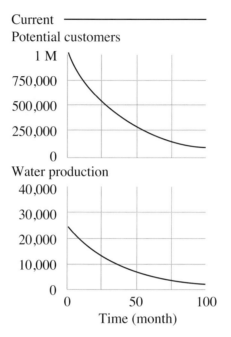

Figure 8.25 *Vensim output time histories for groundwater production*

Variable identification

Identify some key quantities that will need to be included in the model for the model to be able to address the issues at hand. Usually a number of these are very obvious. It can sometimes be useful just to write down all of the variables that might be important and try to rank them in order to identify the most important ones.

Reference modes

A reference mode is a pattern of behaviour over time. Reference modes are drawn as graphs over time for key variables, but are not necessarily graphs of observed behaviour. Rather, they are cartoons that show a particular characteristic of behaviour that is interesting. For example, a water company's water sales history may be growing but bumpy, and the reference mode may be the up and down movement around the growth trend. Reference modes can refer to either past behaviour or future behaviour. They can represent what you expect to have happen, what you fear will happen and what you hope will happen. They should be drawn with an explicitly labelled time axis to help refine, clarify and bound a problem statement.

Reality check

Define some reality check statements about how things must interrelate. These include a basic understanding of what actors are involved and how they interact, along with the consequences for some variables of significant changes in other variables. Reality check information is often simply recorded as notes about what connections need to exist. It is based on knowledge of the system being modelled.

Dynamic hypotheses

A dynamic hypothesis is a theory about what structure exists that generates the reference modes. A dynamic hypothesis can be stated verbally, as a causal loop diagram or as a stock and flow diagram. The dynamic hypotheses you generate can be used to determine what will be kept in models, and what will be excluded. Like all hypotheses, dynamic hypotheses are not always right. Refinement and revision is an important part of developing good models.

Simulation model

A simulation model is the refinement and closure of a set of dynamic hypotheses to an explicit set of mathematical relationships. Simulation models generate behaviour through simulation. A simulation model provides a laboratory in which you can experiment to understand how different elements of structure determine behaviour.

This process is iterative and flexible. As you continue to work with a problem you will gain understanding that changes the way you need to think about the things you have done before.

Vensim provides explicit support for naming variables, writing reality check information, developing dynamic hypotheses and building simulation models. Creating good issue statements and developing reference modes can easily be done with pencil and paper or using other technologies. Dynamic hypotheses can be developed as visual models in Vensim, or simply sketched out with pencil and paper. Simulation is one stage where it is necessary to use the computer for at least part of the process.

This section illustrates the development of a simple system dynamics simulation model (that is, the last step of the modelling process elaborated above). Specifically, we develop and investigate a prototype 'bathtub' model, which is a part of many water resources problems with storage system component. The purpose of this example is to familiarize you with what is required to build a system dynamics simulation model, and how such a model can be used.

Example 10

This prototype *bathtub* example asks you to create a simulation model of taking a bath, with the following characteristics:

- The faucet (tap) has a maximum flow of 2 litres/min, and is turned on until the tub is filled. Assume that the tub is considered filled when the volume in it reaches 40 litres.
- You are to bathe for exactly 20 min. Note that during bathing, no water should leave the tub.
- After bathing, you unplug the drain and the water should leave the tub. Assume the outflow rate depends on the amount of water in the tub. (This makes intuitive sense, as a higher pressure head will induce higher exit velocity.) The maximum outflow of 2 litres/min should occur when the tub is completely filled (i.e., at 40 litres) and the minimum outflow of 0 litres/min should occur when the tub is empty (i.e., at 0 litre).

The questions that we wish to answer are: (a) How does the water volume in the tub change over time? and (b) How long will it take for the water to completely drain from the tub?

Let us start the model development using Vensim by specifying the following simulation model time settings: initial time $t_o = 0$, simulation time horizon $t_f = 200$, $dt = 0.5$. The time is in minutes.

The basic model structure should involve one stock variable *water in tub*, and two flow variables, *faucet flow* and *drain flow*. The volume of water in the tub increases through the flow of water from the faucet and decreases through the drain outflow. Figure 8.26 shows the starting model structure.

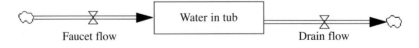

Figure 8.26 *Simple bathtub model flow diagram*

Since the problem statement requires controls of both flows, we shall add two auxiliary variables to our model: *faucet control* and *drain control*. These are the variables that tell the model when to start filling and when to start draining the tub. The modified model structure is in Figure 8.27.

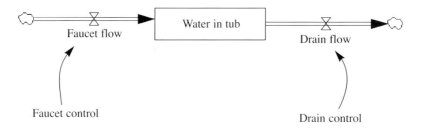

Figure 8.27 *Modified bathtub model structure*

Both faucet and drain control are functions of time. Therefore we shall add one more auxiliary variable, *time*, which will allow time counting for setting our flow controls. Note, that *time* is an existing variable in Vensim, and therefore it will show in our diagram in grey as a shadow variable. (If you attempt to add an existing variable to a model, Vensim will respond with the message that the variable already exists. In that case you should select the shadow variable menu button, <VAR>, which prompts you to select a variable from the list of already existing variables.) *Drain flow* requires a link with the amount of water in the tub. To accommodate that requirement, let us introduce one more auxiliary variable that will assist in expressing *drain flow* as a function of *water in tub*. Let's name this variable *drain function*. The complete bathtub model structure is now shown in Figure 8.28.

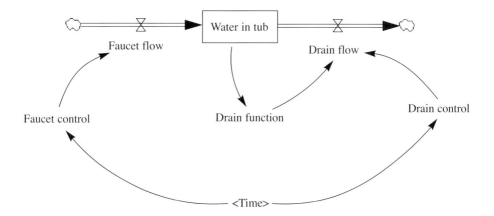

Figure 8.28 *Complete bathtub model structure*

We are now ready to enter the model equations and data. Starting with the stock variable, we shall use an initial value of *water in tub* of 0. *Faucet flow* is constant and equals 2 litres/min. However, in order to enter the flow as required by the problem statement we have to define *faucet control* in the following way:

> IF (*Time* <20) THEN
> *Faucet control* = 1;
> OTHERWISE (8.11)
> *Faucet control* = 0;

Thus, we shall set the *Faucet control* equation as:

> *Faucet flow* = *Faucet control* × 2 (8.12)

A similar logical statement will be used for setting the *drain control* variable:

> IF (*Time* > 40) THEN
> *Drain control* = 1;
> OTHERWISE (8.13)
> *Drain control* = 0;

Next let us define the auxiliary variable called *drain function*, which will be a graphical function of *water in tub*. This function is an increasing function, as more water in the tub implies higher outflow. We shall define the *x*-range from 0 to 40 and the *y*-range from 0 to 2, and enter the two values: (0,0) and (40,2). Note that the input variable is *water in tub*, and that the output variable is *drain function*. This graphical function is shown in Figure 8.29.

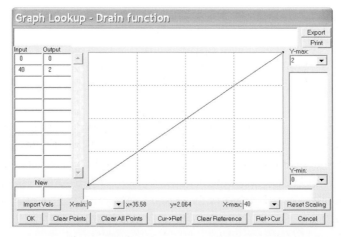

Figure 8.29 *Vensim screen shot of the* Drain function

From the graph in Figure 8.29, when *water in tub* is 0, *drain flow* is 0 litres/min, and when *water in tub* is 40 litres, *drain flow* is 2 litres/min, with all the values in between. At the end we shall define the equation for *drain flow* as:

$$\text{Drain flow} = \text{Drain function} \times \text{Drain control} \tag{8.14}$$

Our model is now complete and ready for simulation. Vensim equations of the complete bathtub model are in Figure 8.30.

(01) Drain control = IF THEN ELSE (Time > 40, 1, 0)
     Units: no units
(02) Drain flow = Drain function * Drain control
     Units: L/min
(03) Drain function = WITH LOOKUP(Water in tub , ([(0,0)-(40,2)],(0,0),(40,2)))
     Units: **undefined**
(04) Faucet control = IF THEN ELSE (Time < 20, 1, 0)
     Units: no units
(05) Faucet flow = 2 * Faucet control
     Units: L/min
(06) FINAL TIME = 200
     Units: Minute
     The final time for the simulation.
(07) INITIAL TIME = 0
     Units: Minute
     The initial time for the simulation.
(08) SAVEPER = TIME STEP
     Units: Minute
     The frequency with which output is stored.
(09) TIME STEP = 0.5
     Units: Minute
     The time step for the simulation.
(10) Water in tub = INTEG(Faucet flow - Drain flow , 0)
     Units: L

Figure 8.30 *Vensim equations of bathtub model*

The first simulation run provides the answer to question (a), How does the water volume in the tub change over time? The answer is shown in Figure 8.31.

The second question (b) How long will it take for the water to completely drain from the tub? can be answered either from reading the graph in Figure 8.31 or by looking at the output table that can be created in Vensim. When rounded to a first decimal place the answer is 159 min. The strip graph in Figure 8.32 provides direct comparison between values of stocks and flows in our example.

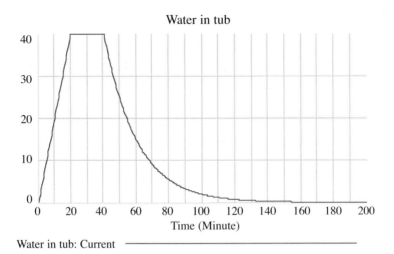

Figure 8.31 *Change of water volume in the tub over time*

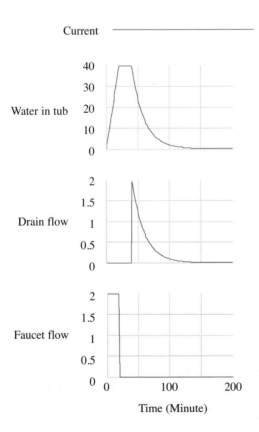

Figure 8.32 *Results of bathtub simulation*

The complete 'bathtub' model is available on the CD-ROM, in directory SYSTEM DYNAMICS, subdirectory Examples, Example 7.

8.2.7 Developing more complex system dynamics models for water resources management

This section presents structures that can be used to represent the water resources decision-making processes within a system dynamics simulation model. The following example is modified after Roberts et al (1983) and Kirkwood (1998).

Example 11: Multi-purpose reservoir operations

This decision problem has two characteristics that make it an appropriate example for models of decision processes. First, the stock and flow variables are easy to determine: the amount of water in the reservoir is a stock, and the flows into and out of the reservoir are flows. Second, it has a structure where the implications of different decision rules can easily be seen. In many water resources management problems, there are several interacting stocks and flows, and thus the impact of changing a single decision rule may be hidden by the complexities of the situation.

Problem description

The Red River Valley has ideal growing conditions for several different types of vegetables, but very little rain. Federal funds were allocated for the construction of the Red Dam in a gorge on the Red River. This dam, together with the Red River Valley Irrigation Project, established an extensive irrigation system throughout the valley, and in the 20 years since the completion of the dam and irrigation system, a prosperous agricultural community has developed there.

The essential features of the reservoir and irrigation system are shown in Figure 8.33. The *Inflow* to the Red Reservoir behind the Red Dam is not under our control. The amount of water in the reservoir is represented by the stock variable *Reservoir volume*. All releases from the reservoir flow into the Red River Valley where the water is primarily used for irrigation of agricultural land. The amount of water available for irrigation at any time is also represented as a stock variable and named *Irrigation supply*. Water is consumed from the Red River Valley in a variety of ways, including evapotranspiration from plants, evaporation and drainage. The flow variable used to describe all of these losses is called *Drainage*. This drainage is not under the control of the Red Dam operator. Thus, there is only one decision variable, the *Release* through the Red Dam.

Figure 8.33 *A simple flow diagram of the Red Reservoir problem*

We shall examine decision policies for managing releases through the Red Dam for use in the Red River Valley. The dam impounds water from a substantial stretch of the Red River, and the average net annual impoundment, after taking into account evaporation losses, is 0.5 million cubic metres (m³). Standard operating procedure at the Red Dam is to maintain a long-term average of 1 million m³ of water behind the dam in the Red Reservoir. However, the actual amount of water in the reservoir may vary over the short term depending on rainfall and other conditions. Not surprisingly, agriculture has expanded in the Red River Valley to consume 0.5 million m³ per year of water. More specifically, the irrigation supply within the valley has 1 mcm of water accessible for agricultural use, and 50 per cent of this is consumed each year.

A reservoir system can have multiple purposes. The rainy season in a region may not coincide with the growing season, and then a reservoir can be used to 'redistribute water in time' from the rainy season to the growing season. If there is flooding in the area, the reservoir can store water during periods of high flow and gradually release it over an extended period of time. If there are periods of drought, the reservoir can store water over several years and release it during dry years.

The primary purpose of the Red Reservoir is to store water which would otherwise flow down the Red River into the ocean, so that this water can be used for agricultural production. Our analysis of the Red Dam operating rules will focus on maintaining sufficient flow to meet the irrigation demand in the Red River Valley, while providing that there is sufficient reserve in the Red Reservoir to meet the needs for irrigation during a drought period. In addition the reservoir releases are managed to prevent spillage of water. Mass conservation results in the requirement that the long-term averages for all flow variables *Inflow*, *Release* and *Drainage* must all be the same.

Reservoir operation options

There are two primary issues that must be addressed in reservoir operations: what variables should be taken into account in the decision model, and how these variables should be combined. Almost all decision processes take into account multiple variables. For the Red Dam, it is clear that any release rule has to consider both the *Reservoir volume* and the *Irrigation supply*. In this case there are explicit or implicit goals with regard to both of these variables. For the *Reservoir volume*, the quantity of water in the reservoir should not be so large that a sudden increase in inflow might lead to spill. Also, the *Reservoir volume* should not be insufficient to provide for irrigation if a drought occurs. In order to prevent flooding in the Red River Valley during high flows and provide secure irrigation during low flows, we maintain a constant value for the *Irrigation supply*.

In this example, our goals are to maintain constant levels for the two stock variables *Reservoir volume* and *Irrigation supply* in Figure 8.33. A variety of different quantitative forms of reservoir operating rules are available to address multiple goals. The two simplest are (a) a weighted-average operating rule and (b) a multiplicative operating rule.

(a) Weighted-average operating rule model

The ideas underlying a weighted-average operating rule model for a flow variable are straightforward:

- A portion of the flow is used to maintain some goal with respect to each of the decision variables, and if the flow deviates from what is needed to maintain that goal, then this portion of the flow should be adjusted.
- Flow adjustments are made over a period of time (that is, averaged) in order to avoid discontinuities in operations, and also to smooth out transient shifts in conditions due to random factors.
- The total flow is made up of a sum of the portions assigned to achieving each goal.
- Different weights are assigned to meeting each goal depending on their relative importance.

Figure 8.34 shows a stock and flow diagram of a weighted-average decision model for the *Release* decision variable. It has been developed from the simple structure shown in Figure 8.33 by adding a variety of auxiliary variables.

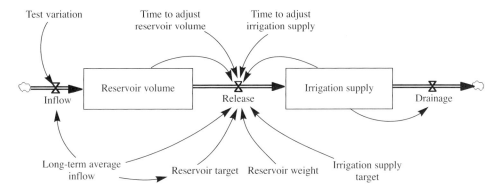

Figure 8.34 *A weighted-average reservoir decision model*

Long-term average inflow is a constant which provides the average flow rate into the Red Reservoir and is equal to 0.5 million m³ per year. This is used to set a target for the amount of water in the Red Reservoir, which is shown in Figure 8.34 as *Reservoir target*. We shall express the target as:

$$\text{Reservoir target} = 2 \times \text{Long-term average inflow} \tag{8.15}$$

Equation (8.15) implies that the reservoir is operated to maintain, on average, two years of inflow.

There is also a target for the amount of accessible irrigation water in the Red River Valley.

$$\text{Irrigation supply target} = 2 \times \text{Long-term average inflow} = 2 \times 0.5 = 1 \text{ million m}^3 \tag{8.16}$$

This target is set to maintain a constant amount of water in the basin over the long term. As noted above, annual irrigation from the valley is 50 per cent of the irrigation supply in the basin. We also know from our earlier discussion that this average irrigation is equal to the *Long-term average inflow*, which is 0.5 million m³.

The two constants *Time to adjust reservoir volume* and *Time to adjust irrigation supply* relate to the averaging period used to address deviations from the goals with respect to the reservoir volume and the irrigation supply, 0.5 and 0.05 respectively. Finally, the constant *Reservoir weight* represents the weight assigned to the goal of maintaining a constant value for *Reservoir volume*, relative to maintaining a constant value for *Irrigation supply*. In our case the weights are equal to 0.5.

In the case of a weighted-average operating rule, the flow is split into several parts which add up to constitute the entire flow. A useful way to develop the decision rule is often to make the flow from a base component needed to maintain stable conditions over the long run, and 'correction' terms needed to address deviations from each of the goals. For the Red Reservoir, the base component is equal to *Long-term average inflow*. The correction term for deviations from the target for the quantity of water in the reservoir is built in three steps. First, note that this correction term should be 0 when the value of *Reservoir volume* is equal to *Reservoir target* (no deviation). Therefore, the correction term is proportional to the difference:

Reservoir volume – Reservoir target

That is, if there is more water in the reservoir than the target, then the release should be increased, while if there is less water than the target, then release should be decreased.

However, if the difference between the *Reservoir volume* and *Reservoir target* is used as the correction term for deviations from the reservoir goal, this would mean that any deviation would be instantly corrected. However, this is not feasible because of physical constraints on the dam, and the correction is averaged over a period of time as follows:

Correction = Reservoir volume – Reservoir target / Time to adjust reservoir volume (8.17)

This means that it will take a length of time equal to *Time to adjust reservoir volume* to completely remove the deviation or:

Reservoir adjustment factor = 1/Time to adjust reservoir volume (8.18)

and then equation (8.17) can be rewritten:

Correction = Reservoir adjustment factor × (Reservoir volume – Reservoir target) (8.19)

By combining equations (8.17), (8.18) and (8.19) we arrive at the correction related to the *Reservoir volume* deviation:

Volume correction = Reservoir weight × ((Reservoir volume – Reservoir target) / Time to adjust reservoir volume) (8.20)

Following the same line of reasoning we can arrive at the correction related to *Irrigation supply* deviation:

Supply correction = (1 – Reservoir weight) × ((Irrigation supply target – Irrigation supply) / Time to adjust irrigation supply) (8.21)

The final expression for the weighted-average release decision rule is obtained by adding the two correction terms in equations (8.20) and (8.21) to the *Long-term average inflow*:

Release = Long-term average inflow + Reservoir weight × (Reservoir volume – Reservoir target) / Time to adjust reservoir volume + (1 – Reservoir weight) × ((Irrigation supply target – Irrigation supply) / Time to adjust irrigation supply) (8.22)

The complete set of Vensim equations for the Red Reservoir management model with a weighted-average additive operating rule is given in Figure 8.35.

The values assumed for the various constants are also shown in Figure 8.35. The initial values of *Reservoir volume* and *Irrigation supply* are set equal to the targets. Thus, so long as *Inflow* continues to be equal to *Long-term average inflow*, the entire process will be in steady state.

As a test input for this simulation model, we can use a step function, as shown by equation (13) in Figure 8.35. The results are shown in Figure 8.36. The set of graphs corresponding to Run 1 shows the results with a *Reservoir weight* equal to 0; the set of graphs of Run 2 shows the results with a *Reservoir weight* equal to 0.5; and the set of graphs of Run 3 shows the results with a *Reservoir weight* equal to 1. Thus, in Runs 1 and 3, only one of the goals is taken into account in setting the Red Reservoir release, while in Run 2 both goals are taken into account.

The dynamics of *Release* are substantially different for the three cases. When there is no weight on the reservoir goal (Run 1) the release remains constant at 0.5 million m^3 per year, and the *Reservoir volume* steadily grows to absorb the extra inflow that is not being released. When there is no weight on the irrigation supply goal (Run 3) the release grows to 0.6 million m^3 per year to stabilize the amount of water in the reservoir, but the *Irrigation supply* grows substantially.

In the case of equal weight (Run 2), the *Irrigation supply* value is closer to the Run 1 case than the Run 3 case. The reason for this can be seen from examining the values for the two constants *Time to adjust irrigation supply* and *Time to adjust reservoir volume* in equations (15) and (16) of Figure 8.35. We see from these equations that *Time to adjust irrigation supply* is one-tenth of *Time to adjust reservoir volume* (0.05 versus 0.5). Thus, adjustments to *Irrigation supply* are made much more quickly than adjustments to *Reservoir volume*, and hence the final results for the equal weight case are closer to the case where all the weight is placed on maintaining a constant value for *Irrigation supply*. This illustrates that the overall performance of a weighted-average decision rule is equally impacted by the weights and the adjustment time constants.

The complete Red Reservoir model with weighted-average operating rule is available on the CD-ROM, in the directory SYSTEM DYNAMICS, subdirectory Examples, Example 8.

(b) Multiplicative operating rule model

Another approach for modelling reservoir operating decision rules is to use a multiplicative form. With the weighted-average form, correction terms are added to a baseflow rate, while with the multiplicative form, correction factors are used to multiply the baseflow rate.

The correction factors are illustrated in Figure 8.37. Figure 8.37a applies to a situation where if the variable of interest is above its target (goal) value the flow

(01) Drainage = 0.5 * Irrigation supply
   Units: mcm
   50% of irrigation supply.
(02) FINAL TIME = 4
   Units: year
   The final time for the simulation.
(03) Inflow = Long-term average inflow + Test variation
   Units: mcm
(04) INITIAL TIME = 0
   Units: year
(05) Irrigation supply = INTEG(Release – Drainage , Irrigation supply target)
   Units: mcm
   Starting with target value.
(06) Irrigation supply target = 1
   Units: mcm
   Twice the annual irrigation use.
(07) Long-term average inflow = 0.5
   Units: mcm
(08) Release = Long-term average inflow + Reservoir weight * (Reservoir volume
   – Reservoir target) / Time to adjust reservoir volume + (1 – Reservoir weight)
   * (Irrigation supply target – Irrigation supply) / Time to adjust irrigation supply
   Units: mcm
   Weighted average operating rule.
(09) Reservoir target = 2 * Long-term average inflow
   Units: mcm
   Two year carry over capacity.
(10) Reservoir volume = INTEG(Inflow - Release , Reservoir target)
   Units: mcm
   Starting with target value
(11) Reservoir weight = 0.5
   Units: no units
   Equal weight given to both stocks.
(12) SAVEPER = TIME STEP
   Units: year
(13) Test variation = STEP (0.1, 0.5)
   Units: mcm
   Step function representing inflow variation.
(14) TIME STEP = 0.01
   Units: year
(15) Time to adjust irrigation supply = 0.05
   Units: year
(16) Time to adjust reservoir volume = 0.5
   Units: year

Note: mcm = million cubic metres.

Figure 8.35 *Vensim equations for the Red Reservoir weighted-average operating rule*

needs to be increased. This is the situation in the Red Reservoir example for *Reservoir volume*. Figure 8.37b applies to a situation where if the variable of interest is above its target value, the flow needs to be reduced. This is the situation in the Red Reservoir example for *Irrigation supply*.

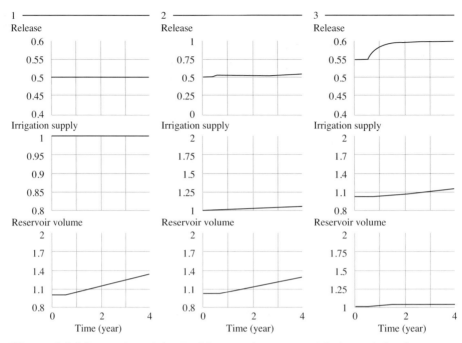

Figure 8.36 *Dynamics of the Red Reservoir system with the weighted-average operating rule*

It is useful to normalize the variables by dividing them by their target values, as shown in Figure 8.37. When this is done, a situation where a normalized variable is equal to one will have a multiplier of 1. Therefore, when the value of a variable is equal to its target value, there will be no correction applied to the base case flow.

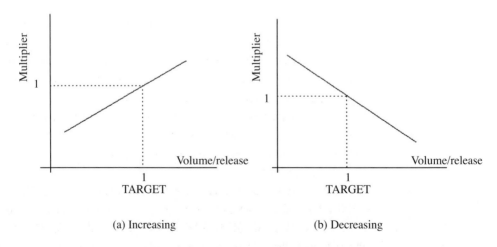

Figure 8.37 *Multipliers*

The slope of the normalized curve sets the strength of the reaction in the flow that occurs for a specified percentage deviation in a variable from its target value. The greater the slope, the greater the response for a specified percentage deviation of a variable.

It is straightforward to derive the equation for the multiplier as a function of the variable, its *Target* and its *Slope*. For the increasing case, this is:

$$Multiplier = (Slope \times Actual/Target) + (1 - Slope) \tag{8.23}$$

and for the decreasing case, this is

$$Multiplier = (1 + Slope) - (Slope \times Actual/Target) \tag{8.24}$$

The results of applying the multiplicative reservoir operating rule approach to the Red Reservoir example are shown in Figures 8.38 (the stock and flow diagram), 8.39 (Vensim equations) and 8.40 (the results of running a simulation with the equations in Figure 8.39).

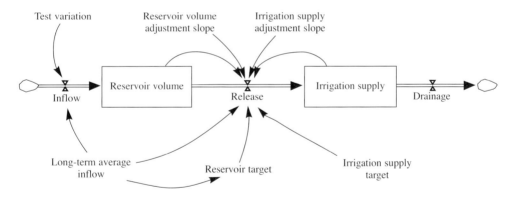

Figure 8.38 *A multiplicative rule reservoir decision model*

The performance of the weighted-average and multiplicative operating rules are similar for small variations from the desired flow, provided the constants in the two models are appropriately adjusted. For larger variations, the multiplicative rule can lead to a more aggressive response than the weighted-average operating rule because the responses for the two variables interact in a multiplicative fashion.

(01) Drainage = 0.5* Irrigation supply
    Units: mcm
(02) FINAL TIME = 4
    Units: year
(03) Inflow = Long-term average inflow + Test variation
    Units: mcm
(04) INITIAL TIME = 0
    Units: year
(05) Irrigation supply = INTEG(Release – Drainage , Irrigation supply target)
    Units: mcm
(06) Irrigation supply adjustment slope = 1
(07) Irrigation supply target = 1
    Units: mcm
(08) Long-term average inflow = 0.5
    Units: mcm
(09) Release = Long-term average inflow * (Reservoir volume adjustment slope
    * (Reservoir volume /Reservoir target) + (1 – Reservoir volume adjustment slope))
    * (1 + Irrigation supply adjustment slope – Irrigation supply adjustment slope
    * (Irrigation supply / Irrigation supply target))
    Units: mcm
(10) Reservoir target = 2* Long-term average inflow
    Units: mcm
(11) Reservoir volume = INTEG(Inflow – Release , Reservoir target)
    Units: mcm
(12) Reservoir volume adjustment slope = 1
(13) SAVEPER = TIME STEP
    Units: year
(14) Test variation = STEP (0.1 , 0.5)
    Units: mcm
(15) TIME STEP = 0.01
    Units: year

Note: mcm = million cubic metres

Figure 8.39 *Vensim equations for the Red Reservoir multiplicative operating rule*

This type of decision rule may be appropriate for modelling some decision-makers. However, the additive model will perform as well as many actual decision-makers. The complete Red Reservoir model with multiplicative operating rule is available on the CD-ROM, in the directory SYSTEM DYNAMICS, subdirectory Examples, Example 9.

8.3 SIMULATION UNDER UNCERTAINTY

The computer simulation model is a formal attempt to construct a computer model of a complex real water resources system to make adequate predictions of its behaviour under different initial and boundary conditions. Deterministic and stochastic

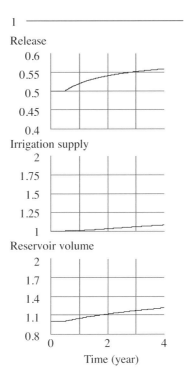

Figure 8.40 *Dynamics of the Red Reservoir system with the multiplicative operating rule*

simulation models are commonly used to simulate the performance of water resources systems, but here we shall consider the less common approach of fuzzy simulation. Fuzzy simulation can be an appropriate approach to include various inherent uncertainties of water resources systems in the simulation process (Pedrycz and Gomide, 1998).

8.3.1 Fuzzy simulation

Among the several commonly used classes of fuzzy simulation models are fuzzy relational equations, fuzzy neural networks, and fuzzy regression models. We shall look at fuzzy regression to simulate the dependency of a water resources system output on its input. Fuzzy regression models are simple tools capable of capturing system uncertainty using fuzzy system parameters. The dependency of an output variable on input variables can be expressed (Klir and Yuan, 1995) as follows:

$$\tilde{Y} = \sum_{i=1}^{n} \tilde{C}_i x_i \tag{8.25}$$

where:
$\tilde{Y}$ = system fuzzy output variable
$\tilde{C}_i$ = fuzzy coefficient
x_i = system real-valued input variable.

Given a set of crisp data observations of system input and output, $(a_1, b_1), (a_2, b_2), \ldots, (a_\mu, b_\mu)$, the fuzzy regression calculates the fuzzy parameters of the assumed model that represents the best fit of these observations. Using a symmetric triangular fuzzy membership function to represent the fuzzy coefficients in the form:

$$\tilde{C}_i(c) = \begin{cases} 1 - \dfrac{|c - c_i|}{s_i}, & \text{if } c_i - s_i \le c \le c_i + s_i \\ 0, & \text{elsewhere} \end{cases} \quad (8.26)$$

where:
c_i = the value at which the parameter $\tilde{C}_i(c)$ membership value = 1 and
s_i = half of the support of $\tilde{C}_i(c)$. The output variable is also a symmetric triangular fuzzy membership number. It can be presented in the following form:

$$\tilde{Y}(y) = \begin{cases} 1 - \dfrac{|y - X^T c|}{s_i |X|}, & \text{if } x \ne 0 \\ 1, & \text{if } x = 0, y \ne 0 \\ 0, & \text{if } x = 0, y = 0 \end{cases} \quad \forall\, y \in R \quad (8.27)$$

where:

$$X = \begin{bmatrix} x_1 \\ x_1 \\ \vdots \\ x_n \end{bmatrix},\; c = \begin{bmatrix} c_1 \\ c_1 \\ \vdots \\ c_n \end{bmatrix},\; s = \begin{bmatrix} s_1 \\ s_1 \\ \vdots \\ s_n \end{bmatrix},\; |X| = \begin{bmatrix} |x_1| \\ |x_1| \\ \vdots \\ |x_n| \end{bmatrix}$$

and T is the transposition operator.

Therefore, the problem is converted into finding vectors c and s such that $\tilde{Y}(y)$ fits the observations as well as possible. The two criteria of goodness of fit are: (i) for each given input observation a_j, the output observation, b_j, should belong to the corresponding fuzzy number $\tilde{Y}_j$ with a grade greater or equal than given h value, as shown in Figure 8.41, where $h \in [0,1]$; i.e. $\tilde{Y}_j(b_j) \ge h$ for each $j, j \in m$; and (ii) the total non-specificity of the fuzzy parameters must be minimized. The non-specificity of parameter $\tilde{C}_i(c)$ is expressed by the value s_i.

The problem of regression parameter selection can be formulated as a simple linear programming (LP) optimization problem (see Chapter 10) as follows:

$$\text{minimize} \sum_{i=1}^{n} s_i$$

subject to: (8.28)

$$(1-h)s^T |a_j| - |b_j - a_j^T c| \geq 0, \quad j \in m$$
$$s_i \geq 0, \quad i \in n$$

Example 12

Here we shall develop a linear fuzzy regression model for evaluating the cost of prefabricated water storage tanks (modified after Terano et al, 1991). The data obtained from the company are as follows:

Input data: x_1 – quality of materials (low quality = 1; average quality = 2; high quality =3); x_2 – bottom level area; x_3 – upper level area; x_4 – number of water chambers; and x_5 – number of pumping chambers.
Output data: y – selling price of the tank (10,000 \$).

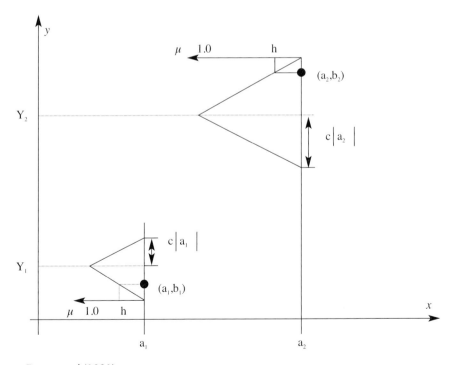

Source: Terano et al (1991)

Figure 8.41 *Typical example of a fuzzy regression model* $\tilde{Y} = \tilde{C}x$

The actual data are presented in Table 8.2. Using these data, let us assume that

$$Y_i = A_0 + A_1 x_{i1} + \cdots + A_5 x_{i5}$$

is the linear possibility function that gives the selling price of company's prefabricated water storage tanks, and A_0 is a fuzzy constant. If, for simplicity, we let fuzzy membership be triangular, that is $L(x) = 1- |x|$, we get the problem for finding the fuzzy coefficient A from equation (8.28):

$$\min_{\alpha,c} J(c) = \sum c^i |x_i|$$
$$y_i \le x_i^t \alpha + (1-h)c^t |x_i|$$
$$y_i \ge x_i^t \alpha + (1-h)c^t |x_i|$$
$$\alpha, c \ge 0$$

Table 8.2 *Input/output data for prefabricated water tanks*

No.	y_i	x_1	x_2	x_3	x_4	x_5
1	606	1	38.09	36.43	5	1
2	710	1	62.10	26.50	6	1
3	808	1	63.76	44.71	7	1
4	826	1	74.52	38.09	8	1
5	865	1	75.38	41.40	7	2
6	852	2	52.99	26.49	4	2
7	917	2	62.93	26.49	5	2
8	1031	2	72.04	33.12	6	3
9	1092	2	76.12	43.06	7	2
10	1203	2	90.26	42.64	7	2
11	1394	3	85.70	31.33	6	3
12	1420	3	95.27	27.64	6	3
13	1601	3	105.98	27.64	6	3
14	1632	3	79.25	66.81	6	3
15	1699	3	120.50	32.25	6	3

The result of the LP problem for $h=0.5$ is:

$A_0^* = (0,0)_L$, $A_1^* = (245.17, 37.63)_L$, $A_2^* = (5.85, 0)_L$, $A_3^* = (4.79, 0)_L$,
$A_4^* = (0,0)_L$, $A_5^* = (0,0)_L$.

You can confirm this calculation using the Linpro program on the CD-ROM (see Chapter 9). The inferred fuzzy values for $Y_i^* = \mathbf{A}^* \times \mathbf{x}_i$ obtained from the odd-numbered data are also shown in Figure 8.42. Since the problem is solved using $h=0.5$, $\mu_{Y_i}(y_i) \geq 0.5$, actual values for $\mu_{Y_i}(y_i)$ are shown in Table 8.3 together with $[Y_i^*]_0 = \{y | \mu_{Y_i^*}(y_i) \geq 0\}$.

We can consider the possible price of prefabricated tank #1 as ranging from $5,671,900 to $7,177,300, and explain the actual $6,070,000 as having been chosen for the actual price. Since A_0, A_4 and A_5 are $(0,0)_L$, and constant, the number of water chambers and number of pumping chambers are not chosen to be variables. Figure 8.42 also shows the relationship between the observed values and the inferred variables. From Table 8.3 we see that samples 4, 6, 13 and 14 for which $\mu_{Y_i}(y_i) = 0.5$ are end points.

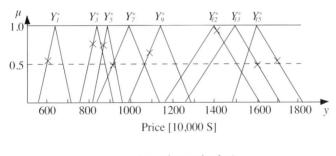

(× : observed value)

Figure 8.42 *Inferred fuzzy output $\tilde{Y}_i^* = \mathbf{A}^* \mathbf{x}_i$ (×: observed value)*

8.4 EXAMPLES OF WATER RESOURCES SIMULATION

Simulation models play an important role in water resources systems management. Many generalized, well-known simulation models were developed primarily for hydrological analyses. Some of the best known are SSARR (streamflow synthesis and reservoir regulation – US Army Corps of Engineers, North Pacific Division), RAS (river analysis system – Hydrologic Engineering Center); QUAL (stream water quality model – Environmental Protection Agency), HEC-5 (simulation of flood control and conservation systems – Hydrologic Engineering Center), SUTRA (saturated–unsaturated transport model – US Geological Survey), and KYPIPE (pipe network analysis – University of Kentucky).

Table 8.3 *Possibility $\mu^*_{Y_i}(y_i)$, lower bounds, centres, upper bounds of 0-level sets for inferred fuzzy output Y^*_i*

No.	Data y_i	$\mu^*_{Y_i}(y_i)$	Lower bound	Centre	Upper bound
1	606	0.516	567,194	642,462	717,730
2	710	0.662	660,199	735,467	810,735
3	808	0.677	757,068	832,336	907,604
4	826	0.500	788,363	863,631	938,899
5	865	0.741	809,239	884,507	959,775
6	852	0.500	776,730	927,266	1,077,800
7	917	0.545	834,908	985,444	1,135,980
8	1031	0.738	919,960	1,070,500	1,221,030
9	1092	0.668	991,414	1,141,950	1,292,490
10	1203	0.869	1,072,160	1,222,700	1,373,240
11	1394	0.969	1,161,240	1,387,050	1,612,850
12	1420	0.976	1,199,600	1,425,400	1,651,210
13	1601	0.500	1,262,280	1,488,090	1,713,890
14	1632	0.500	1,293,300	1,519,100	1,744,910
15	1699	0.540	1,369,330	1,595,140	1,820,940

System dynamics simulation is becoming increasingly popular for modelling water resources systems. Palmer and his colleagues (Palmer et al, 1993; Palmer, 1994) have done work in river basin planning using system dynamics. Keyes and Palmer (1993) used a system dynamics simulation model for drought studies. Matthias and Frederick (1994) have used system dynamics simulation techniques to model the sea-level rise in a coastal area. Fletcher (1998) has used system dynamics as a decision support tool for the management of scarce water resources. Simonovic et al (1997) and Simonovic and Fahmy (1999) have used the system dynamics approach for long-term water resources planning and policy analysis for the Nile River basin in Egypt. Saysel et al (2002) analysed regional agricultural projects based on water resources development using system dynamics simulation, taking into consideration many potential impacts on social and natural environments. A water resources system dynamics model of the Yellow River basin in China (Xu et al, 2002) has been developed for simulating a water resources system and capturing the dynamic character of the main elements affecting water demand and supply. Stave (2003) documented the process of building a strategic-level system dynamics model using the case of water management in Las Vegas, Nevada. The purpose of the model was to increase public understanding of the value of water conservation in Las Vegas. The discussion of a dynamic model of the new irrigated lands around Mazarron and

Aguilas, Spain, which led to the overexploitation of local aquifers and seawater intrusion, is presented in Fernandez and Selma (2004). Comprehensive use of water has been modelled using the system dynamics approach by Fedorovskiy et al (2004). The Idaho National Engineering and Environmental Laboratory has developed a system dynamics model in order to evaluate its utility for modelling large complex hydrological systems like the Bear River basin (Sehlke and Jacobson, 2005).

Here we shall look at five water resources management simulation models, selected from my experience in the application of system dynamics. They illustrate the variety of water resources management situations where system dynamics simulation is a dominant approach. They cover:

- the Shellmouth Reservoir, Manitoba, Canada (Ahmad and Simonovic, 2000)
- a global water assessment model *WorldWater* (Simonovic, 2002)
- a flood evacuation simulation for the Red River basin in Manitoba, Canada (Simonovic and Ahmad, 2005)
- a hydrological simulation for predicting floods from snowmelt (Li and Simonovic, 2002)
- the use of system dynamics simulation in resolving water sharing conflicts (Nandalal and Simonovic, 2003).

8.4.1 Shellmouth Reservoir simulation model

Ahmad and Simonovic (2000) introduced a system dynamics simulation approach for modelling reservoir operations. This approach is more attractive than other systems analysis techniques for modelling reservoir operations because of the ease of modification in response to changes in the system and the ability to perform sensitivity analysis. It has been applied to the Shellmouth Reservoir on the Assiniboine River in Canada. Operating rules are developed for high flow/flood years to minimize flooding. Alternative operating rules are explored by changing the reservoir storage allocation and reservoir outflows. Impacts on the flood management capacity of the reservoir are investigated by simulating a gated spillway in addition to an existing unregulated spillway. Sensitivity analysis is performed on the reservoir levels at the start of the flood season and outflow from the reservoir.

Introduction

The model was developed for a single multi-purpose reservoir, with a focus on its flood management role. It has been used to develop a reservoir operational policy for high-flow years to minimize flooding. It also serves as a tool for studying the impacts of changing reservoir storage allocation and the temporal distribution of reservoir levels and outflows. The general architecture of the model is presented first, followed by a discussion of model sectors and complex dynamic relationships among these sectors.

The model was constructed using the graphical building blocks discussed in Section 8.2.4. Because of the modular nature of the simulation tool, the model is divided into three main sectors: the reservoir, upstream area and downstream area. Figure 8.43 is a schematic diagram of the reservoir and its three sectors.

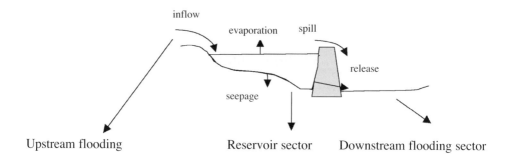

Figure 8.43 *Schematic diagram of the Shellmouth Reservoir*

The reservoir sector is the core sector of the model. Inflows and outflows from the reservoir are the main components of this sector. Flow from all tributaries directly contributing to the reservoir is considered as inflow to the system. Inflow data files, one for each flood year, were provided to the model as input. The total reservoir outflow consists of reservoir releases, spill, evaporation and seepage losses. Reservoir storage is described in terms of a mass balance equation:

$$Storage(t) = Storage(t-1) + (Q_{in} - Q_{out}) \times dt \qquad (8.29)$$

Conduit flow and spillway modules govern the flow through the conduit and the spillway respectively. System constraints, spillway curves and conduit outflow capacity at different gate openings were provided to the model as part of its database. Reservoir operating rules are captured in this sector using if-then-else statements.

The upstream flooding sector calculates the area flooded upstream of the reservoir. Upstream flooding is triggered by a combination of reservoir inflow, reservoir level and reservoir outflow. The number of days when the upstream area is flooded is counted in this sector.

The downstream flooding sector calculates the individual and total flooded area, and duration of flooding as a result of the reservoir operation at selected locations between the dam and the final disposal point of the river. All sources and sinks affecting the flow in the river are introduced in this sector.

Shellmouth Reservoir model

Figure 8.44 (Plate 11) is a schematic diagram of the study area. The Assiniboine River, on which the Shellmouth Reservoir is located, flows through the towns of Russell, St Lazare, Miniota, Griswold, Brandon, Holland and Portage, with the reservoir close to the Manitoba/Saskatchewan border in Canada, and finally joins the Red River at Winnipeg. At Portage, a portion of the river discharge can be diverted to Lake Manitoba through a diversion channel of 710 cubic metres per second (m^3/s) capacity. Headingley is the last station on the river where discharge is measured before Winnipeg.

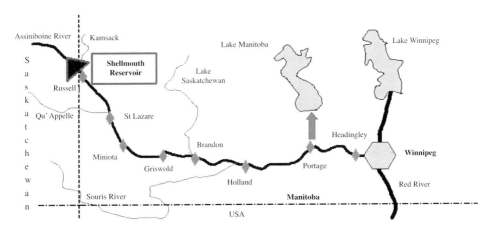

Note: See Plate 11 for a colour version.

Figure 8.44 *Shellmouth reservoir area*

Flooding of the river, mainly caused by heavy spring runoff, has resulted in extensive damage to residential, agricultural and industrial property in the past. The Shellmouth Dam and Reservoir were developed primarily to protect the cities of Brandon and Winnipeg from these floods, and supplementary benefits of the project include flood control to agricultural land in the river valley, but there are still flooding problems in the area. Upstream of the reservoir these are caused by a combination of high water levels in the reservoir and high inflows during flood season. Releases from the reservoir that exceed the channel capacity cause flooding at several locations downstream.

The Shellmouth Dam is a zoned earth-fill embankment, approximately 1319 m long with an average height of 19.8 m. A gated concrete conduit with maximum discharge capacity of 198.2 m^3/s on the east abutment and a concrete chute spillway on the west abutment control outflow from the dam. The reservoir is 56 km in length, 1.28 km in average width and covers a surface area of 61 square kilometres (km^2) when full. The elevation of top of the dam is 435 m above mean sea level, with a dead storage elevation of 417 m. The spillway elevation is 12 m higher, at 429 m. The volume of inactive pool below the conduit invert elevation is 12.3×10^6 m^3. The difference between the volume of the reservoir at the active storage level (370×10^6 m^3) and the crest level of the natural spillway (477×10^6 m^3) is the flood storage capacity of the reservoir, i.e. 107×10^6 m^3. Current operating rules specify that the reservoir should be brought to 185×10^6 m^3 by 31 March to accommodate floods, and a reservoir volume of 370×10^6 m^3 is a goal during the summer months. Maximum reservoir outflow is limited to 42.5 m^3/s to prevent flooding downstream, and the outflow must be greater than 0.71 m^3/s to avoid damage to fish and aquatic life in the river system (Water Resources Branch, 1995). Currently there is no control structure on the spillway to regulate spill from the reservoir (Water Resources Branch, 1992).

Considering these aspects of flooding, the objectives of the simulation modelling study were defined as:

- developing a reservoir operational policy for high-flow years to minimize flooding;
- exploring the impacts on the reservoir flood management capacity of installing gates on an existing unregulated spillway;
- developing a tool for evaluating alternative operating rules by changing the reservoir storage allocation, reservoir levels at the start of the flood season and the reservoir outflows.

The data sets used to set up the reservoir simulation model include:

- reservoir volume curve;
- reservoir area curve;
- reservoir inflow (daily);
- reservoir water levels (daily);
- reservoir operating rules;
- spillway rating curve;
- conduit rating curve;

- relationship between depth of water and area flooded at all points of interest upstream and downstream of the reservoir;
- additional flows joining the Assiniboine at different downstream locations;
- evaporation and seepage losses from the reservoir.

Model development

The main objective of the study was to develop a reservoir operating policy for high-flow years/floods using the system dynamics simulation approach. The five largest flood events in the history of the reservoir, occurring in 1974, 1975, 1976, 1979 and 1995, were selected for simulation. Only inflow was considered as input to the reservoir. Outflow through conduit and spills were considered as total outflow from the reservoir. After defining connections between the three model sectors and components, operating rules were incorporated in the model using logical statements of IF-THEN-ELSE structure:

IF (Res_level>429.3)
AND (Spillway_Control = 0)
AND (TIME >120)
AND (Reservoir_Inflow>Unregulated_Spillway)
THEN (198) (8.30)

This statement in equation (8.30) explains that if the reservoir is full (429.3 m), the unregulated spillway is selected for simulation (it is flooding season, May), and inflow is more than outflow through the unregulated spillway, then the conduit must be operated at its maximum discharge capacity (198 m^3/s). Similarly, if for simulation, a gated spillway option is selected and the reservoir level has reached between 430.5 m and 431.2 m and it is flooding season (late April to mid-June), then outflow should be equal to the inflow to the reservoir:

IF (Spillway_Control = 1)
AND (Res_level >= 430.5)
AND (Res_level <= 431.2)
AND (TIME>110)
AND (TIME <165)
THEN (Reservoir_Inflow) (8.31)

An option is provided in the model to route floods through the reservoir using natural spill or gated spill scenarios. The model uses a spillway rating curve and information on current reservoir level, inflows and time of the year to make decisions about discharges through the spillway. The conduit flow module defines the flow through the gated conduit. Based on which spillway option is active, there are two different sets of operating rules for conduit flow. Once spillway selection is made, this information is automatically passed to conduit control and appropriate conduit operating rules are fired. The current reservoir level, inflows, time of the year and safe channel capacity downstream of the reservoir are criteria on which the quantity of the release through the conduit is based. The quantity of water for diversion at Portage is a function of Lake Manitoba water levels and the capacity of the diversion channel.

Several tests were performed to validate the model and to confirm that the model response matches the response of the system being modelled, including a behaviour replication test, behaviour sensitivity test and behaviour prediction tests. The model development process involved all the steps as summarized in the schematic diagram of a model life cycle (Figure 8.45).

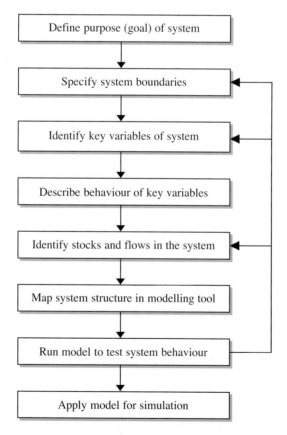

Figure 8.45 *System dynamics model development life cycle*

Model use

The model's main control screen to run the reservoir operation simulations is shown in Figure 8.46. There are five separate input data files for the five largest flood years. The user can select the flood year for simulation using a graphical tool (slider). Choices from 1 to 6 on the slider correspond to different flood years. The spillway module has a slider that provides the user with an option to choose either the unregulated or gated spillway for simulation. Warnings linked to minimum and maximum reservoir levels have been provided in the model in the form of text messages and sounds. A text message '*Spillway will start operating soon*' prompts the user when the reservoir level reaches the spillway crest level. A sound warning in the model is activated when the reservoir reaches the minimum or the maximum level.

While the simulation is running, the user has control over the flow through the conduit and can increase or decrease the discharges as the need arises. As output, the model provides information on variations in the reservoir levels. The model also calculates the number of days when the reservoir is full or at the minimum level, and the number of days the spillway is operated. Other model output includes the number of days of downstream/upstream flooding and the number of days when channel capacity is exceeded as a result of reservoir operation. The model also calculates total and individual areas flooded at several locations along the river as a result of the reservoir operation.

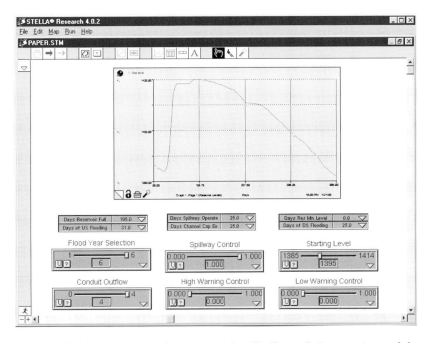

Figure 8.46 *Main control screen of the Shellmouth Reservoir model*

After the model had been developed, and it had been ensured that it was replicating the actual system behaviour, several model runs were carried out. Following each run, the reservoir levels and area flooded through the reservoir operation were carefully studied. Then modifications of operating rules were made with the target of improving the reservoir performance for flood management. The calculation of the flooded area and duration of flooding as a result of reservoir operation provides information on the effectiveness of different operating policies for flood management. Simulation techniques are not capable of generating directly an optimal solution to a reservoir operation problem; however, by going through several runs of a model with alternative policies, an optimal or near-optimal operating policy can be identified.

Simulations of the Shellmouth Reservoir operation were made for the five largest historic floods with natural and gated spill scenarios. Model inputs were the annual series of daily inflows to the reservoir during the five major flood events. Model output included daily variation of the reservoir level, daily discharges from the reservoir, total flooded area upstream of the reservoir, discharges and flooded area at seven downstream locations, and diversion to Lake Manitoba at Portage. Discharges at Headingley were used to estimate the contribution of the Assiniboine River to the flooding of Winnipeg City. Policy alternatives were explored by changing the initial reservoir storage level, both at the start of simulation and at the start of the flooding season. Trials were also made to explore the effects of changing outflow through the conduit on the variation of the reservoir level.

Models are developed to answer questions related to relatively uncertain conditions, and this is especially true in water resources systems management. System dynamics simulation provides a very convenient and powerful tool to explore how changes in one system variable impact other variables. For this study, a sensitivity analysis was performed on the variables *time step* and *delays*. Time step, also called delta time (DT), is the interval of time between model calculations, so DT represents the smallest time interval over which a change in numerical values of any element in the model can occur. The delay function returns a delayed value of input, using a fixed lag time of delay duration. This function is important to capture the timing of the flood peak, as the flow takes five days to reach Winnipeg once it has been released from the dam. With several trials it was found that a time step of one day provided the best trade-off between the speed of calculation and the accuracy of the results. Similarly, a variation of the delay function, used for flood routing, affects the timing and the duration of flooding at downstream locations.

Results and discussion

Daily variations of the reservoir levels for four major flood events (1974, 1975, 1976 and 1979) are shown in Figure 8.47. Selected results are provided in Tables

8.4, 8.5 and 8.6. Revised operating rules with natural spill and gated spill are shown in Figure 8.48 along with existing operating rules. The results show that with revised operating rules it would have been possible to operate the reservoir with only minor flooding upstream and downstream for four out of the five major flood events. When simulating the floods in 1975 and 1976, the spillway was not operated and there was no flooding upstream as well as downstream. For 1974 the spillway was operated for only five days, with a maximum discharge of 9.35 m^3/s, and 70 hectares (ha) of land were flooded. Similarly in 1979 the spillway was operated with a maximum discharge of 37.97 m^3/s and 151 ha of land were flooded. Simulations were made again for the flood events of 1974 and 1979 with the gated spillway option, and it was found that downstream flooding can easily be avoided without increasing flooding upstream of the reservoir.

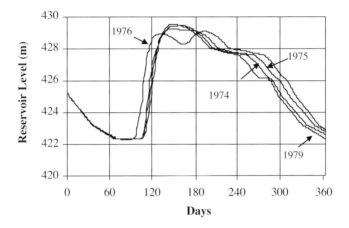

Figure 8.47 *Water levels in the Shellmouth Reservoir for flood years 1974, 1975, 1976 and 1979*

The flood in 1995 has a return period of 100 years, and inflows were well over three times the volume usually experienced. However, this flood event provided an opportunity to look into the advantage of having a gated spillway. With the free spill option, 166 ha upstream and 21,371 ha downstream were flooded for 5 and 38 days respectively (Table 8.4). Peak discharge was reduced from a natural 660.92 m^3/s to 359.45 m^3/s through the reservoir operation. By routing the flood of 1995 through the reservoir with the gated spillway option there was a reduction of about 5000 ha in the flooded area and the flood days were reduced to 23. Maximum outflow was reduced to 223.85 m^3/s, almost a 40 per cent improvement over the unregulated spillway option.

Table 8.4 *Flood management with revised operating rules for selected flood years*

Flood year	Operating rules	Spill	Reservoir full days	Upstream flooding (days)	Downstream flooding (days)	Area flooded (ha)
(1)	(2)	(3)	(4)	(5)	(6)	(7)
1974	Existing	Natural	101	4	11	630
	Revised	Natural	119	6	5	70
	Revised	Gated	125	1	0	0
1976	Existing	Natural	120	2	7	370
	Revised	Natural	158	0	0	0
	Revised	Gated	158	0	0	0
1979	Existing	Natural	106	5	19	1067
	Revised	Natural	121	11	12	151
	Revised	Gated	129	0	0	0
1995	Existing	Natural	161	7	47	24,530
	Revised	Natural	193	5	38	21,537
	Revised	Gated	250	30	23	16,234

Table 8.5 *Impacts on flooding by changing reservoir levels (a) Start of year; (b) Start of flooding season for 1976 flood year without using gated spillway*

	Initial reservoir level (m)	Reservoir full (days)	Upstream flooding (days)	Downstream flooding (days)	Total area flooded (ha)
	(1)	(2)	(3)	(4)	(5)
A	422.2	163	0	0	0
	425.2	163	0	0	0
	428.3	176	2	0	152
	429.2	195	5	0	152
B	422.2	163	0	0	0
	425.2	174	12	17	4790
	428.3	195	13	26	20,160
	429.2	195	13	31	21,030

With the gated spillway the maximum discharge at Headingley was 5.66 m³/s, which is equal to the minimum required flow, compared with 172.2 m³/s with the free spill option. This means that the Assiniboine River's contribution to the flooding of

Table 8.6 *Impacts on flooding by changing flow through the conduit*

Conduit outflow (m³/s(cfs)) (1)	Reservoir full (days) (2)	Upstream flooding (days) (3)	Downstream flooding (days) (4)	Reservoir level at the end of year (m) (5)
45.3 (1600)	157	0	0	422.64
34.0 (1200)	164	0	0	422.79
19.8 (700)	196	10	0	424.37
11.3 (400)	247	10	0	427.33
0 (0)	250	10	0	429.25

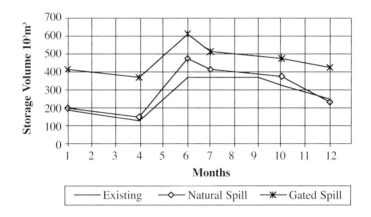

Figure 8.48 *Existing and revised rule curves with natural spill and gated spill*

Winnipeg City was zero. Figure 8.49 shows a comparative graph of the reservoir level variation with and without gates on the spillway for the 1995 flood. The reservoir levels with gated spill are higher than the levels with free spill, as more water is stored during the flood. The reservoir with gated spill reaches a lower level at the end of the simulation because discharges through the conduit, over the falling limb of the hydrograph, are higher to release water stored during the flood.

The 1976 flood year was selected to investigate how initial water levels in the reservoir, at the start of simulation and at the start of flood season, affect the spillway operation and the reservoir levels during the flood; 1976 was selected because this year was the second-largest flood in terms of volume of inflow, and the spillway was not operated during the simulation with the free spill option. Several simulations were carried out by considering different levels at the start of the year

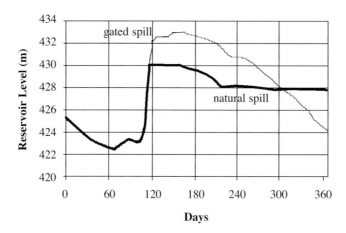

Figure 8.49 *Comparative graph of water levels in the Shellmouth Reservoir (flood year 1995) with natural spill and gated spill*

and at the start of the flooding season. The simulations considered a range of reservoir levels between empty (422.5 m) and full (429.5 m).

The effects of varying the reservoir level at the start of the year are shown in Figure 8.50. It can be noted that if simulation starts in January, the reservoir levels at the start of simulation do not have any serious impact on the reservoir levels during the flood and the total area flooded as a result of reservoir operation (Table 8.5). As floods arrive typically in late April or early May, there is sufficient time to bring the reservoir to a lower level to accommodate incoming floods. Release rules are written in such a way that they adjust outflow based on the information on inflow, time of year and the reservoir level. If the reservoir simulation starts in April, the impact of the initial reservoir level on the reservoir levels during the flood is significant (Figure 8.51). As the flood arrives soon after the simulation starts, there is not enough time to bring the reservoir to a lower level to accommodate the incoming flood. By increasing the initial reservoir level, the number of days when the reservoir is full, the number of days of upstream and downstream flooding and the flooded area are also increased (Table 8.5).

The variation of the reservoir level caused by changing the outflow through the conduit for a range from no flow to maximum flow is shown in Figure 8.52. Data in Table 8.6 and Figure 8.52 support the suggestion that both reservoir levels at the end of simulation and the number of days when the reservoir is full are very sensitive to outflow through the conduit.

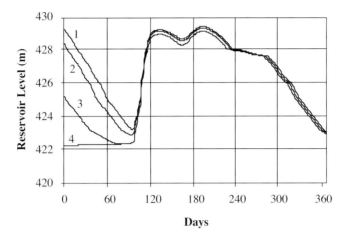

Note: Curve 1 = 422.2 m, Curve 2 = 425.2 m, Curve 3 = 428.3 m and Curve 4 = 429.2 m

Figure 8.50 *Reservoir levels by varying the initial reservoir level at the start of the simulation*

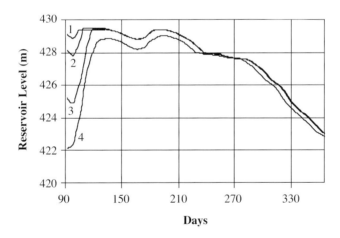

Note: Curve 1 = 422.2 m, Curve 2 = 425.2 m, Curve 3 = 428.3 m and Curve 4 = 429.2 m.

Figure 8.51 *Reservoir levels by varying the initial reservoir level at day 90*

Concluding remarks

The research reported in this case study focused on the simulation of a single multi-purpose reservoir for flood management purposes using the system dynamics simulation approach. Operating rules were revised for high-flow/flood years to minimize flooding. Impacts on the flood management capacity of the reservoir were

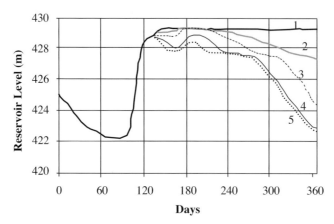

Note: Falling limb of hydrograph at elevation 429.2–428 m. Curve 1 = 0 m³/s, Curve 2 = 11.3 m³/s, Curve 3 = 19.8 m³/s, Curve 4 = 34.0 m³/s and Curve 5 = 45.3 m³/s.

Figure 8.52 *Reservoir levels by varying discharges through the conduit*

explored by simulating a gated spillway in addition to the existing unregulated spillway. Alternative operating rules were explored by changing the reservoir storage allocation, reservoir levels at the start of flooding season and the reservoir outflows.

The Shellmouth Reservoir simulation model can be fine-tuned easily in the light of operating experience, or with the help of insight provided by an expert. The system dynamics simulation approach offers a way for operators to participate in the model-building process, thus increasing their trust in the model. The operator's feedback provides directions for follow-up simulations and modifications in the model structure.

The architecture of the Shellmouth Reservoir model is generic in nature and can be applied to other reservoirs by replacing the Shellmouth data and operating rules with those for another reservoir. Numerous simulation scenarios, in addition to what has been demonstrated in this case study, can be tested using the existing framework. As the current model provides information on the extent and duration of flooding, another sector can be added to calculate damage to crops or economic losses through the lost opportunity of seeding. The model can also be extended from a single multi-purpose reservoir to a system of reservoirs.

8.4.2 Global water resources assessment simulation model

The growing scarcity of clean freshwater is among the most important issues facing civilization in the 21st century. Despite the growing attention to the chronic, pernicious crisis in the world's water resources, our ability to correctly assess and predict global water availability, use and balance is still quite limited. An attempt is presented here to model global world water resources using the system dynamics approach (Simonovic, 2002). In the model, water resources (both quantity and quality) are integrated with five sectors that drive industrial growth: population, agriculture, non-renewable resources, the economy and persistent pollution. The WorldWater model was developed on the basis of the last version of the World3 model (Meadows et al, 1992).

Introduction

Many countries around the world are caught between a growing demand for fresh and clean water, and limited and increasingly polluted water supplies. Many countries face difficult choices. Populations continue to grow rapidly. Yet there is no more water on Earth now than there was a few thousand years ago, when the population was less than 3 per cent of its current size. Today, some 1.1 billion people in developing countries lack access to safe drinking water. In addition, 2.4 billion lack adequate sanitation.

One of the first steps in addressing the crisis is an accurate assessment of water resources and their use, which can then form the basis for future predictions. Methodologically, all previous studies have estimated the quantitative characteristics of renewable water resources using observed river runoff data. The mean value of renewable global water resources is estimated at 42,750 cubic kilometres (km^3) per year, with considerable spatial and temporal variation (Shiklomanov, 2000). Quantitative characteristics of water use on the global scale have been determined using several basic factors, such as the socio-economic development level, population and physiographic (including climatic) features. Combinations of these factors determine the volume and character of water use, its dynamics and future tendencies (Table 8.7). Relationships between the important factors are not explicitly addressed, and their important temporal and spatial dynamics are lost in the integration. Therefore, the prediction of future water use and balance is very difficult and subject to a wide margin of error.

A limited effort has been devoted to the global modelling of water resources in a way that takes into consideration dynamic interactions between the quantitative characteristics of available water resources and water use (Simonovic, 2002). Not one of the existing models is dynamic in nature. Different models treat the water sector with different levels of detail. Dynamic feedback relationships between the physical characteristics of water balance and population growth, development of agriculture and industry, technological development, and use of other resources are not captured explicitly. Therefore, the utility of these models for understanding the impact of water on world development at the global scale is quite limited.

Table 8.7 *Assessment of water use in the world (in km³/year) by sector of economic activity*

Sector	Assessment year							
	1900	1940	1950	1960	1970	1980	1990	1995
Population (million)			2542	3029	3603	4410	5285	5735
Irrigated land area (million ha)	47.3	75.9	101	142	169	198	243	253
Agricultural use	513 / 321	895 / 586	1080 / 722	1481 / 1005	1743 / 1186	2112 / 1445	2425 / 1691	2504 / 1753
Industrial use	21.5 / 4.61	58.9 / 12.5	86.7 / 16.7	118 / 20.6	160 / 28.5	219 / 38.3	305 / 45.0	344 / 49.8
Municipal use	43.7 / 4.81	127 / 11.9	204 / 19.1	339 / 30.6	547 / 51.0	713 / 70.9	735 / 78.8	752 / 82.6
Reservoirs	0.30	7.00	11.1	30.2	76.1	131	167	188
Total (rounded)	579 / 331	1088 / 617	1382 / 768	1968 / 1086	2526 / 1341	3175 / 1686	3633 / 1982	3788 / 2074

Note: First row: water withdrawal; second: water consumption.
Source: IHP (2000)

Global modelling has, however, received attention from many researchers with probably the best-known publication being *The Limits to Growth* (Meadows et al, 1972). This work drew much attention to the global system dynamics modelling of the planet. It reported the results of a study of the future under the present growth conditions. Simulations of the future were performed using a system dynamics model named World3. The main conclusion reached was that if the present growth trends in world population, industrialization, pollution, food production and resource depletion continue unchanged, the limits to growth on this planet will be reached sometime within the next 100 years.

Global world modelling has devoted very little attention to water resources. The main argument is that water is a regional, not a global, resource. This statement is in contrast to the main findings of the water community, which realize the importance of regional differences in solving water-related problems, but clearly indicate the existence of a global water crisis (Cosgrove and Rijbersman, 2000). In the simple accounting of most global modellers (Meadows et al, 1992: 54–57), the renewable flow from which all freshwater inputs to the human economy are taken is estimated to be 40,000 km³/year. When the amount of water flowing to the sea is deducted (28,000 km³/year) and seasonal characteristics of the runoff taken into consideration (5000 km³/year), the human population is left with approximately 7000 km³/year of accessible stable runoff. This figure is compared with the approx-

Box 8.2 Water in Canada

- Almost 9 per cent, or 891,163 km^2, of Canada's total area is covered by freshwater.
- Annually, Canada's rivers discharge 7 per cent of the world's renewable water supply – 105,000 m^3/s.
- Approximately 60 per cent of Canada's freshwater drains north, while 85 per cent of the population live along the southern border with the United States.
- Canada holds 20 per cent of the world's freshwater, but has only 7 per cent of the world's fresh renewable water.
- The highest waterfall in Canada is Della Falls, BC at 440 m.
- The longest Canadian river is the Mackenzie River in the NWT at 4241 km.
- The largest lake entirely in Canada is Great Bear Lake in the NWT at 31,328 km^2.
- The Great Lakes are the largest system of fresh surface water on earth, containing roughly 18 per cent of the world supply.
- One out of every three Canadians and one out of every ten US residents depends on the Great Lakes for water.
- The Great Lakes' coastline accounts for 4 per cent (10,000 km) of the total length of Canada's coasts.

Source: Environment Canada, http://www.ec.gc.ca/water/en/e_ quickfacts.htm (last accessed December 2005)

imate total water use of 3500 km^3/year, leading to the conclusion that we do not have a problem at the global scale. As will be shown later in this section, this accounting is subject to some major flaws. Global world modellers also assume that there are ways to increase the water limit by using technical measures (storage facilities and desalination) as well as some non-technical solutions like bringing people closer to the water or water closer to the people. Many of these concepts are infeasible today. Dams are receiving serious opposition; desalination is still very costly and unaffordable for most of the world (Gleick, 2000); and the other two options of relocating populations or providing long-distance transfer (export) of water are rich with social problems.

This section presents an attempt to model future world development taking into consideration water resources limitations (Simonovic, 2002). A system dynamics simulation approach is used in this work. The two main objectives are (a) to demonstrate the importance of strong feedback links between water availability and the future development of human economy, and (b) to identify the most important water issues on a global scale.

The traditional approach to modelling water balance during the latter part of the 20th century used projections of population growth, unit water demand, agricultural production and industry growth, among other factors (Gleick, 2000). These projections are then used to estimate future water demand and water balance. Gleick demonstrates that future water projections are variants of current trends, and as such are subject to considerable uncertainty (Figure 8.53). The use of different periods for making predictions results in a high variability in the value of the predicted variable. The dynamic character of the main variables and how they will affect water use in the future is not captured through the traditional approach. A system dynamics simulation offers a new way of modelling the future dynamics of complex systems.

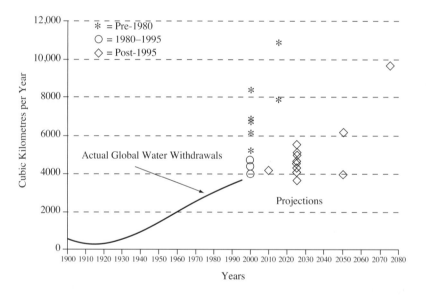

Source: Gleick (2000)

Figure 8.53 *Water scenarios: projected and actual global water withdrawals*

The World3 model

Social systems, technology and the natural environment interact in different ways to produce growth, change and stress. In the past, the main forces of change were dealt with through migration, expansion, economic growth and technology. However, more recently we have become aware of some forces that cannot be resolved through historical solutions. World3 is a system dynamics model of world scope. It comprises five main sectors: population, agriculture (food production, land fertility, and land development and loss), non-renewable resources, the economy (industrial output, services output and jobs) and persistent pollution. The time horizon of the World3 model is 200 years starting from 1900. The period from 1900 to 2000 is used for the calibration of model relationships and verification of model performance.

World3 is a model that attempts to represent the continuous dynamic interaction between the human population and the global resource base. The model contains numerous feedback relations representing demographic and techno-economic means of achieving a balance between the population size and the supply of resources. Two basic characteristics of human populations that are captured by the model are a tendency towards exponential growth, and a long delay in the population's adaptive response to changing external conditions. The major factors affecting population growth that are included in the model are births, deaths, fertility, life expectancy, industrial output, pollution, food, health, service output and crowding, among others.

The main assumption of the agricultural sector of World3 is that the total amount of food that can be produced on Earth each year has some limit. Physical resources that can be allocated to food production are limited. In World3 the available agricultural land is limited, the amount of fertilizer is limited by the total industrial production capacity, and land fertility is limited by pollution absorption mechanisms. The major factors incorporated in this sector include arable land, land yield, land erosion, land fertility, food per capita, agricultural investment and land development.

In World3, non-renewable resources are assumed to be finite, and are defined as mineral or fossil fuel commodities that are essential to industrial production and are regenerated on a timescale longer than the 200-year time horizon of the model. Unknown resources and proven reserves are aggregated into one level that decreases over time as resources are utilized by the industrial sector. Pollution generation is modelled as a function of resource use. The main factors in this sector of the model are non-renewable resources, their usage rate, per capita resource usage and fraction of capital allocated to obtaining resources, among others.

The main objective of the capital sector of the model is to relate the basic components that would demonstrate long-term patterns in the population's access to material goods, services and food. This sector is based on gross national product (GNP) as a measure of historical global productivity. However, industrial output is used in the model instead of GNP. Three categories of capital (service, agricultural, industrial) and four categories of output (service output, agricultural output, production of non-renewable resources and industrial output) are included in World3. Two uses of the model output are defined, consumption and investment. The main factors in this sector of the model are three named categories of capital, investment rates, four named outputs, fractions allocated to consumption and fractions allocated to investment, among others.

Among a broad spectrum of 23 environmental problems associated with demographic and material growth, World3 models a group of material pollutants of potential importance to the world system over the next 100 years. Persistent pollutants in World3 are materials that cause damage to some form of life, are released through many different forms of industrial and agricultural activity, and are sufficiently long-lived that they may be transported through the global environment. These persistent material pollutants include industrial and agricultural chemicals, radioactive isotopes and heavy metals. These are generated at increasing rates, accumulate in the global environment, and there is a delay between their release and the time their full effects on the ecosystem finally appear. The main factors in this sector of the model are pollution generation, appearance and assimilation rates, fraction of resources that are persistent materials, and an industrial and agricultural materials toxicity index, among others.

In the latest version of World3, a concept of adaptive technology is implemented. In the adaptive approach there is a system goal and when the actual system state deviates from the goal in a negative direction, capital is allocated to new technologies.

The WorldWater model

World3 was used as the basis for the development of the WorldWater model (Simonovic, 2002). Two new sectors are introduced (water quantity and quality), together with multiple feedback links between the new sectors and the rest of the model. The graphical presentation of causal relationships between the water and other model sectors is shown in Figure 8.54. The total water stock in the model includes precipitation, ocean resources and non-renewable groundwater resources. The model also takes into account water recycling as a portion of water use (although this is not directly visible in Figure 8.54). The water use side is modelled in a traditional way, to include municipal water use for the needs of the population, industrial and agricultural water needs. However, the most important difference between WorldWater and other global water models is in its ability to address the needs of freshwater resources for the transport and dilution of polluted water.

Most future water balance assessments based on other models are optimistic because they take no account of the qualitative depletion of water resources as a result of the ever-increasing pollution of natural water. This problem is very acute in industrially developed and densely populated regions of the Earth where no efficient wastewater purification takes place. For example in the International Hydrologic Programme (IHP) CD-ROM (IHP, 2000) the following estimates are provided:

By assessments made, in 1995 wastewater volume was 326 km^3/year in Europe, 431 km^3/year in North America, 590 km^3/year in Asia, and 55 km^3/year in Africa. Many countries practice discharging a greater part of wastewater containing harmful substances directly into the hydrographic network. No preliminary purification is carried out. Thus water resources are polluted and their subsequent use becomes unsuitable, especially for water supply to the population. It is estimated that every cubic metre of contaminated wastewater discharged into water bodies and streams make unsuitable 8 to 10 cubic metres of pure water. This means that most regions and countries of the world are already facing the threat of catastrophic qualitative depletion of water resources…

Assumptions used in the development of WorldWater are:

- water is a partially renewable resource;
- water is limiting the growth of population, food production and industry;
- water can be polluted;
- water is a finite resource;
- oceans are an important source of freshwater through desalination;
- pollution consequences of desalination are not incorporated in the model.

One of the most important conceptual assumptions in WorldWater is the hierarchical modelling of water availability. Growing demand in different sectors is being provided for, first of all from renewable surface water resources. The total stock of renewable water resources in the world is estimated at 42,650 km^3/year (Shiklomanov, 2000), and this must be reduced by 67 per cent to account for variability of surface runoff. When the water demand exceeds the available renewable surface water resources, an additional 8.4 km^3/year can be taken from non-renewable groundwater resources. After the demand exceeds the available surface and groundwater resources, water reuse is considered. From all the water used, 55 per cent is being returned into the environment and becomes available for reuse after treatment. A conservative estimate of 20 per cent reuse is used in the model. It is important to note that some countries of the world, such as Israel, are already at the level of 65–80 per cent reuse.

If the demand is still higher than the available supply, desalination of seawater is considered. The general agreement is that desalination is not the solution for the global world water crisis (Gleick, 2000). This process is technologically mature, but energy-intensive and expensive. The total global desalinating capacity of the world is currently 4.82 km^3/year. More than 60 per cent of all capacity is in the oil-rich Middle Eastern countries. The economic attractiveness of desalination is directly tied to the cost of energy. Production of 1 litre of desalinated water requires 2.8 kilo-

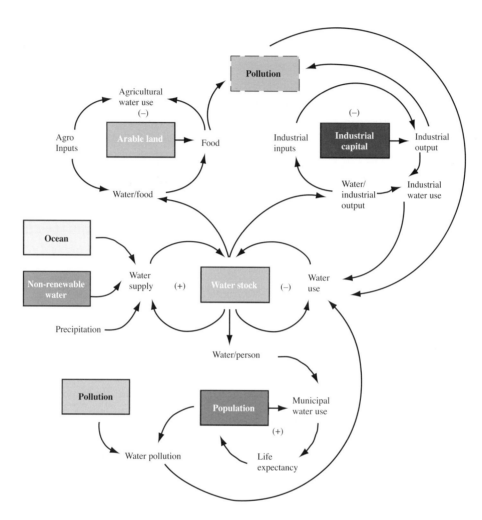

Figure 8.54 *WorldWater model causal diagram*

joules (kJ) of energy. Despite the high current cost ($1–8 per m³) some estimates indicate that the cost could be reduced to $0.55 per m³. There is no consideration of the serious pollution caused by desalination processes. In WorldWater testing, initial runs are made using existing desalination capacity and then this limit is removed, assuming that unlimited seawater will be available for desalination. Therefore, the water quantity limits are replaced with their economic surrogate – capital needs for water production.

Full integration of the water quantity and quality sectors with World3 is established through the development of the following set of relationships (italics – variable in WorldWater; bold – variable in World3):

- *Domestic water use* expressed as a function of **population**.
- *Population without sanitation and water supply* expressed as a function of **population**.
- **Population** expressed as a function of *domestic water supply*.
- **Life expectancy** expressed as a function of *total water quantity*.
- **Life expectancy** expressed as a function of *water quality*.
- *Irrigated land* expressed as a function of **arable land.**
- **Land yield** expressed as a function of *irrigation water supply*.
- *Industrial water use* expressed as a function of **industrial capital.**
- **Industrial output** expressed as a function of *water capital needs*.
- *Urban population* expressed as a function of **industrial output.**
- *Amount of wastewater* expressed as a function of **pollution index.**

Historical data (food production, health sector, industrial production and manufacturing, demographic, etc.) and numerous runs of World3 were used in the development of these relationships. It is important to note that most of the relationships should be strengthened through the enhancement of newly available databases.

Data

The WorldWater model is data-intensive. Most of the data effort was focused on the water quantity and quality sectors, and the relationships used to link these two sectors with the rest of the World3 model. The data required for the five sectors of the original World3 were used, as provided by the authors of the model for twelve different scenarios described in Meadows et al (1992). Historical data on water availability and use are taken from Shiklomanov (2000), Gleick (2000) and IHP (2000). A summary of the basic data is given in Table 8.7.

Using input from the population sector, the model performs a calculation of domestic water needs. The quantification of water use in WorldWater is based on water demand for public and domestic needs, industrial production and agriculture (irrigation). Water losses from large reservoirs are also taken into account. All the future projections have been made ignoring potential anthropogenic global climatic change, i.e. they are for a stationary climatic situation. The volume of public water use depends on the size of an urban population and the services and utilities provided, as well as climate conditions. In many large cities, present water withdrawal amounts to 300–600 litres per day per person. By the end of 2006, the specific per capita urban withdrawal was expected to increase to 500–1000 litres per day per person (Cosgrove and Rijbersman, 2000). On the other hand, in more agricultural developing countries public water withdrawal is a mere 50–100 litres per day. In certain individual regions with insufficient water resources, it is no more that 10–40 litres per day of water per person.

WorldWater uses three municipal water demand scenarios:

- optimistic – assuming growth, between now and the year 2100, in municipal water withdrawal up to the level of today's maximum water supply in the developing world of 2,000 litres per day;
- status quo – maintaining the level of today's water supply in the developing world of 1000 litres per day until the end of the simulation period (year 2100);
- realistic – making a redistribution of municipal water withdrawal between the developed and the developing world, which is seen as ending with overall municipal water withdrawal of 700 litres per day in 2100.

Nonlinear relationships over time have been developed using the available data between 1900 and today.

The relationship between life expectancy and domestic water supply is modelled using a relationship known as the *life expectancy multiplier*. It is represented as an exponential function, indicating that life expectancy is equal to 0 if there is no water available, and reaches a maximum of 1 for maximum supply. Since qualitative data to support this relationship are not available, extensive sensitivity testing of the model results has been performed to evaluate the impacts of changing the shape of this relationship.

The distribution of population between urban and rural is taken from the United Nations Population Fund. These data are then correlated with industrial output (generated by the industrial sector of the model), which is taken as a measure of living standards. WorldWater model simulations are based on the use of this dynamic relationship.

The population sector of the model is linked to the water quality sector through a set of relationships ending in the life expectancy–water quality function. Historic data on rural and urban populations not served by water supply and sanitation services were used to develop this relationship. Available data on mortality as a result of water-related diseases were not used. At the current level of model detail it was impossible to extract more general relationships from the available mortality data.

Water in industry is used for cooling, transportation and washing, as a solvent, and it also sometimes enters the composition of the finished product. Thermal and nuclear power generation lead the list of major users. Analysis of the available data on industrial water use demonstrated a very strong correlation with industrial output and industrial capital. Therefore in the WorldWater model, industrial water use has been expressed as a function of industrial capital.

The development of surface water infrastructure (that is, diversion projects, storage projects and so on), groundwater development and recharge, recycling and desalination require major capital investment. The cost of water supply is taken from Gleick (2000) and expressed in 1980 US dollars. Valid cost estimates and comparisons between different types of technology are extremely difficult to make, given the differences in capital costs, repayment periods, operation and maintenance requirements, and non-economic externalities. The data used in the model are average values for typical facilities built today. There is a wide range of estimates for some costs: desalination costs from $0.55 to $8 per m^3, and recycling from $0.07 to $1.8 per m^3. A sensitivity analysis of model output was performed to evaluate the impact of different cost levels.

Land irrigation has been practised for millennia in order to maximize food production for humanity. At present, about 15 per cent of all cultivated land is being irrigated. Food production from irrigated areas amounts to almost half the total crop production in terms of value. Agriculture accounts for more than 70 per cent of the total water consumption in Europe and in North America. In Asia, Africa and South America, agriculture (and irrigation in particular) is the major component of water withdrawal (65–82 per cent of the total water withdrawal in 1990). The amount of water used for irrigation varies considerably around the world: 7000–11,000 m^3/ha in eastern Europe, 8000–10,000 m^3/ha in the USA, 20,000–25,000 m^3/ha in Africa, and 5000–17,000 m^3/ha in Asia, South and Central America. In the future these values will change considerably as a result of advancements in irrigation systems, and improvements in watering requirements, regimes and techniques. The total irrigation water use data from 1900 to 1995 used in the WorldWater model are from various sources presented in Simonovic (2002).

Two different irrigation scenarios are considered for the future:

- continuation of the current trend: a slow reduction in irrigation water demand to reach on average 9000 m^3/ha by 2100;
- a major irrigation efficiency scenario, which dramatically reduces irrigation water demand to 650 m^3/ha by 2100.

These nonlinear relationships are combined with an increase in the area of irrigated land, which is estimated to be considerably below the level of increase we have witnessed in the past (3 per cent per year between 1970 and 1982; 1.3 per cent between 1982 and 1994). Using data from the Food and Agriculture Organization (FAO) of the United Nations (UN) the increase in agricultural land has been estimated at 0.6 per cent per year for the period 2000–2025, 0.4 per cent for 2025–2050 and 0.3 per cent for 2050–2100. These data are used to define the relationship between available arable and irrigated land. Data from 1900 to 1995 are used for relationship calibration, and those from 1995 to 2000 for relationship verification. Three future scenarios are included in the model:

- no change (ending with 15 per cent of arable land being irrigated in 2100);
- optimistic (ending with 20 per cent of arable land being irrigated in 2100);
- pessimistic (ending with 13.8 per cent of arable land being irrigated in 2100).

The WorldWater model contains an additional relationship to capture the feedback effect of water availability for agriculture on land yield. This relationship is represented as an exponential function that starts with zero land yield (for no water supply) and ends with a maximum value of 1 (for maximum water supply). Qualitative data is not available to describe this relationship. Extensive sensitivity analyses of model output were conducted to evaluate the impacts of different shapes of this relationship.

The development of large surface reservoirs leads to major transformations in the temporal and spatial distribution of river runoff, and an increase in water resources during low-flow limiting periods and dry years. At present, the total volume of the world's reservoirs is about 6000 km^3 and their total surface area is up to 500,000 km^2. Evaporation losses from existing reservoirs are very significant for the global world water balance. Data from Shiklomanov (2000) are used in the WorldWater model. Future reservoir development is thought to be limited as most of the best sites have already been used, and considerable opposition to the future development of reservoirs is being expressed in developed countries. An extension of the data for the next 100 years was performed according to the classical S-shaped growth curve, ending at reservoir losses of just above 300 km^3/year in 2100.

WorldWater differs from most of the available models in its consideration of water pollution. IHP (2000) estimates for 1995 are used to calibrate the relationship between clean water needs to dilute the wastewater being discharged directly into the hydrographic network. A conservative estimate of a total 700 km^3/year of wastewater discharge in 1995 (representing only 50 per cent of the total 1995 estimate in IHP, 2000) is used in WorldWater. This value is then related to the pollution index from the persistent pollution sector to help to develop a predictive relationship between the pollution index and the amount of wastewater generated. A fixed need for clean water for dilution of waste, in the ratio of 9:1, is then applied in the simulations.

Calibration and verification of the WorldWater model with data available for the period between 1900 and the present day ensures there can be confidence in the results of the model and the data being used. Calibration has been performed using the period between 1900 and 1995. Data for a standard run of World3 was used for calibration. The standard run of World3 is based on the 'limits to growth', and assumes that world society proceeds along its historical path as long as possible without major policy change. Since only the period between 1900 and 1995 was of interest for calibration purposes, the standard run assumptions were judged acceptable. The period between 1995 and 2000 was used for the verification of the model and its main relationships. Data for calibration and verification required by the water sectors of WorldWater were taken from IHP (2000) and Cosgrove and Rijbersman (2000). Simulations of future states of the world, for the period from 2000 to 2100, were performed according to different scenarios.

World water dynamics

In the simulated environment of the World3 and WorldWater models, the inherent assumption is one of continuous economic growth. The population in both models will stop growing only when it is rich enough or supporting resources are depleted. The world's resource base is limited and erodable. The feedback loops that connect and inform decisions (Figure 8.54) in WorldWater contain many delays, and the physical processes have considerable momentum. The most common mode of behaviour, as pointed out by Meadows et al (1992), is overshoot and collapse. In *Beyond the Limits*, Meadows et al (1992) investigated a broad range of 12 scenarios. By varying the basic global policy assumptions they showed a range of outcomes, from collapse to sustainability. These options can be used with WorldWater too. However, here we shall discuss the results from just 3 of the 12 scenarios: the standard run, double run and stable run.

The *standard run* is taken from *The Limits to Growth* (Meadows et al, 1972), and assumes that world society proceeds along its historical path as long as possible without major policy change. Population and industry output grow until a combination of environmental and natural resource constraints eliminate the capacity of the capital sector to sustain investment. As capacity falls, the level of food supplies and health services also falls, decreasing life expectancy and raising the death rate.

The global population in this simulation (line 4 in Figure 8.55a, Plate 12) rises from 1.6 billion in the simulated year 1900 to over 6 billion in 2000. Total industrial output expands by a factor of 20 between 1900 and 2000 (line 2 in Figure 8.55a). Between 1900 and 2000 only 27 per cent of the Earth's stock of non-renewable resources is used (line 1 in Figure 8.55a). Pollution in the year 2000 is starting to show a significant rise (line 5 in Figure 8.55a). Food production is increasing and major changes are starting to show in the late 1990s (line 3 in Figure 8.55a).

In this scenario the growth of the economy stops and reverses because a combination of limits are reached. As we can see, after 2000 pollution rises rapidly and begins to seriously affect the fertility of the land. Total food production continues to decline after 2000. This causes the economy to shift more investment into the agriculture sector to try to maintain output. During the first simulated 20 years of the 21st century, the increasing population and industrial output use many more non-renewable resources than in the past. Therefore a shift in capital investment is required to support this development. As both food and non-renewable resources become harder to obtain in this simulated world, capital is diverted to producing more of them.

The WorldWater simulation of this scenario (Figure 8.55b) shows quite a different picture. Water is demonstrated to be the main limitation to future growth. With a rise in pollution and in the demand for clean water after 2010, shortages start to occur in food production (line 3 in Figure 8.55b). Capital resources are drained and industry starts to collapse (line 2 in Figure 8.55b). The total population level is also affected by water and food shortages, decreasing at a faster rate then in the simulated scenario in Figure 8.55a.

The *double run* is a scenario in which the simulated world develops powerful technologies for pollution abatement, land yield enhancement, land protection and the conservation of non-renewable resources. All these technologies are assumed to require capital investment and to take 20 years to be fully implemented. Figure 8.56a shows that their implementation allows growth to continue longer into the future.

The population level rises (line 4 in Figure 8.56a, Plate 13), reaching about 11 billion by 2100. Food production remains adequate but not abundant after 2040 (line 3 in Figure 8.56a). Pollution remains quite low (line 5 in Figure 8.56a). Non-renewable resources do not become scarce despite their constant decrease (line 1 in Figure 8.56a). Industry stagnates and starts to decline after 2020 (line 2 in Figure 8.56a).

Again the WorldWater simulation shows a different future. Because of the major impact of the increasing demand for water and the heavy pollution load, food production starts to decline in 2005 and experiences a major collapse in 2040 (line 3 in Figure 8.56b, Plate 13). This contributes to a total population decline which starts in 2025 and culminates in 2060. With water demand and pollution in decline, more water becomes available for food production, and population levels recover, with the total population reaching 7 billion by 2100 (line 4 in Figure 8.56b). However, capital resources are taken from industry and its decline continues at a slower rate (line 2 in Figure 8.56b).

The *stable run* is taken from *Beyond the Limits* (Meadows et al, 1992), as an illustration of a possible path towards a sustainable world. In this scenario population and industrial growth are moderated and technologies are developed to conserve resources, protect agricultural land, increase land yield and reduce pollution.

In this scenario the population levels off at just under 8 billion by 2060, and people live at a desired standard of living for the rest of the simulation timespan (line 4 in Figure 8.57a, Plate 14). Pollution peaks and falls before it causes irre-

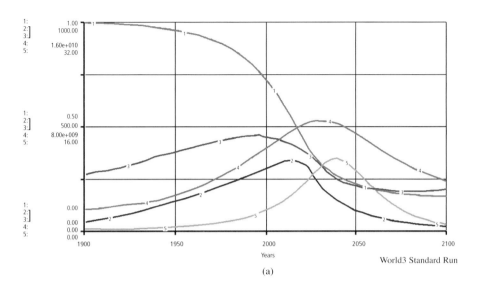

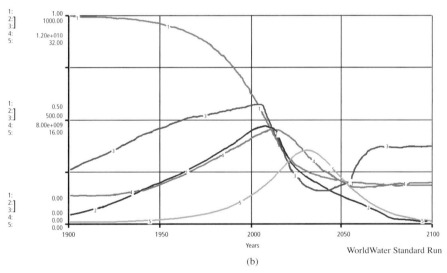

Note: See Plate 12 for a colour version.

Figure 8.55 *State of the World: 'Standard run' results of World3 (a) and WorldWater (b) models*

versible damage (line 5 in Figure 8.57a). The non-renewable resources depletion rate is very slow, and about 50 per cent of original resources are still available in the simulated year 2100. The world avoids an uncontrollable collapse, maintains its standard of living and holds itself nearly in equilibrium.

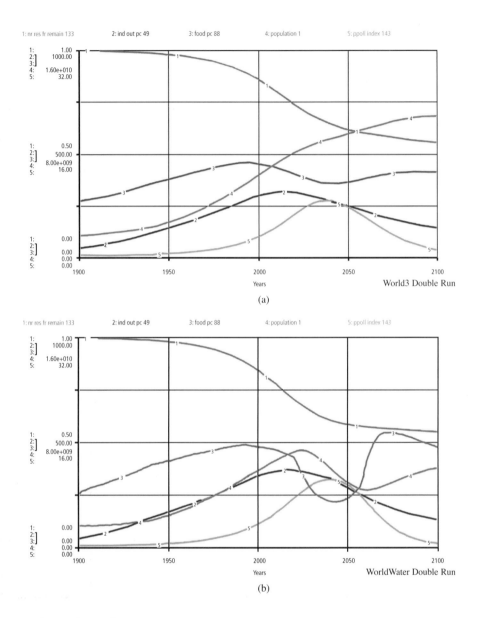

Note: See Plate 13 for a colour version.

Figure 8.56 *State of the World: 'Double run' results of World3 (a) and WorldWater (b) models*

WorldWater again reveals a different picture (Figure 8.57b, Plate 14). The link between persistent pollution and wastewater production creates a tremendous demand for clean water around 2040. This is reflected in a major decline in food

Simulation **383**

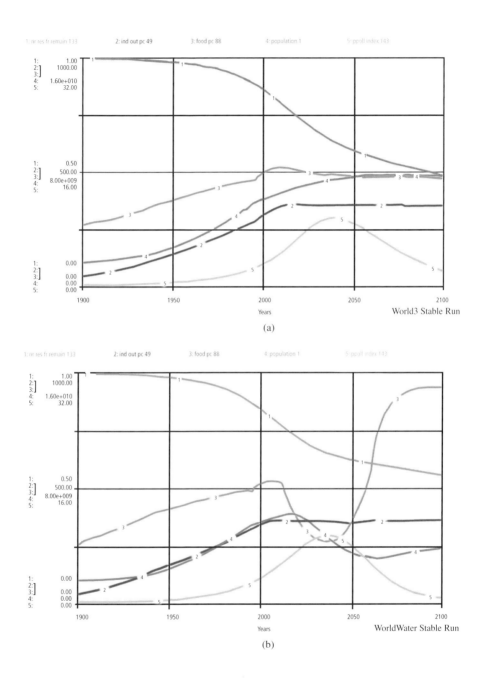

Note: See Plate 14 for a colour version.

Figure 8.57 *State of the World: 'Stable run' results of World3 (a) and WorldWater (b) models*

production (line 3 in Figure 8.57b) and impact on population levels (line 4 in Figure 8.57b), which begin to decline in 2020. However, as in the case of the double run this collapse is reversible. With better control of persistent pollution, water pollution is reduced and the demand for clean water for dilution and transport of wastewater declines, allowing more water for food production and use by the population. In this scenario equilibrium is not reached within the simulation period.

The WorldWater simulations clearly demonstrate the strong feedback relations between water availability and different aspects of world development. The results of numerous simulations contradict the assumption made by the developers of World3 that water is not an issue on a global scale. It is quite clear that water is an important resource on the global scale, and its limits do affect food production, total population growth and industrial development (Figures 8.55b, 8.56b and 8.57b).

Water issues on the global scale

WorldWater provides a detailed insight into the dynamics of water use over the simulation horizon. Figure 8.58 shows predicted water use patterns for the set of data from the standard run. Two major observations can be made from this simulation. First, the use of clean water for the dilution and transport of wastewater, if not modified, will impose a major stress on the global world water balance. Using conservative data on wastewater disposal and rate of dilution from Shiklomanov (2000) and IHP (2000), it is shown that this use exceeds the total water use by six times. Therefore the main conclusion of this research is that water pollution is the most important future water issue on a global scale. Second, the water use by different sectors shows quite different dynamics from those predicted by classical forecasting tools and other water models. Inherent linkages between the water quantity and quality sectors, and the food, industry, persistent pollution, technology and non-renewable resource sectors of the model create an overshoot and collapse behaviour in water use dynamics.

For the standard run simulation, water use is seen to increase in all sectors until 2015. The use of water for agriculture stops growing after 2015 but afterwards remains at approximately the same level, since food production levels start to suffer from the impact of pollution (line 1 in Figure 8.58). Water use for municipal supply follows the total population level: it grows until 2015 and then collapses with the decrease in the total population. After 2060, when the water dilution and transport demand is brought under control, municipal water use begins to rise again (line 3 in Figure 8.58). Industrial water use shows the very same behaviour (line 2 in Figure 8.58). Reservoir losses rise at a moderate pace following the expected development of water storage around the world (line 4 in Figure 8.58, Plate 15).

The use of clean water for dilution and transport of wastewater follows the dynamics of persistent pollution. It peaks around 2040 and then starts to decrease

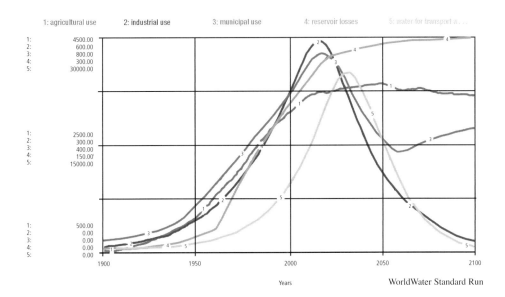

Figure 8.58 *Use of water: 'Standard run' results of WorldWater*

following the reduction in the growth of food production and population. Since this use of water is so important, let me review again the data and assumptions made in modelling it. IHP (2000) estimates that in 1995, the rate of wastewater disposal in the environment was 1402 km^3/year. The assumption of the WorldWater model is that this waste requires a considerable amount of clean water for dilution and transport. An assumption is also made that polluted water cannot be used for other purposes. Both of these assumptions are open to debate. First, more and more wastewater is being treated before being disposed of in the environment (Gleick, 2000). Second, some industrial and agricultural water needs may be satisfied using polluted water. There are no precise global estimates of the percentage of wastewater being treated and used for other purposes, so a conservative estimate is made in WorldWater that in 1995 the amount of wastewater discharged and not treated or reused was about 50 per cent of the total estimate in IHP (2000), or 700 km^3/year. This value is then related to the pollution index from the persistent pollution sector of the model to help develop a predictive relationship between persistent pollution and the amount of generated wastewater. A figure of clean water needed for the dilution and transport of waste was then applied in the simulations by taking this to be a fixed nine times the amount of wastewater. This is in line with IHP (2000), which estimates it at between eight and ten times.

Sensitivity analysis conducted with WorldWater (Simonovic, 2002) did not show any significant change in world water dynamics when this ratio was changed. However, any changes obviously affected the total amount of water taken to be used for dilution and transport of wastewater. Sensitivity analysis performed on another important variable, the percentage of cultivated land being irrigated, showed very similar results. Simonovic (2002) observes that the total use of water for irrigation does not change significantly with a change in the percentage of cultivated land being irrigated. Obviously this change does not affect world water dynamics significantly. A much more detailed sensitivity analysis can be easily performed on all the variables in WorldWater.

Two of the most significant global water studies, IHP (2000) and Cosgrove and Rijbersman (2000), provide static predictions of future water needs for the year 2025. These predictions are shown in Table 8.8 together with the predictions of WorldWater. Some interesting observations can be made by comparing the predictions. First, these classical predictions do not take into consideration the needs for dilution and transport of wastewater. Both studies do describe water pollution as one of the main future issues, but they do not assess its impact in this way. Second, although there is good agreement between the estimates of total water use for purposes other than wastewater handling, static predictions of particular water uses vary substantially from the WorldWater results. For example, the estimated industrial demand in 2025 is 1170 km^3/year according to Shiklomanov (2000) and IHP (2000), 900 km^3/year according to Cosgrove and Rijbersman (2000) and only 520 km^3/year according to WorldWater. The main reason for this difference is that WorldWater considers the dynamic behaviour of the industrial sector, while the models use static predictions. As shown in Figure 8.53, there is considerable variance in these predictions depending on the factors used in making them and the historical period used as a basis.

Table 8.8 *Comparison of standard projections for 2025 with the results of WorldWater (km^3/year)*

Use	Shiklomanov (2000), IHP (2000)	Cosgrove and Rijbersman (2000)	WorldWater
Agriculture	3189	2300	3554
Industry	1170	900	520
Municipal	607	900	723
Reservoir	269	200	276
Total withdrawal	**5235**	**4300**	**5073**

Conclusions

When working with the WorldWater model and examining different visions of the future, it is important to remind ourselves that we are dealing not with the 'real world' but a representation of the world. The exact numbers of model predictions are less important than the overall change in a variable over time (known as dynamic behaviour). The real value of WorldWater lies less in estimating the total municipal water supply of the world in 2063 than in understanding how municipal water supply relates to population growth, food production and future industrial development. The model provides valuable insights into how, for example, the municipal water supply might change dynamically over time, as a function of the numerous assumptions that are used to form a possible world development scenario.

In developing and carrying out numerous simulations with WorldWater, we came to two very important conclusions:

1. Water is one of the limiting factors that needs to be considered in the global modelling of future world development. Attempts made by World3 in the past to seek scenarios leading to a sustainable world future should be carefully reexamined now with the clear understanding that water must be part of the global picture.
2. Water pollution is the most important future issue on the global scale. In spite of the rhetoric of many water experts, the results of WorldWater simulations are explicitly, and for the first time, bringing water pollution to the forefront as the most alarming issue that needs attention from the world population, water experts and policy-makers.

WorldWater is a powerful tool. However, work with the global model is also opening up future directions for its use and improvement. Probably the most important is the transformation of WorldWater into numerous 'RegionalWater' models. Solutions for water problems are generally to be found at the regional level, and the power of dynamic regional models, developed using the same principles as are incorporated in WorldWater, can increase our understanding of water problems and our ability to reach sustainable solutions for them.

Simonovic and Rajasekaram (2004) have developed a CanadaWater regional model. It takes into consideration dynamic interactions between quantitative characteristics of the available water resources and water use, which are determined by the socio-economic development level, population and physiographic features of Canada's territory.

8.4.3 Flood evacuation simulation model

Simonovic and Ahmad (2005) present an interesting application of system dynamics simulation for capturing human behaviour during emergency flood evacuation. Their model simulates the acceptance of evacuation orders by the residents of the area under threat, the number of families in the process of evacuation, and the time required for all evacuees to reach safety. The model is conceptualized around the flooding conditions (both physical and management) and a core set of social and psychological factors that determine human behaviour before and during the flood evacuation. The main purpose of the model is to assess the effectiveness of different flood emergency management procedures. Each procedure consists of the choice of a flood warning method, warning consistency, timing of evacuation order, coherence of the community, upstream flooding conditions and a set of weights assigned to different warning distribution methods. The model use and effectiveness were tested through the evaluation of different flood evacuation options in the Red River basin, Canada.

Introduction

Preparation for emergency action must be taken before a crisis for several reasons. Conditions in a disaster-affected region tend to be chaotic. Communication is difficult and command structures can break down because of logistical or communications failure. Human behaviour during the emergency is hard to control and predict. Complaints cannot normally be addressed during the emergency. Experience with emergency evacuation in the Red River basin (Manitoba, Canada) during a major flood in 1997 unveiled an abundance of problems that the population affected by the disaster had with policies and their implementation. They were not happy with the timing of the evacuation orders, evacuation process implementation, order of command and many related issues. The literature confirms that there is a very similar situation in other kinds of disaster. There is an obvious need to improve:

- our understanding of the social side of emergency management processes;
- our understanding of human behaviour during emergencies;
- the communication between the population affected by the disaster and emergency management authorities;
- preparedness through simulation, or investigation of 'what-if' scenarios.

The proper understanding of human behaviour in response to a disaster, and the ability to capture it in a dynamic model, are valuable additions to emergency management policy analysis. This example develops a theoretical framework for studying flood evacuation emergency planning in a more holistic way, integrating a broad range of social and cultural responses to the evacuation process. It also provides new insights by developing a dynamic model of the process, which is converted into a gaming format for policy analysis and for other practical applications. The model integrates empirical survey data to fit the characteristics of specific communities.

Human behaviour during disasters

The modelling process required a very detailed consideration of major factors that affect human behaviour during disasters (like individual risk perception, disaster recognition and acceptance of the evacuation order). It was found that core factors determining how people cope with floods are their economic status and previous experience with flooding. Stress indicators were measured using fear, desperation, action, depression and family health indexes. Issues covered in modelling flood knowledge were the flood warning system, contributing topographical factors, contributing effects of urbanization and political trends. Both economic status and previous experience with flooding are incorporated in the model and are discussed below.

The amount of human effort involved in coping with natural hazards varies greatly, and there are several different levels and thresholds. A social group moves from one level to another in a cumulative fashion as it acquires experience. Factors that influence this movement are the severity of a hazard, recency of a hazard, intensity and extensiveness of human activities in the area, and the wealth of the society. We can model this by proposing an awareness threshold which precedes an acceptance level. This is followed by an action threshold, which leads people to modify and prevent events, and then an intolerance threshold, marked by 'change use' or 'change location' steps. The Red River basin evacuation model uses a structure that divides the process into three phases: concern, danger recognition and evacuation decision.

The relocation of residents after a natural disaster contributes to environmental, social and psychological stress. Research has showed that people from the same neighbourhood prefer to be evacuated together. Evacuation orders that direct people from close social environments to different temporary accommodation do not meet with ready acceptance, and can delay the general evacuation process.

A system dynamics model for flood evacuation planning

Flood management is aimed at reducing the potential harmful impact of floods on people, the environment and the economy of the region. In Canada the flood management process can be divided into three major phases: planning, emergency management and post-flood recovery. During the *planning* phase, different alternative measures (structural and non-structural) are analysed and compared for possible implementation to reduce flood damage. *Emergency management* involves regular appraisal of the current flood situation and daily operation of flood control structures to minimize damage. Following appraisal of the situation, decisions may be made to evacuate areas. *Post-flood recovery* involves decisions regarding the return to normal everyday life. The main concerns during this phase are provision of assistance to flood victims and rehabilitation of damaged properties. This example focuses on issues related to emergency management, provision of assistance and conduct of the evacuation process. Human behaviour during evacuation, in response to a disaster warning, is captured in a system dynamics model that allows emergency managers to develop the 'best' possible response strategy in order to minimize the negative impacts of a flood disaster. Theoretical knowledge collected from the relevant literature was used to conceptualize the model. Model relationships and all other necessary data were obtained through interviews conducted in the Red River basin immediately after the flood of 1997.

The human decision-making process in response to a disaster warning can be divided into four psychological phases:

- concern;
- danger recognition;
- acceptance;
- evacuation decision.

The factors that play an important role in the decision-making process can also be divided into four groups: initial conditions, social factors, external factors and psychological factors (denoted as IF, SF, EF and PF in Figure 8.59). Figure 8.59 shows the conceptual framework of the behavioural flood evacuation model. The four groups of factors are identified, with their acronyms. The vertical arrow alongside each of the variables indicates the direction of causal relationship between the variable and the psychological phase under consideration. For example, if a family has had previous flood experience, its concern rate will be lower than that of a family without this experience. Therefore, an arrow pointing down is shown along the variable *flood experience*. Variables in italics are the policy variables, and can be changed by emergency managers. A detailed discussion of the links between the phases and the main groups of factors follows.

Model variables: initial conditions and social factors

This concerns social and demographic aspects of the population, such as income, age group and daily life pattern, and includes attributes such as an inhabitant's experience of natural disasters, awareness of being at risk and knowledge about disasters. Each family living in a disaster-prone area has a certain degree of disaster awareness and a life pattern of its own. This disaster and risk awareness forms a set of initial conditions for that family's behaviour when the disaster hits the area. The behavioural patterns of the household are further affected by the information provided about the disaster and by physical parameters such as the intensity of the disaster and the size of the area affected. Initial conditions trigger a *concern*. Based on the data collected in the Red River basin, concern is higher if experience with flooding is missing, the sense of risk is high, and the event (precipitation, flood peak, water levels, etc.) is large. In the model *concern* is defined as the first phase of the decision-making process, when an individual or family is aware of risk, and has basic information on the type of disaster and its impacts (Figure 8.59). This concern is always present, even when there is no imminent threat of a disaster. Initial conditions provide a background for an individual's perception of danger.

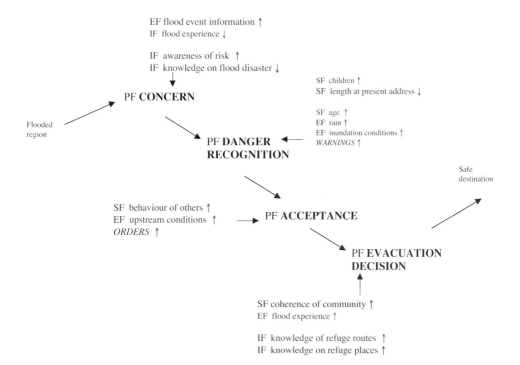

Figure 8.59 *Conceptual framework of a behavioural model for evacuation planning*

The process through which initial conditions affect an individual's behaviour is complicated, and involves a chain of complex psychological reactions. The model considers two categories of households, those that have experienced disaster before, and those that have not. For households with previous experience, it considers the extent of damage they experienced, whether they have evacuated in the past, whether they evacuated in the most recent disaster, the reasons for any decision not to evacuate, and damage to property. Information required from households with no disaster experience includes their knowledge about disasters in the area, the criteria they would use in order to decide whether evacuation is necessary, and awareness of the risk to their property. In the Red River basin this information was obtained through personal interviews.

Depending on the severity of the situation *concern* may develop into *danger recognition*, defined as a new variable and calculated in the model using a different equation. There is a positive causal relationship between these two variables, which is shown in Figure 8.60. In this second phase an individual or family is aware of imminent threat and is on alert.

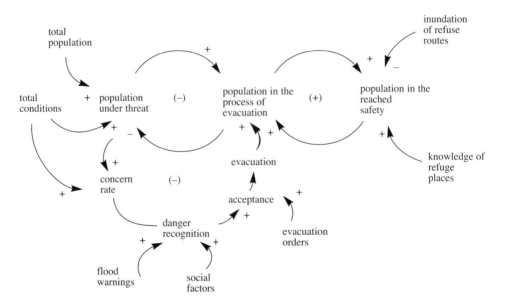

Figure 8.60 *Causal diagram of a behavioural model for evacuation planning*

Model variables: external factors

The external factors that play a vital role in forming responses to disaster situations are information provided by the media, responsible emergency management authorities, and personal experience of the physical conditions. An evacuation order directly affects the acceptance and therefore the evacuation decision, as it initiates the action of evacuation itself (Figure 8.59). Experience of weather conditions affects the decision more indirectly, through *danger recognition*. Each household's *danger recognition* level changes over time, and affects not only the evacuation decision but also broader attitudes to the disaster. People with a high danger recognition rate try to get as much information as possible about a disaster event from all possible sources.

Model variables: psychological factors

Psychological factors are used in this model to represent all phases of the evacuation decision-making process. They cover the phases of *concern*, *danger recognition*, *acceptance*, and *evacuation decision*. An inhabitant's decision to evacuate is a result of external factors influencing the initial conditions. Social factors such as age and presence of dependants in the family (children or elderly people), in combination with external factors like awareness of heavy rain and inundation conditions, give rise to *danger recognition*. Once the *danger recognition* rate reaches a certain threshold level, evacuation orders and the behaviour of others can precipitate an *evacuation decision*. The reaction of each inhabitant to external factors differs. For example, there are households that do not evacuate in spite of receipt of an evacuation order, and there are those who evacuate even before an evacuation order is issued. To incorporate these behaviours in the model, a variable called *acceptance level* is introduced. This measures the extent to which a household accepts the danger. The *evacuation decision* results from the interaction between the *acceptance level* and the trigger information. An evacuation order and the behaviour of other households are considered the trigger information in the model.

After a household decides to evacuate, it has to determine a refuge place and a route to it. Its ability to reach safety depends on household members' knowledge of these things. This knowledge affects the behaviour of the inhabitants after they decide to evacuate. If they lack knowledge it will lose them valuable time in the evacuation process. The behaviour of a family with little knowledge of a route may be affected by the behaviour of other families with better knowledge.

Policy variables

The flood evacuation system dynamics model was developed to investigate different emergency policy options. The two main sets of policy variables concern flood warnings (media used for dissemination and consistency) and evacuation orders (media used for dissemination and timing). Warnings can be disseminated using the television, radio, mail, Internet and visits. For dissemination of evacuation orders only two options are considered, mail and visits to the household.

Model structure

The variables discussed above are interrelated through the model structure. The basic causal diagram (not including all the variables for simplicity of presentation) is shown as Figure 8.60. This identifies the main feedback forces that determine the behaviour of the system captured with our model. There are two feedback loops in the centre of the model connecting three stocks: population under threat, population in process of evacuation and population that reached safety. The loop on the left-hand side represents negative feedback and the loop on the right-hand side represents positive feedback. The reference behaviour mode is S-shaped growth (see Section 8.2.2) (this was confirmed by the results, discussed later). The growth of population that reaches safety is exponential at first, but then gradually slows with the decrease in the population under threat. The upper boundary value of the system state is also the goal, and equals the total population under threat that can be evacuated. Figure 8.60 also shows a negative feedback loop that links the psychological variables (*concern*, *danger recognition*, *acceptance* and *evacuation decision*) with the main stocks.

Mathematical relationships

The data set used to develop this model derived from a field survey of families that evacuated during the 1997 flood. A questionnaire was administered to 52 households involving more than 200 respondents in 6 different community types. These communities represented a broad range of people affected by the 1997 flood, including:

- an urban community (Kingston Row and Crescent in Winnipeg);
- a rural community protected by a ring dike (St Adolphe);
- a rural community without structural protection (Ste Agathe);
- a suburban community (St Norbert);
- an urban fringe community (Grande Pointe);
- rural estates/farmers.

The survey was conducted less than one month after the flood, when many families were still in the process of recovery and under considerable stress. Both closed and open-ended questions were used. The data collected directly by the survey was verified through the process of public hearings organized by the International Joint Commission at five locations, on two occasions: immediately after the flood (autumn 1997) and before submission of the final report (spring 2000). There were more than 2000 participants in these hearings. Note that the relationships developed and used in the model apply only to the communities in south Manitoba. It is expected that major value systems captured by the survey will not change with time, since the population in the flooded regions of the Red River basin tends to be stable in both size and characteristics.

The data collected were processed to establish different relationships among the variables in the evacuation model. For example, the relationship in equation (8.32) describes the relative importance (weight) of each variable used for representation of the *concern* rate.

$$\begin{aligned}Concern = &\ (Awareness\_of\_Flood\_Disaster) \times 0.1 \\ &+ (Previous\_Flood\_Experience) \times 0.7 \\ &+ (Awareness\_of\_Risk) \times 0.2\end{aligned} \quad (8.32)$$

The relationship derived for *danger recognition* is:

$$\begin{aligned}Danger\_Recognition = &\ (Age\_factor) \times 0.05 \\ &+ (Impacts\_of\_Warning) \times 0.3 \\ &+ (Concern) \times 0.3 \\ &+ (Rain\_Factor) \times 0.1 \\ &+ (Inundation\_Factor) \times 0.15 \\ &+ (Children\_Factor) \times 0.05 \\ &+ (Stay\_Factor) \times 0.05\end{aligned} \quad (8.33)$$

This equation describes different variables involved in the calculation of *acceptance* level:

$$\begin{aligned}Acceptance\_Level = &\ (Danger\_Recognition) \times 0.2 \\ &+ (Behaviour\_of\_Others) \times 0.3 \\ &+ (Order\_Impacts) \times 0.2 \\ &+ (Flooding\_Factor) \times 0.3\end{aligned} \quad (8.34)$$

Finally, the *evacuation decision* is expressed as a function of the *acceptance level*, previous experience with evacuation and disaster claims, and support available from the community where the family lives:

$$\begin{aligned}Decision = &\ (Acceptance\_Level) \times 0.7 \\ &+ (Experience\_Factor) \times 0.2 \\ &+ (Support\_Factor) \times 0.1\end{aligned} \quad (8.35)$$

All the variables in equations (8.32) through (8.35) are restricted to values between 0 and 1. Quantification of weights is done through the model calibration procedure. Data collected from the Manitoba Emergency Management Organization (MEMO) provided details on the evacuation process (length, timing and number of people) and were compared with the outcome of the model simulations. The weights that generated model output that matched the observed data the best were selected and used in the model.

Other relationships between model variables require graphical description. These relationships are developed from the data collected through the field survey. For example, a graphical relationship for the *Flooding_Factor*, which is a function of *Upstream_community_flooded*, is shown in Figure 8.61. Relative values of the *Flooding_Factor* are between 0 and 1. The value of 0 indicates no upstream flooding information. The value of 1 indicates the full knowledge of the upstream flooding situation. Relative values of the *Upstream_community_flooded* are also between 0 and 1. They are derived from the survey data by calculating the relative ratio of people aware of upstream flooding, if it existed. The shape of the graph reflects the notion that the more knowledge about upstream flooding was available to people, the higher the attention that was given to this information in their process of making a decision about personal and family evacuation.

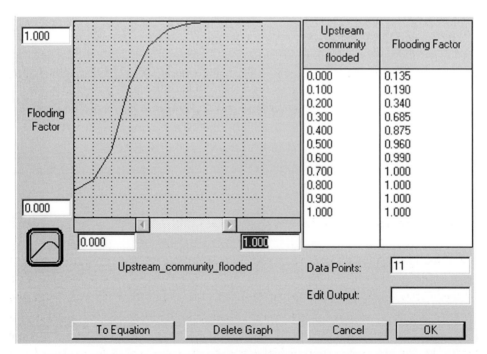

Figure 8.61 *A graphical relationship between* Flooding_ factor *and* Upstream_community_ flooded

Delays and random number generation functions are used for describing different processes in the model. For example, reaching a refuge place after evacuation is conditioned on the individual's knowledge of refuge place location and inundation of access routes. There are 'information' and 'material delays' involved in making a decision to evacuate and in reaching a refuge place. An example of information delay is the time difference between the moment when a flood warning is issued and the time an individual takes to make an evacuation decision. An example of material delay shown in equation 8.36 is the time required by people in the process of evacuation to reach the final destination. It is captured in the model in the following form:

$$Reaching = DELAY((Evacuation\_In\_Process) \times (Knowledge\_of\_Refuge\_Places) \times 0.2 + (Route\_Factor) \times 0.8\ Random(1,5,5)) \quad (8.36)$$

There are three main stocks in the model. They are *Population*, *Evacuation_in_Process* and *Reached_the_Destination*. The *Population* stock represents the total number of households in the area under threat (52 families with more than 200 individuals in seven different communities in the Red River basin). The outflow from this stock is the number of families that decide to evacuate. The population stock in mathematical form is expressed as:

$$d\ Population\ (t)\ /\ dt = -Evacuating\ (t) \quad (8.37)$$

The second important stock in the model is *Evacuation_in_Process*. This stock represents the difference between the number of families that have decided to evacuate and the number of families that have reached a refuge:

$$d\ Evacuation\_in\_Process\ (t)\ /\ dt = Evacuating\ (t) - Reaching\ (t) \quad (8.38)$$

The third stock accumulates the number of families that have reached safety:

$$d\ Reached\_the\_Destination\ (t)\ /\ dt = Reaching\ (t) \quad (8.39)$$

The flood evacuation system dynamics model was developed in four sectors that are linked together: initial conditions, social factors, psychological factors and external factors. After mapping each sector and defining connections between different sectors, decision rules are developed and incorporated in the model using logical statements with an IF-THEN-ELSE structure. The following rule states that if the threshold value for an evacuation decision is less than or equal to, for example, 0.65, then there will be no evacuation; otherwise there will be an evacuation:

IF *Decision* <= 0.65 THEN 0 ELSE 1 (8.40)

The evacuation model depends on the input from other models. For example, the information on dynamically varying water levels is provided by a hydrodynamic model, and a geographic information system (GIS) provides data on the spatial location of each community, its distance from the river, and its relative location to other communities (upstream or downstream).

Model use

The model interface shown in Figure 8.62 is the main control window for the use and navigation of the evacuation model. The main use for this model is to develop different scenarios for assessing the impact of different emergency evacuation policy options. An appropriate interface is required to provide the user with an easy process of scenario development and assessment. The upper section of the interface window provides an introduction to the model and may be browsed by scrolling the text within the window. In order to use the model properly, policy selection is required. Switches and sliders are used to set the value for different variables. All sliders offer the choice of a value between 0 and 1. Zero always indicates the lowest level of importance with 1 always indicating the highest level. The model is ready for simulation when all values are selected. A simulation run is started by clicking on the 'Run Model' button. The graph window on the interface shows in real time the results of model calculations by redrawing the two lines (time series) shown (1 and 2). Completion of the graph indicates the end of the simulation.

Basic model results are presented in the form of a graph, as shown in Figure 8.62 (Plate 16). A line (numbered 2) shows the number of families (out of 50 stored in the model database) that have reached a refuge place. Line number 1 shows the number of families on the way. Both variables are shown as functions of time. The total simulation horizon is 96 hours, or four days. The shape of these two lines is a function of the policy selected, and encompasses the warning distribution, the evacuation orders distribution, characteristics of the community, awareness of incoming flood, and the weights given by community members to different warning and evacuation order distribution modes.

The detailed application of the system dynamics evacuation model to the analyses of flood emergency procedures in the Red River basin, and sensitivity analyses of flood evacuation strategies, are presented in Simonovic and Ahmad (2005).

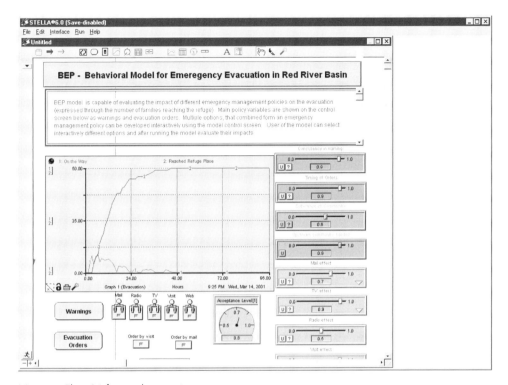

Note: see Plate 16 for a colour version

Figure 8.62 *Evacuation model interface*

8.4.4 Hydrological simulation model for predicting floods from snowmelt

This example uses a system dynamics approach to explore hydrological processes in the geographic locations where the main contribution to flooding is from snowmelt (Li and Simonovic, 2002). Temperature is identified as a critical factor that affects watershed hydrological processes. Based on the dynamic processes of the hydrological cycle in a watershed, the feedback relationships linking the watershed structure and climate factors to streamflow generation were identified prior to the development of a system dynamics model. The model is used to simulate flood patterns generated by snowmelt under temperature change in the spring. Its structure captures a vertical water balance using five tanks representing snow, interception, surface, subsurface and groundwater storage. Calibration and verification results show that temperature change and snowmelt play a key role in flood generation. Data from the Red River basin, which is divided between Canada and the United States, were used in model development and testing.

Existing hydrological models have been applied for either generating streamflow or determining runoff response to an external change. However, an analysis of the endogenous feedback structure of a watershed system that generates and regulates dynamic hydrological behaviour is not addressed in the existing models, and the canopy interception capacity and impact of temperature on soil infiltration rate are assumed to be constant in most models. These assumptions ignore the impact of temperature fluctuations on both vegetation growth dynamics and the change in soil physical states, which has a significant impact on hydrological dynamics. Therefore, this example attempts to develop a dynamic model for addressing flood generation from the snowmelt associated with hydrological processes. The model considers temperature as a critical external factor to determine the canopy interception capacity and physical state of the soil. System dynamics simulation is applied as a methodology that provides an inside view of endogenous feedback structures relating to hydrological processes. The model developed in this example captures the essential dynamic characteristics of surface and subsurface hydrological processes that are nonlinear, occur in the feedback form and include time delays.

Model development

Lumped or integrated approaches have a long tradition in hydrological modelling, because they can effectively use available daily data related to runoff records, long and reliable records of precipitation and temperature. A lumped parameter conceptual model could be capable of simulating the various components of streamflow. This example builds on the existing models and integrates climatic factors and hydrological processes. Model parameters are defined for the whole watershed, and a simulation of vertical water balance is performed using five tanks representing snow storage, canopy storage, surface soil storage, subsurface soil storage and groundwater storage.

Figure 8.63 shows that any precipitation falling as snow accumulates in *snow storage*. Precipitation as rainfall and water from snowmelt first enter *canopy storage*, which represents the interception of moisture by vegetation, and varies with vegetation growth over the seasons. The loss from this storage is due to evaporation. Any moisture in excess of the canopy storage maximum capacity is passed to the *surface soil storage*. There is a limit on the rate at which moisture can enter surface soil storage. This is a function of the surface soil conditions and soil moisture content. Temperature critically determines the soil physical state, and existing soil water saturation in the form of feedback affects the infiltration. The difference between the volume of water from canopy storage and the amount infiltrated into the soil becomes overland flow into rivers.

There are losses from *surface soil storage* through evapotranspiration, interflow and percolation to *subsurface soil storage*. Evapotranspiration flux aggregates the losses through physical (evaporation) and biological (transpiration) processes. It is dependent on moisture saturation and weather conditions. Interflow (i.e. lateral flow) is a very complex function of the effective horizontal permeability, water saturation and availability, the gradient of the layer and the distance to a channel or land drain. Percolation to the lower layer is dependent on the water saturation within the surface and subsurface soil layers. *Subsurface soil storage* is moisture below the surface layer but still in the root zone. Water enters this layer by percolation from the surface soil. Similar losses to those in surface soil storage exist in the subsurface soil storage: evapotranspiration, interflow runoff and percolation to groundwater. Evapotranspiration from the subsurface soil layer depends on vegetation transpiration, and varies with vegetation type, the depth of rooting, density of vegetation cover, and the stage of plant growth, along with the moisture characteristics of the soil zone. Interflow and percolation to *groundwater storage* may depend on moisture saturation. *Groundwater storage* as an infinite linear reservoir continuously contributes to the runoff. Subsurface interflow and baseflow from groundwater are important contributions to the streamflow, especially in a dry or winter season. Their contribution to the streamflow is dependent on the spatio-temporal characteristics of the watershed, especially in the topography, effective horizontal permeability, the gradient of the layer and the distance to a channel or land drain.

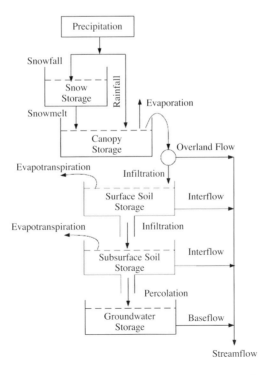

Figure 8.63 *A schematic representation of the vertical water balance*

Based on the above analysis, temperature is presented as an important climate factor that influences snowpack accumulation and snowmelt as well as the physical states of soil and water. Runoff and flood generation from snowmelt may follow a general pattern as temperature changes during the active snowmelt period. In the winter period, precipitation accumulates as a snowpack because of the low temperature, and the runoff contribution mostly comes from the groundwater and subsurface soil storage, because the surface soil is frozen. As temperature reaches an active point in the early spring, the snow starts melting. Most of the snowmelt becomes overland flow because of the small amount of canopy storage and the frozen surface soil. As the temperature increases, the snowmelt generates more water, which rapidly increases the streamflow and gradually leads to flood flows. In the meantime, the active temperature also gradually defrosts the soil, therefore increasing the infiltration rate and the surface soil storage capacity. As a result, the streamflow starts to decline. If heavy rain occurs during the snowmelt period, the streamflow will rise more rapidly and the peak magnitude will be larger. As the accumulated snowpack melts, the streamflow gradually returns to normal level. After the snowmelt period, the main streamflow contributions will come from the groundwater and the soil storage. Fluctuations in the streamflow depend strongly on the rainfall magnitude. This pattern has been clearly observed in different locations along the Assiniboine River and the Red River in Manitoba, Canada.

Dynamic hypothesis

A system dynamics simulation approach was applied to develop the hydrological model, which represents the dynamics of the hydrological processes described above. From the viewpoint of system dynamics, the dynamic behaviour of the hydrological system is dominated by the feedback loop structure that controls change in the system. As external and internal conditions vary, the contribution of each feedback loop may change, and the dominance in controlling internal moisture dynamics may shift from one feedback loop to another. Hence, an integrated analysis of complex feedback relationships could be helpful for a better understanding of the watershed hydrological dynamics.

Based on the hydrological processes in the surface–subsurface layers, a basic hypothesis was developed to generate the hydrological dynamics (Figure 8.64). This shows that the feedback structure of the fundamental state variables is related to hydrological flow processes as well as exogenous factors (Figure 8.64). The strength of each hydrological flow process is represented by a flow variable. Linking state variables to flow variables, feedback loops can be formed to control the hydrological behaviour. When rainfall or snowmelt enters into the system, the hydrological flow processes are regulated by these feedback loops. There is one negative feedback loop which controls canopy capacity and water interception. It shows that water interception by the canopy increases the amount of water in canopy storage, which reduces the interception capacity, and finally limits the water interception rate. Interception capacity is dependent on the vegetation cover, which is subjected to active temperature accumulation during the snowmelt active period.

There are five negative feedback loops controlling surface soil storage. One describes the source of water for surface soil storage. Water infiltration through the surface soil increases the water saturation, which further limits infiltration into the surface soil. Another describes the evapotranspiration process. Two loops explain that the percolation of water to the subsurface storage is dependent on the water saturation in the surface and subsurface soil layers. Finally, one loop describes the surface soil interflow, which is influenced by the saturation level. All these loops are strongly regulated by temperature during snowmelt active periods when the surface soil is frozen. Frozen surface soil limits the water infiltration rate and water availability for evapotranspiration, percolation and interflow.

Three negative loops are identified to control moisture losses from the subsurface soil storage. They show that evapotranspiration, interflow and percolation processes are determined by the subsurface soil storage saturation. Groundwater storage is assumed to behave as a shallow reservoir, and baseflow is determined by one negative loop in Figure 8.64.

The dynamic hypothesis in Figure 8.64 shows that the rainfall and the snowmelt are the most important external water sources affecting the water balance between the soil layers and the groundwater storage. Internal hydrological processes and negative feedback structures among the soil layers and the groundwater reservoir provide internal storage buffers and adjustment mechanisms that reduce or delay the impact of the external disturbance on the streamflow. The main role of the negative feedback structures is to maintain the system balance. Floods occur when the external water volume exceeds internal storage buffers and its adjustment capacity.

Model structure

Any precipitation falling as snowfall accumulates in the snow storage. A critical temperature is used to determine whether the measured or forecasted precipitation is rainfall or snowfall. The snowmelt rate can be calculated by the degree-day factor (Li and Simonovic, 2002). On the basis of water balance, the snow storage change rate can be expressed mathematically as:

$$\frac{dSS}{dt} = SF \times swec - \alpha \times T \tag{8.41}$$

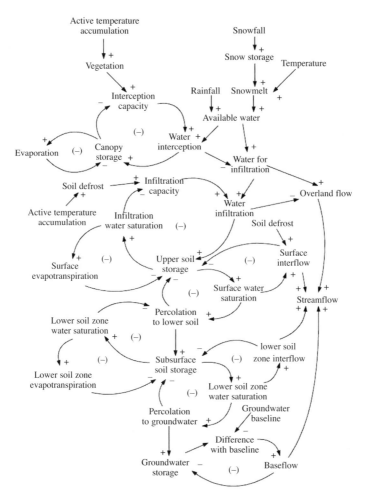

Figure 8.64 *Basic dynamic hypothesis of hydrological dynamics in a watershed*

where:
SS   = water in snow storage (cm)
SF   = precipitation as snowfall (cm day^{-1}) identified by a critical temperature
swec = snow–water equivalent coefficient (cm snow/cm precipitation)
α   = degree-day factor for snowmelt (cm °C^{-1} day^{-1})
T   = daily mean temperature (°C).

The canopy interception rate is dependent on the canopy interception capacity, existing water in the canopy storage, and the availability of water from snowmelt and rainfall. Water loss in canopy storage is due to evaporation, which is assumed to depend on air temperature when intercepted water is available for evaporation. The water balance equation for the canopy storage and the interception rate can be written as:

$$\frac{dCS}{dt} = CI - c_2 T \qquad (8.42)$$

where:
CS = water in the canopy storage (cm)
CI = canopy interception rate (cm day^{-1})
c_2 = evaporation coefficient (cm °C day^{-1}).

CI depends on the canopy interception capacity, which varies with the seasonal growth of vegetation (Li and Simonovic, 2002). During the winter season, the canopy interception capacity is very small, since the leaves have fallen from the trees and grasses are submerged. In a normal biological growth pattern, canopy growth follows an S-curve pattern: that is, with a rise in temperature, plants start growing, and the growth increases as the temperature increases until it reaches a maximum.

Moisture change in the surface soil storage depends on infiltration, evapotranspiration, interflow and percolation. Evapotranspiration, interflow and percolation from the surface soil are determined by the climatic conditions, water saturation and water availability. Water availability is influenced by water available from the canopy storage and can be expressed as a function of active temperature. Therefore, the change of water in the surface soil storage is determined by:

$$\frac{dSW}{dt} = I - SWEP - SWIF - SWP \qquad (8.43)$$

where:
SW = the water in surface soil storage (cm) and
I, SWEP, SWIF and SWP = the rates of infiltration, evapotranspiration, interflow and percolation respectively in surface soil storage (cm day^{-1}).

The effect of temperature on infiltration is a complex phenomenon affected by temperature fluctuation and the length of time the temperature stays above and below the active temperature. This phenomenon results in the soil defrosting and refreezing. It is ignored by most existing models. This model assumes that the soil defrosts exponentially with active temperature accumulation (Li and Simonovic, 2002). However, soil will be refrozen again if the temperature drops below zero for a number of days. The active temperature accumulation will be lost and will start again from zero.

The source of moisture for the subsurface soil storage is surface soil percolation, whereas the moisture losses are from evapotranspiration $SSMEP$, interflow $SSMIF$ and percolation to the groundwater storage $SSMP$. Rate of loss terms are determined by the subsurface soil water saturation. Vegetation cover and climatic conditions also influence evapotranspiration. Similar equations to those that describe the canopy interception capacity are developed for expressing vegetation as a function of active temperature accumulation (Li and Simonovic, 2002). Hence, the following equation is used to describe moisture dynamics in subsurface soil storage:

$$\frac{dSSM}{dt} = SWP - SSMEP - SSMIF - SSMP \qquad (8.44)$$

Groundwater storage is described as a linear shallow reservoir. Water enters the groundwater storage through percolation from the subsurface soil storage, and comes out of this storage as baseflow to streams. It is assumed that there exists a baseline groundwater level. The baseflow rate depends on the difference between the actual groundwater storage level and the baseline groundwater level. The equation for the change in groundwater storage can be written as:

$$\frac{dGWS}{dt} = SSMP - BF \qquad (8.45)$$

where:
GWS = the water in groundwater storage (cm) and
BF = baseflow (cm day^{-1}).

Application of the model

The proposed system dynamics model has been applied for simulation of runoff in two river basins in Southern Manitoba, Canada: the Assiniboine River basin and the Red River basin (Figure 8.65). The Assiniboine River originates in mid-north-west Saskatchewan and drains the area from the eastern part of Saskatchewan to the western part of Manitoba. Its major tributaries include the Qu'Appelle River and Souris River. The Assiniboine River flows from north-west to south-east, and joins the Red River in Winnipeg, Manitoba. The lower reach of the river is below the Shellmouth Dam, which can significantly reduce flow rates and downstream water levels. Therefore, this case study focuses on the Assiniboine River basin from its headwaters to the Shellmouth Reservoir.

The study area covers 16,496 km². Topographically, the basin is gently to moderately undulating, with higher relief evident in the north-east portion. The north-east part of the basin is located within the boreal plains ecozone, with brush and wooded bluffs cover and a steeper flow gradient, whereas the southern part lies within the prairie ecozone, a flatter terrain characterized by less brush and fewer trees. Climatologically, the basin is continental sub-humid, characterized by a long cold winter and short warm summer. The frost-free season varies from 90 to 110 days. Annual precipitation averages about 450 mm (with variations between 140 and 550 mm), of which 27 per cent is snow. The streamflow in the basin is highly variable on a daily basis. During the springtime, water levels in the Assiniboine River are high because of the snowmelt. About 63 per cent of annual total flow is contributed during the months of April and May, whereas there is only 3 per cent during the period from December to March. Yearly flow variation is also high because of climate variations.

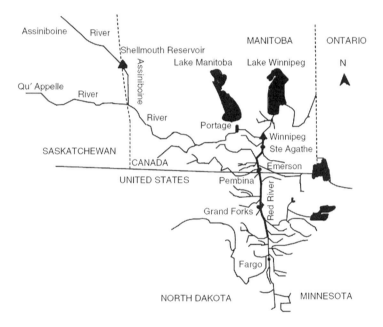

Figure 8.65 *Study area and locations*

The Red River originates in Minnesota and flows north. It is located in the geographic centre of North America. It enters Canada at Emerson, Manitoba, and continues northward to Lake Winnipeg. With the exclusion of the Assiniboine River and its tributaries, the Red River basin covers 116,550 km^2, of which 103,600 km^2 is in the United States. The basin is remarkably flat, and the slope of the river averages less than 15 cm per 1.6 km. The basin has a sub-humid to humid climate with moderately warm summers, cold winters, and rapid changes in daily weather patterns. Annual precipitation is about 500 mm, with almost two-thirds occurring between May and July. Precipitation during the dry months from November to

February averages only about 13 mm per month. Because the river flows from south to north, its southern reaches thaw before the lower river does so. Heavy snowfall and spring rains, coupled with late spring thaws, can cause the river to spill over shallow banks and across the floodplain, and result in major natural disasters.

The input data set for model use includes all calibrated parameters, temperature, precipitation and a set of initial values for the stock variables. Most of the parameters used in the model were given values (through the calibration process) within those available in the literature.

Model results

The simulated and measured streamflow data for the calibration year of 1995 and the verification year of 1979 in the Assiniboine River are shown in Figures 8.66 and 8.67. The results indicate that the simulated streamflow pattern is quite similar to that observed. In the case of the calibration flood year (Figure 8.66), the streamflow is smaller during the winter season because of the frozen surface soil. The active temperature starts in early March, which results in snowmelt and an increase in streamflow. From late March to early April, negative temperature lasted for about two weeks, which led to freezing of the surface soil again, and streamflow receded to the normal low level because of the absence of snowmelt. In mid-April, the temperature rose to the active point and snowmelt started again. As the temperature increased, more water was produced from snowmelt, and streamflow increased rapidly. In the meantime, the active temperature gradually defrosted the surface soil, which increased the infiltration rate and the surface soil storage capacity. More water infiltration into the surface soil increased water saturation, which in turn limited further water infiltration into the surface storage. Although the infiltration rate increased with the increase in temperature, streamflow continued to increase because of the delay in snowmelt. Before the surface soil was fully defrosted, streamflow reached a peak in association with a rainfall in late April. Fully defrosted surface soil infiltrated most of the snowmelt water and reduced the streamflow. A lasting high active temperature gradually depleted accumulated snowpack before mid-May, and streamflow returned to a normal level. After the snowmelt period, groundwater and soil storage again became the main contributors to streamflow, and fluctuations of streamflow were strongly dependent on the rainfall magnitude.

Since the catchment area of the Red River basin is very large, it was divided into three sub-catchments. The streamflow at the lower reach is routed together with the local inflow into the upper reach with a delay. Because the Red River flows north, its southern reaches thaw before the northern stretches do. The flow in the southern reaches significantly influences that in the northern reaches. As a result, flood starting and peak dates in the northern reaches are later than those in the southern

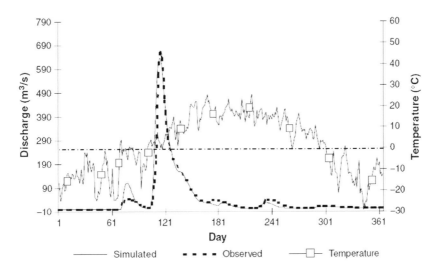

Figure 8.66 *Simulated and measured streamflow in the Assiniboine River basin for 1995 (calibration)*

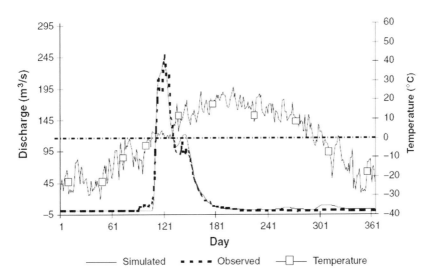

Figure 8.67 *Simulated and measured streamflow in the Assiniboine River basin for 1979 (verification)*

reaches. Calibration and verification of the model for the Red River basin show (Figure 8.68) that this real-life pattern was well reproduced in the model, which captured the essential dynamics of flows occurring in the basin.

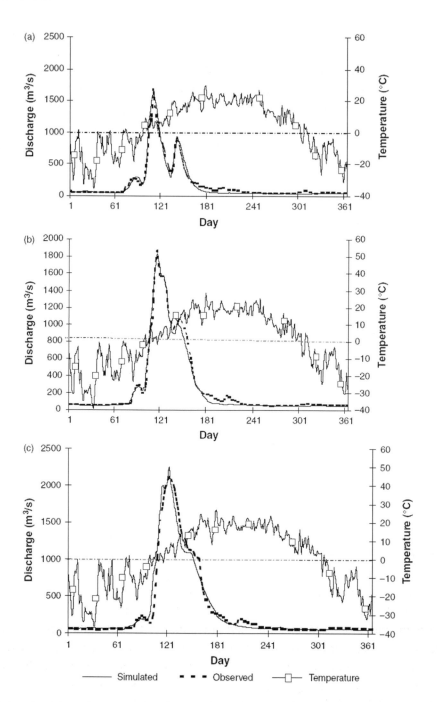

Figure 8.68 *Simulated and measured streamflow in the Red River basin for 1996 (calibration): (a) simulated and measured streamflow at Grand Forks; (b) simulated and measured streamflow at Emerson; (c) simulated and measured streamflow at Ste Agathe*

In the calibration flood year of 1996, simulated streamflow matched the observed flood pattern well during the snowmelt active period at Grand Forks: flood from snowmelt started and reached its peak in mid-April, and lasted about 30 days (Figure 8.68a). In late May, heavy rainfall resulted in a second peak. The flood pattern at Emerson and Ste Agathe is, to a great extent, determined by the pattern in the southern reaches with a time delay. Addition of snowmelt water in the lower reaches meant that flow peaks at Emerson and Ste Agathe were higher than that at Grand Forks, and the flood duration at Emerson and Ste Agathe was much longer than at Grand Forks. This pattern was well reproduced by the model at Emerson and Ste Agathe (Figures 8.68b and c). However, the magnitude of the peak at Emerson was overestimated. At Ste Agathe, a second peak generated in late May was also overestimated by the model. This peak was mainly generated by local heavy rainfall in the lower reaches of the river.

Figure 8.69 compares simulated and measured discharges for the verification flood year of 1997 (the flood of the century) in the Red River basin. At Grand Forks, the predicted flood duration matched the observed one, but the peak magnitude was smaller than observed and the peak time was delayed. At the Emerson and Ste Agathe stations, peak magnitude and time matched those measured very well. After the snowmelt active period, there was a heavy rainfall in early July, which produced another streamflow peak. After July, streamflow remained at the normal level.

The model reproduced the basic dynamics of streamflow occurring in the watershed on a daily basis. It does not capture daily changes in temperature. The Assiniboine River basin is represented in the model in aggregated form. One set of parameters is used for the whole watershed. This aggregation ignores the spatial variation of climate, land use, mantle and soil properties within the watershed. Although three sub-catchments are used in the Red River basin, each sub-catchment contains a large area. The number and size of the sub-catchments for a watershed model depend on catchment characteristics, data availability and quality. For this reason, the combination of a system dynamics model with other tools, such as a GIS, may improve the presentation of spatially varying processes.

The performance of the hydrological model shows that the simulated streamflow reflects the variation in temperature and precipitation as well as the moisture interaction between the surface, subsurface and groundwater storages. Comparison of the results from simulation and observation indicates that the model can reproduce well the observed flood starting time, peak and duration. Statistical analysis (Li and Simonovic, 2002) revealed that the error is unsystematic and the model quantitatively matches the historical data. The model in its present form provides a yearly prediction of streamflow on a daily basis. It can be used to make a long-term prediction of streamflow under different climate change scenarios. Further studies to refine the hydrological dynamics by taking spatial variations into account are warranted, to improve the model's ability to reproduce historic data and to predict future flood events.

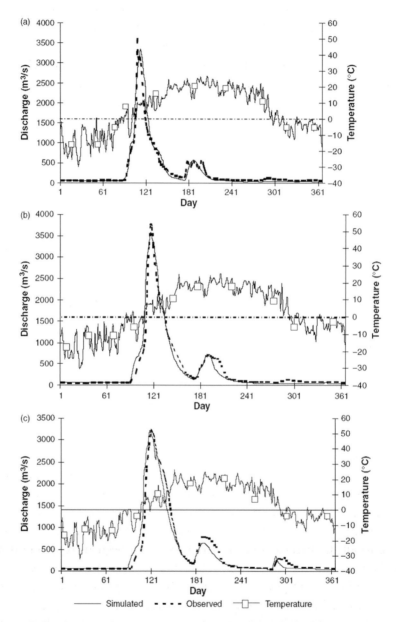

Figure 8.69 *Simulated and measured streamflow in the Red River basin for 1997 (verification): (a) simulated and measured streamflow at Grand Forks; (b) simulated and measured streamflow at Emerson; (c) simulated and measured streamflow at Ste Agathe*

8.4.5 A simulation model for resolution of water sharing conflicts

With increased industrial development and economic growth, conflicts over the use and allocation of water have been increasing. Although diverse efforts have been made towards resolving conflicts through computer-based models, a clear understanding of the processes is a prerequisite for models to be effective. This example presents a system dynamics simulation approach to assist stakeholders in two different jurisdictions in a hypothetical water resources system to resolve a potential water sharing conflict (Nandalal and Simonovic, 2003).

Introduction

Water is essential to sustain life in both human systems and ecosystems. Water is unevenly distributed across the Earth, both temporally and spatially. Frequent and regular rainfall in some regions contrasts sharply with prolonged droughts in others. Some regions are blessed with an abundance of freshwater while others face scarcity. Moreover, the freshwater resources of the world are not partitioned to match political borders. Thus the distribution and use of limited water resources can create conflicts at local, regional and even international level. History shows, and the future may confirm, that water has a strategic role in conflicts between stakeholders. Improved water management, conflict resolution and cooperation could ameliorate such conflicts. The water conflict resolution process has been approached by many disciplines such as law, economics, engineering, political economy, geography, anthropology and systems theory.

Traditional conflict resolution approaches using judicial systems, state legislatures, commissions and similar governmental instruments mostly provide resolutions in which one party gains at the expense of the other. The successful resolution of national as well as international water conflicts requires an understanding of the nature of the conflict and then modelling and analysing the inherent problems. To reach a final agreement concerning how much of the shared water resource is allocated to each party or nation, the assistance of procedures or methodologies acceptable to all the parties concerned is very much needed. A system dynamics study of the nature and conduct of conflict and cooperation between parties, based on new technologies and practices, could assist with the efficient management of water resources, and thereby reduce tension between parties in dispute over water.

This is a new approach for water resource conflict resolution. It uses systems thinking and a system dynamics simulation model to provide a powerful alternative to traditional approaches for conflict resolution, which often rely too much on outside mediation. By helping stakeholders explore and resolve the underlying structural causes of conflict, this approach can transform problems into significant opportunities for cooperation between all parties involved.

This example shows how a system dynamics approach can help to get a better understanding of a water-related conflict during its resolution process, and to identify the key factors to be considered in such a conflict, including their interrelationships. It also shows how system dynamics simulation can assist in helping stakeholders reach a consensus over final water allocations, and determine the time to reach such an agreement.

Dynamic hypothesis of conflict resolution

Most environmental conflicts, including those that are water-related, spring from three sources (Nandalal and Simonovic, 2003). The first is an actual or prospective human intervention in the environment, which provokes changes in natural and societal systems. Conflict arises when one or more of the stakeholder groups sees the activity as disturbing the complex interaction between physical, biological and social processes. The second source is disagreement over the management of a water supply at one location when it affects the use of it elsewhere. The third source is climatic variability and change independent of direct human activity, which places new stresses on the water resources and generates fresh adaptations to available resources.

We shall focus here on the first and second sources of water conflict. We consider the case of a river basin that traverses a boundary, which could be either local, regional or international. The basis of the conflict is a development (a reservoir) and its management by one of the stakeholders in the water resource. Such decisions impact the neighbour during water shortage conditions, and create conditions for a number of water conflicts. The behaviour of two stakeholder groups confronted with sharing a limited water resource is examined using a system dynamics simulation.

The system modelled comprises a reservoir and a downstream service area, as shown in Figure 8.70. The service area is assumed to fall into two administrative authorities. The stakeholders from these two regions (communities A and B) are facing a problem in fulfilling their objective of water sharing, since there is not sufficient water to satisfy the total demand of both. The conflict caused by the water shortage results in the problem of how to share the scarce resource between them.

The two communities may have an influence on water allocation decisions based on a conventional operations research-type model. Either might claim as relevant factors, for example, that the reservoir is located on its territory, or that it contributes more to maintenance expenses. One community's struggle for more water might make the other community increase its own effort to obtain more. That

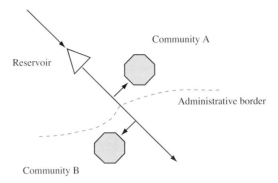

Figure 8.70 *A schematic diagram of the hypothetical water resource system*

is, the actions of A will affect B, and of B will affect A, through a chain of causes and effects. Despite the very simple structure, this hypothetical system is sufficient to demonstrate how system dynamics could assist in studying and resolving water resources conflicts. More complicated systems will obviously require more complex system representation and simulation analyses.

The conflict over water allocation is taken to proceed through several stages:

- Both stakeholders fight for their target water allocations.
- Failure to fulfil the target creates dissatisfaction.
- Displeasure leads to a propensity to fight for more.

This dynamic hypothesis of the conflicting situation between the two stakeholders is identified in Figure 8.71 using the systems language of causal diagrams. The two communities (*A* and *B*) share a limited amount of water, thus the *allocations* are inversely related. More *allocation* to a community reduces its *dissatisfaction*, while its *fight for more water* increases with *dissatisfaction*. If a community enhances its fight, it will get more water while the opposite party will get less. For example, in loop M, the increase of water allocated to A decreases its dissatisfaction. The targeted (*aspiration A*) and actual water allocations to A determine its dissatisfaction. If A suffers an increase in dissatisfaction, it fights for more water (in the diagram, it increases *fight for more A*). This increased fight results in an increased allocation. Loop N shows the similar behaviour of community B. As loop P shows, an increase in A's fight negatively affects B's allocation. Likewise, an increase in B's fight decreases A's allocations, as shown in loop Q.

This causal loop diagram helps us to get a deeper insight in the complex water sharing conflict. All the negative (balancing) feedback loops indicate that the conflicting situation will reach equilibrium or settlement in the end. The allocations and the dissatisfactions of the two communities depend on their initial targets (provided the policy adopted is to allocate water according to the ratio of their initial target requirements). As we noted, both communities may have various reasons to exert more pressure. Recognition of the degree of concern of each could be given by coupling a weight to the community's fight for more water. The weights introduced would then affect the final settlement with respect to water allocations and the time necessary to achieve that settlement. Thinking systemically and taking action to give certain weights to the fights of the stakeholders would enable the conflict to end with a more reasonable solution of the water allocation problem.

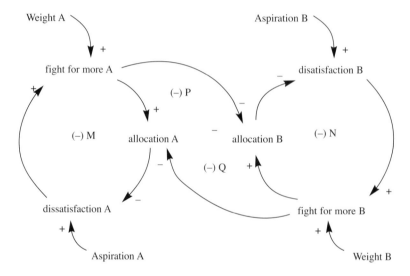

Figure 8.71 *A causal loop diagram representing the water sharing conflict between the two users*

Model development

A system dynamics model was developed to represent the water sharing conflict, based on the insight obtained from the causal loop diagram. The model structure is shown in Figure 8.72. In the middle of the diagram is a stock *availability*, representing the water in the reservoir available for distribution. The flow *maintain* into it keeps the *availability* at a constant level during the simulation. The flows *allocation A* and *allocation B* are the amounts of water supplied to A and B respectively. The converters *aspiration A* and *aspiration B* are the initial demands of the two communities. Failure to assure the supply of total demand creates dissatisfaction in the communities. The values of the converters *dissatisfaction A* and *dissatisfaction B* increase when the difference between allocation and aspiration level increases, and vice versa. When the dissatisfaction of a community rises, it fights for more water. The flows *fight for more A* and *fight for more B* go up. Their efforts accumulate over time, in the stocks *cumulative fight A* and *cumulative fight B*, and influence the allocations made to the two communities. The converters *weight A* and *weight B* are coupled to the struggles of the two communities to influence the allocation and the importance given to their fights.

A deviation of the amount of water received from the desired quantity creates 'dissatisfaction'. This dissatisfaction may be expressed in many different forms using the difference between the allocation and the desired level. In the model, if aspiration exceeds allocation, dissatisfaction is assumed to be equal to the squared difference between the aspiration and the allocation ($n = 2$ in equation (8.46)). There is no dissatisfaction if the allocation exceeds the aspiration. If n takes a higher value, the dissatisfaction will increase more rapidly. The choice of an appropriate value for n should be based on practical experience and knowledge of the average time required for the resolution of conflict. The dissatisfaction of the community could also be expressed in the form of a penalty function. Equation (8.46) shows the dissatisfaction of A, while a similar equation can be used for B:

$$dissatisfaction\ A = \begin{cases} \{aspiration\ A - allocation\ A\}^n; & \text{if } aspiration\ A > allocation\ A \\ 0 & ; \text{if } aspiration\ A \leq allocation\ A \end{cases}$$

(8.46)

where:
dissatisfaction A = dissatisfaction of community A at time t
aspiration A = targeted allocation of community A
allocation A = water allocated to community A at time t
n = constant.

The intensity of the fight for more water for each community goes up with dissatisfaction, and can vary significantly. In the model the fight is taken as equal to the square root of the dissatisfaction, representing a lower-intensity fight. If *fight for more* were taken to equal dissatisfaction, that would indicate a more intense fight. The two communities might have different levels of influence on the water resource, as noted above. For example, one community might claim that its land has been inundated by the reservoir and that this should be taken into consideration when the allocation decision is made. Such influences are introduced to the model as weights that can be given to the fights of different communities. The initial weights to be assigned could be agreed upon by the two communities involved in the conflict at the outset of the resolution process. The participation of a third party as a mediator

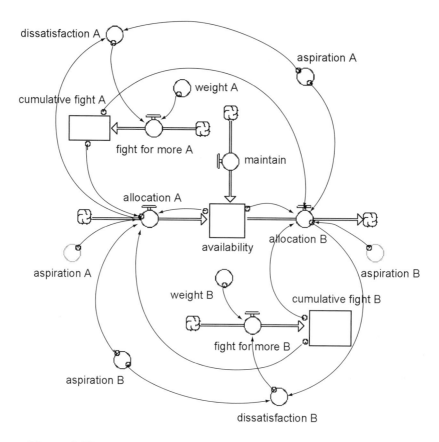

Figure 8.72 *A system dynamic model of the water sharing conflict*

is not expected in the proposed conflict resolution process. However, the communities may obtain the technical assistance of a third party in the application of the model and interpretation of its results. If such a mediator is involved in the resolution process, he or she could assist the communities in deciding and applying the weights in the model. Equation (8.47) presents the fight of community A, and B can be represented in the same manner.

$$\text{fight for more A} = \left\{\sqrt{\text{dissatisfaction A}}\right\} \times \text{weight A} \qquad (8.47)$$

where:
fight for more A = the fight of A for more water at time t, and
weight A = the weight given to that fight.

When dissatisfaction builds up, the *cumulative fight* increases in both communities. The allocation made for a community depends both on its own cumulative fight (or its dissatisfaction) and the cumulative fight of the rival community (or its dissatisfaction). Equation (8.48) shows the allocation to A used in the model, while a similar equation can be used for B.

$$\text{allocation A} = \text{availability} \times \left(\frac{\text{cumulative fight A}}{\text{cumulative fight A} + \text{cumulative fight B}} \right) \quad (8.48)$$

where:
cumulative fight A = cumulative fight of A at time t
cumulative fight B = cumulative fight of B at time t
availability = total water available in the reservoir for allocation.

As shown, the water allocation policy adopted in the model depends on the dissatisfactions of both communities. However, different water allocation policies can be adopted during the modelling process. For example, water could be allocated according to the squared value of the cumulative fight ratio in equation (8.48).

Model application results

The causal loop diagram in Figure 8.71 offers a better understanding of the dynamic behaviour of stakeholders confronted with a water-related conflict during its resolution process. The system dynamics model developed to fit the causal loop diagram was tested using a common pattern I have observed in real-world water conflicts, such as:

- sharing the water resources of the Nile River between riparian countries;
- management of the Sihu basin in China for flood control, water supply and irrigation;
- sharing the hydro potential of the Danube River between Romania, Bulgaria and the former Yugoslavia;
- management of the Shellmouth Reservoir in Manitoba, Canada for flood control, recreation and water supply;
- floodplain management in the Red River basin, shared between the United States and Canada.

A number of simulation experiments were carried out to investigate the conflict resolution process with respect to both the final allocations acceptable to both parties, and the time to reach agreement.

The system dynamics model needs the initial aspirations of the two communities and the weights to be given to their fights as inputs. The initial aspirations are provided by the two communities directly. They reflect each community's water needs. The model outputs of interest for the study include the variations of dissatisfactions of the two parties and allocations to them with time. Figure 8.73 shows the typical outputs from a simulation run. It shows the variation of dissatisfactions of the two communities and the variation of the allocations to them with time when their initial aspirations are 30×10^6 m^3 for A and 60×10^6 m^3 for B. Note that the reservoir is assumed to provide 50×10^6 m^3 water for distribution between the two parties.

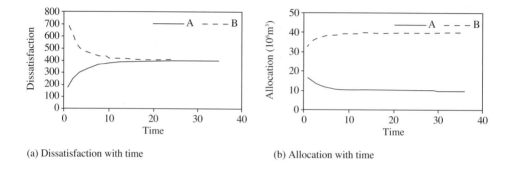

(a) Dissatisfaction with time (b) Allocation with time

Figure 8.73 *Variation of dissatisfactions and water allocations with time*

Simulation starts with the allocation of the water available in the reservoir to A and B according to the ratio of their initial aspirations. As shown in Figure 8.73(b), these initial allocations are 16.7×10^6 m^3 and 33.3×10^6 m^3 respectively. The dissatisfactions these create are 710 and 178 units. The weight given to the fight of community A is 0.2 (that of B is 0.8). The two communities are assumed to have reached an agreement when the dissatisfaction of each community repeats the same value for at least five consecutive time steps. This criterion has been selected on the basis of the similar criteria used for approximate solutions of numerical problems. The time from the beginning to the first of such time steps is taken as the time necessary to reach the agreement. At the agreement, which is achieved after 32 time units, the allocations to A and B are 10×10^6 m^3 and 40×10^6 m^3 respectively. The dissatisfactions of both communities reach 400 units at this agreement. For the same set of initial aspirations of the communities, the time to reach an agreement will vary depending on the weights given to their fights. The communities will also have two different dissatisfactions at each of those agreements.

The impact of the choice of weights, initial conditions and aspiration levels on model performance is investigated in detail in Nandalal and Simonovic (2003).

Conclusions

The example shows how the use of a system dynamics approach can be effective in enabling stakeholders to increase their insight into the nature of a conflict between two groups faced with a water allocation problem. It also shows how the results of system dynamics simulations can assist in making wise decisions regarding the resolution of the conflict.

The detailed analysis of the conflict resolution process based on the proposed model (Nandalal and Simonovic, 2003) demonstrates the impacts of the initial aspirations of the stakeholders, the magnitude of their influence on the water resource and the intensity of the struggle on the final agreement. The initial aspirations of stakeholders will affect the time to reach an agreement. In general, the time required will increase with the difference between their aspirations. The starting allocations considered in the simulations do not affect the final agreement, but can affect the time to reach an agreement. The influence stakeholders have on the water resource affects the final agreement considerably. Both the time to reach an agreement and the final allocations depend on the stakeholder's relative control over the water resource. For example, if a party has a very high influence over the resource, although its initial aspiration is low, it may receive more water than a weak opponent with a higher initial aspiration. If the gap between the initial aspirations of the two stakeholders is larger while their influences remain the same, the difference between the final allocations at the agreement will increase.

8.5 REFERENCES

Ahmad, S. and S.P. Simonovic (2000), 'Modelling Reservoir Operations for Flood Management Using System Dynamics', *ASCE Journal of Computing in Civil Engineering*, (14): 190–198

Atkinson, K. (1985), *Elementary Numerical Analysis*, John Wiley, New York

Cosgrove, W.J. and F.R. Rijbersman (2000), *World Water Vision*, for the World Water Council, Earthscan Publications Ltd, London

Fedorovskiy, A.D., I.Y. Timchenko, L.A. Sirenko and V.G. Yakimchuk (2004), 'Method of System Dynamics in Simulating the Problems in the Comprehensive Use of Water', *Hydrobiological Journal*, 40(2): 87–96

Fernandez, J.M. and M.A.E. Selma (2004), 'The Dynamics of Water Scarcity on Irrigated Landscapes: Mazarron and Aguilas in South-eastern Spain', *System Dynamics Review*, 20(2): 117–137

Fletcher, E. J. (1998), 'The Use of System Dynamics as a Decision Support Tool for the Management of Surface Water Resources', *Proceedings of the First International*

Conference on New Information Technologies for Decision-Making in Civil Engineering, Montreal, Canada, 909–920

Ford, A. and H. Flynn (2005), 'Statistical Screening of System Dynamics Models', *System Dynamics Review*, 21(4): 273–303

Forrester, J.W. (1990), *Principles of Systems*, Productivity Press, Portland, OR, first published 1968

Gleick, P.H. (2000), *The World's Water 2000–2001*, The Biennial Report on Freshwater Resources, Island Press, Washington, DC

IHP (International Hydrologic Programme) (2000), *World Freshwater Resources*, CD-ROM prepared by I.A. Shiklomanov, International Hydrologic Programme, UNESCO, Paris

Keyes, A. M. and Palmer, R. (1993), 'The Role of Object-oriented Simulation Models in the Drought Preparedness Studies', *Proceedings of the 20th Annual National Conference*, Water Resources Planning and Management Division of ASCE, Seattle, WA, 479–482

Kirkwood, C.W. (1998), *System Dynamics Methods: A Quick Introduction*, Arizona State University, AZ

Klir, G. and B. Yuan (1995), *Fuzzy Sets and Fuzzy Logic: Theory and Applications*, Prentice-Hall, Englewood Cliffs, NJ

Li, L. and S.P. Simonovic (2002), 'System Dynamics Model for Predicting Floods from Snowmelt in North American Prairie Watersheds', *Hydrological Processes Journal*, 16: 2645–2666

Malthus, T.R. (1798), *An Essay on the Principle of Population*, J. Johnson, London

Matthias, R. and P. Frederick (1994), 'Modelling Spatial Dynamics of Sea-level Rise in a Coastal Area', *System Dynamics Review*, 10(4): 375–389

Meadows, D.H., D.L. Meadows, J. Randers and W.W. Behrens III (1972), *The Limits to Growth*, A Potomac Associates Book, Washington, DC

Meadows, D.H., D.L. Meadows and J. Randers (1992), *Beyond the Limits*, McClelland & Stewart Inc., Toronto, Ontario, Canada

Nandalal, K.D.W. and S.P. Simonovic (2003), 'Resolving Conflicts in Water Sharing: A Systemic Approach', *Water Resources Research*, 39(12): 1362–1373

Palmer, R. (1994). '(ACT-ACF) River Basin Planning Study', University of Washington, US, http://www.tag.washington.edu/projects/act-acf.html (last accessed November 2005)

Palmer, R., A.M. Keyes and S. Fisher (1993), 'Empowering Stakeholders through Simulation in Water Resources Planning', *Water Management for the '90s*, K. Hon, ed., ASCE, Seattle, WA, 451–454

Pedrycz, W. and F. Gomide (1998), *An Introduction to Fuzzy Sets*, MIT Press, Cambridge, MA

Roberts N., D.F. Anderson, R.M. Deal, M.S. Garet and W.A. Shaffer (1983), *Introduction to Computer Simulation: The System Dynamics Approach*, Addison-Wesley, Reading, MA

Saysel, A.K., Y. Barlas and O. Yenigun (2002), 'Environmental Sustainability in an Agricultural Development Project: A System Dynamics Approach', *Journal of Environmental Management*, 64: 247–260

Sehlke, G. and J. Jacobson (2005), 'System Dynamics Modelling of Transboundary Systems: The Bear River basin model', *Ground Water*, 43(5): 722–730

Senge, P.M. (1990), *Fifth Discipline – The Art and Practice of the Learning Organization*, Doubleday, New York

Shiklomanov, I.A. (2000), 'Appraisal and Assessment of World Water Resources', *Water International*, 25(1): 11–32

Simonovic, S.P. (2002), 'World Water Dynamics: Global Modelling of Water Resources', *Journal of Environmental Management*, 66(3): 249–267

Simonovic, S.P. and H. Fahmy (1999), 'A New Modelling Approach for Water Resources Policy Analysis', *Water Resources Research*, 35(1): 295–304

Simonovic, S.P., H. Fahmy and A. El-Shorbagy (1997), 'The Use of Object-oriented Modelling for Water Resources Planning in Egypt', *Water Resources Management*, 11(4): 243–261

Simonovic, S.P. and S. Ahmad (2005), 'Computer-based Model for Flood Evacuation Emergency Planning,' *Natural Hazards*, 34(1): 25–51

Simonovic, S.P. and V. Rajasekaram (2004), 'Integrated Analyses of Canada's Water Resources: A System Dynamics Model', *Canadian Water Resources Journal*, 29(4): 223–250

Stave, K.A. (2003), 'A System Dynamics Model to Facilitate Public Understanding of Water Management Options in Las Vegas, Nevada', *Journal of Environmental Management*, 67: 303–313

Sterman, J.D. (2000), *Business Dynamics: Systems Thinking and Modelling for a Complex World*, McGraw Hill, New York

Terano, T., K. Asai and M. Sugeno (1991), *Fuzzy Systems Theory and its Applications*, Academic Press, London

Ventana Systems (1995), *Vensim User's Guide*, Ventana Systems Inc., Belmont, MA

Ventana Systems (2003a), *Vensim 5 Reference Manual*, Ventana Systems Inc., Belmont, MA

Ventana Systems (2003b), *Vensim 5 Modelling Guide*, Ventana Systems Inc., Belmont, MA

Water Resources Branch (1992), *Shellmouth Reservoir Study: Preliminary Study of Spillway Control Gates*, Manitoba Department of Natural Resources, Canada

Water Resources Branch (1995), *Assiniboine River Flooding & Operation of Shellmouth Dam*, Summary Material: Presentation to the Shellmouth Flood Review Committee, Manitoba Department of Natural Resources, Canada

Xu, Z.X., K. Takeuchi, H. Ishidaira and X.W. Zwang (2002), 'Sustainability Analysis for Yellow River Water Resources Using the System Dynamics Approach', *Water Resources Management*, 16: 239–261

8.6 EXERCISES

1. In the famous excerpts below from Thomas Malthus's *First Essay on Population* (1798), Malthus implicitly describes feedback loops that influence the dynamics of population. Draw a causal diagram to show his feedback thinking. The first paragraph sets the stage; it is the second and third paragraphs that should be diagrammed:

   > Population, when unchecked, increases in a geometrical ratio. Subsistence increases only in an arithmetical ratio...
   >
   > By that law of our nature which makes food necessary to the life of man, the effects of these two unequal powers must be kept equal. This implies a strong and constantly operating check on population from the difficulty of subsistence. This difficulty must fall somewhere; and must necessarily be severely felt by a large portion of mankind...
   >
   > Population, could it be supplied with food, would go on with unexhausted vigour, and the increase of one period would furnish the power of a greater increase the next, and this without any limit...
   >
   > Foresight of the difficulties attending the rearing of a family acts as a preventative check [acting on the birth rate]; and the actual distress of some of the lower classes, by which they are disabled from giving the proper food and attention to their children, acts as a positive check [acting on the death rate], to the natural increase of population.

   You might choose to include in your causal diagram population, births, deaths, preventative checks, positive checks, food adequacy and food.

2. For the diagrams below: (i) assign polarities to each of the causal links; (ii) assign polarities to each of the feedback loops; (iii) write a brief but insightful paragraph describing the role of the feedback loops in your diagram. Do not describe every link in your diagram (assume your figure and its polarities take care of that), but talk mainly about the loops.

a. Municipal water system causal loop diagram showing the effect of household water use on supply.

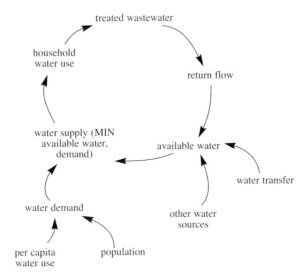

b. Feedback loops in water supply system capacity expansion and water demand.

After assigning polarities, consider what the left-hand loop would do by itself. What do the right-hand loops contribute?

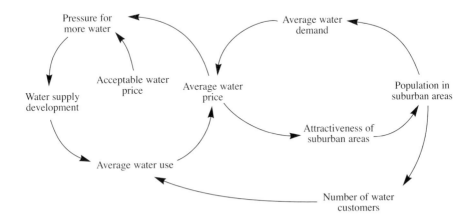

c. Main factors and feedback loops of the irrigated lands model.

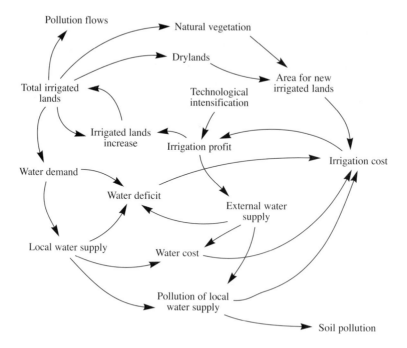

3 Present a causal loop of your own (from your own thinking or from an article, or book of some sort). Explain it sufficiently in words so that your picture and the story it tries to tell are clear.

4 Identify at least one stock within the engineering activity that your actions help to fill or deplete.

5 You have learned about the basic building blocks (objects) of the modelling tool Vensim which are used to develop system dynamics simulation models. With the knowledge you have gained so far, define the objects Auxiliary variable and Arrow in your own words. Comment on their use in Vensim models.

6 In each of the following groups, identify a stock and one or more related flows. Some of the words represent concepts that are not connected stocks or flows – they are just information in the system. Show in a causal diagram how you think those other concepts in the group are related to the stock and flow sets you identify. (Suggestion: don't add any more concepts to these lists.)
   a. Pipes, pipeline construction, water flow.
   b. Births, deaths, population, fecundity, life span.
   c. Knowledge, learning, forgetting, intelligence.
   d. Deficit, debt, income, spending, interest payments on debt, payments on debt principle.

7 Develop a stock and flow diagram for each example in Exercise 2. Explain your choice of stock and flow variables.
8 Consider a reservoir with a single inflow rate Q_1 and outflow rate Q_2. Draw the change in reservoir volume given the two sets of flow and inflow rates below. The initial volume of the reservoir is 100 units in both cases. Do not use the computer. This exercise should help you to develop intuition about stocks and flows, and the ability to relate their behaviour.

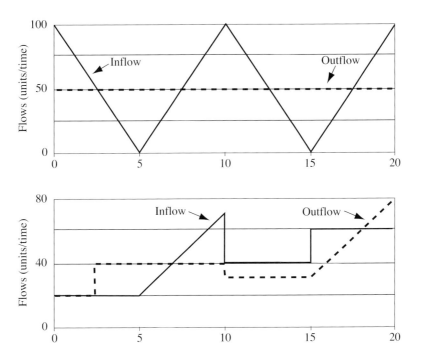

9 For the prototype 'bathtub' problem in Example 10 (Section 8.2) create a simulation model of storing the water during a water shortage in your area (caused by maintenance of the water main).
   a. The faucet has a maximum flow of 2 litres/min, and it is turned on until the tub is filled. The tub volume is 50 litres.
   b. The maintenance work will require five hours.
   c. You will require 15 litres of water after 50 minutes and an additional 20 litres after 90 minutes.
   d. Assume that after three hours your service will be reestablished and you will drain the remaining water from the tub.
   e. Assume the outflow rate depends on the amount of water in the tub. The maximum outflow of 2 litres/min should occur when the tub is completely

filled (i.e., at 50 litres) and the minimum outflow of 0 litres/min should occur when the tub is empty (i.e. at 0 litres).
f. Use Vensim to simulate the model. How does water volume in the tub change over time? How long will it take for the water to completely drain from the tub? Suggestion: use the 'bathtub' model available on the CD-ROM, directory SYSTEM DYNAMICS, subdirectory Examples, Example 7 and modify it according to the needs of the problem.

10 For the Red Reservoir problem in Example 11 (Section 8.2), which is available on the CD-ROM, directory SYSTEM DYNAMICS, subdirectory Examples, Example 8, modify the Vensim simulation model to accommodate the following monthly inflow: 0.2, 0.25, 0.4, 0.7, 0.9, 0.65, 0.5, 0.45, 0.4, 0.55, 0.3, 0.25 million m^3 (starting with January and ending with December). Show the dynamics of the Red Reservoir system with a weighted-average operating rule and with a multiplicative operating rule.

11 Read the following articles:
a. Saysel et al (2002).
b. Sehlke and Jacobson (2005).
c. Stave (2003).
d. Comment on the use of system dynamics simulation for the three problems in the articles. In each case indicate how you would expand the model presented in the article. What feedback loops would you suggest? How would you obtain the data necessary for the suggested expansions? What do you expect your expansions will contribute to the models already developed?

12 Solve the LP problem in Example 12 (Section 8.3) for $h = 0.6$ and $h = 0.4$. For these calculations you are encouraged to use the Linpro program on the CD-ROM (described in Chapter 9).
a. Compare the results from your calculations with those in Table 8.3.
b. Show your results using graphs similar to the graph in Figure 8.42.

9

Optimization

The procedure of selecting the set of decision variables that maximizes/minimizes the objective function subject to the system constraints is called the *optimization* procedure. Numerous optimization techniques are used in water resources systems management. One of the most significant contributions of optimization to the solution of water resources management problems was the introduction of linear programming (LP) in the late 1950s (Dantzig, 1963). LP is applied to problems that are formulated in terms of separable objective functions and linear constraints. However, neither objective functions nor constraints are in a linear form in most practical water management applications. Many modifications have been used in real applications in order to convert nonlinear problems to use LP solvers. Examples include different schemes for the linearization of nonlinear relationships and constraints, and the use of successive approximations.

An expansion of optimization applications followed the introduction of *nonlinear programming*, an optimization approach used to solve problems when the objective function and the constraints are not all in linear form. Successful applications are available for some special classes of nonlinear programming problems such as unconstrained problems, linearly constrained problems, quadratic problems, convex problems, separable problems, non-convex problems and geometric problems. The main limitation in applying nonlinear programming to water management problems is that the method is generally unable to distinguish between a local optimum and a global optimum (except by finding another better local optimum).

Dynamic programming (DP) offers advantages over other optimization tools, since the shape of the objective function and constraints do not affect it, and as such, it has been frequently used in water resources systems management. DP requires the discretization of the problem into a finite set of stages. At every stage a number of possible conditions of the system (states) are identified, and an optimal solution is identified at each individual stage, given that the optimal solution for the next stage is available.

In the very recent past, most researchers have been looking for new approaches that combine efficiency and an ability to find the global optimum. One group of techniques, known as *evolutionary algorithms*, seems to have a high potential since it holds the promise of achieving both. Evolutionary techniques are based on similarities with the biological evolutionary process. In this concept, a population of individuals, each representing a search point in the space of feasible solutions, is exposed to a collective learning process, which proceeds from generation to generation. The population is arbitrarily initialized and subjected to the processes of selection, recombination and mutation through stages known as *generations*, such that newly created generations evolve towards more favourable regions of the search space. In short, the progress in the search is achieved by evaluating the fitness of all individuals in the population, selecting the individuals with the highest fitness value, and combining them to create new individuals with increased likelihood of improved fitness.

From the history of the application of optimization techniques to water management, it has become obvious that more complex analytical optimization algorithms are being replaced with simpler and more robust search tools. This chapter describes two optimization methods in detail, and discusses their practical implementation in the water management context. LP is presented for its academic and practical significance, and evolutionary optimization is discussed as one of the methods used in contemporary water resources systems optimization.

9.1 LINEAR PROGRAMMING

LP is one of the most widely used techniques in water resources systems management. This section introduces the basic concepts of its optimization technique (Wagner, 1975; Jewell, 1986; de Neufville, 1990; Hillier and Lieberman, 1990).

9.1.1 Formulation of linear optimization models

Sections 4.4.1–4.4.3 provide three examples of linear optimization model formulations: for a city water supply, operation of a multi-purpose reservoir and wastewater treatment. Model formulation is the most difficult part of the process. Wagner (1975) offers the following guidelines for this stage of the optimization analysis:

- What are the key decisions to be made? What problem is being solved?
- What makes the real decision environment so complex as to require the use of a

> **Box 9.1** The largest power plant in the world
>
> In the spring of 2005 my work brought me to the Iguazu Falls on the border between Brazil and Argentina. The trip provided me with an opportunity to visit the nearby Itaipú Dam on the Paraná River, the largest hydroelectric dam on the planet. It has been called one of the seven wonders of the modern world by the American Society of Civil Engineering.
>
> Built from 1975 to 1991, in a binational development, Itaipú represents the efforts and accomplishments of two neighbouring countries, Brazil and Paraguay. The power plant's 18 generating units add up to a total production capacity of 12,600 megawatts (MW) and a reliable output of 75 million MWh a year. Itaipú's energy production has broken several records over recent years, since the last generating unit was commissioned in 1991. The generation of 77,212,396 MWh a year in 1995 was surpassed in 1996, when the new record of around 80 million MWh a year was established. The municipalities that had part of their land flooded by the formation of the Itaipú Lake, which feeds the dam, receive a share of the profit, paid monthly in the form of royalties.
>
> In the 1970s and 1980s, when it was built, the Itaipú was considered an imperial project, a product of the megalomania of the military which governed Brazil at that time. Today, three decades later, opinions have changed. In addition to the production of energy, Itaipú is also a symbol of environmental preservation. The company maintains the Bela Vista Biological Refuge and is responsible for permanent reforestation, fish breeding and other wild animal programmes.
>
> I stood at the entrance to the dam, feeling the shaking of the ground caused by the rotation of the turbines, admiring the ability of the human race to create amazing structures in order to support its own existence and at the same time sustain its relationship with the surrounding environment.
>
> (A memory from 2005)

  linear optimization model? What elements of complexity are incorporated in the model? What elements are ignored?
- What distinguishes a practical decision from an unusable one in this environment? What distinguishes a good decision from a poor one?
- As a decision-maker, how would you employ the results of the analysis? What is your interpretation of results? In what ways might you want or need to temper the results because of factors not explicitly considered in the model?

In order to formulate the mathematical model in terms of linear relationships, two conditions must be satisfied:

- *Divisibility.* For each activity, the total amounts of each input and the associated value of the objective are strictly proportional to the level of output – that is, to the activity level. Each activity is capable of continuous proportional expansion or reduction.
- *Additivity.* Given the activity levels for each of the decision variables x_j, the total amounts of each input and the associated value of the objective are the sums of the inputs and objective values for each individual process.

9.1.2 Algebraic representations of linear optimization models

In many water resources management problems, the aim is to maximize or minimize some objective, but there are certain constraints on what can be done to this end. The term *linear programming* (LP) refers to a way of modelling many of these problems so that they have a special structure, and to the way of solving problems with such a structure. It is a technique that can be applied in many different problem domains. The process of formulating a model was discussed in detail earlier in the book. The first step is to decide which are the *decision variables*. These are the quantities that can be varied, and so affect the value of the objective. The second step in the formulation is to express the *objective* in terms of the decision variables. Lastly we must write down the *constraints* that restrict the choices of decision variables. One common-sense constraint is that many variables cannot realistically be negative. We can summarize the mathematical representation of the LP model in the following way. Letting x_j be the level of activity j, for $j = 1, 2, ..., n$, we want to select a value for each x_j such that:

$$C_1 x_1 + C_2 x_2 + ... + C_n x_n$$

is maximized or minimized, depending on the context of the problem. The x_j are constrained by a number of relations, each of which is one of the following type:

$$a_1 x_1 + a_2 x_2 + ... + a_n x_n \leq a$$
$$b_1 x_1 + b_2 x_2 + ... + b_n x_n = b$$
$$c_1 x_1 + c_2 x_2 + ... + c_n x_n \geq c$$

The first relation includes the possible restriction $x_j \geq 0$. Such a constrained optimization problem may have:

- no feasible solution: that is, there may be no values of all the x_j, for $j = 1, 2,..., n$, that satisfy every constraint;

- a unique optimal feasible solution;
- more than one optimal feasible solution;
- a feasible solution such that the objective function is unbounded; that is, the value of the function can be made as large as desired in a maximization problem, or as small in a minimization problem, by selecting an appropriate feasible solution.

Changing the sense of the optimization

Any linear maximization model can be viewed as an equivalent linear minimization model, and vice versa, by accompanying the change in the optimization sense with a change in the signs of the objective function coefficients. Specifically,

$$\max \sum_{j=1}^{n} c_j x_j \text{ can be treated as } \min \sum_{j=1}^{n} (-c_j) x_j \qquad (9.1)$$

and vice versa.

Changing the sense of an inequality

All inequalities in an LP model can be represented with the same directioned inequality since:

$$\sum_{j=1}^{n} a_j x_j \leq b \text{ can be written as } \sum_{j=1}^{n} (-a_j) x_j \geq -b \qquad (9.2)$$

and vice versa.

Converting an inequality to an equality

An inequality in a linear model can be represented as an equality by introducing a non-negative variable as follows:

$$\sum_{j=1}^{n} a_j x_j \leq b \text{ can be written as } \sum_{j=1}^{n} a_j x_j + 1s = b \text{ where } s \geq 0$$

$$\sum_{j=1}^{n} a_j x_j \geq b \text{ can be written as } \sum_{j=1}^{n} a_j x_j - 1t = b \text{ where } t \geq 0 \qquad (9.3)$$

It is common to refer to a variable such as s as a *slack variable*, and t as a *surplus variable*.

Converting equalities to inequalities

Any linear equality or set of linear equalities can be represented as a set of like-directed linear inequalities by imposing one additional constraint. The idea can be generalized as follows:

$$\sum_{j=1}^{n} a_{ij}x_j = b_i \text{ for } i = 1,2,\ldots, m \text{ can be written as}$$

$$\sum_{j=1}^{n} a_{ij}x_j \leq b_i \text{ for } i = 1,2,\ldots, m \text{ and } \sum_{j=1}^{n} \alpha_j x_j \leq \beta \quad (9.4)$$

where

$$\alpha_j = -\sum_{i=1}^{n} a_{ij} \text{ and } \beta = -\sum_{i=1}^{n} b_i. \quad (9.5)$$

Canonical forms for linear optimization models

Sometimes it is convenient to be able to write *any* linear optimization model in a compact and unambiguous form. The various transformations presented above allow us to meet this objective, although it is now apparent that there is considerable freedom in the selection of a particular canonical form to employ. I illustrate one such representation here.

Any linear optimization model can be viewed as:

$$\text{maximize } \sum_{j=1}^{n} c_j x_j \quad (9.6)$$

subject to:

$$\sum_{j=1}^{n} a_{ij}x_j \leq b_i \text{ for } i = 1,2,\ldots, m$$

$$x_j \geq 0 \text{ for } j = 1,2,\ldots, n \quad (9.7)$$

It is typical, although not required, that $n > m$.

The CD-ROM accompanying this book includes the Linpro software and all the LP examples developed in the text. The folder LINPRO contains two sub-folders, Linpro and Examples. The read.me file in the LINPRO folder contains instructions for the installation of the Linpro software and a detailed tutorial for its use as a part of its Help menu.

9.1.3 Geometric interpretation of linear optimization models

There are two geometric representations of linear optimization models. One is called the *solution space representation* and is treated in this book. The other, the *requirements space representation*, is less important and is not treated in this text.

Example 1 – solution space representation

Let us directly consider a numerical example with two dimensions (decision variables):

$$\text{maximize } 12x_1 + 15x_2 \tag{9.8}$$

subject to:

$$4x_1 + 3x_2 \le 12 \tag{9.9}$$

$$2x_1 + 5x_2 \le 10 \tag{9.10}$$

$$x_1 \ge 0 \text{ and } x_2 \ge 0. \tag{9.11}$$

The problem is graphed in Figure 9.1. Observe that both inequalities (9.9) and (9.10) are drawn as equations. Then each inequality is indicated by an arrow on the side of the line representing permissible values of x_1 and x_2. Since the two variables must be non-negative, the region of permissible values is also bounded by the two coordinate axes.

Accordingly, the polygon $\overline{oabc}$ represents the region of values for x_1 and x_2 that satisfy all the constraints. This polygon is called the *solution set*. The set points described by the polygon are *convex*. The vertices $0, a, b$ and c are referred to as the *extreme points* of the polygon, in that they are not on the interior of any line segment connecting two distinct points of the polygon.

The parallel lines in the figure represent various values of the objective function. The arrow points in the direction of increasing values of the objective function. The unique optimal solution is at the extreme point b, where $x_1 = 15/7$, $x_2 = 8/7$ and $12x_1 + 15x_2 = 300/7$.

The example represented by the set of relationships (9.8) to (9.11) is on the CD-ROM, in the directory LINPRO, subdirectory Examples, Example1.

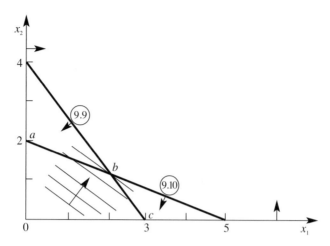

Figure 9.1 *Solution space*

Example 2 – alternative optimal solutions

If the coefficients of the objective function are changed so as to alter the direction of the parallel lines in Figure 9.1, it is clear that the optimal solution may change, but in any case there is always an extreme-point optimal solution. Consider in Figure 9.2 a problem for one specific rotation of the parallel lines:

$$\text{maximize } 4x_1 + 10x_2 \tag{9.12}$$

again subject to the same set of constraints (9.9), (9.10) and (9.11).

Now, all the points (an infinite number) on the segment *ab* are optimal. Thus $x_1 = 15/7$ and $x_2 = 8/7$ are still optimal. But so are $x_1 = 0$ and $x_2 = 2$, as well as any positive-weighted average of these two optimal solutions. The optimal value of the objective function is 20.

The example represented by the set of relationships (9.12), (9.9), (9.10) and (9.11) is on the CD-ROM, in the directory LINPRO, subdirectory Examples, Example2.

Example 3 – unbounded optimal solutions

The third illustration, shown in Figure 9.3, is based on the model:

$$\text{maximize } -2x_1 + 6x_2 \tag{9.13}$$

subject to:

$$-1x_1 - 1x_2 \leq -2 \tag{9.14}$$

$$-1x_1 + 1x_2 \leq 1 \tag{9.15}$$

$$x_1 \geq 0 \text{ and } x_2 \geq 0. \tag{9.16}$$

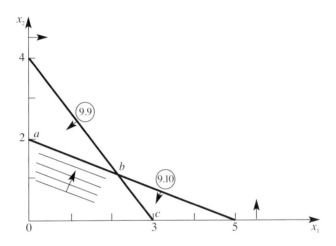

Figure 9.2 *Alternative optimal solutions*

The solution set for this problem is unbounded. The objective function for the problem can be made arbitrarily large. Given any value for the objective function, there is always a solution point with an even greater objective function value, and there is always such a point satisfying (9.15) with equality.

The example represented by the set of relationships (9.13), (9.14), (9.15) and (9.16) is on the CD-ROM, in the directory LINPRO, subdirectory Examples, Example 3.

Example 4 – infeasible problem

Figure 9.4 is based on the problem:

$$\textit{maximize } 1x_1 + 1x_2 \tag{9.17}$$

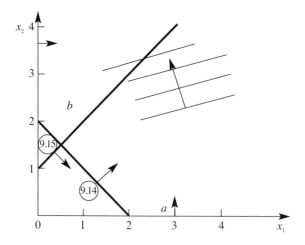

Figure 9.3 *Unbounded solution*

subject to:

$$-1x_1 + 1x_2 \leq -1 \tag{9.18}$$

$$1x_1 - 1x_2 \leq -1 \tag{9.19}$$

$$x_1 \geq 0 \text{ and } x_2 \geq 0. \tag{9.20}$$

Figure 9.4 illustrates that the problem does not have a feasible solution.

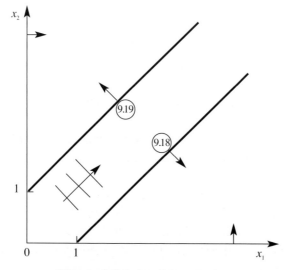

Figure 9.4 *Infeasible solution*

The example represented by the set of relationships (9.17), (9.18), (9.19) and (9.20) is on the CD-ROM, in the directory LINPRO, subdirectory Examples, Example 4.

From the several geometric illustrations presented we can summarize that:

- if the solution set is not empty, it is convex and may be either bounded or unbounded;
- if the solution set is not empty, the optimal value of the objective function may be finite or unbounded. If finite, then an optimal solution exists at an extreme point.

9.1.4 Simplex method of solution

Probably you will never have to calculate manually the solution of an LP model in a real application, since a computer can do the work. Therefore you might ask, 'Why do we need to know the underlying theory of linear optimization models?' In the light of considerable experience in applying LP to water resources systems management problems, I am convinced that a novice to the field must understand the principles explained here in order to make truly effective and sustained use of this optimization tool.

Many different algorithms have been proposed to solve LP problems, but the one below has proved to be the most effective in general. This is the general procedure:

Step 1: Select a set of m variables that yields a feasible starting trial solution. Eliminate the selected m variables from the objective function.
Step 2: Check the objective function to see whether there is a variable that is equal to 0 in the trial solution but would improve the objective function if made positive. If such a variable exists, go to Step 3. Otherwise, stop.
Step 3: Determine how large the variable found in the previous step can be made until one of the m variables in the trial solution becomes 0. Eliminate the latter variable and let the next trial set contain the newly found variable instead.
Step 4: Solve for these m variables, and set the remaining variables equal to 0 in the next trial solution. Return to Step 2.

The resulting algorithm does find an optimal solution to a general LP problem in a finite number of iterations. Often this method is termed *Dantzig's simplex algorithm*, in honour of the mathematician who devised the approach.

Let us examine a 'well-behaved' problem and explain the *simplex method* by means of this example.

Example 5

Consider the mathematical model:

$$\text{maximize } 4x_1 + 5x_2 + 9x_3 + 11x_4$$

subject to:

$$\begin{aligned}
1x_1 + 1x_2 + 1x_3 + 1x_4 &\le 15 \\
7x_1 + 5x_2 + 3x_3 + 2x_4 &\le 120 \\
3x_1 + 5x_2 + 10x_3 + 15x_4 &\le 100 \\
x_1 \ge 0 \quad x_2 \ge 0 \quad x_3 \ge 0 \quad x_4 &\ge 0
\end{aligned} \quad (9.21)$$

Let x_0 be the value of the objective function and add slack variables. Then write the system as:

$$\begin{aligned}
1x_0 - 4x_1 - 5x_2 - 9x_3 - 11x_4 &= 0 & \text{Row 0} \\
1x_1 + 1x_2 + 1x_3 + 1x_4 + 1x_5 &= 15 & \text{Row 1} \\
7x_1 + 5x_2 + 3x_3 + 2x_4 + 1x_6 &= 120 & \text{Row 2} \\
3x_1 + 5x_2 + 10x_3 + 15x_4 + 1x_7 &= 100 & \text{Row 3}
\end{aligned}$$

where all the variables must be non-negative. Notice how the introduction of the variable x_0 in Row 0 permits us to express the objective function in equation form. The example represented by the set of relationships (9.21) is on the CD-ROM, in directory LINPRO, subdirectory Examples, Example 5.

The task of Step 1 is to find a starting feasible solution to (9.21). There are a large number of such solutions, but it is certainly most convenient to begin with $x_0 = 0$, $x_5 = 15$, $x_6 = 120$, $x_7 = 100$, and all other variables equal to 0. In other words, we start with an all-slack solution. We term this an *initial feasible basic solution*, and x_0, x_5, x_6 and x_7 are known as the *basic variables*, sometimes shortened to the *basis*. The remaining variables we call *non-basic*.

Interpretation of coefficients in Row 0

Each coefficient represents the increase (for negative coefficients) or decrease (for positive coefficients) in x_0 with a unit increase of the associated non-basic variable.

Iteration 1

For Step 2 the simplex method adopts the following easy-to-apply rule for deciding the variable to enter the next trial basis.

Simplex Criterion I (maximization)

If there are non-basic variables with a negative coefficient in Row 0, select the one with the most negative coefficient, that is, the best per unit potential gain (say x_j). If all non-basic variables have positive or zero coefficients in Row 0, an optimal solution has been obtained.

To decide which variable should leave the basis we apply the following rule, or Step 3.

Simplex Criterion II

(a) Take the ratios of the current right-hand side to the coefficients of the entering variable x_j (ignore ratios with 0 or negative numbers in the denominator).
(b) Select the minimum ratio – that ratio will equal the value of x_j in the next trial solution. The minimum ratio occurs for a variable x_k in the present solution; set $x_k = 0$ in the solution.

The process of applying Criterion II is known as a *change-of-basis* calculation, or a *pivot operation*. A detailed calculation is presented in Table 9.1.

Iteration 2

At this point the first iteration of the simplex method has been completed. On returning to Step 2, you are ready to determine whether an optimal solution has been obtained or another simplex iteration is required. Criterion I, which examines the non-basic variables, indicates that a still better solution seems to exist. You might profitably enter into the basis x_1, x_2, or x_3. Criterion I selects x_1, since it promises the greatest gain per unit increase. Next perform the Step 3 calculations, using Criterion II. From Table 9.1, notice that x_1 will replace x_5 in the next trial solution.

Iteration 3

Having completed the second simplex iteration, once more examine the coefficients in Row 0 to ascertain whether you have discovered an optimal solution. It now appears favourable to enter x_3 and remove x_4, which was entered at the first iteration.

At this iteration you have just seen another aspect to the computational rule in Criterion II. To sum up, Criterion II ensures that each new basic solution results in only 0 or positive values for the trial values of the basis. Consequently, the solution remains feasible at every iteration.

Table 9.1 Simplex tableau for the Example 5 problem

Iteration values	Basis	Current	x_1	x_2	x_3	x_4	x_5	x_6	x_7	Row
1	x_0	0	−4	−5	−9	−11				0
	x_5	15	1	1	1	1	1			1
	x_6	120	7	5	3	2		1		2
	x_7	100	3	5	10	15			1	3
2	x_0	$\frac{220}{3}$	$\frac{-9}{5}$	$\frac{-4}{3}$	$\frac{-5}{3}$				$\frac{11}{15}$	0
	x_5	$\frac{25}{3}$	$\frac{4}{5}$	$\frac{2}{3}$	$\frac{1}{3}$		1		$\frac{-1}{15}$	1
	x_6	$\frac{320}{3}$	$\frac{33}{5}$	$\frac{13}{3}$	$\frac{5}{3}$			1	$\frac{-2}{15}$	2
	x_4	$\frac{20}{3}$	$\frac{1}{5}$	$\frac{1}{3}$	$\frac{2}{3}$	1			$\frac{1}{15}$	3
3	x_0	$\frac{1105}{12}$		$\frac{1}{6}$	$\frac{-11}{12}$		$\frac{9}{4}$		$\frac{7}{12}$	0
	x_1	$\frac{125}{12}$	1	$\frac{5}{6}$	$\frac{5}{12}$		$\frac{5}{4}$		$\frac{-1}{12}$	1
	x_6	$\frac{455}{12}$		$\frac{-7}{6}$	$\frac{-13}{12}$		$\frac{-33}{4}$	1	$\frac{5}{12}$	2
	x_4	$\frac{55}{12}$		$\frac{1}{6}$	$\frac{7}{12}$	1	$\frac{-1}{4}$		$\frac{1}{12}$	3
4	x_0	$\frac{695}{7}$		$\frac{3}{7}$		$\frac{11}{7}$	$\frac{13}{7}$		$\frac{5}{7}$	0
	x_1	$\frac{50}{7}$	1	$\frac{5}{7}$		$\frac{-5}{7}$	$\frac{10}{7}$		$\frac{-1}{7}$	1
	x_6	$\frac{325}{7}$		$\frac{-6}{7}$		$\frac{13}{7}$	$\frac{-61}{7}$	1	$\frac{4}{7}$	2
	x_3	$\frac{55}{7}$		$\frac{2}{7}$	1	$\frac{12}{7}$	$\frac{-3}{7}$		$\frac{1}{7}$	3

Iteration 4

All the coefficients in Row 0 are non-negative, and consequently Criterion I asserts that we have found an optimal solution. Thus the calculations are terminated in Step 2.

Summary

In brief, the simplex method consists of four steps:

1. Selection of an initial basis.
2. Application of simplex Criterion I. If the solution is not optimal go to Step 3; otherwise, stop.
3. Application of simplex Criterion II.
4. Change of basis, and return to second step.

The progress of the simplex method can easily be interpreted through the geometry of the solution space. Each basis corresponds to a cortex of the convex polyhedral set of feasible solutions. Going from one basis to the next represents going from one extreme point to an adjacent one. Thus the simplex method can be said to seek an optimal solution by *climbing along the edges, from one vertex of the convex polyhedral solution set to a neighbouring one*. Once we master the straightforward logic of the simplex iterations, considerable writing effort can be saved by organizing the computations in a convenient tabular form called a *simplex tableau* (Table 9.1).

9.1.5 Completeness of the simplex algorithm

In the application of Criterion I, when two or more variables appear equally promising, as indicated by the values of their coefficients in Row 0, an arbitrary rule may be adopted for selecting one of these. For example, use the lowest numbered variable, or the one suspected to be in the final basis.

In the application of Criterion II, when two or more variables in the current basis are to fall simultaneously to the level 0 upon introducing the new variable, only one of these is to be removed from the basis. The others remain in the basis at 0 level. The resultant basis is termed *degenerate*. Unless some care is given to the method of deciding which variable to remove from the basis, there is no *proof* that the method always converges. However, long experience with simplex computations leads to the conclusion that for all *practical* purposes, the selection can be arbitrary and the associated danger of non-convergence is negligible.

If at some iteration in applying Criterion II there is no positive coefficient in any row for the entering variable, then there exists an *unbounded* optimal solution. In this event, the entering variable can be made arbitrarily large, the value of x_0 thereby increases without bound, and the current basis variables remain non-negative. Thus we may drop the earlier assumption that the optimal value of the objective function is finite. The simplex algorithm provides an indication of when an unbounded optimal solution occurs. Criterion II is easily reworded to cover this case.

Starting basis

Let me review the selection of an initial basis to begin the algorithm. Because each constraint in the example of the preceding section is of the form:

$$\sum_{j=1}^{n} a_{ij}x_j \le b_i \quad \text{where } b_i \ge 0 \tag{9.22}$$

adding a slack variable to each relation and starting with an all-slack basic solution provided a simple way of initiating the simplex algorithm. However, the constraints in any LP model can be written as:

$$\sum_{j=1}^{n} a_{ij}x_j = b_i \quad \text{for } i = 1,2,\dots, m \quad \text{where } b_i \ge 0 \tag{9.23}$$

In this form, if a variable appears only in constraining relation i and has a coefficient of 1, as would be the case for a slack variable, it can be used as part of the initial basis. But relation i may not have such a variable. This can occur, for example, if the i-th equation is linearly dependent on one or more of the other equations, such as being a sum of two equations. Then we can utilize the following approach.

Write the constraints as:

$$\sum_{j=1}^{n} a_{ij}x_j + 1y_i = b_i \quad \text{for } i = 1,2,\dots, m \quad \text{where } b_i \ge 0 \tag{9.24}$$

and where $y_i \ge 0$, use y_i as the basic variable for relation i. It is assumed here, for simplicity, that every constraint requires the addition of a y_i. The name *artificial variable* is given to y_i because it is added as an artifice in order to obtain an initial-trial solution. Is this approach valid? The answer is yes, provided Condition A is satisfied.

Condition A. To ensure that the final solution is meaningful, every y_i must equal 0 at the terminal iteration of the simplex method.

If there is no feasible solution, it will be impossible to satisfy Condition A. At the final iteration of the simplex algorithm, at least one positive y_i will be in the solution indicating an *infeasible* optimal solution.

This completes the rules of the simplex algorithm. They can be programmed for operation on any computer. Microsoft Excel® includes an LP solver that can be used to solve LP problems. The accompanying CD-ROM includes the LP software Linpro (folder LINPRO, subfolder Linpro) developed on the basis of the simplex algorithm for the solution of LP problems.

The Big M method

There are a number of computational techniques for guaranteeing Condition A. One approach is to add to the maximizing objective function each y_i with a large penalty-cost coefficient:

$$x_0 - \sum_{j=1}^{n} c_j x_j + \sum_{i=1}^{m} M y_i = 0 \tag{9.25}$$

where M is a relatively large number. Thus, each y_i variable is very costly compared with any of the x_j variables. To initiate the algorithm, first we eliminate each y_i from (9.25) by using (9.24). This gives:

$$x_0 - \sum_{j=1}^{n} c_j x_j + M \sum_{i=1}^{m} \sum_{j=1}^{n} a_{ij} x_j = -M \sum_{i=1}^{m} b_i \tag{9.26}$$

which simplifies to:

$$x_0 - \sum_{j=1}^{n} (c_j + M \sum_{i=1}^{m} a_{ij}) x_j = -M \sum_{i=1}^{m} b_i \tag{9.27}$$

Because the y_i variables are so expensive, the optimization technique drives the y_i variables to 0, *provided* there exists a feasible solution. Whenever a y_i drops from a basis at some iteration, we need never consider using it again, and can eliminate it from further computations. The following example will clarify the approach.

Example 6

Consider the problem:

$$\text{maximize } -3x_1 - 2x_2 \tag{9.28}$$

subject to:

$$1x_1 + 1x_2 = 10 \tag{9.29}$$

$$1\, x_1 \geq 4 \tag{9.30}$$

$$x_1 \geq 0 \quad x_2 \geq 0 \tag{9.31}$$

After adding a surplus variable x_3 in (9.30), we can write the model as:

$$\begin{aligned} x_0 + 3x_1 + 2x_2 &= 0 & &\text{Row 0} \\ 1x_1 + 1x_2 &= 10 & &\text{Row 1} \\ 1x_1 - 1x_3 &= 4 & &\text{Row 2} \end{aligned} \qquad (9.32)$$

Next, introduce artificial variables y_1 and y_2, and let $M = 10$, giving:

$$\begin{aligned} x_0 + 3x_1 + 2x_2 + 10y_1 + 10y_2 &= 0 & &\text{Row 0} \\ 1x_1 + 1x_2 + 1y_1 &= 10 & &\text{Row 1} \\ 1x_1 - 1x_3 + 1y_2 &= 4 & &\text{Row 2} \end{aligned} \qquad (9.33)$$

To initiate the algorithm, we subtract M times Row 1 and M times Row 2 from Row 0 to eliminate y_1 and y_2:

$$\begin{aligned} x_0 - 17x_1 - 8x_2 + 10x_3 &= -140 & &\text{Row 0} \\ 1x_1 + 1x_2 + 1y_1 &= 10 & &\text{Row 1} \\ 1x_1 - 1x_3 + 1y_2 &= 4 & &\text{Row 2} \end{aligned} \qquad (9.34)$$

Using Linpro software we can verify that $x_1 = 4$ and $x_2 = 6$ are the optimal solutions. Example 6 is on the CD-ROM, directory LINPRO, subdirectory Examples, Example 6.

9.1.6 Duality in linear programming

LP offers much more than the numerical values of an optimal solution. The mathematical structure of a linear model means, for example, that if a linear optimization model has a finite optimal solution, there exists an optimal *basic* solution. Many important post-optimality questions are easily answered, given the numerical information at the final simplex iteration. However, before we proceed with the discussion of sensitivity analysis let us introduce a unifying concept, known as *duality*, which establishes the interconnections for all of the sensitivity analysis techniques.

Primal and dual problems

Consider the pair of LP models:

$$\text{maximize} \sum_{j=1}^{n} c_j x_j \qquad (9.35)$$

subject to:

$$\sum_{j=1}^{n} a_{ij} x_j \leq b_i \text{ for } i = 1, 2, ..., m \tag{9.36}$$

$$x_j \geq 0 \text{ for } i = 1, 2, ..., n. \tag{9.37}$$

and

$$\text{minimize } \sum_{i=1}^{m} b_i y_i \tag{9.38}$$

subject to:

$$\sum_{i=1}^{m} a_{ij} y_j \geq c_j \text{ for } j = 1, 2, ..., n \tag{9.39}$$

$$y_j \geq 0 \text{ for } i = 1, 2, ..., m. \tag{9.40}$$

We arbitrarily call (9.35), (9.36) and (9.37) the *primal problem* and (9.38), (9.39) and (9.40) its *dual problem*.

Example 7

As an illustration, let us write a dual formulation of the following primal LP model:

$$\text{maximize } 4x_1 + 5x_2 + 9x_3 \tag{9.41}$$

subject to:

$$\begin{aligned} 1x_1 + 1x_2 + 2x_3 &\leq 16 \\ 7x_1 + 5x_2 + 3x_3 &\leq 25 \\ x_1 \geq 0 \quad x_2 \geq 0 \quad x_3 &\geq 0 \end{aligned} \tag{9.42}$$

Its dual formulation is then:

$$\text{minimize } 16y_1 + 25y_2 \tag{9.43}$$

subject to:

$$\begin{aligned} 1y_1 + 7y_2 &\geq 4 \\ 1y_1 + 5y_2 &\geq 5 \\ 2y_1 + 3y_2 &\geq 9 \\ y_1 \geq 0 \quad y_2 &\geq 0 \end{aligned} \tag{9.44}$$

The dual problem can be viewed as the primal model flipped on its side:

- The jth column of coefficients in the primal model is the same as the jth row of coefficients in the dual model.
- The row of coefficients of the primal objective function is the same as the column of constants on the right-hand side of the dual model.
- The column of constants on the right-hand side of the primal model is the same as the row of coefficients of the dual objective function.
- The direction of the inequalities and sense of optimization is reversed in the pair of problems.

Now we can define more closely the significant aspects of the primal-dual relationship:

Dual theorem

(a) In the event that both the primal and dual problems possess feasible solutions, then the primal problem has an optimal solution x^*_j, for $j = 1, 2, ..., n$, the dual problem has an optimal solution y^*_i, for $i = 1, 2, ..., m$, and

$$\sum_{j=1}^{n} c_j x^*_j = \sum_{i=1}^{m} b_i y^*_i. \tag{9.45}$$

(b) If either the primal or dual problem possesses a feasible solution with a finite optimal objective function value, then the other problem possesses a feasible solution with the same optimal objective-function value.

The duality relationships are summarized in Table 9.2.

Table 9.2 *Relationship between primal and dual problems*

Primal (maximize)	Dual (minimize)
Objective function	Right-hand side
Right-hand side	Objective function
jth column of coefficients	jth row of coefficients
ith row of coefficients	ith column of coefficients
jth variable non-negative	jth relation an inequality
jth variable unrestricted in sign	jth relation an equality
ith relation an inequality	ith variable non-negative
ith relation an equality	ith variable unrestricted in sign

In addition we can observe the following relationship:

Optimal values of dual variables
(a) The coefficients of the slack variables in Row 0 of the final simplex iteration of a maximizing problem are the optimal values of the dual variables.
(b) The coefficient of variable x_j in Row 0 of the final simplex iteration represents the difference between the left- and right-hand sides of the jth dual constraint for the associated optimal dual solution.

By reference to the notion of duality, we deepen our understanding of what is really happening in the simplex method. The coefficients of the slack variables in Row 0 of the primal problem at each iteration can be interpreted as trial values of the dual variables. So the simplex method can be seen as an approach that seeks feasibility for the dual problem while maintaining feasibility in the primal problem. As soon as feasible solutions to both problems are obtained, the simplex iterations terminate.

Example 8

For further explanation, let us consider the problem from Example 5 represented by the set of relationships (9.21). Using this example we shall illustrate the process of solving the dual problem.

The primal formulation given by (9.21):

$$\text{maximize } 4x_1 + 5x_2 + 9x_3 + 11x_4$$

subject to:

$$1x_1 + 1x_2 + 1x_3 + 1x_4 \leq 15$$
$$7x_1 + 5x_2 + 3x_3 + 2x_4 \leq 120$$
$$3x_1 + 5x_2 + 10x_3 + 15x_4 \leq 100$$
$$x_1 \geq 0 \; x_2 \geq 0 \; x_3 \geq 0 \; x_4 \geq 0$$

yields the following dual formulation:

$$\text{minimize } 15y_1 + 120y_2 + 100y_3 \tag{9.46}$$

subject to:

$$1y_1 + 7y_2 + 3y_3 \geq 4$$
$$1y_1 + 5y_2 + 5y_3 \geq 5$$
$$1y_1 + 3y_2 + 10y_3 \geq 9$$
$$1y_1 + 2y_2 + 15y_3 \geq 11$$
$$y_1 \geq 0 \; y_2 \geq 0 \; y_3 \geq 0$$

Using Linpro software we can verify, by checking the coefficients of the three slack variables in Row 0 of the final iteration, that

$$y_1 = 13/7, \; y_2 = 0 \text{ and } y_3 = 5/7 \qquad (9.47)$$

are the optimal solutions of the dual problem. Example 8 is on the CD-ROM, directory LINPRO, subdirectory Examples, Example 7.

Copy these values as we shall refer to them below. First, verify that the constraints in (9.46) are satisfied:

$$28/7 \geq 4$$
$$38/7 \geq 5 \qquad (9.48)$$
$$63/7 \geq 9$$
$$88/7 \geq 11$$

Second, check that the value of the dual objective function is the same as the value of the primal objective function:

$$15(13/7) + 120(0) + 100(5/7) = 695/7. \qquad (9.49)$$

The values in (9.47) must be optimal, since they satisfy all the dual constraints and yield an objective-function value equal to the optimal primal value.

Finally, let us calculate the differences between the left- and right-hand sides of (9.48). For example, the second and third constraints give:

$$38/7 - 5 = 3/7$$
$$63/7 - 9 = 0 \qquad (9.50)$$

These are the coefficients of x_2 and x_3, respectively, in Row 0 of the final iteration of the dual-problem solution.

9.1.7 Sensitivity analysis

Sensitivity analysis is the study of how the optimal solution and the value of the optimal solution to a linear program change given changes in the various coefficients of the problem. That is, we are interested in answering questions such as the following:

- What effect will a change in the coefficients in the objective function (c_j) have?
- What effect will a change in the right-hand side values (b_i) have?
- What effect will a change in the coefficients in the constraining equations (a_{ij}) have?

Since sensitivity analysis is concerned with how these changes affect the optimal solution, the analysis begins only after the optimal solution to the original LP problem has been obtained. Hence, sensitivity analysis can be referred to as *post-optimality analysis*.

The mechanics for all post-optimality analyses are straightforward extensions of the simplex arithmetic, but duality is the key idea that ensures the mechanics are correct. Every LP model has a dual formulation. By solving one of these we automatically solve the other.

There are several reasons why sensitivity analysis is considered important from a water resources management point of view. First, consider the fact that water resources management occurs in a dynamic environment. Basic physical variables (precipitation and flow, for example) change over time; demand for water fluctuates; water infrastructure ages and new replaces the old, and so on. If an LP model has been used in a decision-making situation and later we find changes in some of the coefficients associated with the initial LP formulation, we would like to determine how these changes affect the optimal solution to our original LP problem. Sensitivity analysis provides us with this information without requiring us to completely solve a new linear program.

Thus, through sensitivity analysis, we will be able to provide valuable information for the decision-maker. We begin our study of sensitivity analysis with the coefficients of the objective function.

Sensitivity analysis: the coefficients of the objective function

Recall in Example 7 that the dual constraint corresponding to the primal variable x_2 is

$$1y_1 + 5y_2 + 5y_3 \geq 5. \tag{9.51}$$

If the objective function coefficient of x_2 becomes $(5 + p_2)$, then $(5 + p_2)$ appears on the right-hand side of (9.51). Substituting the optimal values of the dual variables into (9.44), where $(5 + p_2)$ is used, yields:

$$1(13/7) + 5(0) + 5(5/7) \geq 5 + p_2 \tag{9.52}$$

or

$$3/7 \geq p_2 \tag{9.53}$$

Thus the current dual solution remains feasible provided p_2 does not exceed 3/7. If p_2 is made larger than this fraction, the dual solution is no longer feasible, and consequently the primal solution is no longer optimal.

In summary, consider the following management interpretation of sensitivity analysis for the objective function coefficients. Think of the basic variables as corresponding to a current product line and the non-basic variables as representing other products a company might produce. Within bounds, changes in the profit associated with one of the products in the current product line would not cause the company to change its product mix or the amounts produced, but the changes would have an effect on its total profit. Of course, if the profit associated with one of the products changed drastically, it would change the product line (i.e. move to a different basic solution). For products that are not currently being produced (non-basic variables), it is obvious that a decrease in per unit profit would not make the company want to produce them. However, if the per unit profit for one of these products became large enough, it would want to consider adding that product to the product line.

Sensitivity analysis: the right-hand sides

From the basic interpretation of the simplex algorithm we know that the coefficient of a slack variable in Row 0 of an optimal solution represents the incremental value of another unit of the resource associated with that variable. In the preceding section we stated that the optimal value of a dual variable is the very same coefficient. Putting the two statements together, we have the following interpretation of the dual variables:

> The optimal value of a dual variable indicates how much the objective function changes with a unit change in the associated right-hand-side constant, provided the current optimal basis remains feasible.

This interpretation is in agreement with the fundamental equality relation in the dual theorem presented in Section 9.1.6, which states:

optimal value of $x_0 = \Sigma$ (right-hand-side constants) × (optimal dual variables)
(9.54)

Since this is such an important property, let me state it again. Associated with every constraint of the primal LP problem is a dual variable. The value of the dual variable indicates how much the objective function of the primal problem will increase as the value of the right-hand side of the associated primal constraint is increased by one unit.

In other words, the value of the dual variable indicates the value of one additional unit of a particular resource. Hence, this value can be interpreted as the maximum value or price we would be willing to pay to obtain one additional unit of the resource. Because of this interpretation, the value of one additional unit of a resource is often called the *shadow price* of the resource. Thus, the optimal values of the dual variables are shadow prices. When the right-hand-side constants represent quantities of scarce resources, the shadow price indicates the unit worth of each resource as predicated on an optimal solution to the primal problem.

For the LP problem from Example 7, the value 13/7 is the shadow price for the first constraint, and similarly 0 holds for the second constraint, and 5/7 for the third constraint. Therefore, an additional unit of resource represented by the first constraint increases the objective function value by 13/7, and an additional unit of resources represented by the third constraint increases the objective function value by 5/7, but an additional unit of resource represented by the second constraint does not improve the objective function value. Why? Because this resource is already in excess supply, as evidenced by the slack variable x_6 being in the optimal basis.

In general, a resource in excess supply is indicated by the slack variable for that resource appearing in the final basis at a positive level. The corresponding shadow price is 0, because additional excess supply is of no value. Since the slack is in the basis, its final Row 0 coefficient is 0.

Interpreting the values of the dual variables as shadow prices leads to an insightful view into the meaning of the dual problem. In the context of Example 7, think of each dual variable as representing the true marginal value of its associated resource, assuming that the decision is made optimally. Then (9.54) indicates that the total value of the objective function is the same as evaluating the total worth of all the resources in scarce supply. Interpret each coefficient a_{ij} as the consumption of the ith resource by the jth activity. The summation:

$$\sum_{i=1}^{m} a_{ij} y_j$$

represents the underlying cost of using the jth activity, evaluated according to the shadow prices. The constraints of the dual problem ensure that an optimal solution never exceeds its true worth.

Sensitivity analysis: the A matrix

To begin consideration of sensitivity tests on a_{ij}, suppose an entirely new activity is added to the model. Is it advantageous to enter it into the basis? The easiest test is to check whether the associated dual constraint is satisfied by the current values of the dual variables. If not, the new activity should be introduced.

Consider the problem from Example 5, as given by (9.21) in Section 9.1.4. Suppose we add another variable:

$+ 1x_8$	Row 1	(9.55)
$+ 2/7 x_8$	Row 2	
$+ 17 x_8$	Row 3	

Let the objective function coefficient of x_8 be c_8. At what value of c_8 is it attractive to enter x_8? The associated dual relation is

$$1y_1 + 2/7 y_2 + 17 y_3 \geq c_8 \tag{9.56}$$

Substitute the current optimal values of the dual variables in (9.49) to obtain

$$1(13/7) + 2/7(0) + 17(5/7) \geq c_8, \tag{9.57}$$

or

$$14 \geq c_8 \tag{9.58}$$

Therefore, if c_8 exceeds 14, we should enter x_8 into the basis.

If x_j is a non-basic variable, we can examine the effect of changing its coefficients a_{ij} in exactly the same fashion. To illustrate, x_4 is non-basic in the optimal solution of the problem in Example 5. Let us alter its coefficient in Row 3 by A. Then the associated dual restriction is

$$1y_1 + 2y_2 + (15 + A)y_3 \geq 11 \tag{9.59}$$

Substitute the current optimal values of the dual variables in (9.59) to obtain:

$$1(13/7) + 2(0) + (15 + A)(5/7) \geq 11 \tag{9.60}$$

or

$$A \geq -11/5. \tag{9.61}$$

Therefore, if A is smaller than $-11/5$, we can enter x_4 into the basis.

If x_j is a basic variable, analysing the effect of changing a matrix coefficient is more complex. The analysis involves a simultaneous consideration of both the primal and dual problems. The derivation goes beyond the scope of this text, but can be found in several advanced books on LP.

9.1.8 Summary

LP is a deterministic optimization technique based on the following set of assumptions:

- The objective function and constraints can be expressed as simple algebraic expressions.
- The objective function must be convex and the constraints must form a convex policy space.
- All decision variables must be positive.

As an optimization tool LP has some advantages and some limitations. It uses standard computer programs (like Linpro and Excel). In the application of LP, the user does not have to know details of the method, only its philosophy and limitations. The approach can handle a large number of state variables. However, on a sequential basis it can handle fewer decision variables than other programming techniques (e.g. DP), but can handle a large dimension of decision variables for one time period. In solving LP problems, constraints tend to increase the computational time. In order to obtain a solution of an LP problem, a convex policy space is required. As was pointed out at the beginning of this chapter, LP requires linear relationships. However, nonlinear relationships can be linearized. The optimal solution to an LP problem is a single set of decisions. No information is made available about second- or third-best solutions.

9.2 FUZZY OPTIMIZATION

Water resources management decisions are characterized by a set of decision alternatives (the *decision space*); a set of states of nature (the *state space*); a relation assigning a result to each pair of a decision and state; and finally, the objective function which orders the results according to their desirability. When deciding under certainty, the water resources decision-maker knows which state to expect and he or she chooses the decision alternative with the highest value of the objective function (in the case of maximization, or the smallest value in the case of minimization), given the prevailing state of nature. When deciding under uncertainty, the decision-maker does not know exactly which state will occur. Only a probability function of the states may be known. Then decision-making becomes more difficult. The theory of stochastic optimization is well developed, and applications in water resources decision-making can be found in the literature and practice.

The discussion in this chapter has so far been restricted to decision-making under certainty. In this instance the model of decision-making is non-symmetric in the following senses: the decision space is described either by enumeration or by a number of constraints, and the objective function orders the decision space using a one-to-one relationship of results to decision alternatives. Hence we can only have *one* objective function supplying the order, but we may have several constraints defining the decision space. Following Bellman and Zadeh (1970), we shall focus here on the expansion of our decision-making model to a fuzzy environment that considers a situation of decision-making under uncertainty, in which the objective function and the constraints are fuzzy. The fuzzy objective function and constraints are characterized by their fuzzy membership functions.

Since our aim is to optimize the objective function as well as the constraints, a decision in a fuzzy environment is defined by analogy with non-fuzzy environments as the selection of activities that simultaneously satisfy an objective function *and* constraints. According to the above definition and assuming that the constraints are 'non-interactive' (not overlapping), the logical '*and*' corresponds to the intersection. The 'decision' in a fuzzy environment is therefore viewed as the intersection of fuzzy constraints and a fuzzy objective function. The relationship between constraints and objective functions in a fuzzy environment is thus fully symmetrical: that is, there is no longer a difference between the former and the latter. An example will illustrate the concept.

Example 9

The objective function 'x should be substantially larger than 10' is given by the membership function:

$$\mu_{\tilde{o}}(x) = \begin{cases} 0 & x \leq 10 \\ (1 + (x - 10)^{-2})^{-1} & x > 10 \end{cases} \tag{9.62}$$

The constraint 'x should be in the vicinity of 11' is given by the membership function:

$$\mu_{\tilde{c}}(x) = (1 + (x - 11)^4)^{-1} \tag{9.63}$$

The membership function of the decision is then:

$$\mu_{\tilde{D}}(x) = \mu_{\tilde{O}}(x) \wedge \mu_{\tilde{C}}(x) \tag{9.64}$$

$$\mu_{\tilde{D}}(x) = \begin{cases} \min\{1 + (x-10)^{-2})^{-1}, (1 + (x-11)^4)^{-1}\} & x > 10 \\ 0 & x \leq 10 \end{cases}$$

$$\mu_{\tilde{D}}(x) = \begin{cases} (1 + (x-11)^4)^{-1} & x > 11.75 \\ (1 + (x-10)^{-2})^{-1} & 10 < x \leq 11.75 \\ 0 & x \leq 10 \end{cases}$$

This example is shown in Figure 9.5.

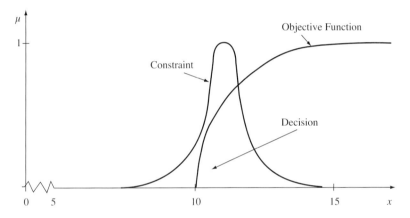

Figure 9.5 *A fuzzy decision*

Bellman and Zadeh (1970) implicitly introduce the following assumptions:

■ The 'and' connecting objective and constraints is the model that corresponds to the 'logical and'.
■ The 'logical and' corresponds to the set of theoretic intersection.
■ The intersection of fuzzy sets is defined by the *min* operator.

Further discussion on how to broaden this concept is available in Zimmermann (1996), and is beyond the scope of this text.

9.2.1 Fuzzy linear programming

The presentation in this section relies on Bellman and Zadeh (1970) and Zimmermann (1996). LP models are considered as a special type of the general decision model discussed above: the decision space is defined by the constraints, the goal is defined by the objective function, and the type of decision is decision-making under certainty. The classical model of LP presented in Section 9.1 and written in matrix form is:

maximize $f(x) = \mathbf{c}^T\mathbf{x}$
subject to:
$$\mathbf{Ax} \leq \mathbf{b} \qquad (9.65)$$
$$\mathbf{x} \geq 0$$
with
$$\mathbf{c}, \mathbf{x} \in \mathbf{R}^n, \quad \mathbf{b} \in \mathbf{R}^m, \quad \mathbf{A} \in \mathbf{R}^{m \times n}$$

We shall now depart from the classical assumptions that all coefficients of $\mathbf{A}$, $\mathbf{b}$ and $\mathbf{c}$ are crisp numbers, that $\leq$ is meant in a crisp sense, and that maximization is a strict imperative.

For transforming LP problems into a fuzzy environment, quite a number of modifications to (9.65) are required. First of all the decision-making problem is transformed from actually maximizing or minimizing the objective function to reaching some aspiration levels established by the decision-maker.

Second, the constraints are treated as vague. The $\leq$ sign is not treated in a strictly mathematical sense, and smaller violations are acceptable. In addition, the coefficients of the vectors $\mathbf{b}$ or $\mathbf{c}$ or of the matrix $\mathbf{A}$ itself can have a fuzzy character, either because they are fuzzy in nature or because perception of them is fuzzy.

Finally, the role of the constraints is different from that in classical LP, where the violation of any single constraint by any amount renders the solution infeasible. Small violations of constraints are acceptable, and may carry different (crisp or fuzzy) degrees of importance. Fuzzy LP offers a number of ways to allow for all those types of vagueness, and we shall discuss some of them below.

In this text we accept Bellman–Zadeh's concept of a symmetrical decision model. We have to decide how a fuzzy 'maximize' is to be interpreted. Our discussion will be limited to one approach for a fuzzy goal, and readers are directed to the literature for different interpretations. Finally we have to decide where and how fuzziness enters the constraints. One way is to consider the coefficients of $\mathbf{A}, \mathbf{b}, \mathbf{c}$ as fuzzy numbers and the constraints as fuzzy functions. Another approach represents the goal and the constraints by fuzzy sets and then aggregates them in order to derive a maximizing decision. We shall use the latter. However, we still have to decide on the type of membership function characterizing the fuzzy sets representing the goal and constraints.

In classical LP the violation of any constraint in (9.64) makes the solution infeasible. Hence all constraints are considered to be of equal weight or importance.

When departing from classical LP this is no longer true, and the relative weights attached to the constraints have to be considered.

Before we develop a specific model of LP in a fuzzy environment, it should be noted once more that by contrast to classical LP, *fuzzy LP* is *not* a uniquely defined type of model. Many variations are possible, depending on the assumptions or features of the real situation being modelled.

Let us now formulate the fuzzy LP problem. Let us assume in model (9.65) that the decision-maker can establish an aspiration level, z, for the value of the objective function to be achieved, and that each of the constraints is modelled as a fuzzy set. The fuzzy LP then finds x such that:

$$\mathbf{c}^T\mathbf{x} \gtrsim z$$
$$\mathbf{A}\mathbf{x} \lesssim \mathbf{b} \qquad (9.66)$$
$$\mathbf{x} \geq 0$$

where $\lesssim$ denotes the fuzzified version of $\leq$ and has the linguistic interpretation 'essentially smaller than or equal to'. Similarly, $\gtrsim$ denotes the fuzzified version of $\geq$ and has the linguistic interpretation 'essentially greater than or equal to'. The objective function in (9.65) should be written as a minimizing goal in order to consider z as an upper bound.

Problem (9.66) is fully symmetric with respect to objective function and constraints, and let us make that even more obvious by substituting $\begin{pmatrix} -\mathbf{c} \\ \mathbf{A} \end{pmatrix} = \mathbf{B}$ and $\begin{pmatrix} -z \\ \mathbf{b} \end{pmatrix} = \mathbf{d}$. Then (9.66) becomes:

$$\mathbf{B}\mathbf{x} \lesssim \mathbf{d} \qquad (9.67)$$
$$\mathbf{x} \geq 0$$

Each of the $(m + 1)$ rows of (9.67) shall now be represented by a fuzzy set, the membership functions of which are $\mu_i(x)$. Then the fuzzy set decision of model (9.67) is:

$$\mu_{\tilde{D}}(\mathbf{x}) = \min_i \{\mu_i(\mathbf{x})\} \qquad (9.68)$$

$\mu_i(x)$ can be interpreted as the degree to which x fulfils (satisfies) the fuzzy inequality $B_i x \leq d_i$ (where B_i denotes the ith row of $\mathbf{B}$).

In order to arrive at the crisp optimal solution from a fuzzy set we shall use the maximizing solution of (9.68), which is the solution to the possibly nonlinear programming problem:

$$\max_{x \geq 0} \min_i \{\mu_i(\mathbf{x})\} = \max_{x \geq 0} \mu_{\tilde{D}}(\mathbf{x}) \tag{9.69}$$

Let us now specify the membership functions $\mu_i(x)$. They should be 0 if the constraints (including the objective function) are strongly violated, and 1 if they are very well satisfied (i.e. satisfied in the crisp sense); and $\mu_i(x)$ should increase monotonously from 0 to 1, that is:

$$\mu_i(\mathbf{x}) = \begin{cases} 1 & \forall \quad B_i\mathbf{x} \leq d_i \\ \in [0,1] & \forall \quad d_i < B_i\mathbf{x} \leq d_i + p_i \quad i = 1,\ldots,m+1 \\ 0 & \forall \quad B_i\mathbf{x} > d_i + p_i \end{cases} \tag{9.70}$$

Using the simplest type of membership function shown in Figure 9.6 for both the objective function and the constraints, we shall assume them to be linearly increasing over the tolerance interval p_i:

$$\mu_i(\mathbf{x}) = \begin{cases} 1 & \forall \quad B_i\mathbf{x} \leq d_i \\ 1 - \dfrac{B_i\mathbf{x} - d_i}{p_i} & \forall \quad d_i < B_i\mathbf{x} \leq d_i + p_i \quad i = 1,\ldots,m+1 \\ 0 & \forall \quad B_i\mathbf{x} > d_i + p_i \end{cases} \tag{9.71}$$

The p_i are subjectively chosen constants of admissible violations of the constraints and the objective function.

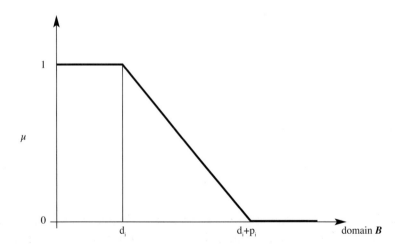

Figure 9.6 *A linear membership function*

Substituting (9.71) into (9.69) yields, after some rearrangements:

$$\max_{x \geq 0} \min_{i} (1 - \frac{B_i x - d_i}{\mu_i}) \qquad (9.72)$$

Introducing one new variable, λ, which corresponds essentially to (9.68), we arrive at:

$$\max \lambda$$

subject to:

$$\lambda p_i + B_i x \leq d_i + p_i \quad i = 1,\ldots,m+1 \qquad (9.73)$$
$$x \geq 0$$

If the optimal solution to (9.73) is the vector (λ, x_0), then x_0 is the maximizing solution (9.69) of model (9.66) assuming membership functions as specified in (9.71).

The elegance of the solution is that this maximizing solution can be found by solving one standard (crisp) LP with only one more variable and one more constraint than model (9.67). This makes this approach computationally very efficient.

We can make (9.72) and (9.73) more general if the membership functions are defined using a variable t_i, $i = 1, \ldots, m+1$, $0 < t_i < p_i$, defined as a measure of the degree of violation of the *i*th constraint. The membership function of the *i*th row is then:

$$\mu_i(x) = 1 - \frac{t_i}{p_i} \qquad (9.74)$$

The crisp equivalent model is then:

$$\max \lambda$$
subject to:
$$\lambda p_i + t_i \leq p_i \quad i = 1,\ldots,m+1 \qquad (9.75)$$
$$B_i x - t_i \leq d_i$$
$$t_i \leq p_i$$
$$x, t \geq 0$$

This model is larger than model (9.73), even though the set of constraints $t_i \leq p_i$ is actually redundant. However, the model form (9.75) has some advantages when performing sensitivity analysis.

The accompanying CD-ROM includes the FuzzyLinpro software. The folder FUZZYLINPRO contains three sub-folders, FuzzyLinpro, FuzzyLinproHelp and Examples. The read.me file in the FUZZYLINPRO sub-folder contains instructions for the installation of the FuzzyLinpro software, and a detailed tutorial for its use is part of the Help menu.

Example 10

A city is deciding on the capacity of its water supply system and choice of sources. Four different sources of water supply (x_1 to x_4) are considered. The objective is to minimize cost, and the constraint is the need to supply all customers (who have a strong seasonally fluctuating demand). That means a certain amount of water has to be delivered from different sources (quantity constraint) and a minimum number of customers per day have to be supplied from different sources (supply constraint). For other reasons, it is required that at least 6 units of water must come from the source x_1. The city management wants to use quantitative analysis and has agreed to the following suggested LP model:

$$\text{Minimize } 41{,}400x_1 + 44{,}300x_2 + 48{,}100x_3 + 49{,}100x_4$$

subject to constraints

$$0.84\,x_1 + 1.44\,x_2 + 2.16\,x_3 + 2.4\,x_4 \geq 170 \tag{9.76}$$
$$16\,x_1 + 16\,x_2 + 16\,x_3 + 16\,x_4 \geq 1{,}300$$
$$x_1 \geq 6$$
$$x_2, x_3, x_4 \geq 0$$

The optimal solution to (9.76) is $x_1 = 6$, $x_2 = 16.29$, $x_3 = 0$, $x_4 = 58.96$. Minimum cost = 3,864,975. Use Linpro to confirm the optimal solution. This part of Example 10 is on the CD-ROM, in the directory LINPRO, subdirectory Examples, Example 8.

When the results are presented to the city management it turns out that they are considered acceptable but that the managers would rather have some flexibility in the constraints. They feel that because demand forecasts are used to formulate the constraints (and because forecasts are not always correct), there is a danger of not being able to meet higher demands by their customers.

The total budget of the city for water supply is 4.2 million, a figure that must not be exceeded. Since the city management feels it should use intervals instead of precise constraints, model (9.67) is selected to model the management's perceptions of the problem satisfactorily. The following parameters are estimated: bounds of the tolerance interval for the objective function d_0 and three constraints d_i, $i = 1,2,3$:

$$d_0 = 3{,}700{,}000 \quad d_1 = 170 \quad d_2 = 1{,}300 \quad d_3 = 6$$

and spreads of tolerance intervals:

$$p_0 = 500{,}000 \quad p_1 = 10 \quad p_2 = 100 \quad p_3 = 6$$

They define the simplest linear type of membership function.

Since the city problem is a minimization problem with 'greater than' constraints, a slight modification of the relationship (9.73), developed for a maximization problem with 'less than' constraints, is required as follows:

max λ

subject to:

$$\lambda p_i - B_i \mathbf{x} \leq p_i - d_i \quad i = 1,\ldots,m+1$$
$$\mathbf{x} \geq 0$$

The new problem definition includes an objective function and four constraints. The first constraint is obtained from the modified formulation by replacing the given values into:

$$\lambda p_O - B_O \mathbf{x} \leq p_O - d_O$$

where B_O are the coefficients of the objective function. The remaining three constraints are simply obtained by replacing given values into:

$$\lambda p_i - B_i \mathbf{x} \leq p_i - d_i$$

So our new problem is now:

Maximize λ

subject to constraints

$$0.083 x_1 + 0.089 x_2 + 0.096 x_3 + 0.098 x_4 - \lambda \geq 6.4$$
$$0.084 x_1 + 0.144 x_2 + 0.216 x_3 + 0.24 x_4 - \lambda \geq 16$$
$$0.16 x_1 + 0.16 x_2 + 0.16 x_3 + 0.16 x_4 - \lambda \geq 12$$
$$0.167 x_1 - \lambda \geq 0$$
$$\lambda, x_1, x_2, x_3, x_4 \geq 0$$

Use FuzzyLinpro to find the optimal solution. This part of Example 10 is on the CD-ROM, directory FUZZYLINPRO, subdirectory Examples, Example 1.

The solution to the city's water supply problem is in Table 9.3. As can be seen from the solution, flexibility has been provided with respect to all constraints at an additional cost just above 2 per cent.

Table 9.3 *Water supply problem solutions*

	Crisp	Fuzzy
Solution		
x_1	6	5.98
x_2	16.29	0
x_3	0	0
x_4	58.96	75.26
Z	3,864,975	3,943,267
Constraints		
1	170	185.66
2	1300	1300
3	6	5.988

The main advantage of the fuzzy formulation over the crisp problem formulation is that the decision-maker is not forced into precision for mathematical reasons. Linear membership functions (Figure 9.6) are obviously only a very rough approximation. Membership functions that monotonically increase or decrease in the interval of $(d_i, d_i + p_i)$ can also be handled quite easily.

9.3 EVOLUTIONARY OPTIMIZATION

Evolutionary algorithms are becoming more prominent in the water management field. Significant advantages of evolutionary algorithms include:

- no need for initial solution;
- easy application to nonlinear problems and to complex systems;
- production of acceptable results over longer time horizons;
- generation of several solutions that are very close to the optimum.

Evolutionary programs are probabilistic optimization algorithms based on similarities with the biological evolutionary process. In this concept, a *population* of individuals, each representing a search point in the space of feasible solutions, is exposed to a collective learning process which proceeds from generation to generation. The population is arbitrarily initialized and subjected to the processes of *selection*, *recombination* and *mutation* such that the new populations created in subsequent generations evolve towards more favourable regions of the search space. This is achieved by the combined use of a *fitness* evaluation of each individual and a selection process which favours individuals with higher fitness values, thus making the entire process resemble the Darwinian rule known as the 'survival of the fittest'.

The terminology, notation and opinions about the importance and the nature of the three underlying processes (selection, recombination and mutation) vary throughout the research community (Goldberg, 1989; Michalewicz, 1999; Pohlheim, 2005). There are three main streams of evolutionary algorithms that have emerged in the last three decades: evolution strategies (ES), algorithms which imitate the principles of natural evolution for parameter optimization problems; evolutionary programming (EP), a technique for searching through a space of small finite state machines; and genetic algorithms (GA), originally proposed to search for the most fit computer program to solve a particular problem. We shall use the term *evolutionary optimization* here for all evolution-based approaches.

Box 9.2 Fish catcher

It was 1964. My family found a spot to rest and enjoy part of the day close to a small creek in the rich grass of the valley. A little later a relative of mine went into the water and slowly positioned himself in the middle of the creek. His hands were long enough to easily reach both banks of the creek, which were covered with lush green grass growing over the edges of the water.

With lightning speed he moved both his hands from the front of his body towards the back, reaching the darker areas under the grass. Each hand surfaced with a fish clutched in it.

We enjoyed the best fish meal I ever had in my life.

(A memory from 1964)

In spite of the lack of strong theoretical background, the evolutionary approach has emerged in the last two decades as a powerful and promising technique that has generated much interest in the scientific and engineering community, mainly as a result of numerous successful applications which far surpassed other search methods in their ability to deliver superior solutions. It is obvious that many different evolution programs can be formulated to solve the same problem. They could differ in the data structure used to represent a single individual, recombination operators used for generating new individuals, the selection process, methods of creating the initial population, methods for handling the constraints of the problem, and search parameters such as population size. Regardless of these differences, they all share the same principle: a population of individuals is subjected to selection and reproduction which is carried out from generation to generation until no further improvement of the fitness function can be achieved.

The accompanying CD-ROM includes the Evolpro software. The folder EVOLPRO contains three sub-folders, Evolpro, EvolproHelp and Examples. The read.me file in the sub-folder Evolpro contains instructions for installation of the Evolpro software, and a detailed tutorial for its use is a part of the Evolpro Help menu.

9.3.1 Introduction

The evolution program is a probabilistic algorithm which maintains a population of individuals, $P(t) = \{x_1^t,...,x_n^t\}$ for iteration t. Each individual represents a potential solution to the problem at hand, and in any evolution program is implemented as a (possibly complex) data structure S. Each solution x_i^t is evaluated to give some measure of its *fitness*. Then a new population (iteration $t + 1$) is formed by selecting the more fit individuals (select step). Some members of the new population undergo transformations (alter step) by means of *genetic operators* to form new solutions. There are low-order transformations m_i (mutation type), which create new individuals by a small change in a single individual $(m_i : S \rightarrow S)$, and higher-order transformations c_j (crossover type), which create new individuals by combining parts from several (two or more) individuals $(c_j : S \times ... \times S \rightarrow S)$. After a number of generations the program converges. It is hoped that the best individual represents a near-optimum (reasonable) solution.

Figure 9.7 is a schematic presentation of the evolutionary optimization process.

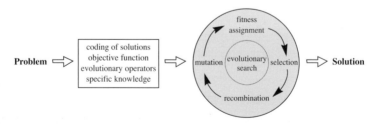

Source: Pohlheim (2005)

Figure 9.7 *Schematic presentation of the evolutionary algorithm*

Let us consider the following example. We search for a network that needs to satisfy some requirements (say, we search for the optimal topology of a water supply network according to criteria such as the cost of pumping and reliability). Each individual in the evolution program represents a potential solution to the problem, i.e. each individual represents a graph. The initial population of graphs $P(0)$ (either generated randomly or created as a result of some heuristic process) is a starting point ($t = 0$) for the evolution program. The evaluation function is usually given – it incorporates the problem requirements. The evaluation function returns the fitness of each network, distinguishing between better and worse individuals. Several mutation operators can be designed which would transform a single network. A few crossover operators can be considered which combine the structure of two (or more) networks into one. Very often such operators incorporate problem-specific knowledge. For example, if the network we search for is connected and acyclic (a tree form), a possible mutation operator would delete an edge from the network and add a new edge to connect two disjoint subnetworks. The other possibility is to design a problem-independent mutation and incorporate this requirement into the evaluation function, penalizing networks that are not trees.

Clearly, many evolution programs can be formulated for a given problem. Such programs may differ in many ways. They can use different data structures for implementing a single individual, genetic operators for transforming individuals, methods for creating an initial population, methods for handling the constraints of the problem, and parameters (population size, probabilities of applying different operators, etc.). However, they all share a common principle: a population of individuals undergoes some transformations, and during the evolution process the individuals do their best to survive.

It should be apparent that evolutionary optimization differs substantially from more traditional optimization methods. The most significant differences are:

- Evolutionary algorithms search a population of points in parallel, not just a single point.
- Evolutionary algorithms do not require derivative information or other auxiliary knowledge. Only the objective function and corresponding fitness levels influence the directions of search.
- Evolutionary algorithms use probabilistic transition rules, not deterministic ones.
- Evolutionary algorithms are generally more straightforward to apply, because no restrictions for the definition of the objective function exist.
- Evolutionary algorithms can provide a number of potential solutions to a given problem. The final choice is left to the user.

The following processes constitute the main part of any evolutionary optimization algorithm.

Selection

Selection determines which individuals are chosen for mating (recombination) and how many offspring each selected individual produces. The first step is fitness assignment by proportional fitness assignment, or rank-based fitness assignment. The actual selection is performed in the next step. Parents are selected according to their fitness by means of one of the following algorithms: roulette-wheel selection, stochastic universal sampling, local selection, truncation selection or tournament selection.

Recombination – crossover

Recombination produces new individuals by combining the information contained in the parents ('parents' are the mating population). Depending on the representation of the variables of the individuals, the following algorithms can be applied: discrete recombination (which is known from the recombination of real-valued variables, and corresponds to uniform crossover of binary-valued variables); intermediate recombination; line recombination; extended line recombination; single-point/ double-point/multi-point crossover; uniform crossover; shuffle crossover; and crossover with reduced surrogate.

Mutation

After recombination every offspring undergoes mutation. Offspring variables are mutated by small perturbations (size of the mutation step), with low probability. The representation of the variables determines the algorithm used. Two operators are of importance: a mutation operator for real-valued variables, and a mutation operator for binary-valued variables.

Reinsertion

After new offspring are produced they must be inserted into the population. This is especially important if fewer offspring are produced than the size of the original population, or not all offspring are to be used at each generation, or more offspring are generated than are needed. A reinsertion scheme determines which individuals should be inserted into the new population and which individuals of the population will be replaced by offspring. The used selection algorithm determines the reinsertion scheme: global reinsertion for an all-population-based selection algorithm (roulette-wheel selection, stochastic universal sampling, and truncation selection) and local reinsertion for local selection.

Now let us look in detail at the processes used in evolutionary optimization.

9.3.2 Selection

In selection the offspring-producing individuals are chosen. The first step is fitness assignment. Each individual in the selection pool receives a reproduction probability depending on its own objective value and the objective value of all other individuals in the selection pool. This fitness is used for the actual selection step which follows. The following definitions are used.

- *Selective pressure* is the probability of the best individual being selected compared to the average probability of selection of all individuals.
- *Bias* is the absolute difference between an individual's normalized fitness and its expected probability of reproduction.
- *Spread* is the range of possible values for the number of offspring of an individual.
- *Loss of diversity* is the proportion of individuals of a population that is not selected during the selection phase.
- *Selection intensity* is the expected average fitness value of the population after applying a selection method to the normalized Gaussian distribution.
- *Selection variance* is the expected variance of the fitness distribution of the population after applying a selection method to the normalized Gaussian distribution.

In rank-based fitness assignment, the population is sorted according to objective values. The fitness assigned to each individual depends only on its ranking and not on an actual objective value. Rank-based fitness assignment overcomes the scaling problems of proportional fitness assignment. The reproductive range is limited, so that no individuals generate an excessive number of offspring. Ranking introduces a uniform scaling across the population, and provides a simple and effective way of controlling selective pressure. Rank-based fitness assignment behaves in a more robust manner than proportional fitness assignment, and as a result it is the method of choice in most evolutionary optimization algorithms (including the one in Evolpro, featured here).

We shall look at both linear and nonlinear ranking methods. Let us use *Nind* for the number of individuals in the population, *Pos* for the position of an individual in this population (the least fit individual has $Pos = 1$, the fittest individual $Pos = Nind$) and *SP* for the selective pressure. The fitness value for an individual is calculated using linear ranking:

$$\text{Fitnes}(Pos) = 2 - SP + 2 \times (SP - 1) \times \frac{(Pos - 1)_i}{Nind - 1} \tag{9.77}$$

Linear ranking allows values of selective pressure in [1.0, 2.0]. The other option is the use of nonlinear ranking:

$$\text{Fitnes}(Pos) = \frac{Nind \times X^{Pos-1}_i}{\sum_{i=1}^{Nind} X^{i-1}} \tag{9.78}$$

where X is computed as the root of the polynomial:

$$0 = (SP - Nind) \times X^{Nind-1} + SP \times X^{Nind-2} + \ldots + SP \times X + SP \tag{9.79}$$

The use of nonlinear ranking permits higher selective pressures than the linear ranking method. Nonlinear ranking allows values of selective pressure in $[1, Nind - 2]$.

The simplest selection scheme is *roulette-wheel selection*, also called stochastic sampling with replacement (Michalewicz, 1999). This is a stochastic algorithm and involves the following technique. The individuals are mapped to adjacent segments of a line, such that each individual's segment is equal in size to its fitness. A random number is generated and the individual whose segment spans the random number is selected. The process is repeated until the desired number of individuals is obtained (which is called the *mating population*). This technique is analogous to a roulette wheel with each slice proportional in size to the fitness (see Figure 9.8).

Example 11

Let us consider the optimization problem with Table 9.4, which shows the election probability for 11 individuals, linear ranking and selective pressure of two together with the fitness value.

Individual 1 is the most fit individual and occupies the largest interval, whereas individual 10 as the second least fit individual has the smallest interval on the line (Figure 9.8). Individual 11, the least fit interval, has a fitness value of 0 and gets no chance for reproduction.

Table 9.4 *Example data*

Number of individuals	1	2	3	4	5	6	7	8	9	10	11
Fitness value	2.0	1.8	1.6	1.4	1.2	1.0	0.8	0.6	0.4	0.2	0
Selection probability	0.18	0.16	0.15	0.13	0.11	0.09	0.07	0.06	0.03	0.02	0

Figure 9.8 *Roulette-wheel selection*

To select the mating population, an appropriate number of uniformly distributed random numbers (between 0.0 and 1.0) is independently generated. A sample of six random numbers is: 0.81, 0.32, 0.96, 0.01, 0.65, 0.42. Figure 9.8 shows the selection process of the individuals for the example in Table 9.4 with these sample random numbers. After selection, the mating population consists of the individuals 1, 2, 3, 5, 6, 9. The roulette-wheel selection algorithm provides a zero bias but does not guarantee minimum spread.

Stochastic universal selection provides zero bias and minimum spread. The individuals are mapped to adjacent segments of a line, such that each individual's segment is equal in size to its fitness, exactly as in roulette-wheel selection. Here equally spaced pointers are placed over the line, as many as there are individuals to be selected. Let us consider *NPointer* to be the number of individuals to be selected, then the distance between the pointers is 1/*NPointer* and the position of the first pointer is given by a randomly generated number in the range [0, 1/*NPointer*]. For six individuals to be selected, the distance between the pointers is 1/6 = 0.167. Figure 9.9 shows the selection for the above example.

Let us select one random number in the range [0, 0.167] to be 0.1. After selection the mating population consists of the individuals 1, 2, 3, 4, 6, 8, which differs from the roulette-wheel selection above. Stochastic universal sampling ensures a selection of offspring which is closer to what is deserved than roulette-wheel selection.

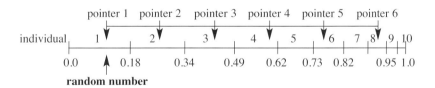

Figure 9.9 *Stochastic universal sampling*

In *local selection*, every individual resides inside a constrained environment called the local neighbourhood. In the other selection methods the whole population is the selection pool or neighbourhood. Individuals interact only with individuals inside this region. The neighbourhood is defined by the structure in which the population is distributed. It can be seen as the group of potential mating partners. The first step is the selection of the first half of the mating population at random (or using one of the other selection algorithms mentioned, for example, stochastic universal sampling or truncation selection). Now a local neighbourhood is defined for every selected individual. Inside this neighbourhood the mating partner is selected (using best, fitness proportional or random methods).

Compared with the previous selection methods modelling natural selection, *truncation selection* is an artificial selection method. It is used for large populations or mass selection. In truncation selection individuals are sorted according to their fitness. Only the best individuals are selected to be parents. These selected parents produce uniform at random offspring. The parameter for truncation selection is the truncation threshold *Trunc*. *Trunc* indicates the proportion of the population to be selected as parents, and takes values ranging from 50 per cent to 10 per cent. Individuals below the truncation threshold do not produce offspring.

In *tournament selection* (Michalewicz, 1999), a number *Tour* of individuals is chosen randomly from the population and the best individual from this group is selected as a parent. This process is repeated as often as individuals must be chosen. These selected parents produce uniform at random offspring. The parameter for tournament selection is the tournament size *Tour*. *Tour* takes values ranging from 2 to *Nind* (number of individuals in population).

9.3.3 Recombination – crossover

Recombination produces new individuals by combining the information contained in two or more parents. This is done by combining the variable values of the parents. Depending on the representation of the variables, different methods can be used. Here we shall look at the discrete recombination method, which can be applied to all variable representations, and intermediate, line and extended line recombination methods for real-valued variables. Methods for binary-valued variables are not presented here, but can be found in the literature (Goldberg, 1989; Michalewicz, 1999; Pohlheim, 2005).

Discrete recombination performs an exchange of variable values between the individuals. For each position the parent that contributes its variable to the offspring is chosen randomly with equal probability according to the rule:

$$Var_i^O = Var_i^{P_1} \times a_i + Var_i^{P_2} \times (1 - a_i) \quad i \in (1,2,...,Nvar) \tag{9.80}$$
$$a_i \in \{0,1\} \quad \text{uniform at random, } a_i \text{ for each } i \text{ new defined}$$

Discrete recombination generates corners of the hypercube defined by the parents. Figure 9.10 shows the geometric effect of discrete recombination.

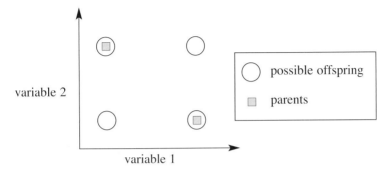

Figure 9.10 *Possible positions of the offspring after discrete recombination*

Example 12

Let us consider the following two individuals with three variables each (that is, three dimensions). These will also be used to illustrate the other types of recombination for real-valued variables:

individual 1	12	25	5
individual 2	123	4	34

For each variable the parent who contributes its variable to the offspring is chosen randomly with equal probability:

sample 1	2	2	1
sample 2	1	2	1

After recombination according to (9.72) the new individuals are created:

offspring 1	123	4	5
offspring 2	12	4	5

Intermediate recombination is a method only applicable to real variables (and not binary variables). Here the variable values of the offspring are chosen somewhere around and between the variable values of the parents. Offspring are produced according to the rule:

$$Var_i^o = Var_i^{P_1} \times a_i + Var_i^{P_2} \times (1 - a_i) \quad i \in (1,2,...,Nvar) \quad (9.81)$$
$a_i \in [-d, 1+d]$ uniform at random, $d = 0.25$, a_i for each i new defined

where a is a scaling factor chosen uniformly at random over an interval $[-d, 1+d]$ for each new variable.

The value of the parameter d defines the size of the area for possible offspring. A value of $d = 0$ defines the area for offspring as the same size as the area spanned by the parents. This method is called (standard) intermediate recombination. Because most variables of the offspring are not generated on the border of the possible area, the area for the variables shrinks over the generations. This shrinkage occurs just by using (standard) intermediate recombination. This effect can be prevented by using a larger value for d. A value of $d = 0.25$ ensures (statistically) that the variable area of the offspring is the same as the variable area spanned by the variables of the parents.

Example 13

Consider the following two individuals with three variables each:

individual 1	12	25	5
individual 2	123	4	34

The chosen a for this example are:

sample 1	0.5	1.1	−0.1
sample 2	0.1	0.8	0.5

The new individuals are calculated according to (9.73) as:

offspring 1	67.5	1.9	2.1
offspring 2	23.1	8.2	19.5

Intermediate recombination is capable of producing any point within a hypercube slightly larger than that defined by the parents.

Line recombination is similar to intermediate recombination, except that only one value of a is used for all variables:

$$Var_i^o = Var_i^{P_1} \times a_i + Var_i^{P_2} \times (1 - a_i) \quad i \in (1,2,...,Nvar), \quad (9.82)$$
$a_i \in [-d, 1+d]$ uniform at random, $d = 0.25$, a_i for all i identical

For the value of d the statements given for intermediate recombination are applicable.

Example 14

Let us consider the following two individuals with three variables each:

individual 1	12	25	5
individual 2	123	4	34

The chosen a for this example are:

sample 1	0.5
sample 2	0.1

The new individuals are calculated according to (9.81) as:

offspring 1	67.5	14.5	19.5
offspring 2	23.1	22.9	7.9

Line recombination can generate any point on the line defined by the parents.

Extended line recombination generates offspring on a line defined by the variable values of the parents. However, it is not restricted to the line between the parents and a small area outside. The parents just define the line where possible offspring may be created. The size of the area for possible offspring is defined by the domain of the variables. Inside this possible area the offspring are not uniformly distributed at random. The probability of creating offspring near the parents is high. There is a low probability of offspring being created far away from the parents. If the fitness of the parents is available, then offspring are more often created in the direction from the worse to the better parent. Offspring are produced according to the following rule:

$$Var_i^O = Var_i^{P_1} + s_i \times r_i \times a_i \times \left\| \frac{Var_i^{P_2} - Var_i^{P_1}}{Var^{P_1} - Var^{P_2}} \right\| \quad i \in (1,2,....Nvar) \tag{9.83}$$

where:
$a_i = 2^{-k \times u}$, k – mutation precision, $u \in [0,1]$ uniform at random
a_i is identical for all i
$r_i = r \times domainr$ – range of recombination steps
$s_i \in \{-1, +1\}$ uniform at random for undirected recombination
$+1$ with probability >0.5 for directed recombination.

The creation of offspring uses features similar to the mutation operator for real-valued variables (see Section 9.3.4). The parameter a defines the relative step size, the parameter r the maximum step size, and the parameter s the direction of the recombination step. Typical values for the precision parameter k are in the area from 4 to 20. A robust value for the parameter r (range of recombination step) is 10 per cent of the domain of the variable. If the parameter s (search direction) is set to -1 or $+1$ with equal probability, an undirected recombination takes place. If the probability of $s = +1$ is higher than 0.5, a directed recombination takes place (offspring are created in the direction from the worse to the better parent – the first parent must be the better parent).

9.3.4 Mutation

Individuals are randomly altered by mutation. The variations (mutation steps) are mostly small. They will be applied to the variables of the individuals with a low probability (mutation probability or mutation rate). Normally, offspring are mutated after being created by recombination. Two approaches exist for the definition of the mutation steps and the mutation rate: either both parameters are constant during a whole evolutionary run, or one or both parameters are adapted according to previous mutations. We focus here only on real-valued mutation. (See the literature for binary mutation.)

Mutation of real variables means that randomly created values are added to the variables with a low probability. Thus, the probability of mutating a variable (*mutation rate*) and the size of the change for each mutated variable (*mutation step*) must be defined. The probability of mutating a variable is inversely proportional to the number of variables (dimensions). The more dimensions one individual has, the smaller is the mutation probability. As long as nothing else is known, a mutation rate of $1/n$ is suggested.

The size of the mutation step is usually difficult to choose. The optimal step size depends on the problem considered, and may even vary during the optimization process. It is known that small steps are often successful, especially when the individual is already well adapted. However, large mutation steps can, when successful, produce good results much more quickly. Thus, a good mutation operator should often produce small step sizes with a high probability and large step sizes with a low probability.

This operator is recommended:

$$Var_i^{Mut} = Var_i + s_i \times r_i \times a_i \quad i \in (1,2,\ldots,n) \text{ uniform at random} \tag{9.84}$$

where:

$a_i = 2^{-k \times u}$, k – mutation precision, uniform at random
$r_i = r \times domainr$ – mutation range (standard 10 per cent)
$s_i \in \{-1, +1\}$ – uniform at random.

This mutation algorithm is able to generate most points in the hypercube defined by the variables of the individual and range of the mutation. (The range of mutation is given by the value of the parameter r and the domain of the variables.) Most mutated individuals will be generated near the individual before mutation. Only a few individuals will mutate farther away. That means the probability of small step sizes is greater than that of bigger steps.

Typical values for the parameters of the mutation operator from equation (9.84) are:

mutation precision k: $k \in \{4, 5, ..., 20\}$
mutation range r: $r \in [0.1, 10^{-6}]$.

9.3.5 Reinsertion

Once the offspring have been produced by selection, recombination and mutation of individuals from the old population, the fitness of the offspring may be determined. If fewer offspring are produced than the size of the original population, then to maintain the size of the population, the offspring have to be reinserted into the old population. Similarly, if not all offspring are to be used at each generation or if more offspring are generated than the size of the old population, a reinsertion scheme must be used to determine which individuals are to exist in the new population. The used selection method (Section 9.3.2) determines the reinsertion scheme: local reinsertion for local selection and global reinsertion for all other selection methods.

Global reinsertion. Different schemes of global reinsertion exist:

- Produce as many offspring as parents and replace all parents by the offspring (pure reinsertion).
- Produce fewer offspring than parents and replace parents uniformly at random (uniform reinsertion).
- Produce fewer offspring than parents and replace the worst parents (elitist reinsertion).
- Produce more offspring than needed for reinsertion and reinsert only the best offspring (fitness-based reinsertion).

Pure reinsertion is the simplest reinsertion scheme. Every individual lives one generation only. This scheme is used in simple evolution algorithms. However, it is likely that very good individuals will be replaced without producing better offspring, and thus good information will be lost.

A combination of elitist and fitness-based reinsertion prevents the loss of information and is the recommended method. At each generation, a given number of the least fit parents is replaced by the same number of the most fit offspring. The fitness-based reinsertion scheme implements a truncation selection between offspring before inserting them into the population. The best individuals can live for many generations. However, with every generation some new individuals are inserted. It is not checked whether the parents are replaced by better or worse offspring. Because parents may be replaced by offspring with a lower fitness, the average fitness of the population can decrease. However, if the inserted offspring are extremely bad, they will be replaced with new offspring in the next generation.

Local selection. In local selection, individuals are selected in a bounded neighbourhood. The reinsertion of offspring takes place in exactly the same neighbourhood. Thus, the locality of the information is preserved. The neighbourhood structures used are the same as in local selection (Section 9.3.2). The parent of an individual is the first selected parent in this neighbourhood. The following schemes are possible for the selection of parents to be replaced and of offspring to reinsert:

- Insert every offspring and replace individuals in the neighbourhood uniformly at random.
- Insert every offspring and replace the weakest individuals in the neighbourhood.
- Insert offspring fitter than the weakest individuals in the neighbourhood and replace the weakest individuals in the neighbourhood.
- Insert offspring fitter than the weakest individuals in the neighbourhood and replace their parents.
- Insert offspring fitter than the weakest individuals in the neighbourhood and replace individuals in the neighbourhood uniformly at random.
- Insert offspring fitter than their parents and replace the parents.

9.3.6 An example of evolutionary optimization

In this example a variant of the evolutionary algorithm is used with the following properties:

- floating-point domain representation, which means that chromosomes are represented with decimal numbers;

- a massive initialization procedure which uses a Monte Carlo random search to find a small initial parent population of high quality;
- multi-parent crossover;
- properties of feasible flows in networks are included in the algorithm such that the search is always restricted to the feasible region by obeying the capacity and flow continuity constraints.

Example 15

Consider the problem of finding the best fit analytical equation of the outflow vs. elevation curve given with ten pairs of (x,y) points in Table 9.5. A typical empirical equation for this curve is:

$$Q = AH^b \qquad (9.85)$$

where:
Q = flow (m³/s)
H = net head (m) above the invert of the outlet
A, b = parameters.

Hence in the case of the curve given in Table 9.5 the net head is reservoir elevation minus 1660 m. Parameters A and b should be determined in such a way that the difference of the sum of squares between the analytic and tabulated values of flow for all ten points is minimized. This problem can be formulated as: find the values of parameters A and b such that the value of the following objective function is minimized:

$$\min \sum_{i=1}^{10} (Q_i - A(H_i - 1660)^b)^2 \qquad (9.86)$$

Flow–elevation values are provided in Table 9.5 for each of the ten points. In addition, from other empirical studies related to similar curve fits, it can be assumed that the most likely range for the values of parameter b is (0,1) and for parameter A is (0,10). To be on the safe side in this example the values of parameter A are inspected in the range of (0,20). The value of parameter b must be less than 1 since it is never a straight line, and it must be greater than 0 since values below 0 would not result in an increasing function, while it is known that the outflow does increase with the increase in net head. Taking into account this simple knowledge about the problem reduces the search space to a value for parameter A in the interval (0,20) and the value of b in the interval (0,1), which has a significant impact on the solution efficiency.

Table 9.5 *Flow elevation data*

Outflow (m³/s)	Elevation (m)
0.000	1660.000
2.350	1661.225
3.678	1662.450
4.954	1663.675
6.029	1664.900
6.977	1666.125
7.834	1667.350
8.622	1668.575
9.355	1669.800
10.044	1671.025

We shall use the Evolpro computer program provided on the CD-ROM to solve the problem. Input data for this problem is in the Examples sub-directory under the name Example1.evolpro.

The algorithm first goes through a process of initialization, where 100 solutions are generated in a pure random manner. Under the Computation menu, the Population size option defines the size of population. After each new solution is created, its objective function is evaluated and compared with the worst objective function of the initial five solutions. We can review the progress of optimization by invoking the Computation>View Iterations Summary menu sequence. The optimal value of $A = 2.10998$ and $b = 0.6537$ with the objective function value $OF = 0.039$. Observe that the problem has been solved with maximum number of iterations set to 1000 and tolerance level set to 0.00001.

9.3.7 The Evolpro computer program

Evolpro facilitates solving an optimization problem using an evolutionary algorithm. The software is capable of handling nonlinear objective functions and constraints with multiple decision variables. The Evolpro algorithm includes the following:

1. Initialization – assignment of a set of random values (genes) between the lower and upper bound for each decision variable (chromosome). The size of population is an input variable (greater than 20).
2. Identification of feasible search region – each chromosome (one set of values for

all decision variables) is checked against the set of constraints. If one of the constraints is not satisfied, the chromosome is discarded. This process is repeated until the required number of chromosomes for the population is obtained.

3   Evaluation of objective function – for each chromosome the value of the objective function is calculated and ranked.
4   Selection – the best-fitted 30 per cent of the population is taken to select parents. Parents are picked randomly.
5   Recombination – using relationship (9.79) new offspring genes are generated.
6   Mutation – new genes are disturbed using a factor $1 + 0.005 \times (0.5 - Rnd())$.
7   Feasibility check – the new chromosome produced from offspring genes is verified against the constraints and bounds. If the constraints are satisfied, the new offspring is ready for migration.
8   Reinsertion – offspring totalling 30 per cent of the population size are inserted to replace the least-fitted 30 per cent of the previous population.
9   Step 3 is repeated with the new population.

Steps 4 to 9 are repeated until the desired accuracy is obtained or the maximum number of iterations is reached. Both the desired accuracy (tolerance level) and maximum number of iterations are program input variables. The Help menus on the CD-ROM give guidance on Evolpro installation and use.

9.4 EXAMPLES OF WATER RESOURCES OPTIMIZATIONS

Optimization models play an important role in water resources systems management. Many applications of linear, nonlinear and dynamic programming are available in the literature. However, the complexity of real water resources management problems today exceeds the capacity of traditional optimization algorithms (Simonovic, 2000a, 2000b; Bhattacharjya and Datta, 2005; Li et al, 2006; Luo et al, 2007; Kumar and Reddy, 2007).

We shall now look at three water resources management optimization models, based on my personal experience. They illustrate the characteristics of current water resources systems management problems and the potential for their solution with the optimization tools presented in this book. The first one (Reznicek and Simonovic, 1990) involves an interconnected hydropower utility. A new algorithm was developed to solve the optimization problem. It was tested using data from the Manitoba Hydro system (Manitoba, Canada) applied to a single-reservoir system. The second example (Teegavarapu and Simonovic, 1999) addresses the imprecision involved in the definition of reservoir loss functions using fuzzy set theory. The model is applied to the

Green Reservoir, Kentucky, United States. The third (Ilich and Simonovic, 1998) deals with minimizing the total cost of pumping in a liquid pipeline. The problem is solved using an evolutionary algorithm with two distinct features: the search is restricted to the feasible region only and it utilizes a floating-point decision variable.

9.4.1 An improved linear programming algorithm for hydropower optimization

A new algorithm, named energy management by successive linear programming (EMSLP), was developed to solve the problem of optimizing hydropower system operation. The EMSLP algorithm has two iteration levels: at the first level a stable solution is sought, and at the second the interior of the feasible region is searched to improve the objective function whenever its value decreases. The EMSLP algorithm was tested using data from the Manitoba Hydro system (Manitoba, Canada) applied to a single-reservoir system.

Computers have made it possible to use optimization techniques for reservoir operation and the replacement of heuristic release rules. Many operations research techniques have been applied: LP, DP, network algorithms, queuing theory, stochastic DP, DP with successive approximations, successive linear programming (SLP), optimal control, and combined linear and dynamic programming algorithms. However, the complexity of the problem requires major simplifying assumptions to be used with most of the methods developed.

Hydropower optimization problem formulation

The task is to optimize the operation of the interconnected hydro utility, with the objective being to maximize system revenue and minimize the costs of satisfying the energy demand described by the given load duration curve, for a given inflow scheme, over the specified planning time period. The solution of the problem for every time step contains values for reservoir storage (ST_t), turbine (R_t), and spilled (S_t) releases and, within the time step for every load duration curve strip, the value of the produced ($HE_{s,t}$), imported ($IE_{s,t}$) and exported ($EE_{s,t}$) energy. These are the decision variables of the problem.

The objective is to maximize the interruptible energy export and the final storage volume while minimizing the production costs of satisfying the system demand (hydro energy production, import, spill costs). The benefit from domestic energy consumption is not included in the objective, since it is constant and defined by the system demand. The mathematical form of the objective function is:

$$\text{Maximize} \left\{ \sum_t \left[\sum_s (-HC_{s,t} \times HE_{s,t} + EB_{s,t} \times EE_{s,t} - IC_{s,t} \times IE_{s,t} - SC_t \times S_t] \right\} + B_T \times ST_T \right. \quad (9.87)$$

where:
$HC_{s,t}$ = hydro energy production cost
$EB_{s,t}$ = export energy benefit
$IC_{s,t}$ = import energy cost
SC_t = cost of spilling water
B_T = benefit from saving the water for future production.

Linear programming (LP) is often used for hydropower optimization. The technique is easy to implement, and the solution can be obtained after a few iterations. A disadvantage of using LP in this context is the difficulty in handling complex objective functions. LP works best when benefits are 'almost' linearly related to releases. The major obstacle to the use of LP is the nonseparable hydro production function:

$$E = \gamma \times Q \times H \times t \times e(Q, H) \quad (9.88)$$

in which the produced energy (E) is the function of discharge (Q), head (H), and efficiency (e) multiplied by the specific weight of water (γ) and time period (t). The multiplication of the two decision variables (Q and H) existing in (9.88) has to be removed in order to use an LP formulation. The simplest way to linearize (9.88) is to assume a constant value for the head (H) and efficiency (e). With this approximation an LP problem can be formulated and solved. The solution is used to update the assumed values. This iterative procedure is repeated until the difference between the assumption and the LP solution is less than the required accuracy.

Some algorithms available in the literature use a more sophisticated linearization procedure. The energy equation (9.88) can be written in the following form:

$$E = ERF \times R \quad (9.89)$$

where ERF stands for energy rate function and is expressed as:

$$ERF = \gamma \times H \times e \quad (9.90)$$

and R designates the release:

$$R = Q \times t \quad (9.91)$$

The assumption is that *ERF* is only a function of the head, that is, of the storage, and is not dependent on the release:

$$ERF = ERF(ST) \tag{9.92}$$

Furthermore, it is assumed that the value of the *ERF* during the time step can be approximated by taking the average of the function values for the initial and final storage:

$$ERF_t = 0.5 \times [ERF(ST_{t-1}) + ERF(ST_t)] \tag{9.93}$$

The energy equation for the *t*th time period has the form:

$$E_t = ERF_t \times R_t \tag{9.94}$$

or

$$E_t = 0.5 \times [ERF(ST_{t-1}) + ERF(ST_t)] \times R_t \tag{9.95}$$

The first-order Taylor expansion of (9.95) around the estimated storage values at the beginning and end of the *t*th time step has the form:

$$\begin{aligned} E_t = 0.5 \times [&ERF(\hat{ST}_{t-1}) + ERF(\hat{ST}_t) \\ &+ DERF(\hat{ST}_{t-1}) \times (ST_{t-1} - \hat{ST}_{t-1}) \\ &+ DERF(\hat{ST}_t) \times (ST_t - \hat{ST}_t)] \times R_t \end{aligned} \tag{9.96}$$

where *DERF* is the first derivative of *ERF* with respect to *ST*.

The linearization of (9.96) is achieved by applying the approximation:

$$ST_t \times R_t = \hat{ST}_t \times \hat{R}_t + (ST_t - \hat{ST}_t) \times \hat{R}_t + \hat{ST}_t \times (R_t - \hat{R}_t) \tag{9.97}$$

Finally, combining (9.96) and (9.97),

$$\begin{aligned} E_t = 0.5 \times \{&[ERF(\hat{ST}_{t-1}) + ERF(\hat{ST}_t)] \times R_t \\ &+ DERF(\hat{ST}_{t-1}) \times (ST_{t-1} - \hat{ST}_{t-1}) \times \hat{R}_t \\ &+ DERF(\hat{ST}_t) \times (ST_t - \hat{ST}_t) \times R_t\} \end{aligned} \tag{9.98}$$

The algorithm using this linearization is iterative and needs assumptions for releases and storages in the solution process.

The energy management by successive linear programming (EMSLP) algorithm uses this linearization procedure of the hydro production equation derived by Taylor expansion. Rewriting (9.98) in terms of the decision variables and sorting the unknown to the left and the constants to the right of the equality sign, the hydro production constraint in the tth time step has the form:

$$-2 \times \sum(HE_{s,t} + [ERF(S\hat{T}_{t-1}) + ERF(S\hat{T}_t) \times R_t \\ + DERF(S\hat{T}_{t-1}) \times \hat{R}_t \times ST_{t-1} + DERF(S\hat{T}_t) \times \hat{R}_t \times ST_t \\ = [DERF(S\hat{T}_{t-1}) \times S\hat{T}_{t-1} + DERF(S\hat{T}_t) \times S\hat{T}_t] \times \hat{R}_t \quad (9.99)$$

In addition to the hydro production constraint, the optimization of the objective function is constrained in each time step by (i) flow continuity for the tth period:

$$ST_t - ST_{t-1} + R_t + S_t = I_t \quad (9.100)$$

where I_t is the reservoir inflow; (ii) the tie line load for every load duration curve strip s and time step t:

$$\frac{IE_{s,t} \times RATIO}{(IEF \times EEF) + EE_{s,t}} \leq \frac{EML_{s,t}}{EEF \times DPS_t} \quad (9.101)$$

where:
RATIO = export and import tie line capacity ratio
IEF = import efficiency
EEF = export efficiency
$EML_{s,t}$ = maximum export load
DPS_t = number of days;

(iii) supply and demand for every load duration curve strip s and time step t:

$$HE_{s,t} - IE_{s,t} + EE_{s,t} = L_{s,t} \times DPS_t \times W_{s,t} \quad (9.102)$$

where $L_{s,t}$ is the system demand and $W_{s,t}$ the load duration curve strip width; (iv) minimum storage in the tth time step:

$$ST_t \geq MAX(STMIN_t, S\hat{T}_t - VARYMX) \quad (9.103)$$

where $STMIN_t$ is the minimum storage and $VARYMX$ the maximum allowed storage variation; (v) maximum storage in the tth time step:

$$ST_t \leq MIN(STMAX_t, S\hat{T}_t - VARYMX) \quad (9.104)$$

where $STMAX_t$ is the maximum storage; and (vi) hydro energy relation to release in the tth time step:

$$\sum_s (HE_{s,t}) - ERF(STMAX_t) \times R_t \leq 0 \qquad (9.105)$$

Constraint (9.99) reflects the SLP approach to the modelling of hydro production. The flow continuity constraint (9.100) ensures the mass conservation, and relates the decision variables of two adjacent time steps to each other. The power demand of the system is specified in every time step by a load duration curve. The load duration curve is discretized into strips (fractions of the time step) where the load is assumed constant. The tie line load constraint (9.101) limits the amount of export and import energy for every strip of the load duration curve. The supply and demand constraint (9.102) ensures that the produced energy, the import minus the export energy, meets the system demand in every strip. Inequalities (9.103) and (9.104) bound the storage and may change from one iteration to the other. The role of (9.105) is to relate the produced hydro energy to the released water volume, since this relation is not explicitly defined in the hydro production constraint (9.99).

Solution algorithm

The EMSLP algorithm developed to solve the above problem has the following steps:

1. Set the LP problem according to the input data, and set the initial storage variability VARYMX. Calculate and accept the initial solution based on the estimated releases (the estimated storages are calculated using the flow continuity equation). Calculate the hydro production constraint coefficients from the solution.
2. Solve the LP problem.
3. Compare the calculated storages with the accepted ones, and if the difference is smaller than the tolerance, stop. The solution is obtained.
4. If the calculated objective function value is better than the accepted one, accept the calculated solution but limit the change in the release policy to 30 per cent of the previous accepted solution. With this release policy used as the estimate, recalculate the coefficients in the hydro production constraint. Reset VARYMX to its initial input value and go to step 2.
5. Otherwise decrease the value of VARYMX, and if it is still greater than the set minimum (VARMIN), go to step 2.
6. If the value of VARYMX is less than the set minimum, then use the first worse objective after the last improvement and the appropriate solution, as if it is better than the accepted one. Go to step 4.

Note that whenever the value of *VARYMX* is changed, actually the bounds on the storage variables are changed. The algorithm has two iteration levels. At the first level a search for a stable solution is performed. At the second level the improvement of the objective function value is sought whenever the objective function value drops between the two iterations. The search is performed by exploring the interior of the feasible region using the decreased storage variability *VARYMX* in the solution procedure. If the search for the better solution at the second level terminates unsuccessfully, the algorithm returns to the first level and accepts the initially identified worse solution. The search terminates on the first iteration level when a stable solution is identified.

The coefficients in the hydro production constraints are recalculated only on the first iteration level. On the second level the lower and upper bounds on the storage volume are changed. The initial wide range is decreased with every iteration at the second level, approaching the accepted storage level.

EMSLP guides the iterative procedure in the following manner. The change in the release policy from one iteration to the other is limited to not more than a fraction (specifically 30 per cent) of the accepted policy. Because of the application of the limited change, the convergence and stability of the iterative process is substantially improved. The storage variability in EMSLP has a somewhat decreased role. It is used only in the search for a better optimum at the second iteration level. The value of the variable does not necessarily decrease during the program execution. It is reinstalled to the original one at the end of the search on the second level. Therefore the storage variability cannot be used as a convergence criteria. Instead, EMSLP checks whether the identified storage trajectory is close enough to the estimated input solution. The search terminates only if this condition is satisfied.

Hydroelectric system description

Manitoba Hydro (Manitoba, Canada) is a power utility responsible for planning the operations of a system of 13 hydro and 3 thermal plants (4250 MW) and associated reservoirs. To evaluate the newly formulated EMSLP algorithm, only a portion of this system was modelled. The system consisted of a single reservoir, power plant, and a tie line which enabled energy import and/or export to satisfy the load. The optimization time horizon consisted of five monthly time steps. The load duration curve was discretized to two segments (on- and off-peak demand). Constant efficiency and constant tailwater level were assumed.

The hydro energy production cost $HC_{s,t}$, energy benefit $EB_{s,t}$ and import cost $IC_{s,t}$ were adopted from Manitoba Hydro practice, where the values are determined on the basis of a separate economic and energy market analysis. The spilling of water was not penalized ($SC_t = 0$). The benefit from the water saved at the end of the planning period, B_t, was one of the parameters in the model, therefore its value was varied in a number of computer experiments. In practice, the benefit B_t is estimated from the value of future energy production. The estimate of the future benefit is discounted by the governing interest rate to obtain the present value.

To account for the fact that reservoirs in Manitoba have a very small operation range, the problem has been adapted to a low head variability case. The algorithm was evaluated by performing a number of computer experiments. Every experiment had a different input data set. The release policy, number of iterations and objective function value were the major output variables. In the input set the ending storage value, generation release limit, and the system load were varied.

Results

Some of the results are presented in Tables 9.6 to 9.8. In the tables each computer experiment is denoted as a 'case' and described in two rows. The first row contains the reservoir levels at the end of each time step, the objective function value, the number of iterations at the first level, and the total number of iterations. The second row contains the releases during the time steps. Table 9.6 illustrates the effect of changing the value of stored water at the end of the planning time period. Reservoir levels and releases m^3/s/day (cubic metres per second-day) or 0.035 kilocubic feet per second (KCFS)-day) are presented for five time steps and eight alternative ending storage values (from \$3,600 to \$4,800 per 2.45 million m^3 or KCFS-day).

Table 9.7 summarizes the results obtained by changing the system load by multiplying the original load using multiples from 0.3 to 2.

Table 9.8 presents the effects of different release limits on the energy production obtained by EMSLP. The effects were examined for releases between 5.67 and 19.86 m^3/s (0.2 to 0.7 KCFS).

The EMSLP results were compared with the results obtained by the optimization program EMMA (energy management and maintenance analysis), which was being used by the utility at that time (Reznicek and Simonovic, 1990). Variation of the input data indicated that the two algorithms identify similar solutions only when they are constrained to do so (e.g. very low release limit), or when the optimization problem is very straightforward (e.g. very low system demand). When the requirement for a trade-off between production, export, import and storage use was noticeable, EMSLP presented better results than EMMA. EMMA was not able to adjust the release policy to the existing price structure as successfully as the EMSLP algorithm. The objective function value was the same in the simple case, but a difference of up to 5 per cent was obtained for the more complex, realistic situations, always in favour of EMSLP. When the storage value or the load had extremely high values, the differences between the objective function values obtained by the two

Table 9.6 *EMSLP results for different ending storage values*

Case		Planning time period (months)					Objective function $1,000	Iterations	
		1	2	3	4	5		First level	Total
Case $3600	Level (m)	91.3	91.5	90.8	88.1	85.4	−126	2	3
	Release (m³/s/day)	218	288	283	422	384			
Case $3700	Level (m)	91.3	91.5	91.5	88.9	86.3	−126	3	4
	Release (m³/s/day)	218	288	214	414	377			
Case $4000	Level (m)	91.3	91.5	91.5	88.9	86.3	−125	6	9
	Release (m³/s/day)	218	288	213	414	377			
Case $4200	Level (m)	91.3	91.5	91.5	91.1	88.7	−124	5	8
	Release (m³/s/day)	218	288	213	208	358			
Case $4500	Level (m)	91.3	91.5	91.5	91.1	88.7	−121	2	3
	Release (m³/s/day)	218	288	229	195	358			
Case $4600	Level (m)	91.3	91.5	91.5	91.5	91.0	−119	6	10
	Release (m³/s/day)	218	288	213	176	176			
Case $4700	Level (m)	91.3	91.5	91.5	91.5	91.5	−117	3	5
	Release (m³/s/day)	218	288	214	178	129			
Case $4800	Level (m)	91.3	91.5	91.5	91.5	91.5	−115	3	4
	Release (m³/s/day)	218	288	214	178	129			

programs decreased. The number of iterations was very similar for both programs. The releases were substantially different. Based on this work, the Manitoba Hydro utility incorporated the EMSLP algorithm into its EMMA program.

9.4.2 Fuzzy optimization of a multi-purpose reservoir

In this example a reservoir operation problem was solved using the concepts of fuzzy mathematical programming. Membership functions from fuzzy set theory were used to represent the decision-maker's preferences in the definition of the shape of loss curves. These functions were assumed to be known, and were used to model the uncertainties. A linear optimization model was developed under a fuzzy environment. This is compared with a nonlinear formulation in Teegavarapu and Simonovic (1999).

Table 9.7 *EMSLP results for different system loads*

Case		Planning time period (months)					Objective function $1,000	Iterations	
		1	2	3	4	5		First level	Total
Case 0.3	Level (m)	91.5	90.3	91.5	91.5	91.5	101	2	2
	Release (m³/s/day)	199	419	101	180	128			
Case 0.5	Level (m)	91.5	91.0	91.5	91.2	90.8	42	1	1
	Release (m³/s/day)	199	353	168	202	174			
Case 0.8	Level (m)	90.7	91.5	90.9	89.2	87.4	−54	3	5
	Release (m³/s/day)	274	232	269	332	297			
Case 1.0	Level (m)	91.3	91.5	91.5	88.9	86.3	−126	5	9
	Release (m³/s/day)	218	288	214	414	377			
Case 1.5	Level (m)	91.5	90.1	91.5	90.1	85.4	−314	6	11
	Release (m³/s/day)	199	438	83	311	565			
Case 2.0	Level (m)	91.5	89.2	91.5	91.5	85.4	−504	6	10
	Release (m³/s/day)	199	518	2	176	699			

Table 9.8 *EMSLP results for varying release limits*

Case		Planning time period (months)					Objective function $1,000	Iterations	
		1	2	3	4	5		First level	Total
Case 5.67	Level (m)	91.5	91.5	91.5	91.5	91.0	−170	1	1
	Release R (m³/s/day)	170	176	170	176	176			
	Release S (m³/s/day)	28	132	43	0	0			
Case 8.57	Level (m)	91.0	91.5	91.0	90.1	88.7	−131	3	5
	Release (m³/s/day)	243	264	255	264	264			
Case 11.35	Level (m)	91.3	91.5	91.5	89.6	87.2	−127	4	7
	Release (m³/s/day)	218	288	217	352	352			
Case 14.19	Level (m)	91.3	91.5	91.5	88.9	86.3	−126	5	9
	Release (m³/s/day)	218	288	214	414	377			
Case 19.86	Level (m)	91.3	91.5	91.5	88.9	86.3	−126	5	9
	Release (m³/s/day)	218	288	214	414	377			

Note: R = turbine; S = spillway.

In the past few decades a variety of optimization models have been developed for long-term, short-term and real-time operation of single- and multiple-reservoir systems. Deterministic and stochastic approaches have been used to handle various issues arising out of the modelling process. Problems in this area have been addressed by many researchers using a wide variety of optimization tools. The emphasis of this example is on short-term reservoir operations, considering the imprecision in the definition of conventional loss functions.

Model formulation

Reservoir operations have typically set a target for release and storage volume values. Deviation from these targets can result in penalties being charged. Short-term reservoir operations are generally optimized (in economic terms) by formulating a model in which a loss function is used that reflects the penalty associated with the deviation from the target value. This type of problem has been addressed by many researchers (Teegavarapu and Simonovic, 1999). Most of the models have used piecewise linearized loss functions. One of the difficult aspects in these models is the quantification of these loss functions. They are usually based on the experience of reservoir operators, and therefore are highly subjective. The values that make up the loss functions are penalty coefficients: that is, the points on the loss function that define the penalty in monetary units corresponding to the penalty zones. Their selection is ultimately the reservoir operator's preference. These values are usually derived from economic information considering the impacts of reservoir operation. Although utility of loss functions have been devised for various reservoir operation models, there still exist unresolved questions about their derivation, shape and the associated penalty coefficients. This imprecision makes the reservoir operation problem difficult to handle.

Operating policies depend on the exact definition of these functions. In practice, the penalty coefficients are not crisp numbers but are aspiration levels which are not well defined. It can be observed that the loss function values reflect the decision-maker's degrees of importance attached to violation of various target values. Therefore the decision-making process involves dealing with the problem in an environment where the objectives and constraints imposed are vague. Fuzzy set theory concepts can be useful in this context, as they can provide an alternative approach to problems in which objectives and constraints are not well defined or information about them is not precise. Here we explore the application of the fuzzy decision-making tools described in Section 9.2 to short-term reservoir operations.

In this case the problem is asymmetric and the procedure (Zimmermann, 1996) includes the following steps:

Step 1: The mathematical programming model is solved, and the objective function value is obtained.

Step 2: The model is again solved with modified constraints which are considered fuzzy.

Step 3: The model is solved with the objective function and constraints (which were earlier assumed as fuzzy) replaced by their fuzzy equivalents using membership functions.

The objective function in steps 1 or 2 becomes a fuzzy constraint in Step 3. The fuzzy constraints in the present study are related to the penalty zones and coefficients, whereas the objective function is the penalty value in monetary units. The procedure described in steps 1–3 can be represented in mathematical form. For a minimization problem the steps are given below:

Step 1: *Minimize CX*
subject to:

$$AX \geq b \tag{9.106}$$

where:
CX = objective function
$X = [x_1, x_2, ...]^T$ = matrix of decision variables
AX = constraint matrix.

Let the objective function value obtained by solving the above problem be f_0.

Step 2: *Minimize CX*
subject to:

$$AX \geq b + t_0 \tag{9.107}$$

Here the tolerance interval t_0, by which the b value can change, is added to the right-hand side of (9.107). Let the objective function value obtained from the solution of step 2 be f_1.

Step 3: *Maximize λ*
subject to:

$$AX - \lambda\, t_0 \geq b \tag{9.108}$$

$$CX + \lambda(f_1 - f_0) \leq f_1 \qquad (9.109)$$

where f_0 and f_1 are the objective function values obtained from the previous steps.

Equations (9.108) and (9.109) represent the fuzzy constraints defined through an appropriate membership function. The objective function from Step 1 becomes a constraint (equation 9.109) in the present formulation. The objective function (max λ) indicates that the objective and the fuzzy constraints are satisfied to the maximum possible degree. This is similar to maximizing the membership function value (λ). The variable λ is referred to as a level of satisfaction, and is represented as L in all the formulations here. The formulations Step 1, Step 2 and Step 3 are referred to as the original, intermediate and final models respectively.

On the basis of the uncertainty issues in this context, two types of problems can be addressed. The first deals with imprecision in the definition of different penalty zones, while the second one is for uncertain penalty coefficients. Figure 9.11 shows one such problem where the location of point A is not precise. This indicates that the penalty zone is not defined exactly. Similarly, point B indicates the uncertainty in penalty coefficients. Here the decision-maker may be interested in decreasing or increasing the length of a particular zone or a penalty coefficient value. The problem becomes difficult to handle if the decision-maker has preferences attached to any change in a particular direction. In this situation a fuzzy set approach would provide a meaningful solution, where membership functions are used to capture the decision-maker's preferences. Here we shall look only at imprecision in the definition of penalty zones. Teegavarapu and Simonovic (1999) give details of the method for the problem of uncertain penalty coefficients.

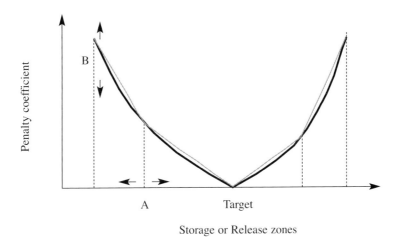

Figure 9.11 *Loss function representing deviations in the penalty zones and coefficients*

A fuzzy LP model was developed to address the problems related to imprecision in definition of penalty zones. The original formulation is presented first and then its fuzzy equivalent.

Imprecision in the penalty zones

The formulation uses piecewise linearization of a nonlinear loss function defined for storage and release. The inflow scheme for the periods for which optimal operation rules are required is assumed to be known. The objective is to minimize the underachievement or overachievement in meeting the storage and release target requirements.

The loss function for release is shown in Figure 9.12. Here, RD_1, RD_2, ... represent the unit penalties (refer to Table 9.9), and a_r, b_r, ... represent various points on the X axis which define the deviation zones, that is, RD_2, RD_1, etc. There is a similar loss function for storage, in which RD_1, RD_2, ... are replaced by SD_1, SD_2, ...; a_r, b_r, ... are replaced by a_s, b_s, ...; the deviation zones RD_2, RD_1, etc. are replaced by SD_2, SD_1, etc.; and RTR is replaced by STR. The objective is to minimize the penalties (equation 9.103), which are incorporated into the deterministic LP formulation. STR and RTR represent the storage and release targets respectively. Constraints (9.101) and (9.112) relate to reservoir target storages and releases; the next ten constraints (equations 9.113–9.122) relate to penalty zones. Obvious constraints related to reservoir mass balance and the upper and lower bounds on release and storage are omitted.

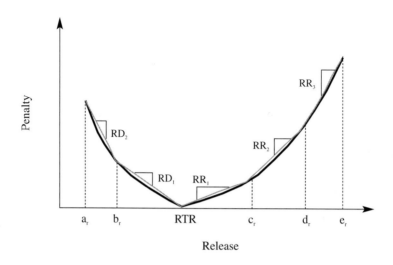

Figure 9.12 *Piecewise linearized loss function for the release*

Table 9.9 *Storage and release zones and corresponding penalties for the winter season*

Storage zone 10^6 m³	Penalty points per 10^6 m³	Release zone 10^6 m³	Penalty points per 10^6 m³
136.53–190.00	5000	0.367–1.50	10
190.00–200.99	1	1.50–2.50	0.1
200.99	target	2.5	Target
200.99–220.00	50	2.50–7.00	150
220–893.58	800	7.00–9.79	900
893.58–1507.09	15,000	9.79–14.67	2000

Model 1

The deterministic model with a piecewise linearized loss function takes the following form:

$$Minimize \sum_{t=1}^{T} (SD_2 SD2_t + SD_1 SD1_t + \\ SS_1 SS1_t + SS_2 SS2_t + SS_3 SS3_t \\ RD_2 RD2_t + RD_1 RD1_t + \\ RR_1 RR1_t + RR_2 RR2_t + RR_3 RR3_t) \quad (9.110)$$

subject to:

$$S_t + SD2_t + SD1_t - SS1_t - SS2_t - SS3_t = STR \quad \forall t \quad (9.111)$$

$$R_t + RD2_t + RD1_t - RR1_t - RR2_t - RR3_t = RTR \quad \forall t \quad (9.112)$$

$$SD2_t \leq b_s - a_s \quad \forall t \quad (9.113)$$

$$SD1_t \leq STR - b_s \quad \forall t \quad (9.114)$$

$$SS1_t \leq c_s - STR \quad \forall t \quad (9.115)$$

$$SS2_t \leq d_s - c_s \quad \forall t \quad (9.116)$$

$$SS3_t \leq e_s - d_s \quad \forall t \quad (9.117)$$

$$RD2_t \leq b_r - a_r \quad \forall t \quad (9.118)$$

$$RD1_t \leq RTR - b_r \quad \forall\, t \quad (9.119)$$

$$RR1_t \leq c_r - RTR \quad \forall\, t \quad (9.120)$$

$$RR2_t \leq d_r - c_r \quad \forall\, t \quad (9.121)$$

$$RR3_t \leq e_r - d_r \quad \forall\, t \quad (9.122)$$

This formulation is first solved to obtain the objective function value f_0 as indicated in Step 1. This type of formulation can be called crisp, as neither the objective function nor the constraints are fuzzy. Tolerances are then added or subtracted to the constraints which are considered fuzzy. This formulation refers to Step 2. Finally, the Step 3 formulation is solved using the membership functions. Membership functions which represent the decision-maker's preferences are assumed to be known in the present study and are used to derive the fuzzy constraints.

Membership functions

Membership functions are appropriate for modelling the preferences of the decision-maker. On the basis of the preferences for reducing or increasing the length of the penalty zones, two types of membership function can be derived. They are shown in Figures 9.13a and 9.13b. These functions indicate the preferences on the 0–1 scale on the Y axis for the length of penalty zone indicated on the X axis.

To keep the scope of the example to linear formulations in the present case, the membership functions are chosen to be linear, but there is no conceptual difficulty in handling nonlinear membership functions if appropriate formulations can be developed. Membership functions for constraint (9.119) can be developed, assuming that the decision-maker wants to reduce the first penalty zone of the release on the left side of the target release and storage. It is assumed that the unit slopes which determine the penalty values for these zones will remain unaltered even after the length of the zone is changed. Using the membership function given in Figure 9.13a, constraint (9.119) can be modified as follows:

$$RD1_t + (H)\, L \leq RTR - b_r \quad \forall\, t \quad (9.123)$$

where:
H = the amount of reduction in the first penalty zone $(RTR - b_r)$, which is equal to $b_0 - b_d$ = given in Figure 9.13a, and
L = the level of satisfaction for the constraint.

For the membership function given in Figure 9.13b, constraint (9.119) is modified as:

$$RD1_t + (H)(1 - L) \leq RTR - b_r \qquad \forall\, t \qquad (9.124)$$

Similar modifications have to be made to constraint (9.114) to represent the imprecision in the storage penalty zones for different membership functions. An additional constraint has to be modified to account for the possible reduction in the first penalty zone, which in turn might increase the second zone, adjacent to the first one, on the left side of the target. The constraint relating to the second zone is modified as:

$$RD2_t - (H)\, L \leq b_r - a_r \qquad \forall\, t \qquad (9.125)$$

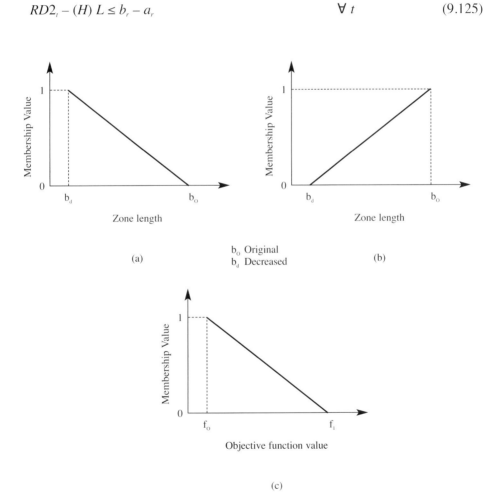

Figure 9.13 *Membership functions for the deviations and the objective function*

If this constraint is not included in the formulation it will result in an infeasible solution. The formulations in Step 1 and Step 2 when solved provide two different objective function values (f_0, f_1), which are used to develop a membership function for the fuzzy objective, which is shown in Figure 9.13c. The objective function (9.110) is modified and is now used as a constraint in the final formulation (Step 3). Model 1A and Model 1B given below refer to final formulations based on the membership functions derived from Figures 9.13a and 9.13b respectively. The objective now is to maximize the level of satisfaction, L. All the models are solved with the complete set of constraints from the original formulation, Model 1, except the constraints which are now modified to their fuzzy equivalents.

Model 1A

In the mathematical form Model 1A can be written as:

Minimize L

subject to:

$$\sum_{t=1}^{T} (SD_2 SD2_t + SD_1 SD1_t + SS_t SS1_t + SS_2 SS2_t + SS_3 SS3_t + \quad (9.126)$$
$$RD_2 RD2_t + RD_1 RD1_t + RR_t RR1_t + RR_2 RR2_t + RR_3 RR3_t) +$$
$$(f_1 - f_0)L \leq f_1$$

$$RD2_t - (H) L \leq b_r - a_r \qquad \forall\, t \qquad (9.127)$$

$$RD1_t + (H)(1 - L) \leq RTR - b_r \qquad \forall\, t \qquad (9.128)$$

H indicates the amount by which the first zone $(RTR - b_r)$ is reduced, thereby moving the point b_r closer to RTR and L has the same notation as that in the earlier fuzzy formulation.

Model 1B

Again, Model 1B in the mathematical form is expressed as:

Minimize L

subject to:

$$\sum_{t=1}^{T} (SD_2 SD2_t + SD_1 SD1_t + SS_1 SS1_t + SS_2 SS2_t + SS_3 SS3_t + \qquad (9.129)$$
$$RD_2 RD2_t + RD_1 RD1_t + RR_1 RR1_t + RR_2 RR2_t + RR_3 RR3_t) +$$
$$(f_1 - f_0)L \leq f_1$$

$$RD2_t - (H)\, L \leq b_r - a_r \qquad \forall\, t \qquad (9.130)$$

$$RD1_t + (H)(1 - L) \leq RTR - b_r \qquad \forall\, t \qquad (9.131)$$

This formulation is the same as the previous one (Model 1A) except that a different membership function (Figure 9.13b) is used for the constraint (9.119). For formulations to reflect the imprecision in the storage zones, (9.130) and (9.131) must be replaced by appropriate constraints which reflect the change in storage zones. Problems addressing the imprecision in both release and storage zones can be handled at the same time.

Case study

A case study of an existing reservoir, Green Reservoir in Kentucky, was chosen to evaluate the sensitivity of reservoir operating policies to the change in the shapes of loss functions. The primary objective of the reservoir is flood control in the Green River basin as well as in the downstream areas of the Ohio River. Secondary objectives include recreation, low-flow augmentation and water quality. The reservoir is the most upstream reservoir in the Green River system, located 489 km above the mouth of the stream, with a maximum storage capacity in excess of 1500×10^6 m^3. The reservoir storage up to the top of spillway is 892.02×10^6 m^3, while the minimum reservoir storage is 136.53×10^6 m^3. The maximum release is based on downstream flood protection and is 14.67×10^6 m^3, whereas the minimum daily requirement of 0.367×10^6 m^3 is based on water quality requirements. The original LP formulation (Model 1) uses the linearized penalty function values given in Table 9.9. These values are based on the intended use of the reservoir. It is evident from Table 9.9 that the primary purpose of the reservoir is for flood protection, indicated by the penalty values (values being high for storage and less for release deviations).

Results and discussion

This work addresses the imprecision in the definition of storage and release zones, and was expanded to address the uncertainty in the available penalty coefficient values in Teegavarapu and Simonovic (1999). The problem was solved using a fuzzy linear optimization model, whereas the expansion in Teegavarapu and Simonovic (1999) was solved using a fuzzy nonlinear optimization model.

All three formulations involving Step 1, Step 2 and Step 3 are solved. The formulations differ depending upon the type of membership function chosen (Figure 9.13a or Figure 9.13b). The membership function used in Figure 9.13c was used for converting the original objective function (9.110) into a fuzzy constraint in all Step 3 formulations. The original, intermediate and final formulations refer to steps 1, 2 and 3 respectively. Model 1 refers to the original formulation, while Models 1A and 1B refer to final formulations based on the membership function used for addressing the imprecision in the release or storage zones. The intermediate formulation refers to the Step 2 procedure, where Model 1 was solved with the appropriate constraints modified. It is apparent from Figure 9.13c that the decision-maker has a higher preference for a lower value of penalty, which is realistic.

The storage and release deviations from the target storage were represented by dividing the entire range of operational storage and release values into different zones. These are generally fixed by decision-makers or reservoir operators. Different simulations were performed to evaluate the sensitivity of reservoir operations to the fuzzy penalty zones. Simulations were performed for a period of 18 days using a known historical inflow scheme and adopting the penalty function values given in Table 9.9. The reservoir storage at the beginning of the first period was taken as 199×10^6 m^3. Model 1 and Model 1A were solved for a case in which the first release zone, on the left side of the target release, was reduced by 0.3×10^6 m^3. The membership function shown in Figure 9.13a was used for this case to represent the decision-maker's preference in reducing the zone. This indicates that the higher penalties were now attached to certain deviations from the target that had earlier had lower values. The storage loss function was unaltered.

Figures 9.14a and 9.14b show the results from all three formulations. The final solution was the optimal release rule for the fuzzy stipulation imposed. In order to reduce the penalties associated with reduced zone, the model opted for increased releases close to the target. Figure 9.14b shows the resulting storage variations. The membership functions used in deriving constraint (9.126) and objective function (max L) incorporate conflicting preferences, thus producing a satisfying solution to the degree of L. Any value of L between 0 and 1 does not necessarily indicate that the final results relevant to storage and releases will lie between the results of the intermediate and original formulations. This is apparent from Figures 9.14a and 9.14b, where the L value obtained is 0.51. This is because the L value is a satisfying value, based on the membership function for the constraints in all the time periods. On the other hand, the final objective function value will always lie between the values f_0 and f_1, obtained from the original and intermediate formulations respectively. In the present case the final objective function value in monetary units was

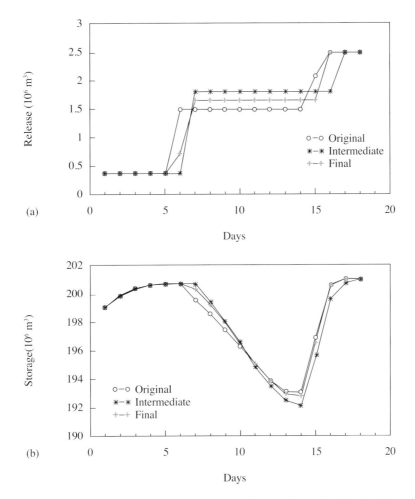

Figure 9.14 *Release and storage variations due to the reduction in the release zone*

124.07, whereas f_0 and f_1 were 108.86 and 139.73 respectively. This can be attributed to the property of the membership function given in Figure 9.13c.

In another experiment, reductions in the penalty zones of release as well as storage were introduced simultaneously. The first storage zone on the left side of the target was reduced by 5×10^6 m^3, and the reduction for release was the same as that of the previous case. The membership function used in this case was again from Figure 9.13a. Figure 9.15a gives the details of the release decisions for all three formulations. The release decisions are different from the one shown in Figure 9.14a. Reservoir storage variations for this case are shown in Figure 9.15b. The L value, or the degree of satisfaction achieved, in this case is 0.63. It can be noted from

Figure 9.15a that the release decisions are higher than the original decisions. A similar trend can be seen in Figure 9.15b for storage variations.

The final objective function value in this case was 131.09, which is higher than the previous value (124.07). This is because the penalty values of both release and storage contribute to the overall penalty value. In order to evaluate the effect of the L value and the type of membership function on the operation schedule, another simulation was performed. In this case the membership function corresponding to Figure 9.13b was used. Only a reduction in the first zone of the release was considered. It is evident from Figure 9.13b that a decreased preference was attached to the reduced zone compared with the original zone.

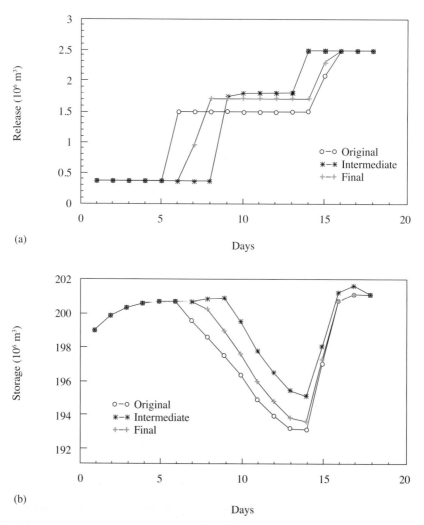

Figure 9.15 *Release and storage values for the reduction in the release and storage zones*

As the objective of Model 1B was to minimize the penalty (reflected indirectly in constraint (9.122)), the formulation forced the result to match with the original formulation results (Model 1). In these types of cases, the L value or the degree of level of satisfaction can be limited to a predetermined constant. This constant value can be interpreted as a minimum level of support in terms of satisfaction. Figure 9.16 shows the storage variations for such an experiment where the L value is limited to 0.75. The variations due to the final formulation are lower than those obtained from all the previous cases. This is because the restriction of the value of L results in higher release decisions and lower storage values.

From the experiments conducted it can be concluded that the reservoir operating rules are sensitive to the decision-maker's preferences attached to the definition of loss functions. This is evident from the results due to the final formulations, which can be called *compromise operating policies*. Sensitivity analysis of operation rules for a variety of conditions is not equivalent to what is achieved by the fuzzy optimization models proposed in the present example. A major limitation of traditional sensitivity analysis is the inability to handle preferences. The conflicting nature of the fuzzy constraints dealing with the penalty zones and coefficients, and the objective, which is the value in monetary units, are captured in the optimization framework. Finally, an important aspect that should be noted is that we cannot expect improved solutions (objective function values) by using fuzzy formulations. The solutions obtained are appropriate for the stipulations imposed. Therefore the results can only be more realistic, meaningful and sensitive to the decision-maker's preferences.

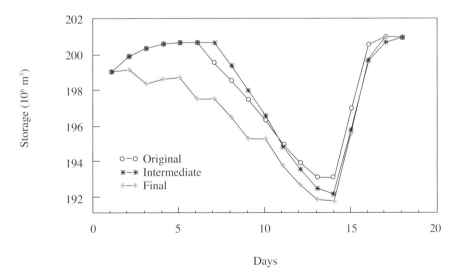

Figure 9.16 *Storage variations based on the reduction in the release zone*

The simulations carried out here were limited to the use of a portion of the release and storage loss functions. This is because the low inflows associated with the winter season would cause the models to generate storage levels or release decisions which lie on the underachievement side of the storage and release targets. Hence any changes made in the loss functions on the other side (right side) might not affect the reservoir operations schedule. The changes made in the penalty zones and coefficients for all the simulations were done in such a way that they increased the overall penalty value in monetary units. This is more realistic, as the reservoir managers tend to impose higher penalties for failure to achieve preset targets. To test the operation models in critical situations, the low inflows associated with the winter season were used. The results due to the final formulations will depend on the inflows, changes made to loss curves, and the membership functions used for preferences. Many variations of the cases mentioned above can be used to generate results for evaluation of the operation rules and their sensitivity to the decision-maker's preferences.

Conclusions

This example presents a new approach in dealing with the problem of uncertainty associated with the definition of conventional loss functions used in reservoir operation models. A fuzzy LP model was developed to handle the decision-maker's preferences in dealing with the imprecision within an optimization framework. The methodology is well suited for applications where the information or the methods through which the loss functions are derived are debatable. Since the penalty zones are considered fuzzy, the decision-maker is no longer compelled to provide a precise definition of loss functions. The optimal operating rules generated are a compromise between the original decisions and the rules when no preferences are attached to the changes made to the loss functions. The model developed is easy to implement in real-life situations if appropriate methods are used to generate the preferences in the form of membership functions. Improvements and extensions to the methodology are possible to incorporate the uncertainty of any other system variables and handle the problems associated with multiple-reservoir systems.

9.4.3 An evolutionary algorithm for minimization of pumping cost

This example deals with minimizing the total cost of pumping in a liquid pipeline (Ilich and Simonovic, 1998). The proposed solution method is an evolutionary algorithm with two distinct features: the search is restricted to the feasible region only, and it utilizes a floating-point decision variable rather than an integer or binary variable, as is the case with most other similar approaches. A numerical example is presented as a basis for verification of the proposed method.

Pipeline optimization is usually associated with minimizing the total cost of operating all pumps in the system. This problem involves complex constraints on flows and pressures, which can exhibit spatial and temporal variability as a result of the transient nature of fluid motion. Added to this is the physical complexity of the pipeline network, with numerous pump units which can be either on or off, units that can operate with either fixed or variable speed, and numerous pressure valves and tanks. A comprehensive analytical solution for pipeline optimization does not exist. Most researchers have resorted to various simplifications, the most common being assuming fixed flow rates over the solution horizon or combining simulation models with nonlinear optimization.

Problem formulation

A pump is a device which converts mechanical energy into pressure, while a pumping system is defined as a combination of pumps and the pipeline which they operate. Pipeline optimization in general refers to finding the most economical way to operate the pipeline. Pump characteristics are described by head–flow–efficiency (H–Q–η) curves. A typical example of an (H–Q–η) curve for a variable speed centrifugal pump is shown in Figure 9.17.

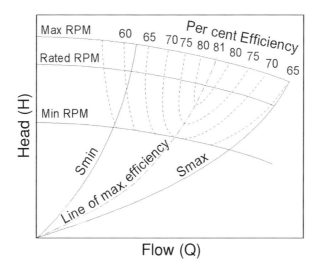

Figure 9.17 H–Q–η *curves*

A variable speed pump has been chosen as the basis for discussion only because it is more general than fixed speed pumps. The discussion is equally applicable to both types of pump. The parabolic dotted lines (also called the iso-efficiency lines) in Figure 9.17 can be used for calculation of the power required to run the pump for any feasible combination of head and flow. The dash-dotted line intersecting them is the line of maximum efficiency, while the solid lines show the typical head–flow relationship for three different speeds – maximum, design and minimum. The head–flow lines define an operating band for the pump in terms of limitations imposed by the size and type of the motor which drives the impeller. The other two lines which define the operating range are the system head curves labelled as S_{min} and S_{max} in Figure 9.17. They show the head system requirements for operating the segment of the pipeline as a function of the target flow rate. The system head characteristics together with the minimum and maximum head flow lines define a feasible range of combinations of head and flow which can be produced by a given pump.

The required power for a given combination of head and flow is then calculated by reading the appropriate efficiency factor for a given head–flow combination using linear interpolation between the nearest iso-efficiency lines. For any fixed flow rate within the operating range a number of heads can be produced, and for each head the corresponding efficiency can be estimated, thus resulting in the graph of power vs. head. Required (or *brake*) power is calculated using the standard relationship:

$$P = \frac{HQ\gamma}{\eta} \tag{9.132}$$

where:
γ = specific weight of the fluid
H = hydraulic head added by the pump
Q = fluid flow across the pump
P = power consumed by the pump.

An example of one such graph is shown in Figure 9.18. Note that the exact shape of the power–head relationship depends on the chosen flow rate, as a result of the distribution of the iso-efficiency lines, which are provided by pump manufacturers. An additional difficulty is that after a certain period of operation the pump efficiency changes and new head–flow–efficiency relationships must be established using measured observations (pressure, flow and power measurements).

For a given flow rate, the power–head relationship becomes the objective function, since each unit of power corresponds to a unit of monetary cost which is charged to the pipeline operators. Hence, it is relatively easy to convert the

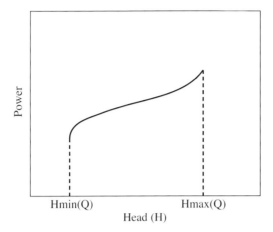

Figure 9.18 *Power vs. head for fixed flow rate*

power–head curves for a given flow rate into the cost–head curve for the same flow rate. The unit cost of power normally varies along a lengthy pipeline which intersects several political regions (provinces or states) which often impose different taxes on power consumption. Also, power is purchased from local suppliers at a local cost. This means that the shape of the cost–head flow curves depends not only on the type of equipment but also on the local cost of power. Therefore, the dollar cost–head function for a given flow rate may or may not have the exact shape as the power–cost function shown in Figure 9.18.

The goal of pipeline optimization is to deliver the target flow rate while simultaneously minimizing pumping cost, subject to the system pressure constraints. Figure 9.19 shows a serial pipeline with five pumping stations. Each station i is assumed to operate pump units with variable speed which can deliver pressure $X(i)$ in the range from $Xmin(i)$ to $Xmax(i)$. Also, each segment of the pipeline between two adjacent stations requires a head equal to $D(i)$ to maintain the target flow rate. The term $D(i)$ refers to the pressure drop for the given pipe between the two adjacent stations. It can be obtained from the appropriate pipeline system curve or the application of either Darcy–Weisbach or Colebrook–White equations (available in any fluid mechanics textbook) assuming that all friction losses (linear and local) are taken into account.

Although head drop happens continually as shown in Figure 9.19, it can be represented as a point loss at the end of the segment between the two stations depicted with upward arrows and $D(i)$, indicating 'demand' for pressure required to drive a given flow rate through the line. The flow rate is the same for each line, since liquid fluids are incompressible and conservation of mass must be preserved. Since a long serial pipeline is a sequential set of smaller subsets consisting of two elements

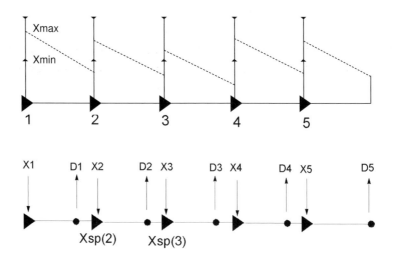

Figure 9.19 *Schematic of a serial pipeline*

(the pumping station i and the pipeline segment which links it to station $i+1$), it is sufficient to consider one such subset in further analysis and extend the observations to the rest.

Consider, for example, pumping station 3 in Figure 9.19. The pressure available in the pipeline immediately above pump 3 is equal to $(X1 - D1 + X2 - D2)$ where $X1$ and $X2$ are pressures produced at stations 1 and 2 while $D1$ and $D2$ are pressure drops between stations 1 and 2, and 2 and 3, respectively. The pressure above pump 3 must exceed the minimum suction pressure $Xsp(3)$ for the pump at station 3, otherwise the pump cannot operate. Similar suction pressure constraints are associated with all other pumping stations.

The problem of optimizing the pipeline can now be stated as:

$$\text{Minimize } \sum_{i=1}^{n} C(X)(i)) \tag{9.133}$$

subject to:

$$Xmin(i) \leq X(i) \leq Xmax(i) \quad i = 1,\ldots,n \tag{9.134}$$

$$\sum_{i=1}^{n} X(i) = \sum_{i=1}^{n} D(i) \tag{9.135}$$

$$X_{sp}^{min}(i) \leq \sum_{l=1}^{i-1} X(l) - \sum_{l=1}^{i-1} D(l) \leq X_{sp}^{max}(i) \quad i = 1,\ldots,n \tag{9.136}$$

where:

$C(X(i))$ = cost of producing (pumping) pressure X at pump i
$Xmin(i)$ = minimum pressure that can be generated for a specific flow rate at pump i
$Xmax(i)$ = maximum pressure that can be generated for a specific flow rate at pump i
$X_{sp}^{min}(i)$ = minimum suction pressure for station i
$X_{sp}^{max}(i)$ = maximum suction pressure for station i
$D(i)$ = head drop between station i and station $i+1$
n = total number of pump stations in a given system
i = station index, i.e. number starting from 1 (the first) to n (the last) counted in the direction of flow
l = counter for all stations and lines located above station i, i.e. preceding station i in terms of the direction of flow.

The first constraint imposes limits on the head that can be produced for a fixed flow rate according to the pump characteristics depicted in Figure 9.17. The second constraint means that the total energy produced by all pumps in the system equals the total energy required to drive the system (the sum of all head drops). Pressure valves were not considered here for simplicity but they could be added to condition (9.133) as an additional loss term, with one major difference: valve throttling would be included as an additional decision variable X with associated cost. The optimization problem would include both pump and valve operation such that valve throttling is minimized. This is of significance when pumps with fixed speed are operated, where valves are essential for controlling the flow rate.

It is convenient to represent the above problem as a minimum cost flow problem in an oriented circulatory network which can be created using the graph theory approach. A graph is defined as a set of nodes and oriented arcs, as shown in Figure 9.20, where each arc is associated with the upper and lower bounds on flow and with a cost of sending a unit of flow from its originating (tail) node to its terminating (head) node along the arc.

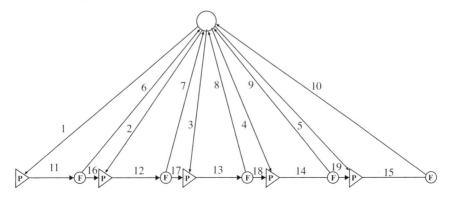

Figure 9.20 *Graph representation of pipeline operation*

Nodes represent the points where pressure is brought into the system (pumping stations), or where it is lost at the end of each segment, denoted with the letter F in Figure 9.20. This node will be addressed in more detail below. There is also a universal source-sink node added to this graph to ensure a circulatory property. The arcs numbered 1 through 5 are oriented from the source-sink node towards the pumping station nodes. They are associated with the decision variable $X(i)$ which represents the pressure produced at each pumping station. These arcs have a lower limit of $Xmin(i)$ and the upper limit of $Xmax(i)$. The arcs numbered 6 to 10 are directed towards the universal source-sink node, and represent pressure drops along each segment of the pipeline. Their lower bound and the upper bounds are set to $D(i)$. Since flow on each arc has to be between the flow bounds, this will ensure that the head loss is taken into account as a hard constraint. Satisfying constraint (9.135) is therefore equivalent to satisfying flow continuity for the source-sink node.

Node F is fictitious. It does not exist in the physical system. It is required to make it possible to subtract the head drop from the pipeline at the end of each segment before the remaining head in the line is made available as the suction head for the next station. A simplified representation of this approach is presented in Figure 9.20. The arcs numbered 11 through 15 represent the segments of the pipeline between the two adjacent pumping stations. Their upper and lower bounds are set to the maximum and minimum permissible pressure that can be tolerated anywhere within a given segment of the pipeline. Note that this representation allows further subdivision of pipeline arcs into shorter segments if a more accurate representation is desired. In the above representation the pressure produced at station 1 is carried out through line 11 with a constant value, until node F where the head loss $D(1)$ is subtracted. In reality pressure drops gradually through the line, and this can be better represented by introducing several fictitious nodes between two pumping stations in the same fashion as node F. However, keeping one node F between two pumping stations is sufficient for demonstrating the ideas in this example. The remaining pressure at the end of segment 1 is $X(1) - D(1)$, and it represents pressure along a short segment of pipeline labelled as arc 16 in Figure 9.20. This pressure must be within the minimum and maximum suction head for a given station.

Consider a vector of pressure values $X(i)$ associated with all arcs in the system $(i = 1,19)$. Expressions (9.133) through (9.136) can now be restated as:

$$\text{Minimize } \sum_{i=1}^{n} C(X)(i) \qquad (9.137)$$

subject to:

$$L(i) \leq X(i) \leq U(i) \quad i = 1,\ldots,n \qquad (9.138)$$

$$\sum_{i \in A} X(i) = \sum_{i \in A'} X(i) \quad k \in N \qquad (9.139)$$

Expressions (9.137) through (9.139) define the *minimum cost flow problem*, which is one of the most widely known topics in network programming. In this formulation constraints (9.138) and (9.139) have replaced constraints (9.134), (9.135) and (9.136). Index i represents the arc number from 1 to n, while set N refers to the set of nodes in the network, which are divided into three types in Figure 9.20 (**P**umping station nodes, **F**ictitious nodes and the **S**ource-sink node). The decision variable $X(i)$ represents the pressure or head available at each arc in the network. Expression (9.138) imposes a lower and upper limit on the pressure for each arc in the network, while expression (9.139) maintains the continuity of flow for each node in the network, with A denoting the subset of incoming arcs into node k while A' represents the subset of outgoing arcs for the same node. Expression (9.139) is written for each node in the network.

When the objective function and constraints (9.137) can be linearized, the above is an LP program with known solution methods (see Section 9.1). The choice and success of solution methods for nonlinear cost functions (9.137) or constraints (9.138) depends on the shape of the cost function and the degree of nonlinearity. Of interest in this example is the case when the cost function (9.137) exhibits a high degree of nonlinearity. For the discussion of various ways of solving this problem and their limitations, see Ilich and Simonovic (1998).

Evolutionary algorithm for pumping cost optimization

In this example we formulate an evolution optimization algorithm for the solution of the pumping cost optimization problem. In this way we are able to overcome the major drawbacks of existing approaches and provide effective convergence to an optimal solution. In particular:

- the search is conducted within the feasible region, which could significantly improve the efficiency compared with other search methods that do not distinguish between feasible and infeasible regions;
- the algorithm applied to the problem uses a floating-point decision variable rather than an integer or binary variable;

- the investigation of the effects of the type of evolutionary generation functions on the convergence process.

An innovative idea without crossover or mutation of artificial chromosomes is implemented in the proposed algorithm. The similarities between the proposed method and the EP presented in Section 9.3 are the evolutionary generation of possible solutions, which proceeds from generation to generation in an iterative fashion, and that the method utilizes the knowledge gained in the current step for improving the chances of finding a better solution in the next generation.

Search within the feasible region

As previously mentioned, many evolutionary algorithms (especially the class of GA) converge to an optimum from both a feasible and infeasible search space. In problems with complex constraints this results in a huge overhead related to the generation of infeasible solutions which have to be eliminated using some type of penalty function added to the objective function of the problem. This means that for many heavily constrained programs, such as for example pipeline optimization, there is a potential for improvement in the efficiency of the solution method if a search can be conducted exclusively within the feasible region.

Consider the minimum cost flow problem defined by expressions (9.137), (9.138) and (9.139). Concepts such as *labelling* and *pivoting*, originally developed in network programming solution techniques (Murty, 1992) are used here to limit the search to a feasible region. It is assumed that the initial starting feasible solution is available, where the feasible solution is vector $X(i)$ which satisfies constraints expressed by (9.138) and (9.139). This can be obtained using a standard LP optimization method, which would either find at least one feasible solution or declare the problem infeasible, regardless of the shape of the objective function. Let the chain of arcs including arcs numbered 1, 11, 16 and 6 be defined as the *augmenting path* or *augmenting cycle*. Consider the initial feasible solution for this circulation vector $V(1)$, and the arc bounds of all arcs in this cycle also form vectors $Vmin(1)$ and $Vmax(1)$, such that:

$$Vmin(1) = \{Xmin(1), Xmin(11), Xmin(16), -Xmin(2)\} \qquad (9.140)$$

$$V(1) = \{X(1), X(11), X(16), -X(2)\} \qquad (9.141)$$

$$Vmax(1) = \{Xmax(1), Xmax(11), Xmax(16), -Xmax(2)\} \qquad (9.142)$$

where the negative sign for $X(2)$ signifies the opposite direction of arc 2 from all other arcs in the chain. One unique chain can be identified for each arc associated with the operation of the pumping station. For example:

$$V(2) = \{X(2), X(12), X(17), -X(3)\} \qquad (9.143)$$

$$V(3) = \{X(3), X(13), X(18), -X(4)\} \qquad (9.144)$$

Note that it is possible to change all values of decision variable $X(i)$ in a single cycle by adding a constant term Δ to all of them such that the continuity of flow for each node in the cycle is preserved while the flow is still within the bounds for every arc. In this process the term Δ can be either positive or negative. There are limits to its positive or negative values, which can be determined from the current values of the elements of vectors $V(i)$, $Vmin(i)$ and $Vmax(i)$. For example, the largest permissible decrease of the initial value of $X(1)$ is defined by:

$$\min \{X(1) - Xmin(1), X(11) - Xmin(11), X(16) - Xmin(16), Xmax(2) - X(2)\}$$

Note that $X(2)$ is not handled in the same way as the other arcs in the cycle due to its orientation. The largest permissible increase for the initial value of $X(1)$ is:

$$\min \{Xmax(1) - X(1), Xmax(11) - X(11), Xmax(16) - X(16), X(2) - Xmin(2)\}$$

It is possible to set up a network database that can monitor the above variables for each augmenting path, thus allowing a search through the feasible region by varying the flows on each cycle. The search begins as a uniformly distributed random walk through the feasible region. The search process is then shaped based on the success of the initial population, such that the direction of the search is improved from generation to generation. This is discussed in more detail below.

Convergence to an optimal solution

All evolution programs utilize the knowledge from previous generations to converge closer to an optimal point in subsequent generations. This is usually done using some type of combinatorial approach (see expressions presented in Section 9.3.3), where the most successful solutions of generation i are combined using techniques like crossover to create solutions for generation $i+1$. In this example, solutions from generation i are not combined directly to generate solutions in generation $i+1$. The model parameters which control the convergence process are those that define the shape of the probability density function used in the generation process. Therefore, the algorithm can be best described as a sequential process of generating feasible solutions using Monte Carlo-type generation, where the knowledge from previous generations is used to adjust the distribution density function of the forthcoming generations, thus increasing the likelihood of an improvement in optimality. The process stops once the improvement in the stated objective stays within a given tolerance criterion from generation to generation. These are the steps in the proposed algorithm.

Step 1
Generate 1000 feasible solutions using the uniform distribution

$$X(i) \sim U[\ Xmin(i), Xmax(i)\ ].$$

Step 2
Rank all 1000 generated solutions in terms of their optimality, i.e. from the best to the worst.

Step 3
Examine the best solutions and fit a distribution for next generation, for example, $\mu(i)$ and $\sigma(i)$ for normal distribution.

Step 4
Generate 1000 feasible solutions using asymmetric distribution, for example:

$$X(i) \sim N[\mu(i), \sigma(i), Xmin(i), Xmax(i)\ ].$$

Step 5
Rank all 1000 generated solutions in terms of their optimality, i.e. from the best to the worst.

Step 6
Test the convergence criteria. If it fails go back to Step 3, otherwise end the search.

The search starts in Step 1 with a Monte Carlo generation of 1000 feasible solutions based on a uniform distribution for each station i within the given bounds. This removes bias and ensures that all corners of the feasible region have an equal chance of being addressed in the search. The number of generated solutions (1000) is arbitrary. The model should generate a sufficient number of initial solutions. Since it is hard to know how many initial solutions are sufficient when there is no previous knowledge related to a given problem, the general rule seems to be, 'the more initial solutions, the better'.

In Step 2 the algorithm stops the generation of feasible solutions. At this point 1000 sets of feasible solutions have been identified. Each solution has five unique values of $X(i)$ and a unique value of the objective function. It is now possible to sort out all solutions in ascending order in terms of the value of their objective functions. Since this is a minimization problem, the first solution in the list is the best.

In Step 3 the best solutions are examined. Typically the first five to ten solutions for each decision variable provide useful information about the direction in which the search should proceed in the next generation. At this point the algorithm uses the best solutions to decide how to transform the initial uniform distribution into an asymmetrical distribution which gives preference to the segment of the feasible region with the best initial solutions. For example, one possible model could be to take the mean μ and the standard deviation σ of the ten best solutions for each decision variable i and use them as parameters for truncated normal distribution in the next generation. The other could be to centre the search on the very best values of $X(i)$ obtained in the previous generation and use the remaining nine best solutions only to estimate the standard deviation σ for the next generation. The best distribution can be found using some type of calibration process.

Step 4 proceeds with the next generation of 1000 feasible solutions using the chosen asymmetrical distribution, such that the most likely generated value is close to the optimal value achieved in the previous generation. For example, if truncated normal distribution is used, then the new population of feasible solutions is generated such that decision variables $X(i)$ comply with:

$$X(i) \sim N\,[\mu(i), \sigma(i), Xmin(i), Xmax(i)\,] \qquad (9.145)$$

Note that $\mu(i)$ and $\sigma(i)$ are different for each decision variable i.

Step 5 is a repetition of Step 2. Finally, in Step 6 the algorithm tests the convergence criteria. The optimal solution is the one with the lowest value of the objective function generated in the process.

Numerical example

Two types of problems were formulated. In the first problem only one pump was considered in each station while in the second there were three pumps in serial connection at each station. The assumed cost function for each pump has the following form:

$$C(X) = Co + Ao\,X(i)^n \qquad (9.146)$$

where Co, Ao and n are user-defined parameters and $X(i)$ is the pressure produced at a given station.

Each pump can have different Co, Ao and n parameters. The approximate shape of the cost functions for the two problems is shown in Figure 9.21.

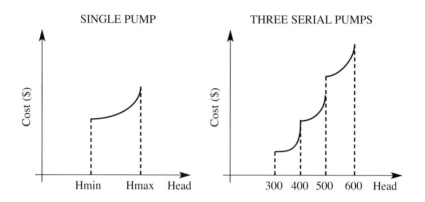

Figure 9.21 *Head vs. cost function for serial pumps*

The first problem is easier to solve and was used mainly as a benchmark. It took only two iterations to get the solution with the proposed method. After the first iteration of the proposed method, the best solution was the one with the total cost of $1189.53. With this solution as a starting point for generation in iteration 2, a total of 11 identical solutions were generated with the same optimal cost of running the pipeline of $1167.49. The process was aborted at the end of iteration 2.

The second problem was much more difficult to solve because of the shape of the cost functions. Table 9.10 shows the calculation of pipeline pressure losses for each segment using the Darcy–Weisbach equation with a friction factor of 0.02. The cost function was also calculated using relationship (9.146) with one distinct difference. All stations operate between the minimum working heads of 300 m and maximum of 600 m (1 m = 3.281 ft). For operation in the range between 300 and 400, the cost function has the form of:

$$C(X) = 0.333 Co + Ao\, X(i)^n \tag{9.147}$$

while for the $X(i)$ operating range of 400–500 this relationship is modified into:

$$C(X) = 0.667Co + Ao\, X(i)^n \tag{9.148}$$

and finally for $X(i)$ between 500 and 600 the function form is:

$$C(X) = Co + Ao\, X(i)^n \tag{9.149}$$

The values of Co, Ao and n are given in Table 9.11. The above rules provide the shape of the cost function similar to the one depicted in Figure 9.21.

Table 9.10 Calculation of pipeline pressure losses

Pipe length	Pipe diameter	Friction factor f	Target flow in	Head drop D(i)	Accumulated head drop Σ D(i)
m	m		m³/s	m	m
(1)	(2)	(3)	(4)	(5)	(6)
14,000	0.31	0.02	0.24	465.66	465.66
15,000	0.31	0.02	0.24	498.92	964.59
13,000	0.31	0.02	0.24	432.40	1396.99
11,500	0.31	0.02	0.24	382.51	1779.50
13,000	0.31	0.02	0.24	432.40	2211.90

Note: 1 m = 3.28ft; 1m³ = 35.3147ft³; f = dimensionless.

Table 9.11 Cost function parameters

Station number	Cost function parameters		
	Co	Ao	n
1	350	0.1	1.24
2	330	0.1	1.33
3	327	0.1	1.12
4	300	0.1	1.13
5	350	0.1	1.27

Table 9.12 shows the best solution for each of the seven generations of populations of 1000 randomly generated feasible solutions. Only the first generation was based on a uniform probability distribution. Subsequent generations were based on the

normal distribution. The total cost value in the final column is the objective function.

Table 9.12 *Solutions for each generation using the proposed evolutionary optimization*

Generation	Head produced at pump station (m)					Total cost
	X(1)	X(2)	X(3)	X(4)	X(5)	$\Sigma C(X)$
1	582.12	388.59	483.75	395.92	361.52	1803.85
2	589.63	374.95	499.07	392.45	355.79	1794.45
3	600.00	364.59	494.92	398.75	353.63	1789.94
4	600.00	364.59	498.37	399.17	349.77	1788.46
5	600.00	364.59	499.58	399.92	347.82	1787.73
6	600.00	364.59	499.96	399.74	347.61	1787.65
7	600.00	364.59	499.95	399.99	347.36	1787.55

The first surprise with the evolutionary search method came after the very first iteration, when it became obvious that there were 11 solutions in the initial pool of 1000 that were very good. The convergence process was then carried out for six more iterations. For each decision variable $X(i)$ a normal distribution was assumed, with the mean equal to the best solution from the previous generation and standard deviation equal to a fraction of the assumed mean. This value of standard deviation was chosen after several trial and error runs. It was found that the standard deviation should be gradually decreased as the process continued from one iteration to another. The standard deviation was gradually reduced in each of them from 0.1 of the mean in the second iteration to 0.001 of the mean in the seventh iteration. Using a constant value for the standard deviation for all iterations would significantly slow down the process of convergence, and if the value were too large there might be no convergence at all. The final solution obtained with the evolutionary search method is $1787.55.

9.5 REFERENCES

Bellman, R. and L.A. Zadeh (1970), 'Decision-making in a Fuzzy Environment', *Management Science*, 17(B): 141–164

Bhattacharjya, R.K. and B. Datta (2005), 'Optimal Management of Coastal Aquifers Using Linked Simulation Optimization Approaches', *Water Resources Management*, 19(3): 295–320

Dantzig, G.B. (1963), *Linear Programming and Extension*, Princeton University Press, Princeton, NJ de Neufville, R. (1990), *Applied Systems Analysis: Engineering Planning and Technology Management*, McGraw Hill, New York

Goldberg, D.E. (1989), *Genetic Algorithms in Search, Optimization, and Machine Learning*, Addison-Wesley, New York

Hillier, F.S. and G.J. Lieberman (1990), *Introduction to Mathematical Programming*, McGraw Hill, New York

Ilich, N. and S.P. Simonovic (1998), 'An Evolution Program for Pipeline Optimization', *ASCE Journal of Computing in Civil Engineering*, 12(4): 232–240

Jewell, T.K. (1986), *A Systems Approach to Civil Engineering Planning and Design*, Harper and Row, New York

Kumar, D.N. and M.J. Reddy (2007), 'Multipurpose Reservoir Operation Using Particle Swarm Optimization', *ASCE Journal of Water Resources Planning and Management*, 133(3): 192–201

Li, Y.P., G.H. Huang and S.L. Nie (2006), 'An Interval Parameter Multi-Stage Stochastic Programming Model for Water Resources Management Under Uncertainty', *Advances in Water Resources*, 29(5): 776–789

Luo, B., I. Maqsood and G.H. Huang (2007), 'Planning Water Resources Systems with Interval Stochastic Dynamic Programming', *Water Resources Management*, 21(6): 997–1014

Michalewicz, Z. (1999), *Genetic Algorithms+Data Structures=Evolution Programs*, 3rd, 4th and extended edn, Springer, Germany

Murty, K.G. (1992), *Network Programming*, Prentice-Hall, Englewood Cliffs, NJ

Pohlheim, H. (2005), *GEATbx Introduction Evolutionary Algorithms: Overview, Methods and Operators*, Genetic and Evolutionary Algorithm Toolbox for use with Matlab GEATbx version 3.7, http://www.geatbx.com (accessed December 2005)

Reznicek, K.K. and S.P. Simonovic (1990), 'An Improved Algorithm for Hydropower Optimization', *Water Resources Research*, 26(2): 189–198

Simonovic, S.P. (2000a), 'Tools for Water Management: One View of the Future', *Water International*, 25(1): 76–88

Simonovic, S.P. (2000b), 'Last Resort Algorithms for Optimization of Water Resources Systems', *CORS – SCRO (Canadian Operational Research Society) Bulletin*, 34(1): 9–19

Teegavarapu, R.S.V. and S.P. Simonovic (1999), 'Modeling Uncertainty in Reservoir Loss Functions Using Fuzzy Sets', *Water Resources Research*, 35(9): 2815–2823

Wagner, H.M. (1975), *Principles of Operations Research*, 2nd edn, Prentice-Hall, Englewood Cliffs, NJ

Zimmermann, H.J. (1996), *Fuzzy Set Theory – And its Applications*, 2nd, rev edn, Kluwer Academic Publishers, Boston

9.6 EXERCISES

1. Consider the following problem:

    maximize $Z = 3x_1 + 2x_2$

    subject to:

    $2x_1 + x_2 \le 6$
    $x_1 + x_2 \le 6$
    $x_1, x_2 \ge 0$

    a. Solve the problem graphically. Identify the corner-point feasible solutions.
    b. Identify all the sets of the two defining equations for this problem. For each one, solve the corner-point solution, and classify it as a corner-point feasible or infeasible solution.
    c. Introduce slack variables in order to write the problem in the canonical form.

2. Consider the following problem:

    maximize $Z = 2x_1 + 4x_2 + 3x_3$

    subject to:

    $x_1 + 3x_2 + 2x_3 \le 30$
    $x_1 + x_2 + x_3 \le 24$
    $3x_1 + 5x_2 + 3x_3 \le 60$
    $x_1, x_2, x_3 \ge 0$

    You are given the information that $x_1 > 0$, $x_2 = 0$, and $x_3 > 0$ in the optimal solution.

    a. How can you use this information to adapt the simplex method to solve the problem in the minimum possible number of iterations?
    b. Solve this problem using the simplex method.

3. Label each of the following statements as True or False, and then justify your answer.

    a. The simplex Criterion I is used because it always leads to the best basic feasible solution.
    b. The simplex Criterion II is used because making another choice normally would yield a basic solution that is not feasible.
    c. When an LP model has an equality constraint, an artificial variable is intro-

duced into this constraint in order to start the simplex method with an obvious initial basic solution that is feasible for the original model.

4 Consider the following LP problem:

maximize $x_1 + x_2$

subject to:

$-x_1 + x_2 \leq -1$
$x_1 - x_2 \leq -1$
$x_1 \geq 0$ and $x_2 \geq 0$

a. Find graphically the solution of the stated problem.
b. Discuss the solution obtained in detail.
c. Modify the problem by changing the sign of both inequalities and adding one more constraint:

$x_1 + x_2 \leq 6$

and solve using the Linpro program on the CD-ROM.

5 Formulate the following linear program using the Big M method and solve it using the Linpro program:

Max $4x_1 + x_2$

subject to:

$2x_1 + 3x_2 \geq 4$
$3x_1 + 6x_2 \leq 9$
$x_1, x_2 \geq 0$

6 A building block manufacturer makes two types of building blocks: type A and type B. For each set of 100 type A blocks the manufacturer can make a profit of $5 whereas for each set of 100 type B blocks he can make $8. Assume that all blocks produced can be sold. It takes one hour to make 100 type A blocks and three hours to make 100 type B blocks. Each day there are 12 hours available for block manufacturing. A set of 100 type A blocks requires 2 units of cement, 3 units of aggregate, and 4 units of water; a set of 100 type B blocks requires 1 unit of cement and 6 units of water. Each day, 18 units of cement and 24 units of water are available for block manufacturing. There is no restriction on the

availability of aggregate. How many type A and type B blocks should be made during the day to maximize profits?
a. Formulate the problem as an LP problem.
b. Name the decision variables, objective function and constraints.
c. Solve the problem graphically.
d. Use the Linpro program on the CD-ROM to solve the problem.
e. What is the value of the dual variable, or shadow price, associated with the 24 units of available water?

7 (Modified after Wagner, 1975) The Sunnyflush Company has two plants located along a stream. Plant 1 is generating 20 units of pollutants daily and plant 2 14 units. Before the wastes are discharged into the river, part of these pollutants is removed by a waste treatment facility in each plant. The costs associated with removing a unit of pollutant are $1000 and $800 for plants 1 and 2. The rates of flow in the streams are $Q_1 = 5$ m³/s and $Q_2 = 2$ m³/s, and the flows contain no pollutants until they pass the plants (see diagram). Stream standards require that the number of units of pollutants per m³ of flow should not exceed 2. Twenty per cent of the pollutants entering the stream at plant 1 will be removed by natural processes before they reach plant 2. The company wants to determine the most economical operation of its waste treatment facilities that will allow it to satisfy the stream standards.
a. Formulate the problem as an LP problem.
b. Name the decision variables, objective function and constraints.
c. Solve the problem graphically.
d. Use the Linpro program on the CD-ROM to solve the problem.

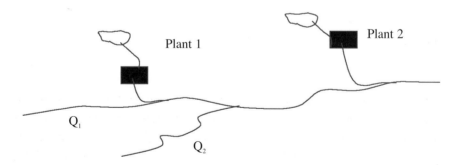

8 (After Wagner, 1975) Reservoir A is used for recreation (swimming, water skiing, canoeing). It is important to keep the average depth of this reservoir within prescribed limits, which vary from one month to the next. The engineers responsible for the operations of reservoir A have estimated a rapid rate of seepage and evaporation from the reservoir. Since rainfall is negligible, reservoir A

must be maintained by spillage from reservoir B. Suppose the planning horizon is 20 months. During month t, let x_t denote the average depth of reservoir A prior to augmenting with the water from reservoir B; $x_1 = 25$ for month 1. Let y_t be the number of metres to be added to the average depth in month t (a positive value for y_t indicates a decision to augment reservoir A). Let L_t and U_t represent the lower and upper prescribed limits, respectively, of the average reservoir depth after augmentation in month t. Assume that x_{t+1} is 0.75 of the average reservoir depth in month t after augmentation. Suppose that the cost of augmenting the reservoir is c_t per metre in month t.

a. Formulate the problem as an LP problem.
b. Name the decision variables, objective function and constraints.

Do not solve the problem.

9 Based on the LP model developed in Exercise 7:
   a. Define the meaning of the dual variables, and their values, associated with each constraint.
   b. Write the dual model of this problem and interpret its objective and constraints.
   c. Solve the dual model using the Linpro program on the CD-ROM, and indicate the meaning of all output data.

10 Consider the following problem:

$$\text{minimize } Z = 4x_1 + 5x_2 + 2x_3$$

subject to:

$$3x_1 + 2x_2 + 2x_3 \leq 60$$
$$3x_1 + x_2 + x_3 \leq 30$$
$$2x_2 + x_3 \geq 10$$
$$x_1, x_2, x_3 \geq 0$$

a. Determine the optimal solution using the Linpro program on the CD-ROM.
b. Now assume the decision-maker has a linear preference function for the objective function between the minimum and 1.5 times the minimum; and the tolerance intervals can be established as $p_1 = 10$, $p_2 = 12$, and $p_3 = 3$. Solve the problem following the methodology from Section 9.2.1 using the FuzzyLinpro program on the CD-ROM.
c. Compare the solutions of the crisp and fuzzy formulations.

11 Consider the water supply capacity expansion problem from Example 10 (Section 9.2.1). Assume the following parameters for the fuzzy formulation:

$$d_0 = 3{,}500{,}000 \qquad d_1 = 170 \quad d_2 = 1\,300 \quad d_3 = 6$$
$$p_0 = 1{,}000{,}000 \qquad p_1 = 120 \quad p_2 = 150 \quad p_3 = 4$$

a. Reformulate the fuzzy problem of water supply capacity expansion.
b. Solve the fuzzy problem using FuzzyLinpro on the CD-ROM.
c. Compare your solutions with those listed in Table 9.3. Discuss the difference. Comment on the impact of uncertainty on this decision-making problem.

12 A rectangular, open-topped reservoir is to be proportioned. The flow rate, and thus the cost of supplying water to the tank, is inversely proportional to the storage volume provided. Typical figures are as follows:

Volume (m³)	Flow (m³/s)	Supply cost ($)
100	0.65	10,000
500	0.50	2000
1000	0.40	1000
2000	0.35	500

The cost of constructing the tank is based on the following rates: base $2/m²; sides $4/m²; and ends $6/m².

a. Find the dimensions for least cost. The supply cost may be approximated by the function $c_1 = 10^6/$ Volume in m³. For solving the problem use the Evolpro computer program from the CD-ROM.
b. Vary the desired accuracy (tolerance level) and the maximum number of iterations. Discuss the solutions.
c. Change the size of population and discuss its impact on the optimal solution and optimization process.

13 Assume that the number of kilometres of water pipe to be laid, H, is a function of both the hours of labour, L, and of machines, M:

$$H = 0.5 L^{0.2} M^{0.8}$$

a. Minimize the cost of installing 20 km of pipe, given that the hourly rates for labour and machines are: $C_L = \$20$; $C_M = \$160$. To solve the problem use the Evolpro computer program from the CD-ROM.
b. Vary the desired accuracy (tolerance level) and the maximum number of iterations. Discuss the solutions.
c. Change the size of population and discuss its impact on the optimal solution and optimization process.

14 A sedimentation tank is circular in plan with vertical sides above ground and a conical hopper bottom below ground, the slope of the conical part being 3 verti-

cally to 4 horizontally. Determine the proportions to hold a volume of 4070 m³ with the minimum area of bottom and sides.
a. To solve the problem use the Evolpro computer program from the CD-ROM.
b. Vary the desired accuracy (tolerance level) and the maximum number of iterations. Discuss the solutions.
c. Change the size of population and discuss its impact on the optimal solution and optimization process.

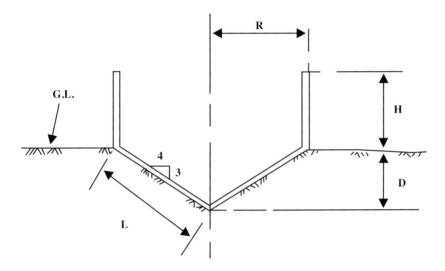

15 Find the optimal reservoir size for irrigation. The total benefit (*TB*) derived from irrigation water in a region has been estimated as:

$$TB = 100\,q - 0.0005q^2$$

where *TB* is expressed as dollars per year ($/yr) and *q* represents dependable water supply, expressed in 10^6 m³ per year (mcm/yr). Estimates of the annual total cost (*TC*) of a reservoir at different sizes result in the expression:

$$TC = 44.42\,q^{0.90} + 0.098\,q^{1.45}$$

where TC has units of dollars per year.
a. The agency responsible for reservoir construction wishes to maximize net benefit (NB), defined as total benefit minus total cost:

$$Maximize\ NB(q) = TB(q) - TC(q).$$

b. To solve the problem use the Evolpro computer program from the CD-ROM.
c. Vary the desired accuracy (tolerance level) and the maximum number of iterations. Discuss the solutions.
d. Change the size of the population and discuss its impact on the optimal solution and optimization process.

10
Multi-objective Analysis

Management of complex water resources systems rarely involves a single objective. In today's modern society, a multitude of complex water resources allocation problems is creating difficulties for planners and policy-makers. Here we look at one example of a problem situation where there is a need to formulate goals, develop alternative plans, and design criteria to be used in the selection of a plan.

One challenging problem area is the long-range development of an existing water resources system. We shall use as an example the Danube River basin in Europe, where my life and professional career started. The queen of Europe's rivers, the Danube is rich in history, enveloped in an aura of legend and myth. From the mountains of the Black Forest, it makes its way through cliffs and surges down wild and romantic gorges to flow by mighty castles, palaces and monasteries, framed by the enchanting landscapes and Baroque splendour along its banks. No other river in the entire world has inspired so many poets, musicians and painters to create masterpieces; along no other river can such a tremendous variety of scenery, historic cities, magnificent architecture and cultural treasures be found.

An artery of trade and communication running through the heart of Europe, the Danube links peoples and nations on its long journey from west to east. Art and music accompany the route of the great river until the melody reaches a gentle climax in a rustling sea of reeds: the Danube delta, one of the last paradises for people and nature.

The Danube is the longest international river in Europe. It flows through Germany, Austria, Slovakia, Hungary, Croatia, Serbia, Romania, Bulgaria, Moldavia and Ukraine. Its spring is in Schwarzwald (in English, the Black Forest), and it empties into the Black Sea. It is 2850 km long, 2414 km of which is navigable. The Danube is at the centre of the European network of navigable inland waterways. The Danube basin, covering 817,000 square kilometres (km^2) – about one-third of continental Europe outside Russia – is the most international river basin

528 Implementation of Water Resources Systems Management Tools

Source: photo courtesy of Danube Tourist Commission

Figure 10.1 *The Danube River – the place where everything started*

in the world, extending over all or part of the territories of 18 countries. The Danube is also Europe's only major river that flows from west to east, from the current Member States of the European Union (EU) through the former eastern bloc countries of central and eastern Europe, many of which are now prospective EU members. The European Commission recognizes the Danube as the 'single most important non-oceanic body of water in Europe' and a 'future central axis for the European Union'.

The main economic uses of the Danube are:

- domestic/drinking water supply;
- water supply for industry;
- water supply for agriculture;
- hydroelectric power generation;
- navigation;
- tourism and recreation;
- disposal of waste (both solid and liquid);
- fisheries.

In addition, the Danube's remaining floodplains provide a range of economically important 'ecological services', such as water quality regulation and flood control. One of the most important factors influencing river basin management activities is the socio-economic contrast between the capitalist and former socialist countries within the basin. Since the changes in the late 1980s, the central and lower Danube region has experienced a rapid shift to free-market democracies in the context of increased globalization, privatization and deregulation, including the loss of much of the formerly guaranteed social security structure. At the same time, as a result of economic restructuring, many countries have lost markets in neighbouring countries. This is especially true of agriculture, which remains the economic mainstay in rural central and eastern Europe, in spite of tough competition from EU-subsidized agricultural products. The result is rural decline, with increased poverty, unemployment and depopulation. Rural environments are being exploited for short-term gain through over-fishing, over-grazing, deforestation and poaching, such that traditional lifestyles and sustainable economic practices are at risk.

The Danube basin is home to a wide variety of natural habitats. Among these are the Alps and Carpathian Mountains, Germany's Black Forest, the Hungarian *puszta* plains, the Lower Danube floodplains and islands, and the vast lakes, reedbeds and marshes of the Danube delta. These habitats are home to a rich and in many cases unique biological diversity, including over 100 different types of fish, among them 6 endangered species of sturgeon. The 600,000-hectare (ha) Danube delta has been designated as a Ramsar site and United Nations Educational, Scientific and Cultural Organization (UNESCO) Biosphere Reserve. It supports more than 280 bird species, including 70 per cent of the world population of white pelicans and 50 per cent of the populations of pygmy cormorant and – in winter – red-breasted goose.

Priority issues for Danube basin management

Until the end of the 19th century, the Danube was a largely natural system with an extensive network of channels, oxbows and backwaters. The river was characterized by constant changes in its course and dynamic natural exchanges with its floodplains. Since then, human interventions in the way of flood protection, agriculture, power production, and navigation have destroyed over 80 per cent of the Danube's wetlands, floodplains and floodplain forests. Major losses in habitats and wildlife have resulted. One example is the considerable reduction of nursery areas for spawning fish, and the blocking of migratory pathways for commercially important species such as sturgeon, which now survive only as small remnant populations. Changes in flow volume and velocity, water temperature and quality as a result of river regulation and pollution have also had negative impacts on biodiversity.

The intensive development of central and eastern Europe has resulted in both positive and negative effects for the middle and lower reaches of the river. On the one hand, many wetland areas were drained to support unsustainable agricultural and forestry practices (e.g. along the Tisza River in Hungary where 2,590,000 ha of floodplains were reduced to 100,000 ha). On the other hand, the main Danube channel itself was not subject to the same level of dam construction as occurred in western Europe, where the upper 1000 km of the river were converted into an artificial waterway by an almost uninterrupted chain of 59 hydropower dams. This contrasts with just two dams on the lower 1800 km of the Danube. Overall, the central and lower reaches possess a generally higher level of biodiversity than do the upper reaches in western Europe. For example, the middle and lower Danube still supports some extensive areas of natural or semi-natural floodplain forest and other wetlands, while more than 95 per cent has been lost further upstream.

Current priority issues at a basin scale include:

- **Proposed shipping developments**. A number of proposals threaten severe ecological damage to the Danube in central and eastern Europe. They include plans to construct a canal through the Ukrainian Danube delta to the Black Sea coast, and another – the Danube-Odra-Elbe canal – linking the Baltic Sea with the Black Sea. In addition to the loss of natural and semi-natural areas that such developments would cause, chronic pollution and the risks of a major oil or chemical spill are also likely to increase.
- **Impacts of EU accession**. Many former eastern bloc countries are now in the process of joining the EU. As part of this accession process, each prospective Member State is required to transpose into national law – and implement – a raft of EU legislation before it is granted entry. The potential impacts on the Danube basin are both positive and negative. While the EU's nature conservation legislation and the Water Framework Directive (which governs water policy and management throughout the EU according to the principles of river basin management) are recognized as positive mechanisms, it is expected that threats to rural economic security in the central and lower Danube will be worsened by the EU's Common Agricultural Policy (CAP). The CAP, although recently reformed, continues to support intensive, unsustainable practices and perverse subsidies. The EU may also provide funding for some shipping development projects through its Trans-European Networks for Transport (TENs-T) programme.
- **Environmental disasters**. The last ten years have seen a number of ecological crises in the Danube basin that have gained worldwide media attention (e.g. the spillage in January 2000 of some 100 tonnes of cyanide into the Tisza River in Romania, following an accident at a gold mining operation). Unless more is done soon to improve environmental security, especially in those parts of the

region where the industrial and urban infrastructure is old and decayed, further catastrophic incidents can be expected.
- **Nutrients and eutrophication**. The main sources of nutrients in the Danube are agriculture (50 per cent), municipal waste (25 per cent) and industry (25 per cent). The total nitrogen load in the Danube is between 537,000 and 551,000 tonnes per year (compared with 50,000 tonnes for the Rhine). The total phosphorus load is 48,900 tonnes per year. The legal limit for nutrient content in groundwater is often exceeded throughout the basin. As a result, the Danube is the biggest contributor of nutrients to the Black Sea, where radical changes to the ecosystem and biodiversity loss have occurred in the last 40 years as a result of eutrophication. There remains insufficient capacity along the Danube to treat municipal and industrial wastewater, and more sewage treatment plants are needed urgently. Restoring wetlands might significantly increase the river's natural 'self-cleansing' capacity.

10.1 MULTI-OBJECTIVE ANALYSIS METHODOLOGY

Multiple-objective, in contrast to single-objective, decisions concerning water resources do not have an optimal solution. As a result, there have been great efforts to develop a methodology for assessing trade-offs between alternatives based on using more than one objective. In the last three decades of multi-objective research, efforts have been made in:

- objective quantification;
- the generation of alternatives;
- selection of the preferred alternative.

10.1.1 Change of concept

Chapter 9 showed that a single-objective programming problem consists of optimizing one objective subject to a constraint set. On the other hand, a multi-objective programming problem is characterized by an r-dimensional vector of objective functions:

$$Z(x) = [Z_1(x), Z_2(x), ..., Z_r(x)] \qquad (10.1)$$

subject to:

$$x \in X$$

where X is a feasible region:

$$X = \{x: x \in R^n, g_i(\mathbf{x}) \leq 0, x_j \geq 0 \;\forall\; i, j\} \tag{10.2}$$

where:
R = set of real numbers
$g_i(x)$ = set of constraints
x = set of decision variables.

The word *optimization* has been deliberately kept out of the definition of a multi-objective programming problem since we cannot, in general, optimize a priori a vector of objective functions. The first step of the multi-objective analysis consists of identifying the set of non-dominated solutions within the feasible region X. So instead of seeking a single optimal solution, a set of non-inferior solutions is sought. The essential difficulty with multi-objective analysis is that the meaning of the optimum is *not defined* as long as we deal with multiple objectives that are truly different. For example, suppose we are trying to determine the best design of a system of dams on a river, with the objectives of promoting national income, reducing deaths by flooding and increasing employment. Some designs will be more profitable, but less effective at reducing deaths. How can we state which is better when the objectives are so different, and measured in such different terms? How can we state with any accuracy what the relative value of a life is in terms of national income? If we resolved that question, then how would we determine the relative value of new jobs and other objectives? The answer is, with extreme difficulty. The attempts to set values on these objectives are, in fact, most controversial.

To obtain a single global optimum over all objectives requires that we either establish or impose some means of specifying the value of each of the different objectives. If all objectives can indeed be valued on a common basis, the optimization can be stated in terms of that single value. The multi-objective problem has then disappeared and the optimization proceeds relatively smoothly in terms of a single objective.

In practice it is frequently awkward if not impossible to give every objective a relative value. The relative worth of profits, lives lost, the environment and other objectives are unlikely to be established easily by anyone, or to be accepted by all concerned. We cannot hope, then, to be able to determine an acceptable optimum analytically.

The focus of multi-objective analysis in practice is to sort out the mass of clearly dominated solutions, rather than determine the single best design. The result is the identification of a small subset of feasible solutions that are worthy of further consideration. Formally, this result is known as the set of *non-dominated solutions*.

10.1.2 Non-dominated solutions

To understand the concept of non-dominated solutions, it is necessary to look closely at the multi-objective problem (10.1) and (10.2). (Note that non-dominated solutions are sometimes referred to by other names: non-inferior, Pareto optimal, efficient, etc. Throughout this text different names are used with the same meaning.) The essential feature of the multi-objective problem is that the feasible region of production of the solutions is much more complex than for a single objective. In single optimization, any set of inputs, x, produces a set of results, z, that can be represented by a straight line going from worst (typically 0 output) to best. In a multi-objective problem, any set of inputs, x, defines a multidimensional space of feasible solutions, as Figure 10.2 indicates. Then there is no exact equivalent of a single optimal solution.

The non-dominated solutions are the conceptual equivalents, in multi-objective problems, of a single optimal solution in a single-objective problem. The main characteristic of the non-dominated set of solutions is that for *each solution outside the set, there is a non-dominated solution for which all objective functions are unchanged or improved and at least one is strictly improved.*

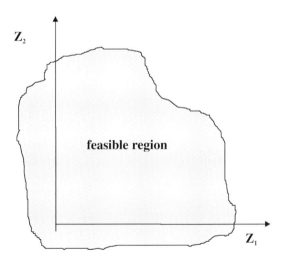

Figure 10.2 *Feasible region of a multi-objective problem presented in the objective space*

The preferred design for any problem should be one of the non-dominated solutions. So long as all objectives worth taking into account have been considered, no design that is not among the non-dominated solutions is worthwhile: it is dominated by some designs that are preferable on all accounts. This is the reason why multi-objective analysis focuses on the determination of non-dominated solutions.

It is often useful to group the non-inferior solutions into major categories. The purpose of this exercise is to facilitate discussions about which solution to select. Indeed, to the extent that it is not possible to specify acceptable relative values for the objectives, and thus impossible to define the best design analytically, it is necessary for the choice of the design to rest on judgement. As individuals find it difficult to consider a large number of possibilities, it is helpful to focus attention on major categories.

If we introduce levels of acceptability for each of the objectives, the non-dominated solutions are best divided into two categories: the major alternatives and the compromises. A *major alternative* group of dominated solutions represents the best performance on some major objective. As Figure 10.3 indicates, the major alternatives represent polar extremes. A *compromise group* lies somewhere in between the major alternatives.

The remainder of the feasible region of solutions is likewise usefully categorized into dominated and excluded solutions. *Dominated* solutions are those that are inferior in all essential aspects to the other solutions. They can thus be set aside from further consideration. *Excluded* solutions are those that perform so badly on one or more objectives that they lie beneath the threshold of acceptability. Thus, they may be dropped from further consideration.

The concepts of non-dominated solutions and of major categories are often highly useful in a practical sense. They organize the feasible designs into a small number of manageable ideas, and draw attention to the choices that must be made. These ideas can be applied even when the feasible region is not defined analytically.

Given a set of feasible solutions X, the set of non-dominated solutions is denoted as S and defined as follows:

$$S = \{x: x \in X, \text{ there exists no other } x' \in X$$
$$\text{such that } z_q(x') > z_q(x) \text{ for some } q \in \{1, 2, ..., r\}$$
$$\text{and } z_k(x') \leq z_k(x) \text{ for all } k \neq q\} \tag{10.3}$$

It is obvious from the definition of S that as we move from one non-dominated solution to another non-dominated solution and one objective function improves, then one or more of the other objective functions must decrease in value.

10.1.3 Participation of decision-makers

From Figure 10.3 it is possible to see that the set of non-dominated solutions is a subset of the initial set of feasible solutions, and that to determine this set the pref-

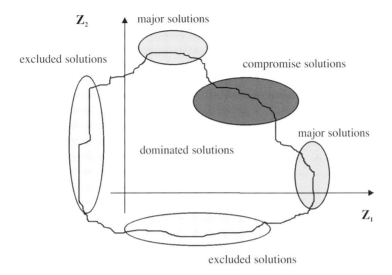

Figure 10.3 *Classification of feasible multi-objective alternative solutions*

erences of the decision-maker are now required. Such partial orderings are characteristic of, but not restricted to, multi-objective planning problems, and imply the need to introduce value judgements into the solution process. At this point in the analysis the decision-maker can be asked to articulate his or her value structure to order the alternative solutions in the non-dominated set (Keeney and Raiffa, 1993). How the value structure of the decision-maker is to be brought into the analysis is not readily apparent. In the theory of the displaced ideal, Zeleny (1974) points out:

> If one obtains an accurate measurement of the net attractiveness (or utility) of each available alternative, one can predict with reasonable accuracy that a person will choose the alternative which is 'most attractive'. So, the problem of prediction of choice becomes the technical problem of measurement and mechanical search. Furthermore, if the alternatives are complex and multi-attributed, then the measurement of utility could be too difficult to be practical. The real question concerns the process by which the decision-maker structures the problem, creates and evaluates the alternatives, identifies relevant criteria, adjusts their priorities and processes information… It is important to realize that whenever we face a single attribute, an objective function, a utility function, or any other single aggregate measure, there is no decision making involved. The decision is implicit in the measurement and it is made by the search… It is only when facing multiple attributes, objectives, criteria, functions, etc., that we can talk about decision making and its theory. (p482)

To define decision-making is not simple. It is a process rather than an act. Although it involves choice from the set of feasible alternatives, it is also concerned with the generation of alternatives. Decision-making is a dynamic process with all its components changing and evolving during its course: alternative solutions are added and removed, the criteria or objectives for their evaluation as well as the relative importance of the criteria are in a dynamic flux, the interpretation of outcomes varies, human values and preferences are reassessed.

Conflict

An element of conflict is inherent in multi-objective problems because resources are limited. Conflict provides the decision-motivating tension, a period of frustration, and dissatisfaction with the status quo of a current situation. Conflict is a property of a situation in which the simultaneous attainment of all the objective functions, at desired levels, is not possible.

Conflict can be resolved in two ways: innovation and adaptation. *Innovation* refers to the development of previously unknown alternatives so that the original goals can be attained. Information plays an important role here, because it can suggest new avenues to search. *Adaptation* refers to changes in the current value structure of the individual so that he or she becomes content with one of the available alternatives. In reality, conflict is often resolved by both methods simultaneously. The decision-maker often chooses to attain that which he or she can attain while still striving to broaden the range of the attainable. This problem of conflict resolution is also dealt with by Zeleny (1974), as he recognizes a predecision situation where the component values of the ideal alternative become clearly perceived, and an effort for conflict resolution is replaced by an attempt at conflict reduction. Also, a post-decision situation may exist where the attractiveness of chosen alternatives is enhanced, while that of the rejected alternatives is reduced.

Goals

A goal can be defined as the 'objective, condition or activity towards which the motive is directed; in short, that which will satisfy or reduce the striving'. A decision-making goal should be something desired, and it should be operational. In the single-objective optimization problem, the search is directed towards an optimal policy vector which possesses well-defined mathematical properties. However, in multi-objective analysis we are dealing with a collection of objective functions and an undefined preference function which, somehow, must be articulated to arrive at a 'solution'. If the concept of an 'optimal solution' is no longer applicable, the concept of a 'satisfactory solution', termed a *satisfactum*, must be examined. Simon (1976) states:

Most human-decision making, whether individual or organizational, is concerned with the discovery and selection of satisfactory alternatives; only in exceptional cases is it concerned with the discovery and selection of optimal alternatives (p272).

Acceptability is a value judgement derived from the individual's preference function. A satisfactum, then, is any value within an interval of acceptability on the range of an objective function. A multiple goal satisfactum implies acceptable values of all objective functions. Since it is assumed that more of each objective function is desirable, a satisfactum is always a member of the set of non-dominated solutions.

10.1.4 Classification of multi-objective techniques

Multi-objective techniques are classified into three groups:

- methods for generating the non-dominated set;
- methods with prior articulation of preferences;
- methods with progressive articulation of preferences (Goicoechea et al, 1982).

(a) Methods for generating the non-dominated set

A generating method does just one thing: it considers a vector of objective functions and uses this vector to identify and generate the subset of non-dominated solutions in the initial feasible region. In doing so, these methods deal strictly with the physical realities of the problem (i.e. the set of constraints), and make no attempt to consider the preferences of a decision-maker. The desired outcome, then, is the identification of the set of non-dominated solutions to help the decision-maker gain insight into the physical realities of the problem at hand.

There are several methods available to generate the set of non-dominated solutions, and four of these methods are widely known. These methods are:

- weighting method;
- ε-constraint method;
- Phillip's linear multi-objective method;
- Zeleny's linear multi-objective method.

The first two methods transform the multi-objective problem into a single-objective programming format and then, by parametric variation of the parameters used to effect the transformation, the set of non-dominated solutions can be generated. The weighting and constraint methods can be used to obtain non-dominated solutions when the objective functions and/or constraints are nonlinear.

The last two methods generate the non-dominated set for linear models only. However, these two approaches do not require the transformation of the problem into a single-objective programming format. These methods operate directly on the vector of objectives to obtain the non-dominated solutions.

(b) Methods with prior articulation of preferences

Methods in this class are further divided into continuous and discrete methods.

Continuous

Once the set of non-dominated solutions for a multi-objective problem has been identified using any of the methods mentioned in (a), the decision-maker is able to select one of those non-dominated solutions as her or his final choice. This solution will be one that meets the physical constraints of the problem and happens to satisfy the value structure of the decision-maker as well. However, a more likely situation is that the decision-maker is unwilling or unable to accept one of those solutions made available. In that case a good alternative is to solicit the decision-maker's preferences regarding the various objective functions in the search for a solution. Various methods are available, where decision-makers are asked to articulate their worth or preference structure, and these preferences are then built into the formulation of the mathematical model for the multi-objective problem.

To assist decision-makers in articulating their preferences, a series of questions may be put to them, where they are asked to consider specific trade-offs among several objectives and so indicate a preference for a particular allocation for each objective. In the process, use is made of basic elements of utility theory (Keeney and Raiffa, 1993) and probability.

The net effect of articulating the preferences of the decision-maker prior to solving the multi-objective problem is to reduce the set of non-dominated solutions to a much smaller set of solutions, facilitating the task of selecting a final choice. Depending on the method used, this smaller set may contain several solutions, one solution or none at all. Preferences, then, provide an ordering of solutions stronger than that provided by the concept of non-dominated solutions.

The best-known methods in the group of continuous methods with prior articulation of preferences are:

- goal programming;
- utility function assessment;
- the surrogate worth trade-off (SWT) method.

Discrete

There are many decision situations in which the decision-maker must choose among a finite number of alternatives which are evaluated on a common set of non-commensurable multiple objectives or criteria. Problems of this sort occur in many practical situations: for example, which of five candidate pipe sizes should be selected, which of ten water supply options should be selected, or which of eight gate operating systems should be chosen and implemented?

In problems of this type, the solution process can be described as follows. First, a statement of the general goals relating to the situation is made. Second, the alternatives must be identified or developed. Third, the common set of relevant criteria for evaluation purposes must be specified. Fourth, the levels of the criteria for each alternative must be determined. Finally a choice is made based on a formal or informal evaluation procedure.

The structure of the discrete problem can be represented in a payoff matrix as shown in Table 10.1. The rating of the ith objective/criteria on the jth alternative solution ($i = 1, 2, ..., m$ and $j = 1, 2, ..., n$) is represented by v_{ij}.

Table 10.1 *Payoff matrix*

Alternatives		1	2	...	n
	1	v_{11}	v_{12}	...	v_{1n}
	2	v_{21}	v_{22}	...	v_{2n}
Criteria	...				
	...				
	m	v_{m1}	v_{m2}	...	v_{mn}

Clearly the choice in a problem such as that represented in Table 10.1 is sufficiently complex to require some type of formal assistance. Determining the worth of alternative solutions that vary on many dimensions presents formidable cognitive difficulties. People faced with such complex decisions react by reducing the task complexity by using various heuristics. Unfortunately, it has been observed that decision-makers who rely on heuristic decision rules systematically violate the expected utility principle. Moreover, decision-makers tend to ignore many relevant variables in order to simplify their problem to a scale consistent with the limitations of the human intellect. While such simplification facilitates the actual decision-making, it can clearly result in a suboptimal decision.

In effect, as the decision-making task increases, researchers have observed systematic discrepancies between rational theory and actual behaviour. Evidence exists that even experts have great difficulty in intuitively combining information in appropriate ways. In fact, these studies and many others indicate that global judgements (i.e. combinations of attributes) are not nearly as accurate as analytical combinations. Because of the severe limitations of the intuitive decision-making process, it is evident that analytical methods are needed to help determine the worth of multi-attributed alternatives.

Ideally, the alternative that maximizes the utility of the decision-maker should be chosen. Therefore, the obvious first step in the application of any discrete multi-attributed method is the elimination of all dominated alternatives. Occasionally, for discrete problems, this dominance analysis will yield only one non-dominated alternative, in which case the problem is solved; no further analysis is needed.

The methods available in this group range from the very simple to the very complex. Some of the methods are:

- exclusionary screening;
- conjunctive ranking;
- weighted average;
- ELECTRE I and II;
- indifference trade-off method;
- direct-rating method.

(c) Methods of progressive articulation of preferences

The characteristic of the methods in this group is the following general algorithmic approach. First, a non-dominated solution is identified. Second, the decision-maker is solicited for trade-off information concerning this solution, and the problem is modified accordingly. These two steps are repeated until the decision-maker indicates the acceptability of a current achievement level, provided one exists.

The methods typically require greater decision-maker involvement in the solution process. This has the advantage of allowing the decision-maker to gain a greater understanding and feel for the structure of the problem. On the other hand, the required interaction has the disadvantage of being time-consuming. The decision-maker may not feel that the investment of the time required provides any better decision-making than ad hoc approaches. That is, the decision-maker may perceive the costs to be greater than any benefits. Some literature points out that decision-makers have less confidence in the interactive algorithms and find them more difficult to use and understand than trial and error methods. Certainly, we cannot ignore these behavioural difficulties. What they seem to indicate is that more research is needed on how analysts can successfully interact with decision-makers to implement improved but complex decision aids.

At any rate, knowledge of some of the advantages and disadvantages should help the managers or analysts to choose the appropriate decision aid. The use of any of the methods as aids to decision-making will depend on the analyst's assessment of the personality and tastes of the decision-maker. The methods of progressive articulation of preferences vary in the degree of sophistication and the degree of required interaction. Some of them are:

- step method (stem);
- Geoffrion's method;
- Compromise programming;
- SEMOPS method.

10.1.5 Water resources management applications

Multi-objective decision-making has a strong presence in everyday water resources management practice. These are some illustrative example applications.

River basin planning

Alternative strategies for the Santa Cruz River basin, as proposed by the Corps of Engineers, were examined with the methods ELECTRE I and II (Gershon et al, 1980). Combinations of flood control actions (e.g. levee construction, channelization, dams, reservoirs and floodplain management) and water supply actions (e.g. wastewater reclamation, new groundwater development, the Central Arizona project and conservation measures) were combined to represent 25 alternative systems.

Conjunctive water uses

The simultaneous utilization of surface and groundwater sources is often a desirable management alternative, particularly in urban water supply. In a case study of Western Skåne, Sweden, consideration was given to using local groundwater and two pipeline systems to supply five municipalities. A two-level hierarchy was structured to aid in the decision process and the STEM method was used to obtain trade-offs among five objectives pertaining to lake water levels, downstream releases and operating costs.

Reservoir operation

In developing a plan of operation for a reservoir, primary consideration may be given to reducing the damaging peak flood stages at principal downstream flood centres. Other objectives may involve recreation and water quality enhancement. A multi-objective dynamic programming (MODP) procedure was applied to the operation of the Shasta Reservoir in California, United States (Tauxe et al, 1979). The three objectives considered were:

1. maximization of cumulative dump energy generated above the level of firm energy;
2. minimization of the cumulative evaporation or loss of the resource;
3. maximization of the firm energy.

Multi-objective Compromise programming (Simonovic and Burn, 1989) was used for the short-term operation of a single multi-purpose reservoir. An improved methodology for short-term reservoir operation was derived which considers the operating horizon as a decision variable which can change in real time. The optimal value of the operating horizon is selected based on the trade-off between a more reliable inflow forecast for shorter horizons and better reservoir operations associated with the use of longer operating horizons.

The methodology, based on the combined use of simulation and multi-objective analysis, was developed by Simonovic (1991). It has been used to modify the existing operating rules of the Shellmouth Reservoir in Manitoba, Canada. Flood control and the water supply objectives are in conflict with the use of reservoir water for dilution of the heated effluent from a thermal generating plant and the improvement of water quality in the river.

Floodplain management

Both structural and non-structural measures can reduce flood damages. Structural means include levees and dams to physically prevent floods from reaching an area. Some of the non-structural means are:

- restrictions on the use of flood-prone areas and reduction of the runoff produced by storms;
- the purchase of floodplain land that is already developed and its conversion to flood-compatible uses to protect life and property.

Novoa and Halff (1977) evaluated eight alternative flooding remedies, ranging from mono-action to stream channelization to complete development of portions of the city of Dallas, Texas, United States. The method of weighted averages was used to evaluate and rank the eight alternative plans. The evaluation criteria reflected:

- relative flood protection;
- relative neighbourhood improvement;

- number of relocated families;
- project cost;
- maintenance cost;
- legal considerations.

Water quality management

The public's increasing concern with water pollution demands that attention be directed to the analysis of water quality determinants, both physical and chemical. These determinants can include dissolved oxygen, coliform count, temperature and concentrations of various pollutants. There are many examples of multi-objective analysis in water quality management.

One regional planning framework that integrates land and water quality management models was developed by Das and Haimes (1979). The SWT method is applied here to investigate trade-offs among four planning objectives:

1. sheet erosion control;
2. phosphorus loading;
3. biological oxygen demand;
4. non-point-source pollution cost control.

Water and related land resources development

The development of water resources frequently affects or relates to the development of land resources, and vice versa. Activities such as coal strip-mining, land reclamation, watershed management and land use planning relate closely to the hydrology of the area and the availability of water resources (e.g. surface and groundwater).

Goicoechea et al (1982) applied multi-objective programming to the problem of reclaiming lands disturbed by coal strip-mining activities in the Black Mesa region of northern Arizona, United States. This study suggested a reclamation programme to enable the land to sustain agricultural, livestock grazing, fish pond harvesting and recreational uses. The PROTRADE method was used to examine levels of attainment and the trade-off for five objectives:

1. livestock production;
2. water runoff augmentation;
3. farming of selected crops;
4. sediment control;
5. fish yield.

This method allows the use of random variables to represent parameters in the objective functions. To generate objective trade-offs the decision-maker is able to consider both the level of attainment and the probability of attainment for each objective function.

10.2 THE WEIGHTING METHOD

The weighting method belongs to the group of techniques for generating a non-dominated set. It is based on the idea of assigning weights to the various objective functions, combining these into a single-objective function, and parametrically varying the weights to generate the non-dominated set. We shall use a presentation of the weighting method to further illustrate the concept of multi-objective analysis and provide a straightforward tool for its implementation.

Mathematically, the weighting method can be stated as follows:

$$\max z(x) = w_1 z_1(x) + w_2 z_2(x) + \ldots + w_r z_r(x) \tag{10.4}$$

subject to:

$$x \in X$$

which can be thought of as an operational form of the formulation:

$$\text{max-dominate } z(x) = [z_1(x), z_2(x) \ldots, z_r(x)]$$

subject to:

$$x \in X$$

In other words, a multi-objective problem has been transformed through (10.4) into a single-objective optimization problem for which solution methods exist. The coefficient w_i operating on the ith objective function, $z_i(x)$, is called a weight and can be interpreted as *the relative weight or worth* of that objective compared with the other objectives. If the weights of the various objectives are interpreted as representing the relative preferences of some decision-maker, then the solution to (10.4) is equivalent to the best compromise solution: that is, the optimal solution relative to a particular preference structure. Additionally, the optimal solution to (10.4) is a non-dominated solution, provided all the weights are positive. The reasoning behind the non-negativity requirement is as follows. Allowing negative weights would be equivalent to transforming the maximization problem into a minimization one, for which a different set of non-dominated solutions will exist. The trivial case where all the weights are 0 will simply identify every $x \in X$ as an optimal solution, and will not distinguish between the dominated and the non-dominated solutions.

Conceptually, the generation of the non-dominated set using the method of weights appears simple. However, in practice the generation procedure is quite demanding. Several weight sets can generate the same non-dominated point (Mishra, 2007). Furthermore, moving from one set of weights to another set of weights may result in skipping a non-dominated extreme point. Subsequent linear combinations of the observed adjacent extreme points would, in many cases, yield a set of points that are only *close* to the non-dominated border. In other words, in practice it is quite possible to miss the non-dominated solution using weights that would lead to an extreme point. Therefore, the most that should be expected from the weighting method is an approximation of the non-dominated set.

The sufficiency of the approximation obviously relates to the proportion of the total number of extreme points that are identified. For example, assume each weight is varied systematically between 0 and some upper limit using a predetermined step size. It seems reasonable to believe that the choice of a large increment will result in more skipped extreme points than the choice of a small increment. However, the smaller the increment, the greater the computational requirements. There is a trade-off between the accuracy of the specification of the non-dominated set and the costs of the computation. Judgement must be exercised by the decision-maker and the analyst to determine the desired balance.

Example 1

A state water agency is responsible for the operation of a multi-purpose reservoir used for:

- municipal water supply;
- groundwater recharge;
- the control of water quality in the river downstream from the dam.

Allocating the water to the first two purposes is, unfortunately, in conflict with the third purpose. The agency would like to minimize the negative effect on the water quality in the river, and at the same time maximize the benefits from the municipal water supply and groundwater recharge.

Thus, there are two objectives: minimize the increase in river pollution and maximize profits. Trade-offs between these two objectives are sought to assist the water agency in the decision-making process. The available data are listed in Table 10.2.

The following assumptions are made:

- Analysis is done for one time period $t = 0, 1$.
- The limiting pump capacity is eight hours per period.
- The limiting labour capacity is four person-hours per period.
- The total amount of water in the reservoir available for allocation is 72 units.
- The pollution in the river increases by 3 units per 1 unit of water used for water supply and 2 units per 1 unit of water used for groundwater recharge.

Table 10.2 *Available data for an illustrative example*

	Water supply	Groundwater recharge
Number of units of water delivered	x_1	x_2
Number of units of water required	1.00	5.00
Pump time required (hr)	0.50	0.25
Labour time required (person-hour)	0.20	0.20
Direct water costs ($)	0.25	0.75
Direct labour costs ($)	2.75	1.25
Sales price of water per unit ($)	4.00	5.00

Based on the preceding information we can formulate the objective functions and constraints of the problem.

Objective functions
The contribution margin (selling price/unit less variable cost/unit) of each allocation is calculated as:

Municipal water supply
$4.00 − $0.25 − $2.75 = $1.00 per unit of water delivered
Sales price — Direct water cost — Direct labour

Groundwater recharge
$5.00 − $0.75 − $1.25 = $3.00 per unit of water delivered
Sales price — Direct water cost — Direct labour

and the objective function for profit becomes:

$$max\ z_1(x) = x_1 + 3x_2$$

The objective function for pollution is:

$$min\ z'_2(x) = 3x_1 + 2x_2$$

This function can be modified to $max\ z_2 = -3x_1 - 2x_2$ so that the maximization criterion applies to both of the objective functions.

Finally, the technical constraints due to pump capacity, labour capacity and water availability are:

$0.5x_1 + 0.25x_2 \le 8$ (pump capacity)
$0.2x_1 + 0.2x_2 \le 4$ (labour capacity)
$x_1 + 5x_2 \le 72$ (water)

Now, using the operational form of the weighting method, the problem to solve is:

$$max\ z(x) = w_1 z_1(x) + w_2 z_2(x)$$
$$= w_1(x_1 + 3x_2) + w_2(-3x_1 - 2x_2) \quad (10.5)$$

subject to:

$$g_1(x) = 0.5x_1 + 0.25x_2 - 8 \le 0 \quad (10.6)$$

$$g_2(x) = 0.2x_1 + 0.2x_2 - 4 \le 0 \quad (10.7)$$

$$g_3(x) = x_1 + 5x_2 - 72 \le 0 \quad (10.8)$$

$$g_4(x) = -x_1 \le 0 \quad (10.9)$$

$$g_5(x) = -x_2 \le 0 \quad (10.10)$$

Let us arbitrarily fix $w_1 = 1$ and increase w_2 at increments of 1 until all the non-dominated extreme points have been identified. For this example, the pairs of values selected for (w_1, w_2) are (1, 0), (1, 1), (1, 2), (1, 3), (0, 1), as shown in Table 10.3. For example, for the pair of weights (1, 0), the objective function to maximize is:

$$\text{Max } z(x) = 1(x_1 + 3x_2) + 0(-3x_1 - 2x_2)$$
$$= x_1 + 3x_2$$

subject to the stated constraints. The solution can be obtained graphically by moving the line $z(x) = x_1 + 3x_2$ out towards the boundary of the feasible region until it just touches the extreme point $x^* = (7, 13)$, yielding $z(x^*) = 46$, $z_1(x^*) = 46$ and $z_2(x^*) = -47$ (Figure 10.4). Since all the objective function and constraints are in the linear form we can use the Linpro computer program on the CD-ROM to confirm our solution. This solution, however, is not unique to the pair of weights $(1, 0)$ as can be observed from Table 10.3.

After a graphical presentation of the solutions in the objective space (Figure 10.5), it is possible to identify non-dominated points by visual inspection. Since all of the non-dominated extreme points are obviously identified, an exact representation of the non-dominated set is achieved. However, for problems with a larger number of variables and constraints, we would probably have to settle for an approximate representation of the non-dominated set.

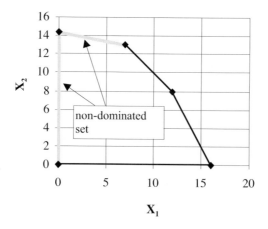

Figure 10.4 *The feasible region and the non-dominated set in the decision space*

Once the non-dominated set is specified, the decision-maker can use the information to select a preferred solution. The trade-offs are now readily apparent. For example, a decision to move from the solution $(7, 13)$ to the solution $(0, 25)$ results in a decrease of 2.8 units of profit with the resulting benefit of a reduction of 18.2 units of pollution. This might be perceived as too great a sacrifice, and feasible production vectors in between these two extremes could be examined, with the corresponding trade-offs.

Table 10.3 *Pairs of weights and associated non-dominated solutions*

Weights (w_1, w_2)	Non-dominated extreme point $x^* = (x_1, x_2)$	$z_1(x^*)$	$z_2(x^*)$	$z(x)$
(1, 0)	(7, 13)	46	−47	46
(1, 1)	$(0, \frac{72}{5})$	$\frac{216}{5}$	$-\frac{144}{5}$	$\frac{72}{5}$
(1, 2)	(0, 0)	0.0	0.0	0.0
(1, 3)	(0, 0)	0.0	0.0	0.0
(0, 1)	(0, 0)	0.0	0.0	0.0

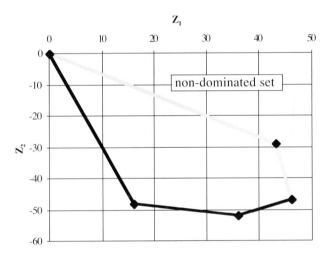

Figure 10.5 *The feasible region and the non-dominated set in the objective space*

10.3 THE COMPROMISE PROGRAMMING METHOD

Compromise programming was originally developed as an interactive method with progressive articulation of decision-maker preferences. It is appropriately used in a multiple linear objective context (Zeleny, 1974). However, many variations of this method have also been used in the analysis of discrete objective problems with prior or progressive articulation of preferences. Compromise programming identifies solutions that are closest to the ideal solution, as determined by some measure of distance. Due to its simplicity, transparency and easy adaptation to both continuous and discrete settings, Compromise programming is recommended as the multi-objective analysis method of choice for application to water resources systems management. This section presents the method in deterministic form, and subsequent sections introduce its extension to multi-objective analysis under uncertainty with single and multiple decision-makers.

10.3.1 Compromise programming

Let us consider a two-objective problem, illustrated in Figure 10.6. The solution for which both objectives (z_1, z_2) are maximized is point I (z_1^*, z_2^*) where the z_i^* is the solution obtained by maximizing the objective i. It is clear that the solution I (named ideal point) belongs to the set of infeasible solutions. Let us consider a discrete case with four solutions available as a non-dominated set: A, B, C and D. The solutions identified as being closest to the ideal point (according to some measure of distance) are called *compromise solutions*, and constitute the *compromise set*. If we use a geometric distance, the set of compromise solutions may include a subset of the non-dominated set A and B.

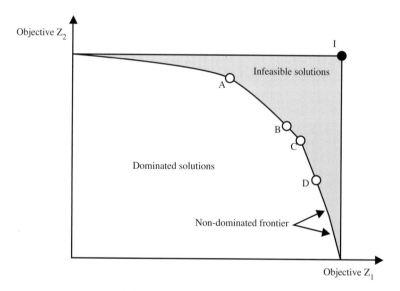

Figure 10.6 *Illustration of compromise solutions*

For a more explicit understanding of what is meant by a compromise solution we must define what is meant by an ideal solution, and specify the particular distance measure to be used.

The ideal solution is defined as the vector $z^* = (z_1^*, z_2^*, \ldots, z_r^*)$ where the z_1^*, known as positive ideals, are the solutions of the following problems:

$$\max z_i(\mathbf{x}) \tag{10.11}$$

subject to:

$$x \in X \quad i = 1, 2, \ldots, r$$

If there is a feasible solution vector, x^*, common to all r problems, then this solution would be the optimal one since the non-dominated set (in objective space) would consist of only one point, namely $z^*(x^*) = (z_1^*(\mathbf{x}^*), z_2^*(\mathbf{x}^*), \ldots, z_r^*(\mathbf{x}^*))$. Obviously this is most unlikely, and the ideal solution is generally not feasible. However, it can serve as a standard for evaluation of the possible non-dominated solutions. Since all would prefer the ideal point if it were attainable (as long as the individual underlying utility functions are increasing), then it can be argued that finding solutions that are as close as possible to the ideal solution is a reasonable surrogate for utility function maximization.

The procedure for evaluation of the set of non-dominated solutions is to measure how close these points come to the ideal solution. One of the most frequently used measures of closeness, and the one we shall use, is a family of L_p, metrics, defined in either of two operationally equivalent ways:

$$L_p = \left[\sum_{i=1}^{r} \alpha_i^p (z_i^* - z_i^*(\mathbf{x}))^p \right]^{1/p} \tag{10.12}$$

or

$$L_p = \sum_{i=1}^{r} \alpha_i^p (z_i^* - z_i^*(\mathbf{x}))^p \tag{10.13}$$

where:

$$1 \leq p \leq \infty$$

Finally, a compromise solution with respect to p is defined as such that:

$$\min L_p(\mathbf{x}) = L_p(\mathbf{x}_p^*) \tag{10.14}$$

subject to:

$$x \in X$$

The compromise set is simply the set of all compromise solutions determined by solving (10.14) for a given set of weights, $\{\alpha_1, \alpha_2, \ldots, \alpha_r\}$ and for all $1 \leq p \leq \infty$.

Operationally, three points of the compromise set are usually calculated, that is, those corresponding to $p = 1, 2$ and ∞. Varying the parameter p from 1 to infinity allows us to move from minimizing the sum of individual regrets (i.e. having a perfect compensation among the objectives) to minimizing the maximum regret (i.e. having no compensation among the objectives) in the decision-making process. The choice of a particular value of this compensation parameter p depends on the type of problem and desired solution. In general, the greater the conflict between objectives, the smaller the possible compensation becomes.

To understand the motivation for this statement and appreciate the role of the α_i and p parameters, consider the following special cases.

Let $\alpha_1 = \alpha_2 = \ldots \alpha_r = 1$ and let $w_i = z_i^* - z_i(x)$. With this, (10.13) becomes:

$$L_p = \sum_{i=1}^{r} w_i^{p-1} (z_i^* - z_i^*(\mathbf{x})) \qquad (10.15)$$

For $p = 1$, $w_i^{p-1} = 1$ we obtain:

$$L_p = L_1 = \sum_{i=1}^{r} (z_i^* - z_i^*(\mathbf{x}))$$

Thus all deviations from the ideal point are weighted equally. This distance is called the *Hamming distance*.

For $p = 2$, (10.13) assumes the form:

$$L_p = L_2 = \sum_{i=1}^{r} w_i (z_i^* - z_i^*(\mathbf{x}))$$

Now, each deviation is weighted in proportion to its magnitude. This distance is also known as *Euclidean distance*. The larger the deviation, the larger the weight. As p becomes larger and larger, the largest deviation receives more and more weight, until finally at $p = \infty$ (Chebychev distance) we observe that:

$$L_\infty = \max_{\text{all } i} (z_i^* - z_i^*(\mathbf{x}))$$

Clearly, the choice of p reflects the decision-maker's concern with respect to the maximal deviation from the ideal solution. The larger the value of p, the greater the concern.

Introduction of α_i allows the expression of the decision-maker's feelings concerning the relative importance of the various objectives. Thus in Compromise programming a double-weighting scheme exists. The parameter p reflects the importance of the maximal deviation and the parameter α_i reflects the relative importance of the ith objective. Consider the following version of (10.13):

$$L_p = \sum \alpha_i^p w_i^{p-1} (z_i^* - z_i^*(\mathbf{x}))$$

The deviation $z_i^* - z_i^*(\mathbf{x})$ is weighted proportionately by the choice of p and then weighted by the pth power of the objective weights. Again, as p increases, the maximal α_i and the maximal deviation receive more and more emphasis until:

$$L_\infty = \max_{\text{all } i} \alpha_i \, (z_i^* - z_i^*(\mathbf{x}))$$

If the objective functions are not expressed in commensurable terms, then a scaling function $S_i(D_i)$, with $D_i = z_i^* - z_i(\mathbf{x})$, is defined to ensure the same range for each objective function. Usually, this range corresponds to the interval (0, 1). This scaling is accomplished by defining the scaling function as:

$$S_i(D_i) = \frac{z_i^* - z_i(\mathbf{x})}{z_i^* - z_i^{**}} \tag{10.16}$$

where z_i^{**} is defined as:

$$z_i^{**} = \min_{x \in X} z_i, \quad i = 1,2,\ldots,r$$

With the indicated transformation, (10.14) is modified by substituting (10.16) for $D_i = z_i^* - z_i(\mathbf{x})$, that is:

$$L_p(\mathbf{x}_p^*) = \min\left[L_p(x) = \sum_{i=1}^r \alpha_i^p \left(\frac{z_i^* - z_i(\mathbf{x})}{z_i^* - z_i^{**}} \right)^p \right] \tag{10.17}$$

subject to:

$$x \in X$$

If the transformation is applied to (10.12) after substitution of (10.16) we arrive at:

$$L_p(\mathbf{x}_p^*) = \min\left\{ L_p(x) = \sum_{i=1}^r \alpha_i^p \left(\frac{z_i^* - z_i(\mathbf{x})}{z_i^* - z_i^{**}} \right)^p \right\}^{1/p} \tag{10.18}$$

subject to:

$$x \in X$$

Expressions (10.17) and (10.18) are equivalent operational definitions of a compromise solution for a given p. Interestingly, solution of either (10.17) or (10.18) always produces a non-dominated point for $1 \leq p \leq \infty$. For $p = \infty$, there is at least one non-dominated solution, x_p^*.

The implementation of Compromise programming results in a reduction of the non-dominated set. If the compromise set is small enough to allow the decision-maker to choose a satisfactory solution, then the algorithm stops. If not, the decision-maker is asked to redefine the ideal point and the process is repeated. Accordingly, the interaction requirement of Compromise programming is not very demanding.

Previous mathematical derivations are not applicable only to continuous settings: Compromise programming can be adapted to discrete settings as well. In a discrete setting the ideal solution is defined as the best value in a finite set of values of $z_i(x)$ (see Figure 10.5). Essentially, the ideal solution in a discrete setting would be defined as the vector of best values selected from a payoff table like the one shown in Table 10.1. The vector of worst values, known as the *negative ideal*, defines the minimum objective function values, that is, the z_i^{**}. With these values defined and α_i and p given, the compromise solution can be determined by calculating the distance of each alternative from the ideal solution, and selecting the alternative with the minimum distance as the compromise solution.

In most cases water resources management multi-objective problems are of a discrete nature, such as:

- Select the appropriate height for a dike from a finite set of choices.
- Determine the size of a pipe from the set of prefabricated diameters.
- Find the number of spillway gates to be open.

Therefore, our further discussion will be limited to discrete settings.

Example 2

We shall modify the reservoir allocation problem described in Example 1. Assume that the objectives from Example 1 are replaced with two new objectives:

$$z_1(x) = x_1 + 7x_2 \tag{10.19}$$

and

$$z_2(x) = 10x_1 + 4x_2 \tag{10.20}$$

representing the maximization of benefits from reservoir water allocation (z_1) and maximization of positive environmental impacts (z_2), respectively. Figure 10.7 shows the non-dominated alternative solutions (A1 to A4) and the location of the ideal point. Considering the discrete setting, our problem is now to identify compromise solutions from the set of four non-dominated solutions presented in Table 10.4 (A1–A4).

As one would expect, the ideal point I (shown as a star in Figure 10.7) corresponds to the maximum value of x_1 (point A4) and x_2 (point A1). To arrive at an approximation of the compromise set, we shall solve problem (10.18) for $p = 1, p = 2$ and $p = \infty$, and the α_i weights must be specified.

Table 10.4 *Non-dominated solutions*

Alternative	Non-dominated extreme point (x_1, x_2)	Objectives	
		z_1	z_2
A1	(0, 14.4)	100.8	57.6
A2	(7, 13)	98.0	122.0
A3	(12, 8)	68.0	152.0
A4	(16, 0)	16.0	160.0

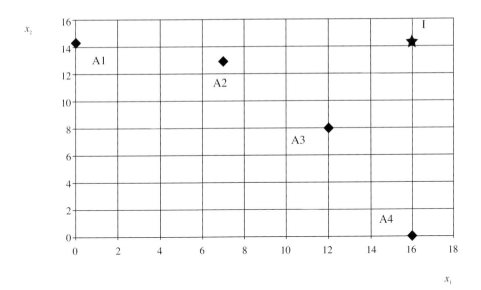

Figure 10.7 *The decision space of the reservoir allocation problem*

Assuming the decision-maker views the objectives as equally important, we shall solve:

$$\min \left\{ L_p(x) = \left[\alpha_1^p \left(\frac{z_1^* - z_1(x)}{z_1^* - z_1^{**}} \right)^p + \alpha_2^p \left(\frac{z_2^* - z_2(x)}{z_2^* - z_2^{**}} \right)^p \right]^{1/p} \right\} \qquad (10.21)$$

subject to:

$$x \in X$$

The compromise solution for $p = \infty$ may or may not be a non-dominated extreme point. To determine the compromise solution in practice, ∞ is replaced by a large number (e.g. 100). The scaled non-dominated solutions are presented in Table 10.5. They are obtained by calculating the scaling function (10.16) for all four non-dominated points.

Table 10.5 *Scaled non-dominated solutions*

Alternative	Non-dominated extreme points (x_1, x_2)	$\dfrac{z_1^* - z_1(x)}{z_1^* - z_1^{**}}$	$\dfrac{z_1^* - z_1(x)}{z_2^* - z_2^{**}}$
A1	(0, 14.4)	0.00	1.00
A2	(7, 13)	0.03	0.37
A3	(12, 8)	0.39	0.08
A4	(16, 0)	1.00	0.00

Using the information from Table 10.5 and solving (10.21) for $\alpha_1 = \alpha_2 = 0.5$ and $p = 1$, $p = 2$ and $p = 100$, the compromise set is identified. The final results are summarized in Table 10.6. The compromise set is a set of solutions closest to the ideal solution for different values of the parameter p. In our case for all three values of $p = 1$, $p = 2$ and $p = 100$, the alternative A2 is ranked first (closest to the ideal point). Therefore, our compromise set of solutions is reduced to alternative A2. In other words this alternative is also the best compromise solution for the reservoir allocation problem.

Table 10.6 *Final compromise solutions*

Alternative	$p = 1$		$p = 2$		$p = 100$	
	$L_p(x)$	Rank	$L_p(x)$	Rank	$L_p(x)$	Rank
A1	0.50	3	0.500	3	0.500	3
A2	0.20	1	0.186	1	0.185	1
A3	0.23	2	0.197	2	0.193	2
A4	0.50	4	0.500	4	0.500	4

You can use the Compro program provided on the CD-ROM (and described later in this chapter) to confirm the solution in Table 10.6. Example 2 data are in the folder COMPRO, sub-folder Examples, file Example1.compro.

Box 10.1 Fishing on the Yangtze River

In 1998 I was on a large cruise ship slowly going through one of the Yangtze gorges when, in front of us, closer to the bank, I noticed a very simple fishing boat with two fishermen throwing large nets in the water and pulling their catch into the boat. The empty nets were going back into the water with almost rhythmical timing.

When our ship came closer I was able to see the fishing boat, fishermen and the river surface very clearly. To my surprise their catch which was in the boat was not fish. Both fishermen were standing, with huge smiles on their faces, knee-deep in used aluminum cans. One look at the river surface explained the rich source of the catch. The surface was covered with glistening aluminum garbage.

The nets continued to fly and the catch was filling the boat. A day's work on the Yangtze River.

(A memory from 1998)

10.3.2 Some practical recommendations

The original purpose of Compromise programming is to reduce the non-dominated set to the compromise set in direct interaction with decision-makers. However, very often water management problems do need to identify one solution that should be a recommended solution. Based on extensive use of the Compromise programming multi-objective method in practice, I suggest that the solution with the smallest Euclidean distance, corresponding to $p = 2$, be used as the first approximation of the 'best compromise solution'. In this case an extensive sensitivity analysis of the final solution selection to the change in parameter p is advised.

Quite often in practice, the preferences of decision-makers are not readily available. In some situations they are not able to articulate them easily; in others, they may not be willing to openly express their values. In order to assist the decision-making process I have developed a concept of *most robust compromise solution* as a replacement for the *best compromise solution* (see the example in Section 10.6.1).

The best compromise solution is one closest to the ideal point for the fixed set of decision-maker preferences and one value of the distance parameter p. My recommendation is that $p = 2$ is used in identification of the best compromise solution. The most robust compromise solution is one that occupies a high rank (not always the highest), the most often for various sets of decision-maker preferences. So in this way we arrive at the solution that is not very sensitive to change in preferences, and therefore has a chance of a higher level of acceptance by the decision-makers. The most robust solution is calculated through systematic sensitivity analysis, or repetitive solutions of (10.17) or (10.18) for different values of α_i and one value of the distance parameter p (again $p = 2$ is suggested).

10.3.3 The Compro computer program

The CD-ROM accompanying this book includes the Compro software and all the examples developed in the text. The folder COMPRO contains three sub-folders, Compro, ComproHelp and Examples. The read.me file contains instructions for installation of the Compro software, and a detailed tutorial for its use is a part of the Help menu.

The Compro software package facilitates multi-objective analysis of discrete problems using Compromise programming techniques. Compromise programming is the method for reducing the set of non-dominated solutions according to their distance from the ideal solution. The distance from the ideal solution for each alternative is measured by a distance metric. An operational definition (10.18) is used for the computation of the distance metric. This value, which is calculated for each alternative, is a function of the criteria values themselves, the relative importance of the various criteria to the decision-makers, and the importance of the maximum deviation from the ideal solution.

The Compro package can be used in two ways: to narrow down the set of non-dominated solutions (find a compromise set), or to identify the *best compromise alternative solution*. Narrowing down the compromise set is achieved by running the software for a set of parameter p values. Identification of the best compromise alternative is achieved by running the program for one preset value of the deviation parameter, $p = 2$.

10.4 FUZZY MULTI-OBJECTIVE ANALYSIS

Up to now we have discussed objective functions, sets of constraints and goal functions. All of these are expressed using fixed quantities for coefficients. This is definitively not a realistic assumption in multi-objective water resources systems management, where many of the coefficients and parameters are subject to various sources of uncertainty, as was discussed in Chapter 6. Some very limited work has been done (and published) on stochastic multi-objective programming (e.g. the probabilistic trade-off development method of Goicoechea et al, 1982; an evolutionary methodology of Jimenez et al, 2006; and a multi-objective evolutionary approach of Medaglia et al, 2007).

We shall continue to focus here on the use of fuzzy sets in extending multi-objective analysis to uncertain conditions. An intuitive, and interactive, decision tool for discrete alternative selection, under various forms of uncertainty, would be valuable for water resources systems management, especially for applications with groups of decision-makers (Karnib, 2004). This section explores the application of fuzzy sets in conjunction with Compromise programming. The adaptation of standard techniques to the fuzzy framework demands a different set of operators. We shall explore how the Compromise programming multi-objective method is transformed into a fuzzy environment (Bender and Simonovic, 2000), and develop and demonstrate the application and use of fuzzy distance metrics as a decision-making tool.

Fuzzy decision-making techniques have addressed some uncertainties, such as the vagueness and conflict of preferences common in group decision-making. Their application, however, demands some level of intuitiveness for the decision-makers, and encourages interaction or experimentation. Fuzzy decision-making is not always intuitive to many people involved in practical decisions because the decision space may be some abstract measure of fuzziness, instead of a tangible measure of alternative performance. The alternatives to be evaluated are rarely fuzzy: it is their performance that is fuzzy. In other words, a fuzzy decision-making environment may not be as generically relevant as a fuzzy evaluation of a decision-making problem.

Most fuzzy multi-objective methods either concentrate on multi-objective linear programming (LP) techniques, or experiment with methods based on fuzzy relations. Carlsson and Fuller (1996) and Ribeiro (1996) provide a review of fuzzy multiple-criteria decision-making.

10.4.1 FUZZY COMPROMISE PROGRAMMING

For Compromise programming to address the vagueness in the decision-maker's value system and criteria value uncertainty, a general fuzzy approach may be appropriate. Simply changing all inputs from crisp to fuzzy produces a definition for fuzzy Compromise programming analogous to the crisp original. The multi-objective problem in Figure 10.6 can no longer be considered a single point for the ideal solution, and each alternative now occupies a small region to various degrees. Measurements of distances between the fuzzy ideal and the fuzzy performance of alternatives can no longer be given a single value, because many distances are at least somewhat valid. Choosing the shortest distance to the ideal is no longer a straightforward ordering of distance metrics, because of overlaps and varying degrees of possibility. The fuzzy multi-objective problem, however, contains a great amount of additional information about the consequences of a decision compared with the non-fuzzy counterpart.

A fuzzy distance metric possesses a valid range of values, each with a characteristic degree of possibility or membership, such that all possible values are a positive distance from the ideal solution (which also becomes fuzzy). Fuzzy inputs include the vagueness of criteria weights, $\tilde{\alpha}_i$, vagueness of both positive, $\tilde{z}_i^*$, and negative ideals, $\tilde{z}_i^{**}$, and vagueness in the appropriate distance metric exponent, $\tilde{p}$. Of course, if any of the inputs are known with certainty, then $\tilde{L}$ becomes less fuzzy.

$$\tilde{L}_p(x) = \left[\sum \tilde{\alpha}_i^{\tilde{p}} \left(\frac{\tilde{z}_i^* - \tilde{z}_i(\mathbf{x})}{\tilde{z}_i^* - \tilde{z}_i^{**}}\right)^{\tilde{p}}\right]^{1/\tilde{p}} \tag{10.22}$$

The process of generating fuzzy sets for input is not trivial. Certainly, arbitrary assignment is simple and may cover the range of possibility, but it is possible to encode a lot of information and knowledge in a fuzzy set. The process of generating an appropriate fuzzy set, accommodating available data, heuristic knowledge or conflicting opinions, should be capable of preserving and presenting information accurately both in detail and in general form. This topic is addressed in Chapter 6. I consider appropriate techniques for fuzzy set generation to be specific to the type of problem being addressed, the availability of different types of information, and the presence of different decision-makers.

In assuming fuzzy set membership functions for the various inputs to a distance metric calculation (10.22), a decision-maker must make a number of assumptions. Normal fuzzy sets are considered. They acknowledge that there is at least one completely valid value, analogous to the expected value case for probabilistic experiments. In circumstances where at least one modal point cannot be found, it is usually better to assign multiple modal points than to assign low membership values across the range of possible values (the universe of discourse), partly for the sake of interpreting evaluations. Multimodal fuzzy sets may consist of multiple modal points or a continuous range of modes. The choice of boundaries for the universe of discourse also makes assumptions about available knowledge on the universe of discourse. Boundary and modal point selection, along with the shape of the fuzzy sets, define a degree of fuzziness which hopefully represents the characteristic fuzziness of real-world behaviour.

In dealing with real-world water problems, fuzzy sets describe a degree of possibility for valid values of a parameter. They do not possess properties such as conditional probabilities for stochastic applications, at least for simple applications. This is acceptable because typical sensitivity analyses explore all combinations of values anyway, and there is usually not enough information to form conditional properties. In an advanced fuzzy application, there is no reason not to provide conditional fuzzy sets.

10.4.2 Properties of fuzzy distance metrics

We shall focus here on maximization problems. In other words, larger values for criteria are assumed to be better than smaller values, and the ideal solution tends to have a larger value than the alternatives.

It is possible, and may be desirable, to fuzzify all parameters in multi-objective problems formulated with a framework. Figure 10.8 shows typical shapes of input fuzzy sets to be used for criteria values, weights, positive ideals and negative ideals for Compromise programming. The fuzzy sets shown are piecewise linear as: (a) and (b) one-sided linear, (c) triangular, (d) trapezoidal, or (e) conflicting, which combines two triangular sets. Nonlinear fuzzy sets can also be used, but this selection typifies the different modal features. The FuzzyCompro computer program provided on the CD-ROM and discussed later in this chapter provides for the use of a wide range of linear and nonlinear input fuzzy shapes.

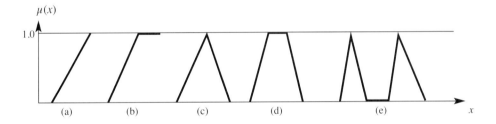

Figure 10.8 *Typical fuzzy input shapes*

Fuzzification of criteria/objectives values is probably the most obvious use of fuzzy sets in decision-making problems. To capture the subtleties of relative performance of different alternatives from the perspective of a decision-maker, there may not be enough choices. Likewise, if a large number of choices are provided, the appreciation of subjectivity in linguistic terms disappears. Fuzzy sets are able to capture many qualities of relative differences in the perceived values of criteria among alternatives. The placement of modal values, along with the curvature and skew of membership functions, can allow decision-makers to retain what they consider the degree of possibility for subjective criterion values.

Quantitative criteria present some slightly different properties from qualitative criteria. It can be assumed that quantitative criteria are measured in some way, either directly or through calculation based on some model. They have stochastic properties which describe the probability of occurrence for values, based on future uncertainties for example. They also have some degree of imprecision in their measurement or modelling. In this way, quantitative criteria may have both stochastic and fuzzy properties. To prevent the complication of many decision-making problems, various uncertainties may be adequately represented with fuzzy sets. In general, the application of quantitative criteria within a fuzzy approach may assume that quantitative criteria are less fuzzy than qualitative criteria.

Criterion weights are an important aspect of the Compromise programming method. Their assignment is completely subjective, usually with a rating on an interval scale. As a subjective value, criterion weights may be more accurately represented by fuzzy sets. Generating these fuzzy sets is also a subjective process. It may be difficult to get honest opinions about the degree of fuzziness of a decision. It might actually be more straightforward to generate fuzzy sets for weights when multiple DMs are involved. Then, at least, voting methods and other techniques are available for producing a composite collective opinion. Regardless of this, more information can be provided about valid weights from fuzzy sets than from crisp weights.

Membership functions for criteria values and criteria weights can both be expressed in three distinct forms (Figure 10.9). They are:

- uncertain (where: known with certainty is a special case with a small degree of fuzziness);
- unknown;
- conflicting.

Both the last two produce a somewhat conflicting interpretation of valid behaviour.

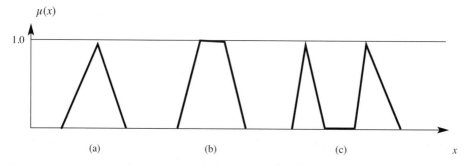

Figure 10.9 *Fuzzy criterion values and weights*

The incorporation of vagueness into the ideal solution is an element which impacts the rankings of alternatives. When we incorporate fuzziness into the location of the ideal solution (both positive and negative), the valid area for the ideal point – in criteria space – affects the measurement of distance to the alternatives. For example, if flood damage reduction is a criterion, then what is the ideal amount of damage reduction? Typically, (crisp) Compromise programming applications use the largest criterion value among the alternatives as the ideal value. This arbitrary placement is probably not valid, and also affects the relative distances to the overall ideal. In another example, if a subjective criterion is rated on a scale of {1,2,3,4,5}, with linguistic interpretations for each, and all alternatives are rated as {3} or {4}, then positive and negative ideals of {4} and {3} respectively will not produce distance metrics indicative of overall alternative performance.

Figure 10.10 shows how positive and negative ideals can be expressed as one-sided fuzzy sets. The three choices are (a) certain, (b) uncertain and (c) unknown. The uncertain case can also be considered as a fuzzy goal. An improvement in criterion value does not improve the level of satisfaction because the goal has already been completely achieved at the initial modal value. The degree of certainty to which the ideals are known is expressed by the range of valid values. Positive and negative ideals may also be triangular, nonlinear or obey any other complex membership function, but they typically assume that a larger value is less valid as the positive ideal solution, or that a smaller value is less valid as the negative ideal.

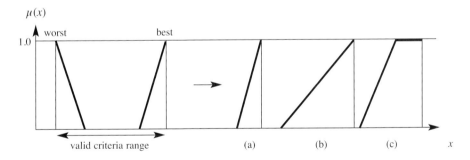

Figure 10.10 *Range of valid criterion values as defined by fuzzy positive and negative ideals*

The distance metric exponent, p, is likely to be the most imprecise or vague element of the distance metric calculation. There is no single acceptable value of p for almost any type of problem, and it can easily be misunderstood. Also, it is not related to problem information in any way except that it provides parametric control over the interpretation of distance. Fuzzification of the distance metric exponent, $\tilde{p}$, can take many forms, but in a practical way it might be defined in one of the five choices shown in Figure 10.11. Options (a) and (c) suggest the common practice of using $p = 2$. However, in (a) it is acknowledged that the distance metric exponent has a possibility of being as small as 1. Options (b) and (d) are the $p = 1$ equivalent. Option (e) offers a theoretical advantage of fuzzy compromise formulation over the crisp one. Using, for example, a triangular membership shape, the principal formulation of fuzzy Compromise programming provides a replacement for setting the range of values for parameter $1 \leq p \leq \infty$. The shape of the triangle and the values of three parameters that define it (in the case of Figure 10.11, 1, 2 and 10 respectively) allow for the solution of the fuzzy Compromise programming problem using only one solution of (10.22). It is important to note that larger values of $\tilde{p}$ may lead to an unmanageable degree of fuzziness (range of possible values), making interpretation of the distance metric difficult.

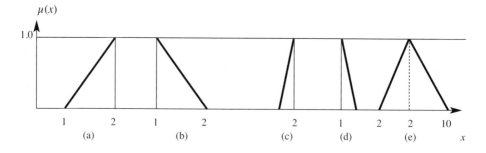

Figure 10.11 *Fuzzy distance metric exponent*

The impact of fuzzy inputs on the shape of the resulting fuzzy distance metric is illustrated in Figures 10.12 and 10.13. Figure 10.12 shows typical shapes for $\tilde{L}$ given triangular weights and criterion values, using different interpretations for $\tilde{p}$.

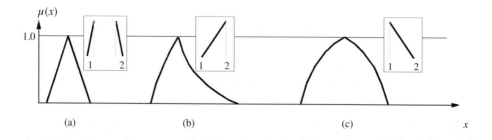

Figure 10.12 *Fuzzy distance metrics for different fuzzy definitions of $\tilde{p}$*

Figure 10.13 shows the impact of unknown and uncertain weights or criteria values. For linear membership functions, areas about the modal value are impacted. The mode may be spread out, split or both.

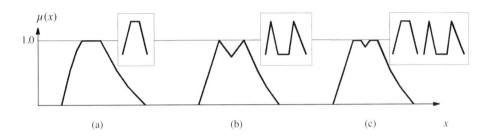

Figure 10.13 *Fuzzy distance metrics for different multimodal criterion values and weights*

The benefits of adopting the general fuzzy approach to Compromise programming are many. Probably the most obvious is the overall examination of decision uncertainty. Sources of uncertainty in criterion values are easily identified (especially for subjective qualitative criteria), but there is also uncertainty or vagueness in other inputs. This is typically ignored, or assessed using sensitivity analysis. Each scenario is normally treated with equal probability unless there are special considerations. Fuzzy compromise treats each potential scenario according to its degree of possibility.

Expressing possibility values with fuzzy inputs enables experience to play a significant role in the expression of input information. The shape of a fuzzy set expresses the experience or the interpretation of a decision-maker. Conflicting data or preferences can also be easily expressed using multimodal fuzzy sets, making fuzzy compromise very flexible in adapting to group decision-making.

Fuzzy criterion values reflect knowledge and confidence regarding the quality of data and models used to calculate criterion values. One assumption in using fuzzy criterion values is that quantitative criteria are generally less fuzzy than subjective criteria. This results in a major enhancement over many multi-objective methods. One tendency in evaluating problems containing both quantitative and qualitative criteria (especially for water resources specialists) is to assign less weight to subjective criteria. Even criterion values known with certainty, in a single decision-maker problem, may be uncertain when additional decision-makers are considered who might disagree on the assessment of the criteria (see Section 10.5). Fuzzy Compromise programming provides better options for expressing differences in subjectivity, because uncertainty in relative importance is supplied by the weights, and uncertainty in relative value is supplied by criterion values.

Fuzzy criterion weights can serve several purposes. The consideration of context-sensitive weighting produces a valid range of possible values. Ideally, the proper weight is conditional on the context of the decision and available information, but this is normally ill-defined and poorly understood. Weights can then be chosen and considered as a fuzzy set of potentially valid weights for a single decision-maker. Considering multiple decision-makers results in fuzzy weighting, even if all the decision-makers are confident about their preferences.

The exponent used to define the distance metric indicates the level of compensation between criteria. The overall level of compensation may be fuzzy. Also, there is no single accepted distance metric. In many cases a decision-maker will be unsure how to penalize difference from an ideal solution. Therefore the definition of p over the range of possible values fits a naturally fuzzy formulation.

10.4.3 Comparing fuzzy distance metrics

Traditional (non-fuzzy) Compromise programming distance metrics measure the distance from an ideal point (Section 10.3), where the ideal alternative would result in a distance metric, $L = 0$. In fuzzy Compromise programming, the distance is fuzzy, such that it represents all of the possible valid evaluations, indicated by the degree of possibility or membership value (Bender and Simonovic, 2000). Alternatives that tend to be closer to the ideal may be selected. This fuzzified distance metric is analogous to a sensitivity analysis for the non-fuzzy Compromise programming case.

Figure 10.14 shows two $\tilde{L}$s. If one alternative is to be selected, the best alternative might be A, a reasonably intuitive choice. Simply consider that A and B have the same shape and degree of fuzziness, but A is shifted towards the origin, which is the ideal solution, assuming that a high membership value near $x = 0$ is desirable. Choosing an alternative is not usually so straightforward, however. If the degree of fuzziness or characteristic shape is different for the available alternatives, choosing the best compromise solution may be difficult.

As an attempt to standardize a procedure for judging which $\tilde{L}$ is best among a set of alternatives, desirable properties can be defined. The most important properties to consider are:

- Possibility values tend to be close to the ideal, $x = 0$, distance.
- Possibility values have a relatively small degree of fuzziness.
- Modal values are close to the ideal.
- Possibility values tend to be far from poor solutions.

An experienced person may be able to visually distinguish the relative acceptability of alternatives, but in cases with many alternatives where each $\tilde{L}$ displays similar

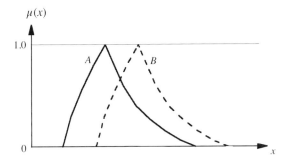

Figure 10.14 *Fuzzy distance metric comparison for two alternatives (A and B)*

characteristics, it may be impractical or even undesirable to make a selection visually. A method for ranking alternatives, based on comparisons of $\tilde{L}$, will make summary ranking information more accessible, automating many of the visual interpretations and creating reproducible results.

There is a theoretical presentation of various methods for comparing fuzzy sets in Section 6.5.4. We introduce the weighted centre of gravity, compatibility-based fuzzy acceptability, the method of Chang and Lee (1994), and the method of Chen (1985). Horizontal methods are those related to the practice of defuzzifying a fuzzy set by testing for a range of validity at a threshold membership value. Vertical methods tend to use the area under a membership function as the basis for evaluation, such as the centre of gravity. Comparative methods are those that introduce other artificial criteria for judging the performance of a fuzzy set, such as a fuzzy goal.

Bender and Simonovic (2000) investigated the use of the weighted centre of gravity and fuzzy acceptability with fuzzy Compromise programming. Prodanovic and Simonovic (2002) compared nine different methods, and fully tested the methods of Chen and of Chang and Lee. The overwhelming conclusions were that most methods behave reasonably well in non-questionable cases, meaning that most methods produce identical rankings for non-questionable cases. However, in difficult cases, i.e. when intuition is not evident, the results are mainly scattered. This means that in difficult cases different methods produce different rankings. Because of the inherent intricacy of fuzzy sets for such difficult cases, degrees of risk have been incorporated into the methods themselves to account for the variability of rankings. Note that for non-questionable cases, the ranking order is not sensitive to changes in the degree of risk.

Based on the results of these two studies, we can conclude that the overall existence ranking index of Chang and Lee (equation (6.57) in Section 6.5.4) is adequate in distinguishing alternatives in both questionable and non-questionable cases. As such, it is recommended as the method of choice in evaluating water resources management alternatives using the fuzzy Compromise programming technique.

10.4.4 An example of a fuzzy Compromise programming application

Let us look at management of the Tisza River basin (which was mentioned earlier) as an example of the implementation of fuzzy Compromise programming.

Example 3

Data from David and Duckstein (1976) and Goicoechea et al (1982) were used for the purpose of comparing five alternative water resources systems for long-range goals in the Tisza River basin according to twelve criteria. Table 10.7 lists the original input data. The last eight criteria in the table are subjective, and have linguistic evaluations assigned to them. The criteria for water quality, recreation, flood protection, manpower impact, environmental architecture and development possibility are all considered on a scale with five linguistic options {excellent, very good, good, fair, bad}. The last two criteria are judged by different linguistic scales. First of all, international cooperation has a subjective scale {very easy, easy, fairly difficult, difficult}. Finally, the sensitivity criterion also uses a subjective scale with four categories (although one of them is not chosen) {very sensitive, sensitive, fairly sensitive, not sensitive}.

No numeric values are provided by David and Duckstein (1976), but numeric differences along an interval scale are given. Issues of uncertainty are not addressed. Subjective criteria are assigned numeric values. The quantitative criteria do not address any stochastic uncertainties normally associated with modelling adequacy, data accuracy or temporal instability. Additional criteria are listed, but are assumed to be handled implicitly.

The weighting of relative importance is also an issue of uncertainty. David and Duckstein (1976) provided criteria weights from the set of {1, 2}. All criteria were weighted as 2 except land and forest use, manpower impact, development possibility and sensitivity, which were given a weight of 1.

In planning water resources systems, a single decision-maker, regulatory agency or interest group is rarely able to represent the interests of others that are impacted by changes in the system. The range of criteria used to evaluate alternatives is admirable, but the relative importance and the relationships between those criteria – defined by different perspectives of stakeholders – are important, if not controlling, aspects of managing water resources systems.

Table 10.7 *Original values used in David and Duckstein (1976)*

Criteria	Alternatives				
	I	II	III	IV	V
1 Total annual cost	99.6	85.7	101.1	95.1	101.8
2 Probability of water shortage	4	19	50	50	50
3 Energy (reuse factor)	0.70	0.50	0.01	0.10	0.01
4 Land and forest use (1000ha)	90	80	80	60	70
5 Water quality	very good	good	bad	very good	fair
6 Recreation	very good	good	fair	bad	bad
7 Flood protection %	good	excellent	fair	excellent	bad
8 Manpower impact	very good	very good	good	fair	fair
9 Environmental architecture	very good	good	bad	good	fair
10 Development possibility	very good	good	fair	bad	fair
11 International cooperation	very easy	easy	fairly difficult	difficult	fairly difficult
12 Sensitivity	not sensitive	not sensitive	very sensitive	sensitive	very sensitive

As a conclusion, without pursuing a sensitivity analysis, David and Duckstein suggest that a mix of systems I and II would be appropriate, since they appear to somewhat dominate the other alternatives and show no overall domination over each other. A sensitivity analysis is implied to be the next logical step in the planning of the Tisza River basin. Changes to the data, weights and time horizon are suggested. Although changes to the data may have probabilistic implications, criterion weights and certainly the impact of the time horizon are more vague because many values may be possible and entirely valid.

Let us introduce a useful improvement for evaluating this water resources system, by treating uncertainties as fuzzy. Although fuzzy applications may not usually exhibit the same explicit definitions as stochastic uncertainties, they should suffice for long-range planning problems.

Bender and Simonovic (2000) introduced the fuzzy definitions shown in Figure 10.15 for linguistic terms used in assessing subjective criteria. Quantitative criteria are also fuzzified, but generally are less fuzzy.

Other fuzzy inputs include the expected ranges of criterion values (Figure 10.16) and the form of distance metric or degree of compensation, $\tilde{p}$, among criteria for different alternatives (Figure 10.16b). Criterion weights, $\tilde{\alpha}_i$, are fuzzified on a range of [0,1] (Figure 10.16c). All of the fuzzy inputs are treated in a simple form, exclusively normal and unimodal. They have either triangular or one-sided membership functions.

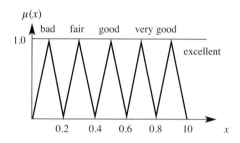

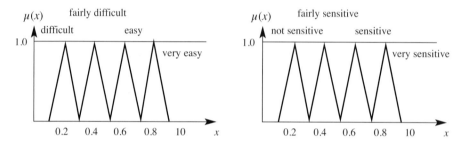

Figure 10.15 *Fuzzy subjective criteria interpretation for the Tisza River problem*

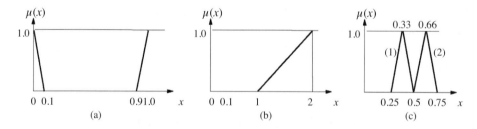

Figure 10.16 *Fuzzy input for the Tisza River problem*

Assuming the fuzzy definition for the distance metric exponent ($\tilde{p}$), and knowing the form of criterion values and weights to be triangular, the resulting fuzzy distance metrics ($\tilde{L}_i$) possess the characteristic shape (Figure 10.17) of near linearity below the mode, and a somewhat quadratic polynomial curvature above the mode. Although the degree of fuzziness (range of valid distances from the ideal solution) is similar for all five alternatives, some of the alternatives are clearly inferior.

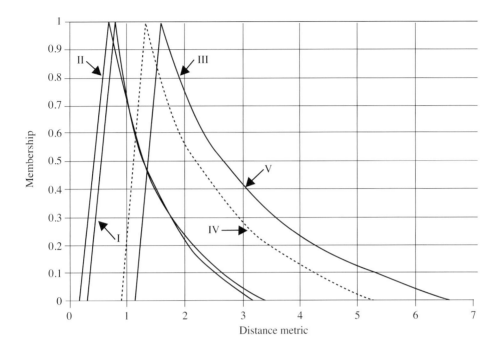

Figure 10.17 *Distance metrics for the Tisza River problem*

Ranking these alternatives is reasonably straightforward because of the simplicity of the shapes, and similarity in degree of fuzziness. We shall use the weighted centroid measure (equation (6.50) in Section 6.5.4). The results are shown in Table 10.8. Rankings are insensitive to changes in levels of risk aversion, as is expected from visual inspection. The resulting ranks confirm that alternatives I and II dominate III, IV and V.

Table 10.8 *Tisza River alternative rankings*

Rank	Alt	WCoG ($q=1$)
2	I	1.045
1	II	0.985
4	IV	2.231
3	III	1.841
5	V	2.241

Note: WCoG = weighted centre of gravity.

You can use the FuzzyCompro program provided on the CD-ROM to confirm the solution in Table 10.8. Example 3 data are in the sub-folder Examples, file Example1.fcompro.

In a live case study with multiple decision-makers, there are opportunities for a group emphasis to collectively adjust fuzzy input to the Tisza River problem. The rankings may change considerably because the values defined for this experiment are predominately simple triangular membership functions, given the form of non-fuzzy input data. Adjustments in relative fuzziness, and the emergence of conflicting opinions about valid criterion values or weights might produce an entirely new outlook – one that might be sensitive to the level of risk aversion characterized by the decision-maker.

10.4.5 The FuzzyCompro computer program

The accompanying CD-ROM includes the FuzzyCompro software. The folder FUZZYCOMPRO contains three sub-folders, FuzzyCompro, FuzzyComproHelp and Examples. The read.me file contains instructions for installation of the FuzzyCompro software, and a detailed tutorial for its use is a part of the Help menu.

FuzzyCompro facilitates the multi-objective analysis of discrete problems using the fuzzy Compromise programming technique. The approach allows various sources of uncertainty, and is intended to provide a flexible form of group decision support. Fuzzy Compromise programming allows a family of possible conditions to be reviewed, and supports group decisions through fuzzy sets designed to reflect collective opinions and conflicting judgements. The transformation of a distance metric to a fuzzy set is accomplished by changing all inputs from crisp to fuzzy and applying the fuzzy extension principle. The measure of distance between an ideal solution and the perceived performance of an alternative is used for comparison among alternatives. An operational definition (10.22) is used for computation of the distance metrics. Choosing the shortest distance to the ideal is no longer a straightforward ordering of distance metrics, because of overlaps and varying degrees of possibility. Ranking of alternatives is accomplished with fuzzy ranking measures designed to illustrate the effect of risk tolerance differences among decision-makers. FuzzyCompro uses a centroid ranking measure (equation (6.50) in Section 6.5.4) to rank alternatives according to the fuzzy distance metric value. It can be used to identify the best compromise alternative under various uncertainties.

10.5 FUZZY MULTI-PARTICIPANT MULTI-OBJECTIVE DECISION-MAKING

Stakeholder participation is a key issue in the planning and management of complex water resources systems. An operational framework that involves managing the evolving relations between regulators, decision-makers, stakeholders and the general public, together with their value systems, is still missing in water resources systems management. This section introduces a methodology that includes the active participation of stakeholders in multi-objective decision-making. A particular emphasis is placed on modelling uncertainties, as well as the preferences of decision-makers, with the aid of the theory of fuzzy sets. Water resources decision-making demands interaction between stakeholders with conflicting interests and/or stakeholders and the environment. We have shown earlier that by using fuzzy sets these types of interactions can be understood and modelled relatively accurately.

An original technique for water resources group decision-making (with multiple objectives) was developed by Prodanovic and Simonovic (2003). The technique integrates a methodology named *group decision-making under fuzziness* (Kacprzyk and Nurmi, 1998), and the fuzzy Compromise programming presented in Section 10.4.

10.5.1 Fuzzy Compromise programming for multi-participant decision-making

Classical Compromise programming is presented as a multi-objective decision analysis technique used to identify the best compromise solution from a set of solutions. We have shown that the measure of distance expressed by equation (10.18)

$$L_p(x) = \left[\sum_{i=1}^{r} \alpha_i^p \left(\frac{z_i^* - z_i(\mathbf{x})}{z_i^* - z_i^{**}} \right)^p \right]^{1/p}$$

referred to as a distance metric, determines the closeness of a particular solution to a generally infeasible (ideal) solution. The advantages of adopting the fuzzy Compromise programming approach as defined by equation (10.22)

$$\tilde{L}_p(x) = \left[\sum_{i=1}^{r} \tilde{\alpha}_i^{\tilde{p}} \left(\frac{\tilde{z}_i^* - \tilde{z}_i(\mathbf{x})}{\tilde{z}_i^* - \tilde{z}_i^{**}} \right)^{\tilde{p}} \right]^{1/\tilde{p}}$$

are plentiful, particularly when dealing with vague criterion weights, $\tilde{\alpha}_i$, the vagueness of both positive, $\tilde{z}_i^*$, and negative ideals, $\tilde{z}_i^{**}$, and vagueness in the appropriate distance metric exponent, $\tilde{p}$.

To pick out a smallest fuzzy distance metric from a group of distance metrics, fuzzy set ranking methods have to be used. The previous section recommended using the method of Chang and Lee, based on the study by Prodanovic and Simonovic (2002). This recommendation was founded on the fact that Chang and Lee's method gives most control in the ranking process, with degree of membership weighting and the weighting of the subjective type. The Overall Existence Ranking Index (OERI) is expressed by (6.57) as

$$OERI(j) = \int_0^1 \omega(\beta) \left[\chi_1 \mu_{jL}^{-1}(\beta) + \chi_2 \mu_{jR}^{-1}(\beta) \right] d\beta$$

where the subscript j stands for alternative j, while β represents the degree of membership. χ_1 and χ_2 are the subjective type weighting indicating neutral, optimistic or pessimistic preferences of the decision-maker, with the restriction that $\chi_1 + \chi = 1$. Parameter $\omega(\beta)$ is used to specify the weights that are to be given to certain degrees of membership (if any). For example, if it is wished to have certain membership values count for more than others, an equation for $\omega(\beta)$ could be formulated to reflect that. For this study, all degrees of membership were weighed equally, so $\omega(\beta) = 1$. Lastly, $\mu_{jL}^{-1}(\beta)$ represents an inverse of the left part, and $\mu_{jR}^{-1}(\beta)$ the inverse of the right part of the membership function.

The risk preferences are: if $\chi_1 = 0.5$, the user is a pessimist (risk averse); if $\chi_1 > 0.5$, the user is neutral; and if $\chi_1 > 0.5$, the user is an optimist (risk taker). Simply stated, Chang and Lee's OERI is a sum of the weighted areas between the membership axis and the left and right inverses of a fuzzy membership.

Group decision-making under fuzziness

Kacprzyk and Nurmi (1998) present a methodology which takes in the opinions of m individuals concerning n crisp alternatives, and then outputs an alternative, or a set of alternatives, preferred by most individuals. Each individual is required to make a pairwise comparison between the alternatives; then a fuzzy preference relation matrix is constructed for each participant, the results are aggregated and a group decision made.

The number of alternatives is denoted by the subscripts $i, j = 1, 2, 3, \ldots n$ and the number of participants by the subscripts $k = 1, 2, 3, \ldots m$. In order to construct a fuzzy preference relation matrix for each individual, we must ask that person to compare every two alternatives in the system. For example, if there are three alternatives in the system (A_1, A_2 and A_3), the participant must compare A_1 with A_2, A_1 with A_3 and A_2 with A_3, and express for each comparison, the degree of preference. The available options include:

$$\mu^k = \begin{cases} 1.0 & \text{if } A_i \text{ is definitely preferred to } A_j \\ c \in (0.5, 1) & \text{if } A_i \text{ is slightly preferred to } A_j \\ 0.5 & \text{in the case of indifference} \\ d \in (0, 0.5) & \text{if } A_j \text{ is slightly preferred to } A_i \\ 0.0 & \text{if } A_j \text{ is definitely preferred to } A_i \end{cases} \quad (10.23)$$

With the restrictions above, each participant is to construct a fuzzy preference relation matrix. For a three-alternative example, a sample matrix for participant 1 might be:

$$d^{k=1} = \begin{cases} \begin{array}{c|ccc} & j=1 & 2 & 3 \\ \hline i=1 & 0 & 0.6 & 0.8 \\ 2 & 0.4 & 0 & 0.4 \\ 3 & 0.2 & 0.6 & 0 \end{array} \end{cases} \quad (10.24)$$

Note that participant 1 prefers A_1 over both A_2 and A_3, and A_3 over A_2, only slightly. Clearly the participant thinks that A_1 is the best option.

Once the fuzzy preference relation matrix is determined for each participant, the aggregation of the results is performed in the following way. First, h_{ij} is calculated to see whether A_i defeats (in pairwise comparison) A_j ($h_{ij} = 1$) or not ($h_{ij} = 0$).

$$h_{ij}^k = \begin{cases} 1 & \text{if } d_{ij}^k < 0.5 \\ 0 & \text{otherwise} \end{cases} \quad (10.25)$$

Then, we calculate

$$h_j^k = \frac{1}{n-1} \sum_{i=1, i \neq j}^{n} h_{ij}^k \quad (10.26)$$

which is the extent, from 0 to 1, to which participant k is not against alternative A_j, where 0 stands for definitely not against and 1 stands for definitely against, through all intermediate values.

Next, we calculate:

$$h_j = \frac{1}{m} \sum_{k=1}^{m} h_j^k \quad (10.27)$$

which expresses to what extent, from 0 to 1, all participants are not against alternative A_j.

Then, we compute

$$v_Q^j = \mu_Q(h_j) \quad (10.28)$$

which represents to what extent, from 0 to 1 as before, Q (most) participants are not against alternative A_j. Q is a fuzzy linguistic quantifier, (in our case meaning 'most') which is defined, after Zadeh (1983):

$$\mu_Q(x) = \begin{cases} 1 & if\ x \geq 0.8 \\ 2x-0.6 & if\ 0.3 < x < 0.8 \\ 0 & if\ x \leq 0.3 \end{cases} \tag{10.29}$$

Note that there are alternative ways to evaluate fuzzy linguistic quantified statements. In Section 6.5.4 we introduced the ordered weighted averaging operator of Yager, for example.

Lastly, the final result (fuzzy Q-core) is expressed as:

$$C_Q = \{(A_1,v_Q^1),(A_2,v_Q^2),(A_3,v_Q^3),...,(A_n,v_Q^n)\} \tag{10.30}$$

and is interpreted as a fuzzy set of alternatives that are not defeated by Q (most) participants.

Similarly, the fuzzy β/Q-core and fuzzy s/Q-core can be determined. The former is obtained by changing equation (10.25) into:

$$h_{ij}^k(\alpha) = \begin{cases} 1 & if\ d_{ij}^k < \alpha \leq 0.5 \\ 0 & otherwise \end{cases} \tag{10.31}$$

and then performing all above steps as before. $(1 - \beta)$ represents the degree of defeat by which A_i defeats A_j; as such it takes values between [0,0.5]. The final result in this case is interpreted as a fuzzy set of alternatives that is not sufficiently (at least to a degree $(1 - \beta)$) defeated by Q (most) participants. The parameter β is arbitrarily chosen at 0.3. The fuzzy s/Q-core is determined by changing equation (10.25) to:

$$\hat{h}_{ij}^k = \begin{cases} 2(0.5 - d_{ij}^k) & if\ d_{ij}^k < 0.5 \\ 0 & otherwise \end{cases} \tag{10.32}$$

and again performing all the above steps as before. With (10.32) above, strength is introduced into the defeat (parameter s stands for strength), and the final result interpreted as a fuzzy set of alternatives that is not strongly defeated by Q (most) participants.

It should be noted that there exists a modification of the above algorithm which can assign different levels of importance to different participants (and/or alternatives). For the purposes of this text, all experts (and all alternatives) are assigned an equal level of importance.

Combining fuzzy Compromise programming with group decision-making under fuzziness

Here is a proposed algorithm for including multiple participants in the multi-objective decision-making process, which uses fuzzy Compromise programming.

1. Each participant is to specify fuzzy weights, $\tilde{\alpha}$, the deviation parameter, $\tilde{p}$, as well as positive, $\tilde{z}_i^*$, and negative ideals, $\tilde{z}_i^{**}$, concerning the objectives/criteria of the problem. The participant's overall degree of risk is to be specified here as well (parameter χ_1). It should be noted that these parameters are entirely subjective and are based on the preferences of a participant.
2. For each participant, a set of fuzzy alternatives is generated via a fuzzy Compromise programming equation. This means that the fuzzy Compromise programming equation takes in r (fuzzy) criteria (for each alternative, for each participant), and produces one (fuzzy) distance metric for each alternative of the problem, for each participant.
3. For each participant, a fuzzy preference relation matrix is generated.
4. Finally, Q-core, β/Q-core and s/Q-core algorithms are performed, and a group decision is made.

A participant fuzzy preference relation matrix is obtained via available ranking methods. These matrices are obtained in the following way. First, a ranking method is called to rank the alternatives for each participant; then, from all the ranking values for that participant, a difference is found for every two alternatives compared. From these differences in the ranking values, a fuzzy preference relation matrix is constructed. If the difference is large and negative, this means that A_1 is much more preferable than A_2, and a fuzzy preference relation for this pair is given a value close to or just less than 1.0. Similarly, if the difference is large and positive, meaning that A_2 is more widely preferred than A_1, a value close to 0 is assigned for that particular pair. However, we must be cautioned when defining the meaning of small and large differences in the ranking values. They may have a profound effect on the results produced by the methodology.

Note that in spite of the need for pairwise comparisons between the alternatives, the participants themselves do not have to perform them directly. They are an integral part of the proposed methodology.

10.5.2 An example of fuzzy Compromise programming for a group decision-making application

The Tisza River basin presented in Section 10.4.4, Example 3, will be used to illustrate the implementation of fuzzy Compromise programming for group decision-making.

Example 4

For the purposes of this example, four participants were asked to provide their input concerning the criteria of the problem. Using a scale from 1 to 5, each participant was asked to express the importance of each criterion, with 1 indicating the least important and 5 the most important. These weights are shown in Table 10.9.

Table 10.9 *Preferences of participants*

Criterion		Participant			
		1	2	3	4
1	Total annual cost	3	1	5	5
2	Probability of water shortage	3	3	4	3
3	Energy (reuse factor) difficult	4	5	3	5
4	Land and forest use (1000 ha)	4	3	5	2
5	Water quality	4	2	5	2
6	Recreation	4	2	5	5
7	Flood protection	4	5	1	4
8	Manpower impact	4	1	2	2
9	Environmental architecture	3.5	5	2	4
10	Development possibility	4	5	2	3
11	International cooperation	3.5	2	5	3
12	Sensitivity	4	2.5	5	2

Fuzzy weights were constructed from the responses in Table 10.9, giving everyone the same level of fuzziness and introducing the following simplifications:

- Triangular fuzzy memberships are used.
- Participants are asked to give only their criterion weights for the problem, while keeping the deviation parameter, $\tilde{p}$, as well as positive, $\tilde{z}_i^*$, and negative ideals, $\tilde{z}_i^{**}$, constant (for all experts).
- Each expert has an equal level of importance and neutral risk preference.
- The Chang and Lee method is used to rank fuzzy distance metrics.

The first two participants considered in the example would like a best compromise solution to be found, while the other two are not as considerate. The third participant is concerned with the protection of the environment, with little consideration of such issues as development possibility. The fourth one has exactly the opposite priorities. Such diverse participants were selected to simulate a conflict among the decision-makers, as this is usually the case in real decision situations.

Figures 10.18–10.21 show fuzzy distance metrics for each participant, as a result of the fuzzy Compromise programming equation applications. The differences in the distance metrics are caused by the differences in weights provided by the participants.

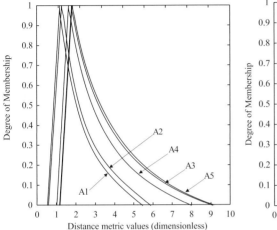

Figure 10.18 *Participant 1 fuzzy distance metrics*

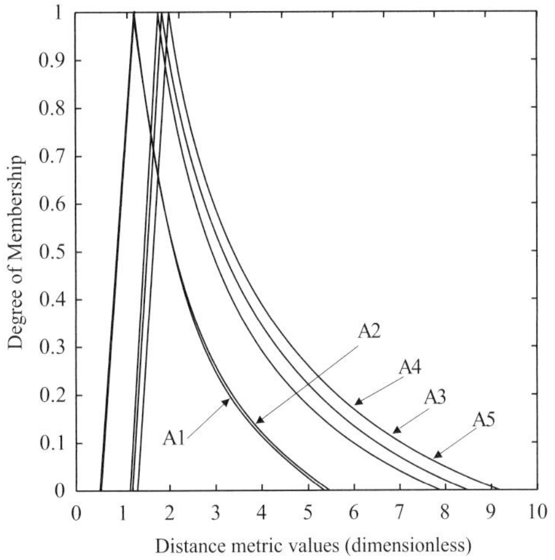

Figure 10.19 *Participant 2 fuzzy distance metrics*

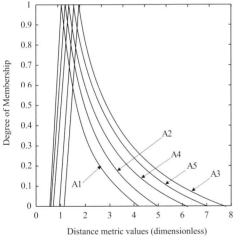

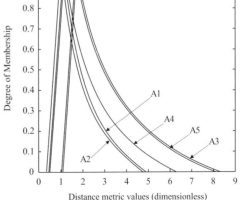

Figure 10.20 *Participant 3 fuzzy distance metrics*

Figure 10.21 *Participant 4 fuzzy distance metrics*

By applying Chang and Lee's fuzzy set ranking method (with neutral viewpoints) to the distance metrics for each participant (see Table 10.10), a set of participant fuzzy preference relation matrices is generated (see Table 10.11).

Table 10.10 *Participants' ranking values, using the Chang and Lee method with* $\chi_I = 0.5$

Alternative	Participant			
	1	2	3	4
1	1.63	1.67	1.37	1.53
2	1.75	1.69	1.61	1.39
3	2.61	2.61	2.43	2.49
4	2.33	2.45	1.87	1.73
5	2.66	2.82	2.16	2.43

These fuzzy preference relation matrices are processed using the algorithm presented above, to produce the following results:

Q-core: {1.0, 1.0, 0, 0.4, 0}; represents the degrees to which alternatives A_1 and A_2 are not at all defeated (in pairwise comparison).

β/Q-core: {0.7750, 0.6500, 0, 0, 0}; gives the degrees to which alternatives A_1 and A_2 are not sufficiently defeated (to a degree of 0.7).

s/Q-core: {0.7250, 0.5625, 0, 0, 0}; expresses the degrees to which alternatives A_1 and A_2 are not strongly defeated.

The results obtained by this methodology concern only the best compromise alternatives, or those that are not defeated in pairwise comparison.

You can use the FuzzyComproGDM program provided on the CD-ROM to confirm the solution in Table 10.10. The data for Example 4 are in the sub-folder Examples, file Example1.fcomproms.

10.5.3 The FuzzyComproGDM computer program

The accompanying CD-ROM includes FuzzyComproGDM software. The folder FUZZYCOMPROGDM contains three sub-folders, FuzzyComproGDM, FuzzyComproGDMHelp and Examples. The read.me file contains instructions for installation, and a detailed tutorial for its use is a part of the Help menu.

This program provides an operational framework that can assist in managing the

Table 10.11 *Individual preference relation matrices*

Participant 1							Participant 3				
	i = 1	2	3	4	5			i = 1	2	3	4	5
j = 1	0.00	0.60	0.95	0.9	1.00	j = 1	0	0.7	1.0	0.8	0.9
2	0.40	0.00	0.95	0.8	0.95	2	0.3	0.0	0.9	0.7	0.8
3	0.05	0.05	0.00	0.3	0.60	3	0.0	0.1	0.0	0.2	0.3
4	0.10	0.20	0.70	0.0	0.70	4	0.2	0.3	0.8	0.0	0.7
5	0.00	0.05	0.40	0.3	0.00	5	0.1	0.2	0.7	0.3	0.0
Participant 2							Participant 4				
	i = 1	2	3	4	5		i = 1	2	3	4	5
j = 1	0.00	0.60	0.95	0.9	1.00	j = 1	0	0.40	0.95	0.6	0.95
2	0.40	0.00	0.95	0.9	0.95	2	0.60	0.00	1.00	0.7	0.95
3	0.05	0.05	0.00	0.4	0.60	3	0.05	0.00	0.00	0.1	0.40
4	0.10	0.10	0.60	0.0	0.70	4	0.40	0.30	0.90	0.0	0.90
5	0.00	0.05	0.40	0.3	0.00	5	0.05	0.05	0.60	0.1	0.00

evolving relations between regulators, decision-makers, stakeholders and the general public, together with their value systems. It is based on a methodology that includes the active participation of stakeholders in multi-objective decision-making. A particular emphasis is placed on modelling uncertainties, as well as the preferences of decision-makers, with the aid of the theory of fuzzy sets.

10.6 EXAMPLES OF WATER RESOURCES MULTI-OBJECTIVE ANALYSES

Multi-objective models play an important role in water resources systems management. Some characteristic applications are introduced in Section 10.1.

Here we shall look at two water resources multi-objective models based on my personal experience. They illustrate the characteristics of current water resources systems management problems, and the potential for their solution with the multi-objective tools presented in this book. The first (Simonovic, 1989) shows the use of Compromise programming multi-objective analysis in the broad context of water resources master planning. The four-step planning procedure is strongly oriented to the use of mathematical models of various complexities. It includes:

- evaluation of the available water resources;
- estimation of the water demand;

- generation of technical alternatives for satisfying water demand from available water resources;
- ranking of the alternative solutions in accordance with a set of objectives.

In the final step, because of a lack of decision-maker preferences and reliable data, the concept of a robust solution as opposed to the best compromise solution is introduced. The example shows the application of the methodology to the development of a water resources master plan for Serbia.

The second example (Simonovic, 2007) presents a methodology developed to capture the views of multiple stakeholders using fuzzy set theory and fuzzy logic. Three possible different response types, scale (crisp), linguistic (fuzzy) and conditional (fuzzy), are analysed to obtain the resultant input by using fuzzy expected value (FEV). FEV input is used with fuzzy Compromise programming and applied to flood management in the Red River basin, Manitoba, Canada. The basin faces periodical flooding, and flood management decision-making problems involve multiple criteria and multiple stakeholders.

10.6.1 Multi-objective analysis for water resources master planning

Water resources planning can be defined as seeking a balance between water demands and the available water resources. It is a search for a solution to meet the demand with the available resources. The solution ranges from reducing demand so that existing resources can meet it, to increases in supply to meet the demand. Water resources planning is as old as humanity's use of water. Of course the planning procedure improves with knowledge and technology development.

The water resources planning steps for developing conditions are classified into:

- single-purpose phase (demand is small; projects are isolated; flood control is very unreliable; water quality control does not exist);
- multi-purpose phase with conservation (increasing demand; increase in required reliability; water quality control is introduced);
- complex water resources systems phase (water becomes the limiting factor in the development; multi-purpose systems are connected; water management is improving; the reservoirs are used for spatial and time redistribution of water resources, and water transfer between river basins is introduced).

Planning principles for developing countries

For developing countries, the United Nations Industrial Development Organization

(UNIDO) recommends that planning analysis considers the following objectives (Dasgupta et al, 1993):

- aggregate consumption;
- income redistribution;
- growth rates of national income;
- employment level;
- self-reliance;
- merit wants.

As was pointed out by Goodman (1984), the United Nations (UN) publication treats the problem of evaluating the extent to which projects advance each of the objectives, and presents their combination as a measure of 'aggregate national economic profitability'. In this approach water consumption is used as the basic unit of account. It is shown that the objectives outlined above, based on UN guidelines, may be expressed in other ways. Another possible formulation is:

- economic sector development (agricultural, industrial, electric power);
- balanced regional development;
- engineering and economic feasibility;
- financial viability.

In developing a system of objectives, the UNIDO guidelines do not lay any stress on the quality of the environment or other intangibles such as the quality of human life.

This example describes some numerical aspects of long-term comprehensive and integrated planning, and illustrates the benefits of using multi-objective analysis in long term water resources planning (Simonovic, 1989). In the application of Compromise programming to planning in developing countries, there is a very important change: the concept of the best compromise solution is replaced by the *most robust compromise solution* (introduced in Section 10.3.2).

As an initial step, the water resources master plan needs to be defined. According to Goodman (1984), master planning is the formulation of a phased development plan to either meet the estimated requirements for a single water resource purpose over a specified period of time, or exploit opportunities for single- and multi-purpose water resources projects in a defined geographic area over a specific period of time, or until all justified projects are completed. In Serbia, no strict definition of master planning has been established. Various approaches have been used in the development of different master plans. However, the Water Law of Serbia (issued in 1975 and revised in 1990, with further revisions in progress) specifies what the Water Resources Master Plan for Serbia (WRMS) must include. According to the Water Law, the WRMS is a long-term planning document used for managing the development of water resources of specific regions. It is obligatory for it to represent the existing condition of water resources and water-related infrastructure in a certain region; and to determine the basic elements, including the water balance and the condition of water resources management, that secure the overall optimal technical and economic solutions for management of water resources, protection against the devastating impact of water, protection of water resources, and water use in that region.

The four-step planning procedure

Chapter 3 presented a systems water resources planning framework which consisted of:

- defining the problem;
- setting objectives and developing evaluation criteria;
- developing alternatives;
- modelling alternatives;
- evaluating the alternatives;
- selecting an alternative;
- planning for implementation.

This is a generic framework (applicable to many different fields and disciplines), but it is precise enough to lend clarity to water resources systems analysis. Step 4 requires a particularly strong grasp of both modelling purposes and modelling techniques.

The planning concept applied in Serbia (Simonovic, 1989) contained almost all the activities presented above, but aggregated into four steps:

1 Inventory, forecast and analysis of available water resources.
2 Inventory, forecast and analysis of water demand.
3 Formulation of alternative solutions for satisfying water demands from available water resources.
4 Comparison and ranking of alternative solutions.

The specification of water problems, goals and objectives was assumed to be done prior to the first step, or before the work on the plan started. A general description of the four steps used in the planning framework follows.

1 Inventory, forecast and analysis of available water resources

Analysis performed at this level included five major activities. The process started with an estimation of available groundwater. The analysis was performed for each territorial unit of water consumption (in the case of the WRMS, the unit considered was a municipality). The estimation of available groundwater included identification of aquifer systems and their physical characteristics and hydraulic properties. These characteristics were summarized on maps which showed aquifer boundaries, piezometric contours and the location of water wells in the aquifer, with indicators of the depth to static water level and the depth to the top of the aquifer. The final results of this phase were maps of available water from aquifer systems in the region analysed (the yield for every well and specific capacity of the well, representing the amount of water available from the well per unit of available drawdown).

The process continued with surface water analysis. The analysis of surface water included a classical regional hydrological study, providing information on rainfall distribution and intensity, evaporation rates and runoff distribution. All of the analyses were performed on the basis of available data. For every stream-gauge as well as for all possible intake sites from the rivers, at this stage, there was a statistical analysis of mean, low and flood flows. In addition a time series of monthly flows was generated for all possible dam sites within the region.

Next, reservoirs were considered. The existing reservoirs were listed with all the available data: total reservoir storage, dead volume, conservation storage, flood control storage, reservoir purposes, reservoir operating rules, etc. For potential reservoir sites there was an investigation of topographical and geological conditions to produce a maximum dam height and reservoir volume curves. During this step unconventional sources of water were also addressed. This involved water that might be obtained through recycling, desalination or similar procedures. It is important to note that the amount of water obtained from unconventional sources is not an active part of the water balance, and is only used to reduce the demand in the following steps of the plan.

The planning process continued with reservoir yield optimization. At this stage, the potential optimal yield was estimated for every reservoir in the region. An original procedure was used for this (Simonovic, 1987). Table 10.12 shows the activities at the first step of the planning concept, the modelling purpose and suggested modelling techniques.

Table 10.12 *Activities, modelling purpose and suggested modelling techniques at the first step of the planning process*

Inventory, forecast and analysis of available water resources

Activities	Available groundwater	Surface water analysis	Reservoirs	Unconventional sources	Reservoir yield determination
Modelling purpose	Determination of groundwater wells yield	Determination of runoff distribution Statistical analysis of mean, low and flood flows Generation of mean flows for ungauged locations	Presentation of existing reservoir data Determination of maximum reservoir storage for potential sites	Recycling Desalinization	Estimation of optimal reservoir yield
Suggested modelling techniques	Simulation	Rainfall-runoff models Simulation Statistical distribution fitting: ■ regression ■ linear and nonlinear interpolation	Optimization Input–output diagrams	Parameter estimation	Implicit stochastic optimization

2 Inventory, forecast and analysis of water demand

Categories of water demand include public water uses (domestic, commercial, industrial and public), rural (domestic, livestock), irrigation, and self-supplied industrial (cooling and processing, thermoelectric and hydroelectric power). All of the categories mentioned are of the withdrawal type. However, there are non-withdrawal uses such as water quality control by dilution, recreation, navigation and environmental uses by natural vegetation and wildlife. The major activities together with the modelling purpose and the modelling techniques are shown in Table 10.13. The major efforts in this planning step were devoted to the correct estimation of population growth, industrial development, energy consumption and production, and the effect on the water quality in the streams. The final result of this step was a map representing the water demand for different purposes at every territorial unit. It is important to note that Tables 10.12 and 10.13 detail a number of modelling techniques which are sometimes difficult to apply because of a lack of reliable data, insufficient funding or other reasons. Obviously, from time to time simplifications were made in the planning process.

Table 10.13 *Activities, modelling purpose and suggested modelling techniques at the second step of the planning process*

Inventory, forecast and analysis of available water resources

Activities	Municipal and industrial water supply	Irrigation	Power production	Water quality control	Other uses
Modelling purpose	Estimation of population growth Prediction of industrial development	Estimation of the water demand Derivation of crop yield–soil moisture relation	Estimation of power demand Thermal power and hydropower production	Estimation of clean water amount necessary for dilution	Recreation requirements Fish production Wildlife
Suggested modelling techniques	Systems dynamics model (diagram of flows, people, resources and products) Input–output diagrams Trends estimation	Simulation Mathematical formula (Blaney Criddle) Optimization of water allocation (stochastic dynamic programming) Optimization of crop structure (linear programming)	Input–output modelling Simulation Optimization Interpretive structural modelling	Quality simulation Optimization of wastewater discharge sites Optimization of waste load	Water level computation (simulation) Water quality simulation

3 Formulating alternative solutions for satisfying water demand from available water resources

Using the results of the first two steps, a water shortage/surplus map was drawn up for the region under consideration. A balance was made between the available resources and the demand for each *water system*. A water system represents the territorial unit inside which all the demand can be satisfied from the available resources. Water systems are determined by aggregating the initial territorial units for which the demand and the available water resources (including water transfer) are defined. The major effort at this planning step was to create a number of alternative technical solutions for satisfying water demand from available water resources within each water system. The alternative technical solutions were developed through preliminary design using 'typical structures', such as water wells, pump stations, dams and water intake structures, and relating the cost of every structure type to flow capacity. The detailed technical design is considered at a more detailed planning level.

As a result of this step, a number of alternatives were presented for each water system. The alternatives presented accounted for interconnections between the water systems, including water transfer. It is important to point out that water systems are dependent neither on administrative boundaries nor on physical catchment boundaries. They are spatial planning units applicable only within the planning region and the time horizon considered (30 years in this example).

4 Comparison and ranking of alternative solutions

Long-term planning over a 30-year horizon in a developing country is a complex process involving different economic, social, environmental, political and other concerns. Since most of the objectives are not quantifiable in monetary terms, the need for multi-objective analysis is clear. The set of objectives is dependent on the particular problem structure and complexity.

Eight objectives were considered for the WRMS:

1. minimization of total costs (local currency);
2. minimization of energy consumption (GWh);
3. maximization of positive effects of alternative plans on water quality (using a relative scale from 1 to 5, 1 being bad and 5 being excellent);
4. minimization of negative effects on resettlement of people (using a relative scale from 1 to 5);
5. maximization of positive environmental effects (using a relative scale from 1 to 5);
6. maximization of regional political interests (with a relative scale from 1 to 5);
7. maximization of local interests (communities) (with a relative scale from 1 to 5);
8. maximization of system reliability (relative scale from 1 to 5).

Only the first two criteria are quantitative, while the remaining six are qualitative.

The ranking of alternative solutions was performed with the assistance of Compromise programming. Since cooperation between planners and decision-makers was not very effective (there were many political hidden agendas, non-cooperative modes, etc.), a comprehensive sensitivity analysis of the rankings was performed. Table 10.14 shows the activities at the fourth step of the planning concept, modelling purpose and suggested modelling techniques.

Table 10.14 *Activities, modelling purpose and suggested modelling techniques at the fourth step of the planning process*

Comparison and ranking of the alternative solutions

Activities	Setting objectives and developing evaluation criteria	Evaluating the alternatives	Selecting an alternative	Sensitivity analysis
Modelling purpose	Model of purposes for every 'water system'	Evaluation of alternative sets according to all criteria used	Decision-making models for ranking alternatives	Analysing the influence of decision-makers' preferences and model parameters on alternatives ranking
Suggested modelling techniques	Objectives tree Hierarchical diagram Interaction matrices	Cost estimation model Cost–benefit model Cost-effectiveness model	Discrete multi-objective techniques Decision table (showing ranking of alternatives for each criterion) Criterion function (mathematical expression for establishing the overall ranking of alternatives)	Preference relations Compromise programming discrete multi-objective technique

Multi-objective analyses of alternative solutions

In the example of the WRMS, eight criteria (as specified above) were used. The preferences of the decision-makers were collected through a set of public meetings and meetings with regional water authorities. Unfortunately, the process did not result in an explicit set of weights to apply using Compromise programming. In response an original approach was developed for the resolution of the problem:

- selection of the distance coefficient p;
- generation of different sets of weights α_i;

- ranking of alternatives for all generated sets from the second step with a selected coefficient p;
- evaluation of the ranks and selection of the suggested solution.

Based on the previous experience and the available literature, $p = 2$ was selected. The next step involved the generation of weights. The introduction of α_i makes it possible to express decision-makers' preferences concerning the relative importance of various objectives. Since the decision-makers were not able (or not willing) to express their preferences, the planning team generated a number of different sets of weights to cover a broad range of potential decision-making positions. After the preferences had been generated, they were ranked using Compromise programming. Planners used the number of ranks generated to search for an alternative that was not always at the top but close to the top. The usual procedure for identifying the best compromise solution was replaced by a search for the most robust compromise solution. The most robust compromise alternative was considered to be the one that was the least sensitive to changes in preferences.

Results

Some of the results from the WRMS are used to illustrate the advantages of the presented planning process. In the comprehensive integrated plan, the following purposes are considered:

- municipal water supply;
- industrial water supply;
- irrigation;
- hydropower generation;
- flood protection;
- water quality control.

The municipal water supply was the major problem considered in the WRMS. All the alternative technical solutions were built around regional water supply systems. The industrial water supply mostly uses water directly from the rivers. Industrial water quality requirements are satisfied. Irrigation development for the planning time horizon was considered only within the bigger river valleys. If development goes as far as upper basin limits, new reservoirs (not included in the WRMS) must be considered.

Hydropower production was analysed in a very specific way. The alternative solutions were created with the same total power production level. That allowed for hydropower production to be excluded from the evaluation of alternative solutions and ranking of the alternatives. Flood protection was considered in the analysis through the formulation of constraints, i.e. river channel flow capacities and water levels. Following the results of some previous investigations, it was accepted as a fact that providing flood control storage in the reservoirs was not economically justifiable. Water quality control was considered directly by providing the minimum release from all the reservoirs. The release is considered to be a function of the river water quality downstream from the reservoir. The time horizon considered was from 1985 to 2015. Because of the complexity of water resources problems in the area, and the size of it, the results, or the plan, were published in ten volumes.

At the first planning step, 120 reservoir sites were considered. Most of the sites were eliminated before the yield optimization because of the geological conditions of the profiles, distance from users and large resettlement requirements. Finally, optimization was performed for 49 sites. Potential reservoir uses were determined by the planners, considering the reservoir location and the demand structure in the nearby region.

For WRMS, the demand analysis (step 2) was done for 98 territorial units (a starting point in the aggregation process into water systems). A very characteristic spatial variability in water demand and water availability guided the aggregation process, which resulted in defining two large water systems (S1 and S2). For these two systems, technical alternative solutions were generated at the third step of the planning process, six for system S1 and eight for system S2.

Table 10.15 lists the input data for S1 and Table 10.16 for S2. Negative signs are an indication that the criterion is being minimized. Six sets of weights (out of 18 used in the original study) are shown in Table 10.17. The same weights are applied to both systems.

Table 10.15 *Input payoff matrix for system S1*

Alternative	Criterion							
	1	2	3	4	5	6	7	8
1	−307.6	−19.3	1	2	3	2	5	3
2	−313.5	−17.6	2	2	4	2	5	3
3	−395.9	−14.5	5	4	5	5	3	4
4	−379.0	−13.7	5	3	5	4	4	2
5	−371.8	−14.0	5	4	5	4	4	2
6	−393.1	−14.7	5	3	5	5	3	5

Table 10.16 *Input payoff matrix for system S2*

Alternative	Criterion							
	1	2	3	4	5	6	7	8
1	−83.3	−3.4	3	4	3	5	3	4
2	−88.3	−3.5	3	2	4	4	4	4
3	−82.9	−3.3	5	4	3	4	3	4
4	−87.5	−3.4	4	3	3	3	5	4
5	−95.4	−3.7	2	5	4	5	3	4
6	−94.1	−3.4	5	3	3	4	4	4
7	−85.9	−3.4	4	4	3	5	3	3
8	−83.8	−3.4	5	2	5	3	5	3

Table 10.17 *Alternative weight sets*

Weight set	Criterion							
	1	2	3	4	5	6	7	8
1	1.2	1.0	0.7	0.7	0.5	1.8	1.1	1.0
2	1.5	1.5	0.8	0.5	0.3	1.2	1.3	0.8
3	1.0	1.0	1.0	1.0	1.0	1.0	1.0	1.0
4	1.0	1.0	1.4	1.3	1.0	1.2	0.3	0.7
5	1.2	1.0	0.7	0.7	0.5	1.1	1.8	1.0
6	1.0	1.2	0.7	0.7	0.5	1.8	1.1	1.0

Table 10.18 *Ranking results for system S1*

Alternative	Weight set					
	1	2	3	4	5	6
1	2.767E-01 [6]	2.080E-01 [2]	2.635E-01 [6]	3.183E-01 [6]	2.121E-01 [2]	2.767E-01 [6]
2	2.655E-01 [5]	1.948E-01 [1]	2.258E-01 [5]	2.750E-01 [5]	1.973E-01 [1]	2.655E-01 [5]
3	2.423E-01 [4]	3.164E-01 [6]	2.201E-01 [2]	1.855E-01 [1]	3.005E-01 [6]	2.422E-01 [3]
4	2.394E-01 [2]	2.796E-01 [4]	2.239E-01 [4]	2.083E-01 [4]	2.484E-01 [4]	2.435E-01 [4]
5	2.300E-01 [1]	2.707E-01 [3]	2.110E-01 [1]	1.867E-01 [2]	2.393E-01 [3]	2.363E-01 [1]
6	2.401E-01 [3]	3.133E-01 [5]	2.232E-01 [3]	1.988E-01 [3]	2.987E-01 [5]	2.410E-01 [2]

Note: First number = distance metric value; second number = rank.

The sensitivity analysis for S1, shown in Table 10.18, suggests that alternative {5} is the most robust compromise solution. Other alternatives that ranked high were {3,6}.

Table 10.19 *Ranking results for system S2*

Alternative	Weight set					
	1	2	3	4	5	6
1	1.676E-01 [1]	1.896E-01 [3]	2.023E-01 [2]	1.885E-01 [3]	2.443E-01 [5]	1.689E-01 [1]
2	1.934E-01 [3]	1.929E-01 [4]	2.028E-01 [3]	2.416E-01 [6]	1.934E-01 [2]	1.946E-01 [3]
3	1.907E-01 [2]	1.868E-01 [2]	1.921E-01 [1]	1.624E-01 [1]	2.451E-01 [6]	1.907E-01 [2]
4	2.504E-01 [6]	1.860E-01 [1]	2.074E-01 [4]	2.408E-01 [5]	1.764E-01 [1]	2.494E-01 [6]
5	2.563E-01 [7]	3.316E-01 [8]	2.577E-01 [8]	2.628E-01 [8]	3.119E-01 [8]	2.563E-01 [7]
6	2.090E-01 [5]	2.169E-01 [7]	2.095E-01 [5]	2.198E-01 [4]	2.090E-01 [4]	1.964E-01 [4]
7	2.061E-01 [4]	2.116E-01 [6]	2.285E-01 [7]	1.842E-01 [2]	2.721E-01 [7]	2.062E-01 [5]
8	2.737E-01 [8]	1.993E-01 [5]	2.189E-01 [6]	2.436E-01 [7]	2.082E-01 [3]	2.744E-01 [8]

Note: First number = distance metric value; second number = rank.

For the S2 alternatives {1,3} are the most robust (Table 10.19). Other alternatives that ranked high were {2,4}.

10.6.2 Participatory planning for sustainable floodplain management

Floodplain management problems in water resources are often associated with multiple objectives and multiple stakeholders. To produce a more effective and acceptable decision outcome, more participation needs to be ensured in the decision-making process. This is particularly relevant for flood management problems where the number of stakeholders can be very large. Although the application of multi-objective decision-making tools in water resources is very wide, application with the consideration of multiple stakeholders is much more limited. The solution methodologies adapted for multi-criteria multi-participant decision problems are generally based on the aggregation of decisions obtained for individual decision-makers. This approach seems somewhat inadequate when the number of stakeholders is very large.

In this example a methodology was developed to capture the views of multiple stakeholders using fuzzy set theory and fuzzy logic (Akter and Simonovic, 2005; Simonovic, 2007). Three possible different response types, scale (crisp), linguistic (fuzzy) and conditional (fuzzy), were analysed to obtain the resultant input by using FEV. The FEV input is then used in fuzzy Compromise programming.

The methodology was applied to flood management in the Red River basin in Manitoba, Canada, which faces periodical flooding and flood management decision-making problems associated with multiple criteria and multiple stakeholders. While the results show the successful application of the methodology, they also show significant differences in the opinions of stakeholders within the basin.

Box 10.2 Emerson flooding

During the flood of the century in the Red River basin, the small town of Emerson on the Canada–United States border told an interesting story. The town is well protected by a dike that kept the water of the Red River outside the community during the 1997 flood. However, because of the very high risk of flooding all the residents were evacuated. Many people took temporary measures to protect their properties before leaving the community. Some moved their valuables to a higher elevation in the house; some covered all the lower-level openings to prevent water from entering the house; some even created smaller temporary sand dikes around their houses.

One resident, however, went even further and filled his basement with clean water, thinking that the clean water would reduce the damage that could be caused if the basement was to flood with polluted water from the wastewater drainage system.

After the great flood of 1997, this was the only flooded property in the town of Emerson.

(A memory from 1997)

Introduction

Flood management in general comprises different water resources activities aimed at reducing the potential harmful impact of floods on the people, environment and economy of a region. Sustainable flood management decision-making requires integrated consideration of the economic, ecological and social consequences of disastrous floods. While economic considerations are given priority in traditional decision-making approaches, the empowerment of stakeholders is an issue that is now demanding increased attention. Flood management activities (i.e. disaster mitigation, preparedness and emergency management) may be designed and achieved without the direct participation of stakeholders. However they cannot be implemented without them. In order to decide about the flood control measures to be adapted in a floodplain, the decision-making process should include different stakeholders. Government policy-makers and professional planners are central to the process, but others such as the general public, communities affected by the decision outcomes, non-governmental organizations and different interest groups should be included as well.

In the wake of the 1997 flood that devastated communities along the Red River in Canada and the United States, work continues to minimize the impact of future flooding on flood victims. A common criticism among the communities in Canada affected by the Red River flooding is the lack of their involvement in decisions on flood control and flood protection measures implemented by the government (Simonovic and Carson, 2003). An International Joint Commission (IJC) was formed by the US and Canadian governments to evaluate the existing flood management plan, after the 1997 flood. In 2000 it published a report (IJC, 2000) with the following recommendation:

> The city of Winnipeg [the largest community in the floodplain], province of Manitoba, and the Canadian federal government should cooperatively develop and finance a long-term flood protection plan that fully considers all social, environmental and human effects of any flood protection measures and respects both the needs of Winnipeg and the interests of those outside the city who might be affected by such a plan.

The objective of the collaborative work presented here and performed by the University of Western Ontario and the communities in the Red River basin is to develop a multi-objective decision-making methodology for a participatory process governing flood management in the Red River basin. This methodology is able to:

- evaluate potential alternatives based on multiple criteria under uncertainty;
- accommodate the high diversity and uncertainty inherent in human preferences;
- handle a large amount of data collected from stakeholders in the Red River basin.

Methodology

The flood management process in Canada, as elaborated for the Red River basin by Simonovic (1999), has three major stages: planning, flood emergency management and post-flood recovery. Appropriate decision-making in each of these stages is very important to establish an efficient flood management process. During the planning stage, different alternative measures (both structural and non-structural) are analysed and compared for possible implementation in order to minimize future flood damage. Flood emergency management includes regular evaluation of the current flood situation and daily operation of flood control works. The evaluation process includes the identification of potential events that could affect the flood situation (such as dike breaches, wind set-up, heavy rainfall, etc.) and the identification of corresponding measures for fighting floods (including building temporary structures or upgrading existing ones). Also, from the evaluation of the situation, decisions are made regarding the evacuation and re-population of flood-affected areas. Post-flood recovery involves numerous decisions regarding the return to normal life. The main issues during this stage include the assessment and rehabilitation of flood damage, and provision of flood assistance to flood victims. In all three stages, the decision-making process takes place in a multidisciplinary and multi-participatory environment.

Flood management decision-making problems are complex because of their multi-objective nature. For a given goal, many alternative solutions may exist that provide different levels of satisfaction for different issues, such as the environmental, social, institutional and political. These concerns naturally lead to the use of multi-objective decision-making techniques in which there is a trade-off between the objectives to find the most desirable solution. Multiple-objective decision-making becomes more complicated with an increase in number of individuals/groups involved in the decision-making process (see Section 10.5.1). The decision problem is no longer limited to the selection of the most preferred alternative among the non-dominated solutions according to a single set of preferences: the analysis must be extended to account for the conflicts among different stakeholders with different objectives. Therefore, it is a real challenge to generate a group decision outcome that can satisfy all those involved in the decision-making process.

In general, the process of decision-making basically involves deriving the best option from a feasible set of alternatives. Most of the existing approaches in multiple criteria decision-making with a single stakeholder/decision-maker consist of two phases: first, the aggregation of the judgements with respect to all criteria and decision alternatives, and second, the ranking of the decision alternatives according to the aggregated judgement. In the case of multiple stakeholders, an additional aggregation is necessary with respect to the judgements of all the stakeholders. Group decision-making under multiple criteria involves diverse and interconnected fields such as preference analysis, utility theory, social choice theory, voting, game theory, expert evaluation analysis, aggregation and economic equilibrium theory.

In the development of the methodology we shall start with Compromise programming (equation 10.18 from Section 10.3.1). Flood management decision-making is always associated with some degree of uncertainty. This uncertainty can be categorized (see Section 6.1) into two basic types: uncertainty caused by inherent hydrologic variability and uncertainty caused by lack of knowledge. The second type of decision uncertainty is more profound in areas of public decision-making, such as flood management. Capturing the views of individuals presents the problem of uncertainty. The major challenge in collecting these views is to find a technique that will capture those uncertainties, and will also be usable in a multi-objective tool like fuzzy Compromise programming (equation 10.22 in Section 10.4.1).

Participation of multiple stakeholders

An aggregation procedure is one of the ways to include information from the participating decision-makers in the decision matrix. The available methods do not seem to be appropriate for flood management for two reasons. First, all the available methods collect information from multiple participants using relatively complicated procedures. Where the participating decision-makers are from both technical and non-technical backgrounds, as in the case of flood management, it is not feasible to apply complicated procedures. The second reason is that when the responses are collected from a large number of participants, there may be a number of common responses. This overlap will not be reflected in the results of traditional (direct aggregation) methods.

The methodology of the present study (Akter and Simonovic, 2005) includes the representation of inputs from a large number of participants and the analysis of the inputs to make them usable in this context. Fuzzy set theory and fuzzy logic are used to represent the uncertainties in stakeholders' opinions. Three possible types of fuzzy input have been considered to capture the subjectivity of the responses from stakeholders (Akter et al, 2004). When a stakeholder is asked to evaluate an alternative against a particular objective/criterion, the answer may take one of the following forms:

(a) a numeric-scale response;
(b) a linguistic answer (e.g. poor, fair, good, very good);
(c) an argument (e.g. 'if some other condition is satisfied then it is good').

For the first type, the input is quite straightforward. For type (b), it is necessary to develop membership functions for the linguistic terms. Type (c) input can be described by using a fuzzy inference system, which includes membership functions, fuzzy logic operators and an if–then rule. For this, the membership functions for the input arguments need to be developed first, then fuzzy operator and fuzzy logic are applied to obtain the output. It should be noted that the interpretation of input values of the last two types is highly dependent on the shape of the membership functions and the degree of severity chosen by the expert for a particular application.

After receiving the inputs from all stakeholders, the next step is to aggregate those inputs to find a representative value. It is obvious that for all input types considered above, the responses are sure to be influenced by the number of repetitions. Many respondents may provide the same response. This implies that the general methodologies of fuzzy aggregation cannot be applied for deriving the resultant input from a large number of decision-makers. The FEV method can be used instead.

This is the definition of an FEV: Let χ_A be a B-measurable function such that $\chi_A \in [0,1]$. The FEV of χ_A over the set A, with respect to the fuzzy measure μ, is defined as:

$$\text{FEV}(\chi_A) = \sup_{T \in [0,1]} \{\min[T, \mu(\xi_T)]\} \qquad (10.33)$$

where

$$\xi_T = \{x \mid \chi_A(x) \geq T\} \qquad (10.34)$$

and

$$\mu\{x \mid \chi_A(x) \geq T\} = f_A(T) \text{ is a function of the threshold } T \qquad (10.35)$$

Figure 10.22 provides a geometric interpretation of the FEV. Performing the minimum operator, the two curves create the boundaries for the remaining triangular curve. The supremum operator returns the highest value of which graphically represents the highest point of the triangular curve. This corresponds to the intersection of the two curves where $T = H$.

The FEV can be computed for all three types of input. For type (a) input, the FEV should be a numeric value between 0 and 1. For both type (b) and type (c) inputs, the FEVs are membership functions. The crisp numeric equivalents of these membership functions can be obtained by applying a defuzzification method, and they can then be compared with type (a) answers.

Box 10.3 Super dike

During a visit to Japan in 1996 I had the opportunity to visit the Tone River Water Authority. My Japanese colleagues were explaining to me the complex flood protection system in their main office when my eyes wandered through the window.

The office was on the third floor and I was still looking into the body of the dike. Behind it was a population of approximately 7 million people.

(A memory from 1996)

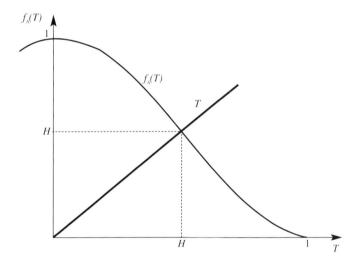

Figure 10.22 *A geometric interpretation of the FEV*

The centroid of the area defuzzification method (equation 6.50 in Section 6.5.4) was used here. It returns a value obtained by averaging the moment area of a given fuzzy set. Mathematically, the centroid, $\bar{x}$, of a fuzzy set, A, is defined in equation (6.50) as:

$$\bar{x} = \frac{\int_0^1 x \cdot \mu_A(x) \, dx}{\int_0^1 \mu_A(x) \, dx}$$

where:
$\mu_A(x)$ = the membership function of the fuzzy set A.

The resultant FEVs are now the aggregated evaluation of the alternatives from all the stakeholders. They can be used as the input value in the payoff matrix (Table 10.1 in Section 10.1.4) for the multi-objective analysis.

Participatory multi-objective decision-making under uncertainty

This work uses fuzzy Compromise programming. The driving force for the transformation from a classical to a fuzzy environment was that there is a need for accurate representation of subjective data in flood control decision-making. The theory of fuzzy sets can represent the subjective data well. Thus, instead of using crisp numbers in the Compromise programming distance metric equation, fuzzy numbers are used; instead of using classical arithmetic, fuzzy arithmetic is applied; instead of simply sorting distance metrics, fuzzy set ranking methods are applied to sort fuzzy distance metrics. In other words, the fuzzy transformation complicates the interpretation of the results but models the decision-making process more realistically.

Mathematically, the Compromise programming distance metric in its discrete form is given in equation (10.18) as:

$$L_j = \left[\sum_{z=1}^{t} \left\{ w_z^P \left(\frac{f_z^* - f_z}{f_z^* - f_z^-} \right)^P \right\} \right]^{\frac{1}{P}}$$

where, using new notation:

- z = 1, 2, 3 ... t criteria
- j = 1, 2, 3 ... n alternatives
- L_j = distance metric of alternative j
- w_z = weight of a particular criteria
- p = a parameter ($p = 1, 2, \infty$)
- f_z^* and f_z^- = the best and the worst value for criteria z, respectively (also referred to as positive and negative ideals)
- f_z = actual value of criterion z.

Bender and Simonovic (2000) fuzzified Compromise programming and thus formulated fuzzy Compromise programming as presented in Section 10.4 and relationship (10.22).

In fuzzy Compromise programming, obtaining the smallest distance metric values is not easy, because the distance metrics are also fuzzy. To pick out the smallest fuzzy distance metric from a group of distance metrics, fuzzy set ranking methods have to be used. Section 10.4 recommended the method of Chang and Lee (1994), because it gives most control in the ranking process, with degree of membership weighting and weighting of the subjective type. Recall (from (6.53), Section 6.5.4) that their OERI has the following mathematical form:

$$OERI(j) = \int_0^1 \omega(\alpha)[\chi_1 \mu_{jL}^{-1}(\alpha) + \chi_2 \mu_{jR}^{-1}(\alpha)]d\alpha$$

where the subscript j stands for alternative j, α represents the degree of membership, χ_1 and χ_2 are the subjective type weighting indicating neutral, optimistic or pessimistic preferences of the decision-maker, with the restriction that $\chi_1 + \chi_2 = 1$; parameter $\omega(\alpha)$ is used to specify weights that are to be given certain degrees of distance metric membership (if any); and $\mu_{jL}^{-1}(\alpha)$ represents an inverse of the left part, and $\mu_{jR}^{-1}(\alpha)$ the inverse of the right part of the distance metric membership function.

For χ_1 values greater than 0.5, the left side of the membership function is weighted more than the right side, which in turn makes the decision-maker more optimistic. Of course, if the right side is weighted more, the decision-maker is more of a pessimist (this is because he or she prefers larger distance metric values, which means the farther solution from the ideal solution). In summary, the risk preferences are: if $\chi_1 < 0.5$, the user is a pessimist (risk averse); if $\chi_1 = 0.5$, the user is neutral; and if $\chi_1 > 0.5$, the user is an optimist (risk taker). Simply stated, OERI is a sum of the weighted areas between the distance metric membership axis and the left and right inverses of a fuzzy number.

Red River basin flood management

One of the problems at the planning stage in the Red River basin was the complex, large-scale problem of ranking potential flood management alternatives. During the evaluation of alternatives, it is necessary to consider multiple criteria which may be both quantitative and qualitative, and the views of numerous stakeholders.

At present the government of Manitoba, Canada is responsible for making decisions about flood management measures (Simonovic, 2004). The decision-making process involves consulting different organizations for their technical input. The concerns of the general public about the alternatives are gathered through public hearings and workshops. Economic analysis plays an important role in formulating plans for reducing flood damages and making operational decisions during the emergency. One of the main limitations of the previous flood management methodology was its high emphasis on the economic criterion. Very minor attention was given to the environmental and social impacts of floods.

The general public have shown increasing concern about decisions on flood control measures. During the 1997 flood some stakeholders in the basin, particularly the floodplain residents, felt they did not have adequate involvement in flood management decision-making. They expressed particular dissatisfaction about evacuation decisions during the emergency management, and about compensation decisions during post-flood recovery (IJC, 2000).

The new methodology was used to collect information from the stakeholders across the Canadian portion of the Red River basin. In order to evaluate the utility of the methodology, it was tested with three generic alternatives for improved flood management. A flood management payoff (decision) matrix with relevant criteria and theoretical alternatives was developed, as shown in Table 10.20.

Table 10.20 *Flood management payoff (decision) matrix*

	Economic criteria			Environmental criteria			Social criteria	
	Cost	Damage	Benefit	Chemical contamination	Alien species	Environment	Community involvement	Personal loss
Structural alternative	e_{11}	e_{12}	e_{13}	e_{14}	e_{15}	e_{16}	e_{17}	e_{18}
Non-structural alternative	e_{21}	e_{22}	e_{23}	e_{24}	e_{25}	e_{26}	e_{27}	e_{28}
Combination alternative	e_{31}	e_{32}	e_{33}	e_{34}	e_{35}	e_{36}	e_{37}	e_{38}
Weight coefficient	W_1	W_2	W_3	W_4	W_5	W_6	W_7	W_8

Note: e_{mn} = stakeholder preference.

The three generic options considered were structural alternatives, non-structural alternatives and a combination of both. The selection of criteria against which the alternatives are ranked is one of the most difficult but important tasks of any multi-objective decision analysis. Here the selection was based mainly on prior studies of Red River flooding (Morris-Oswald et al, 1998; IJC, 2000). Economic objectives (cost, damage, benefit, etc.) are in general the most important, and are also straightforward to quantify. Environmental objectives (chemical contamination; inter-basin transfer of alien invasive species, and protection and enhancement of the floodplain environment) are highly important too. Generally, most flood management decision-making processes exclude or ignore social objectives. This is mainly because of the difficulties inherent in selecting and quantifying them. However, both studies of Red River floods and numerous interviews with stakeholders made it clear that it was of prime importance to consider social impacts if a flood management policy in the Red River basin was to be implemented successfully. Two social objectives were considered in the case study: the level of community involvement, and the amount of personal losses (include financial, health and psychological losses).

A detailed survey was conducted in the basin to collect information on these social criteria (Akter et al, 2004). It used a survey questionnaire to allow stakeholders to express their views in an easy way. Therefore, the remainder of this presentation focuses on the application of the methodology using the three generic alternatives and real data on the two social criteria.

Each objective was expanded through a set of questions rather than a single preference value:

Objective 1: community involvement

1. What is the level of opportunity provided by each alternative to get involved during the planning stage of flood protection?
2. What is the level of opportunity provided by each alternative to get involved during the time of flooding?
3. To what degree does each alternative induce a sense of complacency to rely heavily on the governmental project?
4. What is the level of technical contribution that you would be able to provide for each alternative?
5. How much training is required for each alternative to be actively involved in flood management activities?
6. What is your level of willingness to participate in such activity for (a) your personal estate; (b) your local community; and (c) the city of Winnipeg?
7. What is the role of leadership to the successful execution and implementation of each alternative?
8. Rate the alternatives according to the degree to which they promote local leadership and community tightness.

Objective 2: personal loss

1. What is the severity of economic loss (land, homestead and business) at a personal level for each alternative?
2. Rate the degree of impact on personal health each alternative would expose the public to during a flood.
3. Rate the level of stress induced in the daily lives of the public by each alternative (a) during the planning and preparation; and (b) during a flood.
4. What is the level of personal safely provided by each alternative?
5. Rate the level of control an individual has over the flood protection measures to be implemented.

Thirty-five respondents were interviewed. Each was asked to answer each question in three forms: (a) using a numeric scale with the range 0–1; (b) using linguistic answers (very low, low, medium, high, very high); and (c) using conditional answers of the form [IF flooding is moderate THEN (very low, low, medium, high, very high)] [IF flooding is severe THEN (very low, low, medium, high, very high)]. The purpose of collecting these three data sets was to compare the effect of uncertainty on the resultant values.

The three response types were:

Type (a): a crisp scale response. This response type simplifies data processing but forces participant to rank their preference on a discrete scale. Crisp values fall short in capturing the uncertainty in human preferences and approximate answers. For this study, the scale chosen was from 0 to 10, where 0 means very low, and 10 very high. Note that this scale was selected for a generic example. In a real decision-making situation, it is necessary to investigate before selecting the scale.

Type (b): a linguistic fuzzy response. This response type complicates data processing but allows participants to respond with familiar preferences in the form of words. Linguistic fuzzy sets allow stakeholder preferences to vary in degree of membership, better modelling human views and opinions. The participants could use one of the following five linguistic terms: very low, low, medium, high and very high, to answer each question.

Type (c): a conditional fuzzy response. This response type adds the dimension of circumstance to stakeholder preference. Participants' opinions are often influenced by the nature of the situation or outcome of other events. Through a set of logical rules, the varying preference is captured by preconditioning survey questions with different circumstances. The participants were asked to answer the questions in one of the linguistic terms (very low, low, medium, high and very high) for two flooding conditions: if a severe flood is expected, or if a moderate flood is expected.

Table 10.21 summarizes the three response types. All three types of inputs obtained from all the stakeholders were processed (Akter et al, 2004) using the FEV method. For the conditional response, the response from each person was first processed to get the crisp value, and then all the responses were further processed to obtain the FEV using the method for scale responses.

Table 10.22 summarizes the results of all three types of inputs (here termed A, B and C) as the evaluation of three alternatives (structural, non-structural, combination) against two criteria (community development, personal loss). The results show good correlation between the numeric scale and linguistic types of inputs, with an average difference of only 0.029. The conditional-type results show a consistently slightly lower value. This is because to obtain the linguistic input from the conditional statements, a level of severity had to be assigned to the flooding, and we rated

Table 10.21 *Three response types as they appear in the survey*

Type (a) Crisp scale	A value on a scale of 0 to 10 is chosen to represent the degree to which the interviewee's view coincides with the scale measure	0　　　　　　　　　　　　　10
Type (b) Linguistic fuzzy	A word response is chosen to represent a 'fuzzy' view – a range on the scale rather than an exact answer	Very low low medium high very high
Type (c) Conditional fuzzy	A conditional response includes an if–then statement. The response given requires an event to occur first before it is true. This survey uses two specific conditions or evaluation.	1 IF severe flood expected THEN 2 IF moderate flood expected THEN 　very low low medium high very high

the 1997 flood at 0.7 on the scale from 0 to 1. This value is subject to change according to expert opinion, and if a higher value were chosen the results would be closer to the other type values.

All three methods used in this study appeared equally accurate in representing the stakeholders' views, and no attempt was made to measure the degree to which one was superior to another.

The FEVs in Table 10.22 were then used to rank the three generic alternatives. All questions were considered to carry the same weight. A set of ranking experiments was conducted to evaluate the impact of different stakeholder groups on the final rank of alternatives:

- experiment 1 – all stakeholders interviewed;
- experiment 2 – stakeholders from the city of Winnipeg;
- experiment 3 – stakeholders from the Morris area (south of Winnipeg;
- experiment 4 – stakeholders from the Selkirk area (north of Winnipeg).

Figure 10.23 shows for illustrative purposes criterion 1, criterion 2 and the resultant distance metric membership functions obtained in evaluation of alternative 1 (structural flood management option) for these four stakeholder groups.

The final results of four ranking experiments with three generic alternatives and two social criteria are shown in Table 10.23 (the defuzzified distance metric value with the rank in brackets). It is obvious that the final rank varies with the experiment, therefore confirming that preferences of different stakeholders are captured by the developed methodology.

Table 10.22 *Resultant FEVs*

Alternative	Structural			Non-structural			Combination		
Type	A	B	C	A	B	C	A	B	C
Question No.	FEV	FEV	FEV	FEV	FEV	FEV	FEV	FEV	FEV
Community Involvement									
1	0.600	0.650	0.544	0.647	0.650	0.544	0.600	0.625	0.544
2	0.529	0.517	0.500	0.500	0.517	0.491	0.500	0.570	0.544
3	0.618	0.700	0.529	0.559	0.625	0.529	0.600	0.625	0.544
4	0.600	0.650	0.544	0.657	0.650	0.559	0.686	0.650	0.544
5	0.700	0.700	0.559	0.629	0.650	0.544	0.700	0.650	0.544
6a	0.800	0.825	0.677	0.704	0.770	0.588	0.800	0.825	0.647
6b	0.771	0.770	0.588	0.714	0.717	0.574	0.743	0.770	0.574
6c	0.700	0.700	0.574	0.629	0.650	0.574	0.686	0.700	0.574
7	0.800	0.825	0.735	0.829	0.850	0.718	0.857	0.825	0.718
8	0.700	0.717	0.574	0.700	0.650	0.574	0.700	0.700	0.574
Personal Loss									
1	0.800	0.770	0.718	0.700	0.700	0.574	0.700	0.717	0.671
2	0.588	0.570	0.544	0.600	0.650	0.544	0.600	0.625	0.574
3a	0.500	0.570	0.574	0.559	0.625	0.574	0.559	0.570	0.574
3b	0.700	0.717	0.625	0.700	0.717	0.588	0.706	0.717	0.588
4	0.771	0.770	0.574	0.700	0.650	0.574	0.700	0.717	0.544
5	0.500	0.570	0.529	0.700	0.570	0.544	0.571	0.570	0.544

Table 10.23 *Final rank of flood management alternatives*

Participants	Alternative 1	Alternative 2	Alternative 3
All stakeholders	13.224 (1)	13.717 (3)	13.280 (2)
Morris area	15.435 (2)	16.086 (3)	13.636 (1)
Selkirk area	14.635 (3)	14.425 (1)	14.585 (2)
City of Winnipeg	13.746 (1)	15.259 (3)	13.923 (2)

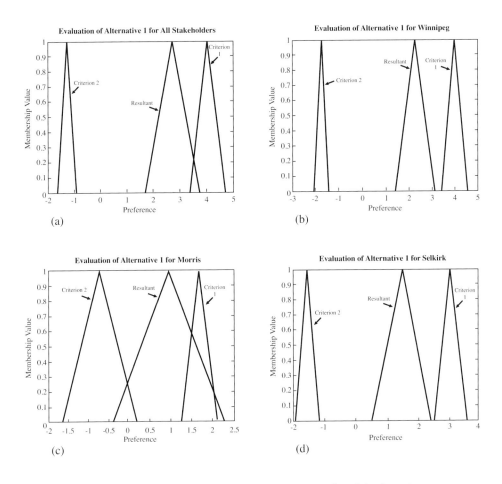

Figure 10.23 *The distance metric fuzzy membership functions*

Conclusions

Although flood control plans can be designed and achieved without stakeholders' participation, they cannot be implemented without it. So, flood management decision-making can be defined as a multi-objective group decision-making problem where alternatives are evaluated against a number of criteria/objectives considering the concerns of all stakeholders. As most of the decision-making processes take place in situations where the goals, the constraints and the consequences of the possible actions are not known precisely, it is necessary to include these types of uncertainty in the decision-making methodology. Fuzzy set and fuzzy logic techniques have been used successfully to represent the imprecise and vague information in many fields, and so are considered an effective way to represent uncertainties in this study.

This work documents an innovative methodology that provides alternative ways to extract and aggregate the inputs from a large number of stakeholders for flood management decision-making. FEV was used as a method to aggregate those inputs and generate the elements of the multi-criteria decision matrix for further analysis. Three possible types of responses for flood management were considered: numeric input, linguistic input and conditional input. The fuzzy Compromise programming technique (Section 10.4) was combined with the group fuzzy membership ranking (Section 10.5) to analyse the alternative flood management options.

The analyses of flood management options in the Red River basin show the applicability of the methodology for a real flood management decision-making problem. The stakeholders can now express their concerns regarding flood hazard in an informal way, and that can be incorporated into the multi-criteria decision-making model. This methodology helps solve the problem of including a large number of stakeholders in the flood decision-making process.

10.7 REFERENCES

Akter, T. and S.P. Simonovic (2005), 'Aggregation of Fuzzy Views of a Large Number of Stakeholders for Multiobjective Flood Management Decision-making', *Journal of Environmental Management*, 77(2): 133–143

Akter, T., S.P. Simonovic and J. Salonga (2004), 'Aggregation of Inputs from a Large Number of Stakeholders for Flood Management Decision-making in the Red River Basin', *Canadian Water Resources Journal*, 29(4): 251–266

Bender, M.J. and S.P. Simonovic (2000), 'A Fuzzy Compromise Approach to Water Resource Systems Planning under Uncertainty', *Fuzzy Sets and Systems*, 115: 35–44

Carlsson, C. and R. Fuller (1996), 'Fuzzy Multiple Criteria Decision-making: Recent Developments', *Fuzzy Sets and Systems*, 78: 139–153

Chang, P.T. and E.S. Lee (1994), 'Ranking of Fuzzy Sets Based on the Concept of Existence', *Computers and Mathematics with Applications*, 27: 1–21

Chen, S.H. (1985), 'Ranking Fuzzy Numbers with Maximizing Set and Minimizing Set', *Fuzzy Sets and Systems*, 17: 113–129

Das, P. and Y.Y. Haimes (1979), 'Multiobjective Optimization in Water Quality and Land Management', *Water Resources Research*, 15(6): 1313–1322

Dasgupta, P., A. Sen and S. Marglin (1993), *Guidelines for Project Evaluation*, United National Industrial Development Organization, Vienna, 383

David, L. and L. Duckstein (1976), 'Multi-Criterion Ranking of Alternative Long Range Water Resources Systems', *Water Resources Bulletin*, 12(4): 731–754

Gershon, M., R. McAniff and L. Duckstein (1980), 'A Multiobjective Approach to River Basin Planning', *Proceedings*, Arizona Section, American Water Resources Association, and The Hydrology Section, Arizona-Nevada Academy of Science, Las Vegas, NV, 41–50

Goicoechea, A., D.R. Hansen and L. Duckstein (1982), *Multiobjective Decision Analysis with Engineering and Business Applications*, John Wiley and Sons, New York

Goodman, A.S. (1984), *Principles of Water Resources Planning*, Prentice-Hall, Englewood Cliffs, NJ

IJC (International Joint Commission) (2000), *Living with the Red*, Ottawa, WA, http://www.ijc.org/php/publications/html/living.html (accessed December 2005)

Jimenez, F., J.M. Cardenas, G. Sanchez, A.F. Gomez-Skarmeta and J.L. Verdegay (2006), 'Multi-Objective Evolutionary Computation and Fuzzy Optimization', *International Journal of Approximate Reasoning*, 43(1): 59–75

Kacprzyk, J. and S.H. Nurmi (1998), 'Group Decision-making under Fuzziness', in *Fuzzy Sets in Decision Analysis, Operations Research and Statistics*, R. Slowinski, ed., Kluwer Academic Publishers, Boston, MA, 103–136

Karnib, A. (2004), 'An Approach to Elaborate Priority Preorders of Water Resources Projects Based on Multi-Criteria Evaluation and Fuzzy Sets Analysis', *Water Resources Management*, 18(1): 13–33

Keeney, R.L. and H. Raiffa (1993), *Decisions with Multiple Objectives: Preferences and Value Tradeoffs*, Cambridge University Press, Cambridge

Medaglia, A.L., S.B. Graves and J.L. Ringuest (2007), 'A Multi-Objective Evolutionary Approach for Linearly Constrained Project Selection Under Uncertainty', *European Journal of Operational Research*, 179(3): 869–894

Mishra, S. (2007), 'Weighting Method for Bi-level Linear Fractional Programming Problems', *European Journal of Operational Research*, 183(1): 296–302

Morris-Oswald T., S.P. Simonovic and J. Sinclair (1998), 'Efforts in Flood Damage Reduction in the Red River Basin: Practical Considerations', Report prepared for the Environment Canada, Environmental Adaptation Group, Institute for Environmental Studies, University of Toronto, Canada, 67

Novoa, J.I. and A.H. Halff (1977), 'Management of Flooding in a Fully-Developed Low-Cost Housing Neighbourhood', *Water Resources Bulletin*, 13(6).

Prodanovic, P. and S.P. Simonovic (2002), 'Comparison of Fuzzy Set Ranking Methods for Implementation in Water Resources Decision-making', *Canadian Journal of Civil Engineering*, 29: 692–701

Prodanovic, P. and S.P. Simonovic (2003), 'Fuzzy Compromise Programming for Group Decision-making', *IEEE Transactions on Systems, Man, and Cybernetics—Part a: Systems and Humans*, 33(3): 358–365

Ribeiro, R. (1996), 'Fuzzy Multiple Attribute Decision-making: A Review and New Preference Elicitation Techniques', *Fuzzy Sets and Systems*, 78: 155–181

Simon, H.A. (1976), *Administrative Behaviour*, 3rd edition, MacMillan, New York

Simonovic, S.P. (1987), 'The Implicit Stochastic Model for Reservoir Yield Optimization', *Water Resources Research*, 23: 2159–2166

Simonovic, S.P. (1989), 'Application of Water Resources Systems Concept to the Formulation of a Water Resources Master Plan', *Water International*, 14: 37–50

Simonovic, S.P. (1991), 'Coping with Changing Objectives in Managing an Existing Multipurpose Reservoir', *IAHS Publication*, 206: 181–191

Simonovic, S.P. (1999), 'Decision Support System for Flood Management in the Red River Basin', *Canadian Water Resources Journal*, 24(3): 203–223

Simonovic, S.P. (2004), 'CANADA – Flood Management in the Red River, Manitoba', *Integrated Flood Management Case Study*, World Meteorological Organization – The Associated Programme on Flood Management, Geneva, http://www.apfm.info/case_studies.htm (accessed December 2005)

Simonovic, S.P. (2007), 'Sustainable Floodplain Management – Participatory Planning in the Red River Basin, Canada', Chapter 11 in *Topics on System Analysis and Integrated Water Resource Management* (IWRM), A. Castelletti and R. Soncini-Sessa, eds, Elsevier, The Netherlands

Simonovic, S.P. and D.H. Burn (1989), 'An Improved Methodology for Short-Term Operation of a Single Multipurpose Reservoir', *Water Resources Research*, 25(1): 1–8

Simonovic, S.P. and R.W. Carson (2003), 'Flooding in the Red River Basin – Lessons From Post Flood Activities', *Natural Hazards*, 28: 345–365

Tauxe, F.W., R.R. Inman and D.M. Mades (1979), 'Multiobjective Dynamic Programming with Application to a Reservoir', *Water Resources Research*, 15(6): 1403–1408

Zadeh, L.A. (1983), 'A Computational Approach to Fuzzy Quantifiers in Natural Languages', *Computers and Mathematics with Applications*, 9: 149–184

Zeleny, M. (1974), 'A Concept of Compromise Solutions and the Method of the Displaced Ideal', *Computers and Operations Research*, 1(4): 479–496

10.8 EXERCISES

1 An SPS consultant has evaluated proposed sites for a multi-purpose water facility using two objectives: economic benefits (EB) and environmental quality improvement (EQUI). The optimal performance of each site is given by the following (EB, EQUI) pairs: A(20,135) B(75,135) C(90,100) D(35,1050) E(82,250) F(60,–50) G(60,550) H(75,500) I(70,620) J(10,500) K(40,350) L(30,800) N(55,250) O(40,–80) P(30,500) Q(30,900) R(60,950) S(80,–150) T(45,550) U(25,1080) V(70,800) W(63,450).

   a. Solve the problem graphically. List the excluded, dominated and non-dominated alternative sites.

b. Identify the non-dominated solutions by maximizing EB subject to EQUI ≥ b where b = 200, 400, 600 and 800.
c. Identify a non-dominated solution by the weighting method for a relative set of weights to (EB, EQUI) of (20,8).

2. Consider the multi-objective problem:

$$max\ (5x_1 - 2x_2)\ and\ max\ (-x_1 + 4x_2)$$

subject to:

$$-x_1 + x_2 \leq 3$$
$$x_1 + x_2 \leq 8$$
$$x_1 \leq 6$$
$$x_2 \leq 4$$
$$x_i \geq 0$$

a. Graph the non-dominated solutions in the decision space.
b. Graph the non-dominated combinations of the objectives in the objective space.

3. Reformulate the problem in Exercise 2 to illustrate the weighting method for defining all non-dominated solutions of part (a), and illustrate this method in decision and objective space.

4. A new civil engineering graduate has job offers from five different companies and is faced with the problem of selecting a company within a week. The individual lists all of the important information in the form of a payoff matrix:

Criterion	Company				
	1	2	3	4	5
Salary ($)	34,000	33,000	36,000	37,500	32,000
Workload (hr/day)	9	7.5	10.5	12	6
Location*	2	3	5	2	4
Additional income	4800	4600	2600	5500	6000
Continuing education*	4	4	3	1	5
Additional benefits*	3	2	4	5	1

Note: *Larger numbers represent more desirable outcomes.

Acting as the new graduate, apply the weighting method and explain your decision.

5. For the wastewater treatment problem in Section 4.4.3:
a. Examine the problem using linear programming (LP) and solve it using

Linpro in order that profits for the steel plant may be optimized according to the city's effluent charges.
   b. Apply the weighting method for w = 1,2 to assist the steel company in making profit projections that reflect the city's current effluent charges and a contemplated doubling of those charges.
6 Under what circumstances will the weighting method fail to identify efficient solutions?
7 Evaluate a plan for disposing of wastewater from a treatment plant. For the given situation you want to use the best combination of two methods: (i) a filter for tertiary treatment and (ii) irrigation with the effluent. The net benefits are: irrigation $5 per cubic metre per second (m^3/s); tertiary treatment –$2 per m^3/s. Another objective is to maximize the dissolved oxygen (DO) in the river bordering the treatment plant and irrigated area. The DO will decrease by 1 mg/litre for each m^3/s of effluent irrigated and increase by 4 mg/litre for each m^3/s given tertiary treatment. The amount of effluent filtered must be less than 3 m^3/s plus the amount of effluent used for irrigation (measured in m^3/s). The total system (tertiary with irrigation) cannot exceed 8 m^3/s; moreover, physical constraints on the capacity of the available land limit the amount for irrigation to 6 m^3/s and the amount of effluent for tertiary treatment to 4 m^3/s.
   a. Considering both the economic and the water quality objectives, conduct a multi-objective analysis using the weighting method.
   b. What is your recommendation?
   c. Why?
8 Evaluate the three alternatives presented below using a multi-objective analysis method of your choice. In your evaluation use three objectives: economic (NED), environmental enhancement (EQ) and regional development (RD). Three alternatives may be used in various combinations: flood control, hydropower and water quality control (mostly low-flow augmentation measure in m^3/s). The flood control alternative, a levee that cannot exceed 2 m in height, will yield $1000/m in annual benefits for both the NED and RD objectives and destroy 20 environmental units per metre of levee height. The hydropower alternative cannot exceed 2 MW of power and will yield $1000/MW for the NED account but only $500/MW for the RD. It will add 10 environmental units per MW. The water quality alternative will yield $500 per m^3/sec of flow NED. It cannot exceed 2 m^3/s and will yield $1000 per m^3/s for RD. It will yield 10 environmental units per m^3/s. The sum of each metre of levee height plus each MW of power plus 1.25 m^3/s of flow augmentation must not exceed 5.0. Your goal is to maximize NED, RD and EQ.
9 For Example 2 in Section 10.3.1:
   a. Identify a compromise set of solutions for $\alpha_1 = 0.1$ $\alpha_2 = 0.9$ and $p = 1, 2, 3, 20, 100$.

b. Identify a compromise set of solutions for $\alpha_1 = 0.9$ $\alpha_2 = 0.1$ and $p = 1, 2, 3, 20, 100$.
c. Compare your solutions to the solution in Table 10.6. Discuss the difference.
d. Use Compro to check your solutions.

10 Use the Compro program to solve the problem from Exercise 5.
   a. Select weights for your criteria and explain your choice.
   b. How sensitive is your solution to the change in weighting of criteria?
   c. What is the most robust compromise solution of the problem?

11 Use Compro to solve the Tisza River problem from David and Duckstein (1976) shown in Table 10.7:
   a. Select weights for the criteria and explain your choice.
   b. Perform an extensive sensitivity analysis to changes in qualitative data, weights, and time horizon.
   c. What is the most robust compromise solution of the problem?

12 For the Tisza River problem from Bender and Simonovic (2000) presented in Section 10.4.4:
   a. Change the fuzzy definitions for linguistic terms and quantitative criteria. Keep the rest of the inputs as given in Section 10.4.4.
   b. Explain the reasoning for the proposed change.
   c. Use FuzzyCompro to solve the problem.
   d. Compare your solution to the one in Table 10.8. Discuss the difference.

13 For the Tisza River problem from Bender and Simonovic (2000) presented in Section 10.4.4:
   a. Develop a new fuzzy definition of degree of compensation, $\tilde{p}$, which is different from the one presented in Figure 10.16b. Keep the rest of the inputs as given in Section 10.4.4.
   b. Explain the reasoning for the proposed change.
   c. Use FuzzyCompro to solve the problem.
   d. Compare your solution to the one in Table 10.8. Discuss the difference.

14 For the Tisza River problem from Bender and Simonovic (2000) presented in Section 10.4.4:
   a. Develop a new fuzzy definition of weights, $\tilde{\alpha}_i$, that is different from the one presented in Figure 10.16c. Keep the rest of the inputs as given in Section 10.4.4.
   b. Explain the reasoning for the proposed change.
   c. Use FuzzyCompro to solve the problem.
   d. Compare your solution with the one in Table 10.8. Discuss the difference.

15 For the water supply example from section 10.6.1:
   a. Apply a fuzzy definition of degree of compensation, $\tilde{p}$, as presented in Figure 10.16b. Keep the rest of the inputs crisp, as given in Section 10.6.1.

b. Use FuzzyCompro to solve the problem S1 and S2 for the third set of weights (Table 10.17).
   c. Compare your solutions with those in Table 10.18 (column for the weight set 3) and Table 10.19 (column for the weight set 3). Discuss the differences.
16 Perform an experiment similar to Example 4 in Section 10.5.2:
   a. Identify five colleagues who will serve as decision-makers.
   b. Obtain their preferences regarding the 12 criteria used in the Tisza River basin example.
   c. Using the results from the previous step, develop fuzzy weights giving every decision-maker the same level of fuzziness and using triangular fuzzy membership functions. Assume that each decision-maker has an equal level of importance and neutral risk preferences.
   d. Use FuzzyComproGDM to solve the problem.
   e. Discuss your solution.
17 For the water supply example from section 10.6.1:
   a. Apply a fuzzy definition of degree of compensation, $\tilde{p}$, as presented in Figure 10.16b. Keep the rest of the inputs crisp, as given in Section 10.6.1.
   b. Identify three colleagues who will serve as decision-makers.
   c. Obtain their preferences regarding the eight criteria for system S1 and system S2.
   d. Using the results from the previous step, develop fuzzy weights, $\tilde{\alpha}_j$, giving every decision-maker the same level of fuzziness and using triangular fuzzy membership functions. Assume that each decision-maker has an equal level of importance and neutral risk preferences.
   e. Use FuzzyComproGDM to solve the problem.

PART V
Water for Our Children

11

The Future of Water Resources Systems Management

The management of water resources systems is as old as human civilization. Its future will have a lot to do with the future development of civilization. The premise of this book and the ultimate goal is water for everyone. Achievement of that goal is going to be difficult. Why is it so difficult? In brief, it is an exceedingly complex problem and one that transcends the knowledge of many individual disciplines. To guide future water resources management towards the goal of 'Water for everyone' requires a 'real' interdisciplinary approach to be developed and implemented. Solutions of complex problems are on the boundaries between various disciplines. I wrote this book primarily for water resources engineers. The challenge for them is whether they will take more of a position of provider of appropriate technical information and solutions, or whether they can assimilate enough understanding of all the issues to enable them to lead and bring human civilization closer to the goal.

In this final chapter I briefly point out a number of issues that I believe will challenge us in the future, then venture into identifying some of the emerging sciences that may provide the ground for innovative solutions.

11.1 EMERGING ISSUES

This will be a very limited view of the future deeply rooted in already evident trends. The world's growing population needs more water. The global move from rural to urban settlements continues, and the challenge is to provide a safe, sufficient and sustainable water supply and sanitation. The climate is changing and affects spatial and temporal patterns of precipitation. Both the frequency and magnitude of floods and droughts are changing too. This set of physical trends is occurring in the middle of changing economics of water. The ability of governments to manage existing complex water resources systems and continue their future development seems to be limited, and therefore new partnerships with the private sector are being investigated. Pressure is growing to treat water as a commodity. At the same time the general population are becoming more and more aware of the seriousness of water issues, and are looking for an opportunity to be more relevant in making future water resources management decisions.

11.1.1 Climate variability and change

The most recent discussions of the Intergovernmental Panel on Climate Change (IPCC) expert group on water have identified from the literature the following emerging impacts of climate change on freshwater resources (IPCC, 2007).

- Climate-driven changes in river flow and other components of the water cycle have already been observed. Even stronger changes are projected. For example, very strong winter climate-related runoff increase (typically between 50 and 70 per cent within the last two decades) has been detected in the most pristine Russian rivers.
- Floods and droughts have become more severe in some regions and are very likely to increase in severity still further. It is well established that precipitation characteristics have changed, and they will continue to change towards more intense and intermittent spells. This translates into more frequent and more severe water-related extremes (floods and droughts). For example, in England and continental Europe the changes in high river flow detected from long-term gauge records are already statistically significant. Flood and drought damage depends on the exposed populations, economies and societies and their adaptive capacity.
- Water demand is likely to grow as a result of climate change. In some regions, especially those with high demographic growth, the water needs for irrigation will considerably increase while available water resources may decrease. The impact of climate change on water demand is strongly dependent on adaptation, but an increase in conflicts between water uses (domestic, agriculture and industry) may be expected.
- Climate change impacts on water quality are likely to be serious. In general, water quality may be degraded through higher water temperatures. Where flows decrease, water quality decreases. But even if flow increase causes increasing dilution, floods may lead to water quality problems, such as pollutants and sewage flushed by runoff and overflowing sewage treatment plants. A rise in sea

level will increase the risk of saltwater intrusion to groundwater. Together with over-pumping of fresh groundwater, saltwater intrusion is endangering the water supply of many people populating large urban areas located close to the sea.
- Climate change is one of multiple pressures on water resources. In many areas and in particular in water shortage areas, anthropogenic pressures, such as population and economic growth and land use change, and not climate change, are the decisive factors behind adverse changes in freshwater resources. However, climate change will make the situation more difficult.
- Quantitative projections of changes in hydrological characteristics at the watershed scale are very uncertain. Precipitation, the principal input to water systems, is not reliably simulated in climate models. One important implication is that adaptation procedures need to be developed that do not rely on precise projections of changes in river discharge, groundwater, etc.
- Integrated water management must be extended to include the effects of climate change. The past can no longer be the key to the future. Integrated water management, necessary for solving increasingly complex water problems, needs to take into account climate change and consider adaptation options. Evidence so far suggests that climate change affects the water resources decision-making process. Technology is only one of the tools that can help to control the effects of climate change on water quantity and quality. However, other tools in the economic and social domains are necessary, for both developed and developing countries.

11.1.2 Water as a social and economic good

Many past failures in water resources management are attributable to the fact that water is viewed as a free good, or at least that the full value of water has not been recognized (GWP, 2000). In a situation of competition for scarce water resources, such a notion may lead to water being allocated to low-value uses and provides no incentives to treat water as a limited asset. One view is that perceptions about water values must change in order to extract the maximum benefits from the available water resources. Treating water as an economic good may help to balance the supply and demand of water, thereby sustaining the flow of goods and services from this important natural asset. When water becomes increasingly scarce, continuing the traditional policy of extending supply is no longer a feasible option. There is a clear need for operational economic concepts and instruments that can contribute to management by limiting the demand for water. Importantly, if charges for water goods and services reflect the full cost involved, managers will be in a better position to judge when the demand for different water products justifies the expenditure of scarce capital resources to expand supply.

An opposing view raises the social consequences of the 'economic good' concept. How would this affect poor people's access to water? The full value of water consists of its use value, or economic value, and intrinsic value. The economic, value which depends on the user and the way it is used, includes the value to (direct) users of water, net benefits from water that is lost through evapotranspiration or other sinks (e.g. return flows), and the contribution of water to the attainment of social objectives.

The future will judge which view has more merits. I would like to support the opposing view. Senge (2003) very eloquently brings to the forefront some of the issues with long-term consequences:

> I think the Industrial Age is a historic bubble, just like the 'dot com' financial bubble. I do not think it will continue, because I don't think it can continue. The Industrial Age has ignored the reality that human beings are part of nature; instead, it has operated based on the idea that nature is a resource waiting to be used by us. If we go back to the idea of independency, human beings depend on nature in many ways for our survival. This is where traditional economics breaks down. Economics says that if the price of a commodity rises, demand for it will go down and a less expensive substitute will replace it. But there are no substitutes for air and water. There is no substitute for a healthy climate. These are common elements shared by everybody. Systems of management that do not value 'commons' cannot continue indefinitely. We don't know when we will hit the wall – we're probably hitting it right now. By some estimates, private soft-drink companies now own rights to more than 10% of the drinkable water in the world. If these companies are allowed to continue their current system of management, which focuses on exponential growth of their products, this percentage will grow even further. We have not yet seen the implications of some of our patterns of development. (p2)

11.1.3 Urbanization

The world has become increasingly urban over the last century, as cities have developed and expanded as centres of commerce, industry and communication. Today, it is estimated that about half of the world's population lives in urban areas, a figure that is expected to rise to 60 per cent by 2030. For example, in Canada, roughly 60 per cent of the population lives in urban areas of 100,000 people or more and about 80 per cent in urban areas of 10,000 or more. Cities are important drivers of the economy: for example, the seven largest cities in Canada generate almost 45 per cent of the national gross domestic product (GDP) (Simonovic, 2005).

In many places around the world municipal water supply is already a problem.

In addition, cities are facing very high levels of threat from extreme events (floods and droughts). In the developing world, the lack of water infrastructure is affecting future development. In the developed world, an ageing water infrastructure is generating a heavy burden on local, regional and federal economies.

The goal of building resilient communities (water infrastructure being one of the most important elements) shares much with the principles of intergenerational equity espoused under the rubric of *sustainable development*, discussed in Chapter 7. In many ways, the decisions we make regarding the siting, design and construction of a community's water infrastructure will affect its sustainability over the long term. Moreover, decisions made today may augment vulnerability in the future, creating problems for future generations.

In order to achieve sustainability of critical urban infrastructure (including water supply and drainage), connections between a wide range of variables must be considered in the design of local community resiliency. We can define sustainable, resilient communities as 'societies which are structurally organized to minimize the effects of abrupt change, and, at the same time, have the ability to recover quickly by restoring the socio-economic vitality of the community'.

Urban water resources management in the future will be part of a paradigm called comprehensive vulnerability management, which is defined as holistic and integrated activities directed to the reduction of water-related emergencies and potential disasters by diminishing risk and susceptibility, and building resistance and resiliency. Comprehensive vulnerability management can be seen as a sort of meta-paradigm, drawing on the strengths of the other concepts. Specific elements include:

- *An inclusive, holistic approach*. Policies for comprehensive vulnerability management would be based on the consideration of risks and vulnerability in the physical, social and organizational environments.
- *A primary focus on vulnerability*. The approach would involve a concerted effort to identify and reduce all types of water-related vulnerabilities.
- *An all-encompassing approach*. The concept recognizes the need to address all types of triggering agents, natural or otherwise, related to both water quantity and quality, as well as water supply, protection from water and protection of water.
- *Participation by a wide range of actors*. The concept requires participation of and collaboration between a diverse set of actors, including public sector organizations, citizens, businesses and non-profit organizations.

By incorporating the many positive elements of other models and strongly emphasizing the need to reduce all forms of water vulnerability, the paradigm provides a framework for developing proactive, tangible strategies to create disaster resilient communities.

11.1.4 Participatory water resources management

Water is a subject in which everyone is a stakeholder. Real participation only takes place when stakeholders are part of the decision-making process. This can occur directly when local communities come together to make water supply, management and use choices. Participation also occurs if democratically elected or otherwise accountable agencies or groups represent stakeholders (GWP, 2000). The type of participation will depend on the spatial scale relevant to particular water management decisions, and on the nature of the political system in which such decisions take place.

Real participation is more than consultation, and requires that stakeholders at all levels of the social structure have an *impact* on decisions at different levels of water management. A participatory approach is the only means for achieving long-lasting consensus and common agreement. However, for this to occur, stakeholders and officials from water management agencies have to recognize that the sustainability of the resource is a common problem, and that all parties must sacrifice some desires for the common good. Participation is about taking responsibility, recognizing the effect of actions within different sectors on other water users and aquatic ecosystems, and accepting the need for change to improve the efficiency of water use and allow the sustainable development of the resource.

There is common responsibility for making participation possible. This involves the creation of mechanisms for stakeholder consultation at various scales (from local, through watershed to national), and the creation of participatory capacity, particularly among marginalized social groups. This may include awareness-raising, confidence-building and education, as well as the provision of the economic resources needed to facilitate participation and the establishment of good and transparent sources of information.

Participation is an instrument that can be used to pursue an appropriate balance between a top-down and a bottom-up approach to integrated water resources management (Castelletti and Soncini-Sessa, 2006).

11.2 EMERGING SCIENCES

I believe two emerging sciences will play an important role in the water management of tomorrow. One is nanotechnology and the other is quantum computing. Both are revolutionary in their own ways.

11.2.1 Nanowater

In this book my focus was on water resources systems management. However, let us not forget that the origin of management needs lies in the core problem – a lack of safe, clean and affordable water. Technology is one of the tools that can help in the solution of water-related problems.

The only technological solution I would like to discuss could be quite small – nanotechnology (www.nanowater.org). The science of the small has the potential to tilt the economic balance of many existing water-related technologies in favour of large-scale use. Traditional remedies, such as filters, desalination and water recovery systems, are limited in scope because they cost too much, are inefficient, require lots of maintenance or use too much energy.

Nanotechnology is not likely to provide much in the way of radical new technologies for desalination, purification or wastewater recovery. But adapting existing nanotechnologies for use in the water industry could provide huge benefits.

Nanotechnology has been developed, but not for water. Nanoscale 'needles' that puncture bacteria could potentially be applied to water treatment applications. So could nanoparticulate silver, which is currently used in medicine to fight infection. Then there are nanofibres, which are already used in many industrial applications. NASA is evaluating ceramic nanofibres for water purification in space because of their ability to increase throughput and reduce clogging compared with traditional filtration methods. Here on Earth, though, nanotechnology is just starting to show what it can do for water.

The use of nanofiltration, which is common in most industrial filtration processes, is the first application to trickle into the water sector. Nanoscale filters can be used to screen out items as small as bacteria and viruses for the specific purpose of eradicating waterborne disease, one of the main killers in developing countries.

But in other segments of water treatment, nanotechnology's potential has yet to be truly investigated. One of the greatest potential areas is desalination, an area where nanotechnology could cut costs, save energy, and improve the lifetime and efficiency of membranes. Today, seawater is most often turned into drinking water through a 40-year-old process called reverse osmosis, which is slow, expensive and energy-intensive. One way of improving the process is using a modern-day version of forward osmosis. There are already developed semi-permeable membranes that act as a molecular sieve, allowing water to pass through while rejecting impurities such as viruses, anthrax spores, E.Coli bacteria, heavy metals and other health threats. The first applications of the technology have already been introduced, to clean up industrial water and to produce some food concentrates.

Another potential application is sensitive-sensor technology, which involves the use of carbon nanotubes, nanowires, and micro- and nanoscale cantilevers to detect contaminants. There are products on the market that can detect waterborne nanoparticles and viruses in real time.

Nanotechnology can also help tackle the decontamination of groundwater from industrial and natural sources. The first tests of a surface-modified gel designed to selectively absorb heavy metal ions from wastewater using a novel nanopore structure are in progress.

If nanotechnology really can bring down the cost of water treatment or desalination, clean water could be within the grasp of a larger portion of the world. And that will be no small feat.

11.2.2 Quantum computing

My generation is already witnessing a revolution that will transform our world. It has drawn together neurobiologists, psychiatrists, computer scientists, physicists and mathematicians in a competition to achieve a more precise understanding of the human brain and to create an even more powerful synthetic brain. Both competitions lead us into the world of quantum mechanics. As with other revolutions, this one will bring about major advances in science and technology. Scientists are already planning for an Internet that functions as a single, worldwide quantum computer.

The basic idea behind a quantum computer (DiVincenzo, 1995; Satinover, 2001) is that data can be entered and placed in a superposition which transforms a certain number of input bits into an arbitrarily large number of superposed quantum bits upon which computation can occur simultaneously. Quantum computation is therefore like parallel processing in many networked universes (dimensions) at once. Alternatively it can be seen as a cellular automaton in which each cell exists in many different states simultaneously, in many different universes (dimensions). In this view it is the universes that exist in parallel. I wish I could go further, but my understanding of quantum mechanics is limited.

However, taking this potential into water management field makes my mind spin. Imagine the situation where computational limits are not part of the reality; where the world unfolds in multiple directions; where space and time are only two out of many dimensions. 'Everything flows' (Heraclitus).

Complex water management problems will be solved easily, and presentations of space–time dynamics will appear on our screens (or in our minds) instantly. However, I would like to draw from the past one message for the future. More powerful computing tools of the future will change the way we formulate and solve water problems. Let me go back to the work of the 'father of the computer', John von Neumann. In 1932 he wrote *Mathematical Foundations of Quantum Mechanics*. In his amazing career he tackled even the pedestrian-sounding subject of turbulent flow. To deal with the problem, von Neumann developed an entirely new approach – an approach with very deep consequences that are still spreading today. Instead of trying to figure out mathematical solutions to the very complex equations that describe turbulence, he devised a method of iterative computation – guess, measure the error, adjust the answer, feed it back in again. The results of iterative calculations were never perfect, but rather were precise to whatever degree was desired. Not only did the method work for turbulence, it proved the only way of solving the problem that led to the creation of the atomic bomb. This was the invention of the modern computer as we know it. 'Guess, measure the error, adjust the answer, feed it back in again' – is the basic method of neural networks, and arguably of natural intelligence (the way our brains process information).

With the power of quantum computing, water-related problems will be approached as dynamic multidimensional systems for which the solution is assembled by stringing together a bunch of simple 'processors' interacting with their neighbours.

11.2.3 Closing comments

The future has not yet been made. In very large part, with many possibilities yet untapped, we will make it what it will be. I hope it will bring water to everyone.

At the end, a word to my children, Dijana and Damjan. I did the best I could to bring you the highest good of all, water.

> **Box 11.1** To the Medway Creek
>
> The creek is my calendar, and sometimes an answer to a question I often think about: What is the meaning of time? It is there every day. My long daily walks bring me to see the swelling spring waters or summer trickle among the pebbles at the bottom. In the autumn the maple colours paint the creek surface and in the winter the battle rages between the free water and the ice.
>
> The creek is different every day, and it talks. The sounds are so powerful when it moves the large spring flow and so deep when it is under the ice. The sounds tell the story of travel through time and space – from the hellish depths of the Earth to the heavenly sky. And we talk. We tell each other what is bothering us and ask each other's help.
>
> She would stand just above the rapids, a little bit down the creek from our home. Red and blue and green. The dog would sniff the creek and look around for small forest creatures. She is one with the creek – she is the creek. She is the love of my life.
>
> (A memory from 2006)

11.3 REFERENCES

Castelletti, A. and R. Soncini-Sessa (2006), *Topics on Systems Analysis and Integrated Water Rersources Management*, Elsevier, Amsterdam, The Netherlands

DiVincenzo, D. (1995), 'Quantum Computation', *Science*, 270(5234): 255–261

GWP (Global Water Partnership) (2000), *Integrated Water Resources Management*, Technical Advisory Committee Background Papers 4, Stockholm, Sweden, 4: 71

IPCC (Intergovernmental Panel on Climate Change) (2007), *Climate Change 2007: The Scientific Basis. Summary for Policymakers. Contribution of Working Group I to the Fourth Assessment Report of the Intergovernmental Panel on Climate Change*, Cambridge University Press, Cambridge and New York

Satinover, J. (2001), *The Quantum Brain*, John Wiley and Sons, New York

Senge, P. (2003), 'The Inescapable Need to Change Our Organization', An Interview with Peter Senge, *Systems Thinker*, 14(3): 2–3

Simonovic, S.P. (2005), 'The Disaster Resilient City: A Water Management Challenge', in *Sustainable Water Management: Solutions for Large Cities*, D. Savic et al, eds, IAHS Publ, 293: 3–13

Index

α-level sets 146, 147–148, 153, 154, 155, 156, 157, 158
abnormal situations 131
acceptable levels
   failure 198–199
   performance 201, 203, 209–210, 211, 218–219, 223
   simulation 393, 395
acceptability 162, 534, 537
adaptive approaches 372, 536, 619
addition of fuzzy numbers 154
additivity 432
advanced decision support systems 9, 10–12
aesthetics 240, 273, 283
aggregation
   membership functions 222–223
   multi-component systems 205, 206–208
   multiple stakeholders 596, 597, 598–599
   operators 169–178, 192–194, 195–196
agreement 420, 421
agriculture
   aquifer sustainability 266
   Danube River 529
   global models 371, 375–376, 377–378, 384, 386
   integrated approaches 6
   reservoirs 337, 338
   revenue 272–273
arguments 597, 598
all-slack basic solutions 440, 444
alternative optimal solutions 436, 437

alternatives
   aquifer policy 259, 262–289
   fuzzy Compromise programming 566–567, 568–569, 570, 571, 572, 574–575, 577
   multi-objective analysis 125, 534, 535, 536, 587–590
   ranking of 160–166
   sustainability 237, 238, 596–597, 601, 602–608
   systems approaches 58–60, 65–66
A-matrix 454–455
ambiguity 130, 131, 132, 133
analytical procedures 44
AND-OR operators 152
applied systems analysis 39–40, 63-112
approximation uncertainty 131
aqueducts 35–36
aquifers 257–289, 326–329, 330, 585, 586
arable land 281, 378
Archimedes 35
artificial variables 444, 446
aspirations 416, 417, 419–420, 421
Assiniboine River, Canada 257–289, 353–366, 406, 407, 409
Assyrians 33
attractiveness 535
augmenting cycles 513–514
auxiliary variables 316–317, 332, 334, 339
availability, water
   aquifer sustainability 283
   conflict resolution 419, 420
   global 3–5, 367–387

multi-objective analysis 585–590
uncertainty 24, 25, 26
awareness, disaster 389, 391

balancing feedback 96
Bangkok, Thailand 119
bathtub simulations 331–337
Bayes's theory 142
behaviour 51–54, 118, 119–120, 299, 300–309, 388–399
Bellman-Zadek's concept 458
benefits *see* cost–benefit analyses
Bernoulli's equation 18
best compromise solutions 557, 558
Beyond the Limits 379
The Big M method 445–446
biodiversity 52–53, 529, 530
brake power 507
Brundtland Commission (1987) 233
Buddhism 235
budgets, water 262–266
Byzantine Empire 36

calibration 378–379, 409, 410, 411, 412
Canada
  aquifer sustainability 257–289
  floods 12–17, 388–412, 594–608
  hydroelectric systems 487–490
  infrastructure 53
  regional performance measures 212–214
  reservoir simulations 353–366
  systems dynamics models 337–346
  water resources data 25, 369
CanadaWater model 387
canals 10, 32, 36, 37, 313
  see also dykes
canonical linear optimization 434
canopy storage 400, 403, 404–405
capital 371–372, 380
caring behaviour 52–53
causal influence diagrams 107, 108
causal loop diagrams 310–315, 327, 374, 392, 394, 415–416, 419
cause-and-effect relations 70, 72, 95, 115, 297
  see also simulation
centralization 92
centre of gravity 162, 205, 220
centroid method 161, 572

change-of-basis calculation 441, 442
Chang and Lee's method 164, 567–568, 574, 580, 600
Chao Praya river, Thailand 119
Chebychev distance 552
chemical treatment 214–215, 217, 220, 221
Chen's method 164–166
China 9–12, 36, 69, 557
chlorine 214–215
Christianity 235
cities 97–99, 212–214, 320, 375, 376, 462–464, 620
  see also municipal level
City of London, Ontario, Canada 212–214
climate change 4–5, 26, 618–619
closed-loop thinking 70
closed systems 80, 93
closeness measures 551
coincidence measures 255, 256, 286
collaborative flood management 12–17
collective learning 23, 123, 430, 464–465
combined configuration systems 206, 208
combined fuzzy reliability–vulnerability measures 202–203
communication, disasters 388
commutative operators 171, 172, 174–175
comparative approaches 160–166, 200, 566–568, 588–589
compatibility 162–164, 200–202, 203, 204, 223
compensation 171–172, 173, 552
competition 89–90, 90
complement membership functions 150
complexity
  competition 90
  fuzzy sets 167, 168, 171
  interdisciplinary approaches 617
  multiple objective decision-making 596
  optimization 121–122
  paradigm 20–24
  pipelines 505
  sustainability 233–234, 240–241, 242
  systems 42, 43, 65, 83
compliance 201–202
composition under pseudomeasures (CUP) 174–175, 177
comprehensive vulnerability management 621
Compro computer program 558

compromise groups/solutions 534, 535
Compromise programming 253–254, 288, 549–558, 559–560, 573–580, 590
    see also fuzzy Compromise programming
computer programs
    Compro 558
    Evolpro 466, 480–481
    FuzzyCompro 572, 580–581
    Linpro 434
    SUSTAINPRO 257, 286
    see also Vensim
computers 20, 22, 26, 38–39, 73, 118, 300, 624–625
    see also technology
concern 391–392, 395
conditional answers 603–604, 604–608
conduit flow 357, 358, 359, 363, 366
conflict 413–421, 536, 578
conjunctive water uses 541
consensus 236, 242, 252–257, 261, 286–288
conservation scenarios
    aquifer policy 264–265, 266
    consensus 287, 288
    fairness 267, 268, 269, 270
    reversibility 283, 284
    risk 277, 278
constraints
    city water supply 98, 99, 462–463
    general systems theory 76
    linear programming 432, 450, 458–459
    multi-purpose reservoirs 101
    pipeline optimization 505, 509–510, 512
    short-term reservoir operations 492–493, 497–498, 501
    uncertainty 456
    wastewater treatment 104–105
construction 41, 320, 322
continuous methods 538, 554
continuum thinking 71
control screens 359, 398, 399
correction factors 342, 344
cost–benefit analyses 58, 239
costs
    city water supply 98, 462, 464
    desalination 373–374
    effectiveness 58
    hydropower 488

    infrastructure 273
    pumping 505–519
    systems engineering 75
    wastewater treatment 103–104
    WorldWater 377
credibility level 146
    see also α-level sets
criteria
    development 58
    fuzzy 562, 563, 565
    sustainability 236, 240–243
    see also objectives
critical components 224
crossover 468, 472–476
    see also recombination
cumulative fight 418–419
CUP see composition under pseudomeasures

damage, flood 180, 183, 184, 188
dams 33, 69, 79, 124, 337, 338, 356, 431, 530, 532
danger recognition 391, 392, 393, 395
Dantzig's simplex algorithm 439–446
Danube River 527–531
Darcy–Weisbach equation 517
data
    aquifer sustainability 259–261
    collection 58
    flood 186, 191, 394–398
    hydrometric 25–26
    uncertainty 166–167
    water resources systems management 56
databases 14, 15, 17, 26–27
decision-making
    complexity 20
    disaster response 388–399
    floods 9, 10–12, 15–16, 16–17
    multi-objective analysis 534–581
    past system performance 51–52
    sustainable water resources 233–257
    system dynamics simulations 337–346
    uncertainty 132–135
decision variables
    city water supply 98, 99
    mathematical models 76, 77
    multi-purpose reservoirs 100
    optimization 429–526
    wastewater treatment 102–103, 105

decomposition 167, 169–178, 178–180
definitions
   conceptual risk 135–138
   consensus 253
   flood control 178–180
   fuzzy approaches 144–148, 197–202
   general systems theory 66–82, 83–86
   master planning 583–584
   operational risk 248–250
   selection 469
   simulation 297–298, 299
   water resources systems management 50, 115
defuzzification 567, 598–599
degree of incompleteness 170
degree of orness 173
delay functions 309, 360, 397
delta time (DT) 360
demand, water
   climate change 618
   complexity 42, 43
   constraints 98
   fuzzy approaches 209–210, 211, 219
   global 366–387
   hydropower optimization 483, 485, 486
   integrated approaches 6–7, 8, 20
   multi-objective analysis 581–582, 586–590, 591
   municipal 462
   water values 619
demographic level 83–84, 391
derivatives 317
   see also flows
desalination 373–374, 623, 624
design 41–42, 46, 60, 74–75
deterministic models 81–82
developing countries 582–584
development 40–47, 264, 529–530, 543–544
development scenarios
   aquifer policy 264, 266
   consensus 287, 288
   fairness 267, 268, 269, 270
   reversibility 283, 284
   risk 277, 278
diagramming languages 68
diagramming notation 315–317
disasters 388–399, 530–531
discharges 409, 410, 411, 412

discrepancy measures 255–256, 286
discrete methods 539–540
disorganization 81
displaced ideal theory 535
dissatisfaction 416–417, 418, 419–420
distance-based fairness measures 244, 245
distance metrics
   Compromise programming 552
   fuzzy Compromise programming 559–560, 561–568, 570, 571, 572, 573, 574, 579, 600–601
   sustainable floodplain management 607
distributed parameter models 82
divisibility 432
division of fuzzy numbers 159–160
domain specific lessons 19
dominated alternatives 540
dominated solutions 125, 532, 534, 535, 550
double run scenarios 380, 382, 384
DP *see* dynamic programming
drainage 6, 9, 10, 12, 53, 332–337
dreamdisintegration arms race 320
droughts 618
DT *see* delta time
duality 446–450, 452–454
Dublin principles 47
dykes 9, 37, 52
   see also canals
dynamic programming (DP) 23, 121, 122, 429–430
dynamics
   definitions 82, 299
   hypotheses 331, 402–403, 404, 414–416
   systems 80, 117–120
   world water 379–384
dynamic thinking 70

EAPWSS *see* Elgin Area Primary Water Supply System
ecological level 280, 285, 286, 529
   see also environmental level
economic level
   Danube River 528–529
   efficiency 56
   global water use 368, 371–372, 379–380
   reservoir loss 491
   reversibility 280, 285
   risk 248, 249, 271, 275

sustainability 239, 601, 602
systems engineering 75
values 619–620
water resources management 40–47
education 46–47, 68–70, 94
efficiency 56, 506–507
efficient solutions *see* non-dominated solutions
effluents 8, 102–105
Egypt 5–9, 32–33, 79
election probabilities 470–472
elements 83, 86
elevation, outflow 479–480
Elgin Area Primary Water Supply System (EAPWSS) 213, 214, 216–218, 223–224
emergency flood management 388–399, 596
emerging technologies 45, 46
Emerson, Canada 594
employment 273, 282
EMSLP *see* energy management by successive linear programming
energy management by successive linear programming (EMSLP) 482, 485, 486–490
energy rate function (ERF) 484–485
engineering 31–36, 40, 73, 74–75
entropy 81
environmental level
   Danube River 529, 530–531
   human activities 52
   impact assessments 240
   reversibility 285
   risk 248, 249, 271, 274, 275–279
   sustainable floodplain management 602
   World3 model 372
   see also ecological level
environment of the problem 57
equality 243, 433–434
equalization periods 269–271
equally likely concept 140
equilibrium 78–79, 80, 86
equity *see* fairness
ERF *see* energy rate function
erosion 272, 282
error 131–132
Euclidean distance 552, 557
Euler integration 325–326

Europe 528, 529–530
eutrophication 108–110, 314, 531
evacuation 185, 189, 288, 388–399
evaporation 337, 338, 400–401, 404–405
evapotranspiration 3, 400–401, 403
events 139, 140, 142
Evolpro computer program 466, 480–481
evolutionary algorithms 23–24, 123, 430, 464–481, 512
excluded solutions 534, 535
excluded variables 131
experience 389, 390, 391, 392, 393
expert systems 45, 46, 180–192, 194–196, 249, 260
exponential growth 87–88, 301, 306–308
extended line recombination 475–476
extension principle 152–153
extremes 180

failure 197–199, 202–203, 204–205, 208, 209, 219, 220–221, 222–223
fairness 236, 241, 242, 243–247, 261, 267–271, 287, 621
faucet flow 310–311, 331–337
FCP *see* fuzzy Compromise programming
feasible regions/solutions
   multi-objective analysis 532, 533, 535, 548, 549
   optimization 432–433, 512–514, 515, 518
feedback
   definition 66
   evacuation planning 392, 394
   flood from snowmelt models 402–403
   general systems theory 92–96
   irrigation flow control 106–107
   lake eutrophication 108, 109
   systems dynamics 301–306, 310–314, 317–318
   water resources systems management 39, 57
   WorldWater model 372, 384
FEV *see* fuzzy expected value
fight 417, 418–419
filtration 217, 623
financial sustainability 239
fishing 465, 557
fitness 23, 123, 465, 466, 467, 469–472, 477, 478

flash mix components 225
flocculation 214
floods
  advanced decision support systems 9, 10–11, 12
  climate change 618
  definitions 178–180
  evacuation models 388–399
  fuzzy membership function for control 178–196
  hydrological simulation models 399–412
  integrated management approaches 12–17
  multi-objective analysis 542–543, 591, 594, 595
  protection levees 146–148, 151
  reservoir simulation models 353, 354–356, 359, 360, 361, 362–364, 365–366
  sustainable floodplain management 593–608
flow
  bathtub simulations 332, 336
  climate change 618–619
  elevation 479–480
  flood from snowmelt models 402–403
  groundwater aquifer production 327, 328
  irrigation 105–107, 302–306, 307
  pipeline optimization 505–519
  reservoir 337, 339
  system dynamics simulation 314–323
  turbulent 625
fluoride 217
food production 377, 380, 382–384
forecasts 585–587
four-step water resources planning frameworks 584
frequency interpretation of probability 140
freshwater availability 3–4
future aspects 47–48, 238, 242, 246–247, 253, 617–626
  see also inter-generational fairness; predictions; projections
fuzzy approaches
  multi-objective analysis 558–581
  optimization 455–464, 491–505
  simulation 347–351
  uncertainty 129–232
FuzzyCompro computer program 572, 580–581

fuzzy Compromise programming (FCP) 559–581, 600–601, 608
  see also Compromise programming
fuzzy expected value (FEV) 598, 599, 604, 605, 606, 608

γ-family of operators 171–172, 192
general systems theory 65–112
generic patterns of behaviour 306–309
generic thinking 70
genetic operators 466, 467
global reinsertion 477–478
Global Run-off Data Centre (GRDC) 26
global water resources 3–5, 366–387, 576
goals 75, 163, 308, 536–537
government employees 274, 275–279
grade of compensation 171–172, 172, 173
graph theory approach 510
GRDC see Global Run-off Data Centre
Greek civilizations 33–34
Green Reservoir, Kentucky 500–505
groundwater
  aquifers 257–289, 326–329, 330
  availability 585, 586
  nanowater 624
  recharge 100, 546
  storage 401, 406, 408, 411
group decision-making 574–577
growth 86–89, 301, 306–309, 326–329, 367

habitat loss 272
Hamming distance 552
hard water 97
head-cost functions 516, 518
head–flow–efficiency 506–507
Hellenistic period, Greece 34–35
hierarchical approaches 92, 170, 373
High Aswan Dam, Egypt 79
Hinduism 235
history of water resources engineering 32–40
holistic approaches 621
homeostasis 53–54, 94
  see also self-regulation
human activities 51–54, 78, 79, 81
hydraulic metaphor 323–324
hydrological simulation models 399–412
hydrology 38
hydrometric data 25–26

hydropower 121–122, 431, 482–490, 587, 590–591
hydrostatics 35
hypotheses
　conflict resolution 414–416
　flood from snowmelt simulation models 402–403, 404
　system dynamics simulation 331
　testing 71

ideal solutions 550–551, 554–555, 558, 559–560, 562–563, 572
IF-THEN-ELSE structures 357, 397–398
ill-defined sets 169, 170, 171, 173
impacts
　abnormal situations 131
　dams 69, 79
　fairness 246, 247
　floods 180, 183, 184, 186, 188, 190
　reversibility 251–252, 280, 284, 285, 286
　risk policy 271–273
　social 239–240
　sustainability decisions 238
importance transform function 173
inclusive approaches 621
incorrect form 131
independence 90
Indian early civilizations 36
indicators, sustainability 236
individual heterogeneity 24, 130
industrial level 375, 376, 384, 386, 587, 590, 620
inequality 433–434
infeasible problems/solutions 432, 437–439, 444, 512, 550
infiltration 400, 405, 408
inflow 337, 340, 354, 360, 361
information 166–167, 186, 190, 309, 391, 397
　see also knowledge
infrastructure 15, 43, 53, 273, 377, 621
initial aspirations 416, 417, 419–420, 421
innovation 536
input 66, 116, 298, 350, 352, 598, 604
INTEGRAL function 324
integrals 317
　see also stocks
integrated approaches 5–9, 12–17, 30, 221, 323, 324–326, 328, 374–375, 619

interception rate 403, 404–405
interdisciplinary approaches 617
inter-generational fairness 244, 245–247, 267–269, 621
intermediate recombination 473–474
International Conference on Water and the Environment, Dublin (1992) 47
Internet 27, 39
intersection 150, 151, 152
inter-temporal fairness 241, 242
intra-generational fairness 244, 245–247, 267–268, 269–270
inventories 585–587
Iranian early civilizations 36
irrigation
　aquifer sustainability 262, 263, 265, 266, 267, 282, 287
　canal intake systems 313
　climate change 618
　early civilizations 32–33
　flow control 105–107, 302–306, 307
　four-step planning framework 587
　global models 377, 386
　risk 274, 275–279
　system dynamics simulation 337–346
Islam 235
issue statements 329–330
Itaipú Dam 431

Japan 599
Judaism 235

knowledge 24, 45, 46, 131, 141–142, 318–319, 393, 514
　see also information
Ktesibios 34

labelling 513
lake eutrophication 108–110, 314
Lake Huron Primary Water Supply System (LHPWSS) 213, 214–216, 223, 224, 225
land reclamation 37
landscape aesthetics 283
learner-directed learning 68–70
learning, collective 23, 123, 430, 464–465
levee height 146–147, 151
LIIPWSS see Lake Huron Primary Water Supply System

licensing 264, 267
life expectancy 375, 376
The Limits to Growth 368, 379
linear aspects
   definition 81
   fuzzy distance metrics 561
   multi-objective analysis 101, 538
   optimization 99, 105, 121, 122–123, 434, 440
   ranking 470
linear programming (LP) 18–19, 22, 121, 122–123, 429, 430–455, 457–464, 482–490
line recombination 474–475
linguistic
   answers 597, 598, 603, 604, 605, 608
   formulations 167
   options 568
   quantifiers 576
Linpro software 434
liquid pipelines 505–519
livestock 262, 263, 264, 282
load 136–137, 138, 197, 200, 248, 486, 488, 490
local reinsertion 478–479
local selection 471–472
logistic curves 88
loss functions 491, 493–494, 495–496, 500–505
Lower Whitemud East, Canada 267, 268, 269, 270
LP *see* linear programming
lumped parameter models 82, 400

major alternative groups 534, 535
margin of safety 198–199, 200, 219, 223
master planning 581–593
material delays 309, 397
maximization
   fuzzy distance metrics 561–566
   fuzzy sets 164–165, 166
   optimization 432–433, 447, 448, 449, 458, 459–460, 461, 463, 493
maximum allowed storage variation (VARYMX) 486, 487
mediators 417–418
medieval ages 37
membership functions
   aggregation methods 195

compatibility 200–202, 203
definitions 144–146
failure 198–199, 222–223
fuzzy 147–148, 149, 150, 157, 164, 166–196
fuzzy Compromise programming 560, 562, 574, 598, 601
fuzzy optimization 456–457, 459, 460, 461, 464
fuzzy simulations 348
multi-component systems 205, 207–208, 209
regional systems performance 221–222
reliability analysis 210
short-term reservoir operations 497–499, 500, 501, 504–505
sustainable floodplain management 607
system-state 219, 221–222
Mesopotamia 32
Millennium Development Goals 48
minimization
   fuzzy optimization 458, 462, 463
   fuzzy sets 164–165, 166
   optimization 432–433, 447, 448, 449, 492–493
minimum cost flow problem 511, 513
MINLP *see* mixed-integer nonlinear programming
misleading similarities 143–144
mixed-integer nonlinear programming (MINLP) 122
modal points 560, 565
Monte Carlo-type generations 514–516
most robust compromise solutions 557–558, 583, 590
multi-component systems 205–212, 220–221
multidisciplinary approaches 57
multi-objective analysis 82, 124–125, 237–238, 242–243, 289, 527–614
multiple reservoir operations 121–122
multiple stakeholders 582, 594–608
multiplication of fuzzy numbers 157–159
multiplicative operating rule models 342–346
multipliers 344, 345
multi-purpose reservoirs 99–101, 121–122, 337–346, 491–505, 545–549
municipal level 384, 386, 387, 462, 546, 587, 590

see also cities
mutation 465, 466, 467, 468, 476–477

nanowater 623–624
natural systems 78–79
necessity measures 163, 201
negative causal links 311
negative feedback 96, 302, 327, 403
   see also balancing feedback
negative ideals 554, 563, 573
negative impacts 275, 279
negative weights 544–545
Nepal 39
The Netherlands 37
von Neumann, John 624–625
neural networks 45, 46
Nile River 5–9, 32–33, 79
nitrogen 531
nodes 510, 511
noise pollution 322
non-dominated alternatives 540, 554
non-dominated sets 537–541, 544–549
non-dominated solutions 125, 532, 533–534, 550, 551, 554, 555, 556, 558
nonlinear aspects 81, 470, 512, 561
nonlinear programming 22–23, 121, 122, 429
non-physical relationships 116, 298
non-renewable resources 371, 380, 381
non-structural measures 182, 185, 189, 542, 606
numerical integration 324–325
numerical procedures 44
numeric scales 597, 598, 603
nutrients 531

objective functions
   city water supply 98, 99
   general systems theory 76, 82
   hydropower 483, 485, 489
   multi-objective analysis 531–532, 546–547
   optimization 121, 429, 440, 456, 459, 512
   reservoirs 100–101, 492, 493, 497–498, 501–502
   sensitivity analysis 451–452, 453, 454, 455
   wastewater treatment 103–104
   weighting 544, 546–547
objective risk 137, 138
objectives 58, 75, 237–238, 432
   see also criteria; multi-objective analysis
object-oriented programming 118, 324
OERI *see* Overall Existence Ranking Index
open systems 80–81, 92–93
operational level
   floods 11
   multi-objective analysis 541–542, 545–549
   practice and technological development 46–47
   research and systems analysis 73
   reservoirs 337–346, 353–366, 500–505
   risk definitions 248–250
   simulation 116, 298
   thinking 71
   water resources systems management 43, 60
operators, aggregation 169–178, 192–194, 195–196
opportunities 129–130
optimal allocation 18–19
optimization 22–23, 77, 81–82, 120–124, 124, 429–526, 532, 585, 586, 591
ordered weighted averaging (OWA) 172–174, 193
orness, degree of 173
oscillation 308
Our Common Future 233
outflow 354, 360, 361, 364, 479–480
   see also spillways
output 66, 116, 298, 350, 351, 352, 371, 375
Overall Existence Ranking Index (OERI) 164, 574, 600–601
overlap analysis 202
overshoot and collapse 308–309, 379
OWA *see* ordered weighted averaging

PAC transfer pumps 225–226
paradoxical situations 143
parallel configuration systems 206, 207–208
parameters of the model 76
parameter uncertainty 131–132
parastatic water clock 34

Pareto optimal solutions *see* non-dominated solutions
partial failure 197–199
participation
 consensus 242
 flood management 14
 integrated water resources 7–8, 8–9
 multi-objective analysis 534–537, 573–581
 survival 51
 sustainable management 274–279, 593–608, 621
participatory water resources management 622
past system performance 51–52
payoff matrices 539, 591, 592, 602
payoff tables 242, 243
P-CUP *see* polynomial composition under pseudomeasure
penalty zones and coefficients 491, 493–496, 497, 500–501, 504, 505
perceived risk 137
performance 51–52, 76, 196–212, 212–226, 320, 321
phosphorous 531
physical relationships 116, 298
physical summativity 90
physical systems 79
piecewise linearization 491, 493–494, 495–496
Pine Creak North, Canada 267, 268, 269, 270
pipelines 208–211, 505–519
pivoting 513
pivot operation 441, 442
planning
 floods 11, 389–390, 391, 392
 integrated approaches 7–8
 multi-objective analysis 541, 569, 581–582, 582–593
 optimization 489
 sustainable floodplain management 596
 systems engineering 75
 urban 97
 water resources systems management 42, 46, 60
point source water pollution 272
policy level 5, 8, 110, 234, 259–289, 393–394, 500–505

pollution
 construction noise 322
 freshwater 4, 136
 global models 371, 372, 373, 378, 380–381, 382–384, 385, 387
 point source water 272
 weighting method 545, 547, 548
polynomial composition under pseudomeasure (P-CUP) 175–178, 193–194, 195–196
population
 evolutionary algorithms 430, 464–465
 flood evacuation 397
 global models 371, 375, 376, 379, 380, 384
 growth 326–329, 367
 policy scenarios 266, 287
 urbanization 620
 water availability 4
positive causal links 311
positive feedback 96, 301
 see also reinforcing feedback
positive ideals 563, 573
positive impacts 275, 279
possibility measures 162, 201
post-load recovery 390, 596
post-optimality analysis 451
 see also sensitivity analysis
power–head relationships 507, 508
precipitation 3, 400, 403–404, 407–408, 411, 618, 619
predictions 367, 386, 399–412
 see also projections
prefabricated water storage tanks 349–351
preferences
 multi-objective analysis 574–575, 577, 578–580, 581, 590
 optimization 497, 503–504, 504–505
 prior articulation of 538–541
 risk 274–279
 see also membership functions
pressure 507, 508, 509–510, 511, 517
primal problems 446–448, 449, 450, 453
principle of transfers 244
prior appropriation rights 264, 267
prior articulation 125, 538–540
prior knowledge 141–142
probabilistic approaches 80, 81–82, 139–144, 197

Index **637**

problem definition 57, 58, 133–134
procedural policies 234
profit 452, 545, 547, 548
progressive articulation of preferences 540–541
progressive segregation 91
projections 370, 619
  see also predictions
PROTRADE method 543–544
pseudomeasures 174–178, 193–194, 195–196
psychological factors 393
public hearings 394–395, 601
pumps 208–211, 505–519
purification 372–373

Q-core 576, 577, 580
qualitative level 178–196, 276, 281, 561, 565
quality, water
  city water supply 98
  climate change 618–619
  environmental 240
  global water 372–373, 576
  monitoring 281
  multi-objective analysis 543, 587, 591
  pollution 136
quantitative level 38, 265–266, 276, 561–562, 565, 568
quantity time graphs 304, 305
quantum computing 624–625
query forms 180, 182, 183

ranking
  consensus 253–256, 286–287
  fitness assignment 469–472
  fuzzy approaches 160–166, 567–568, 571, 572, 574, 577, 580
  multi-objective analysis 588–589, 592, 593
  sustainable floodplain management 605, 606
rates *see* flows
rational equation 17–18
reality checks 330
reclamation programmes 543–544
recombination 465, 466, 468, 472–476
recovery times 204–205, 208, 220
recreation 273

Red River, Canada 12–17, 337–346, 388–399, 399–412, 594–596, 601–608
reference modes 330
refuge reaching 397
regional level 212–226, 368, 387
RegionalWater models 387
regression models 347–351
regrets 552
reinforcing feedback 96
reinsertion 468, 477–479
relation matrices 574, 575, 577, 581
release
  hydropower 488, 489, 490
  multi-purpose reservoirs 338, 339, 342, 364
  short-term reservoir operations 494, 495, 500, 501, 502, 503, 504
reliability analysis 141–142, 196–212, 218–226
religion 234–235
Renaissance period 37–38
reports 60
required power 507
reservoirs
  capacity 221–222
  conflict resolution 414–421
  global water resources 378
  hydropower 488
  multi-objective analysis 541–542, 554–557, 585, 586, 591
  multi-purpose 99–101, 121–122, 337–346, 491–505, 545–549
  simulation models 353–366
  standards projections 386
resiliency 204–205, 209, 621
resistance 137, 138, 197, 200, 248
resolution, conflict 413–421
responsibility 622
reversibility 236, 247, 250–252, 261, 280–286, 287
rights, prior appropriation 264, 267
risk
  aquifer sustainability 261, 271–280, 287
  conceptual definitions 135–138
  floods 179, 183, 187, 191, 192
  fuzzy Compromise programming 574, 577, 601
  operational 248–250

sustainability criteria 236, 241
   uncertainty 129, 130, 133, 164, 196–197
R-metric 251, 283, 284, 285
robustness 204, 210–211, 212, 219,
   223–224, 225, 226
rolling mills 102, 103
Roman Empire 35–36
roulette-wheel selection 470–472
Runge-Kutta methods 325, 326
Russell's paradox 143

safety
   acceptable performance 209–210
   flood evacuation 288, 393, 394, 397
   margin of 197, 198–199, 200, 219, 223
   uncertainty 141, 196
sample spaces 139–140
satisfaction level 502, 536–537
satisficing 78
scaled non-dominated solutions 556
scaled risk preferences 276
scaling functions 553
scientific level 31, 38–40, 56, 71, 622–625
Second Industrial Revolution 66
Second World War 39, 40
segregation 91
selection 465, 466, 468, 469–472
self-confidence 320, 321
self-regulation 93–94
   see also homeostasis
sensitive-sensor technology 623–624
sensitivity analysis
   alternatives evaluation 60
   linear programming 451–455
   planning 569
   reservoir models 360, 503–504
   reversibility 252, 283–284, 285, 286
   weighting 238
   WorldWater model 386
Serbia 583–584
serial components 224
serial configuration systems 206, 207
serial pipelines 507–509
serial pumps 516
set-theoretic operations 148–152
sewerage 17–18, 37
   see also wastewater
shadow prices 453
shared vision modeling 120

shared water resources 413–421
Shellmouth reservoir, Canada 353–366
Shinto 235
shipping development 530
short-term reservoir operations 491–505,
   542
Sihu basin, China 9–12
Simon, Herbert 78
simplex algorithm 439–446
simulation 8, 21–22, 68, 77, 115–120,
   297–428
simultaneous differential equations 83–84
slack variables 433, 440, 444, 449, 450,
   452, 453
SLP *see* successive linear programming
snowmelt simulation 399–412
social level
   flood evacuation 389, 391–392, 393
   impact assessments 239–240
   reversibility 280, 285, 286
   risk 248, 249, 271, 275
   structures 52
   sustainable floodplain management
      602–608
   water values 619–620
socio-economic level 7, 529
soil storage 400–401, 402, 403, 405–406,
   408, 411
solid waste generation 323
solution space representation 435–436
spatial aspects 24, 26, 55–56, 130, 241
spillways 353, 357–358, 359, 361,
   362–363, 364, 366
S-shaped growth 308
stable run scenarios 380–384, 383
stable states 86
stakeholders
   conflict resolution 414–421
   consensus 253
   risk 249–250, 274
   sustainable floodplain management
      597–599, 601–602, 602–608
stale variables 317
   see also stocks
standard intermediate recombination 474
standard run scenarios 379–380, 381, 384,
   385
static systems 80, 82
stationary states 84–86

status quo scenarios
    aquifer policy 262–264, 266
    consensus 287, 288
    fairness 267, 268, 269, 270
    reversibility 283, 284
    risk 277
stochastic models 81–82
stochastic multi-objective programming 558–559
stochastic universal sampling 470–472
stocks
    bathtub simulations 333, 336
    flood evacuation 397
    groundwater aquifer production 327, 328
    hydraulic metaphor 323–324
    reservoirs 337, 339
    system dynamics simulation 314–323
    WorldWater model 372, 373
storage
    flood management 14–15
    hydropower 485–486, 487, 489
    reservoirs 338, 354, 494, 495, 499, 500, 501, 502, 503, 504
    snowmelt 400–401, 402, 403–404, 405–406, 408, 411
    tanks 310–311, 348–351
streamflow 272, 402, 407, 408, 409, 411, 412
structural aspects 15, 70, 117, 118, 119, 131, 299, 300–309, 310–314, 606
subjective aspects 137, 138, 140, 141–142, 249, 251, 568, 570
subsurface soil storage 401, 403, 405–406, 411
subtraction of fuzzy numbers 155–157
successive linear programming (SLP) 486
summativity 90–91
support of a fuzzy number 148, 149
surface soil storage 400–401, 402, 403, 405, 408, 411
surface water analysis 585, 586
surplus variables 433, 446
surrogate variables 131
surrogate worth trade-off (SWT) method 543
survival 51–52, 620
survival of the fittest 23, 123, 465
sustainability 233–293, 593–608, 621
SUSTAINPRO computer program 257, 286

SWT *see* surrogate worth trade-off method
system dynamics simulation 117–120, 298–428
systems analysis 20–21, 38–40, 44–45, 54, 72–73, 73–74, 299
system-state functions 206–207, 219, 221–222, 225

teaching 319
technical lessons 19
technology 39, 43–47, 380, 622–625
temperature
    floods from snowmelt 399, 400, 402, 403, 405, 408
    streamflow simulation 409, 410, 411, 412
    thermostat control 322–323
temporal distribution-based fairness measures 244–246
temporal variability 24, 26, 130
Thailand 119
theory of the displaced ideal 535
thermostats 93, 322–323
thinking, systems 67–72
threats 129–130
Three Gorges Dam, China 69
timescales
    bathtub simulations 333
    engineering 33
    fairness 246–247
    groundwater aquifer production 329, 330
    hydropower 488, 489, 490
    reservoir models 340, 341, 342, 360
    stocks and flows 317, 318
    sustainability 234, 238, 241–242
Tisza River 568–572, 577–580
Tone River, Japan 599
tournament selection 472
tradable water share systems 265
trade-offs 237, 238, 239–240, 531, 541, 543–544, 545, 548
transfers, principle of 244
trapezoidal fuzzy membership 209, 210, 211, 224
treatment systems 212–226, 623, 624
triangular fuzzy functions 154–160, 209, 210, 211, 224, 348, 572
truncated normal distribution 515–516
truncation selection 472

turbulent flow 625

unbounded optimal solutions 436–437, 438, 443
UNCED *see* United Nations Conference on the Environment and Development
uncertainty
　consensus 253
　fuzzy approaches 129–232, 565
　impact distributions 247
　multi-objective decision-making 600–601
　optimization 456
　paradigm 24–26
　short-term reservoir operations 493
　simulation 346–351
　sustainable floodplain management 597, 604
　water resources systems management 54–55
UNIDO *see* United Nations Industrial Development Organization
union membership functions 150, 151, 152
United Nations Conference on the Environment and Development (UNCED) (1992) 47
United Nations Industrial Development Organization (UNIDO) guidelines 582–583
United States of America 500–505
universe of discourse 146
urbanization 620–621
use, water
　conjunctive water 541
　four-step planning framework 586, 587
　global models 287, 299, 368, 372, 375–377, 384–385

vagueness 560, 562, 573
value 75, 246, 535–536, 619–620
valve operation 510
variability 24, 26, 130–131, 132, 272
variable identification 330
VARYMX *see* maximum allowed storage variation
vector optimization 82

Vensim 300, 315, 316, 328, 330, 331, 334, 335, 343, 346
verification 378, 379, 409, 410, 412
vertical water balance 400–401
vulnerability 202–203, 212, 223, 224, 225, 226, 621

warning systems 393–394
wastewater 102–105, 372–373, 382, 384–386, 623
watermills 37
water-sharing conflict resolution 413–421
WCoG *see* weighted centre of gravity measure
weighted-average operating rule model 339–342, 344
weighted centre of gravity (WCoG) measure 161–162, 567
weighting
　Compromise programming 551, 552–553, 562, 565, 566
　conflict resolution 416, 417–418, 420
　consensus 254, 287
　flood evacuation 396
　fuzzy Compromise programming 568, 570, 574, 578, 579, 601
　multi-objective analysis 544–549, 590, 591, 592
　OWA 172–174, 193
　reversibility 252, 280, 284
　sensitivity analysis 238
　sustainable development 276
weir position 106, 107, 313
wetland drainage 53
wholeness 91
wildlife habitat loss 272
workshops 7–8, 260, 274–275, 601
World3 368, 370–372, 378–379, 381, 382, 383, 387
WorldWater model 372–379

Yangtze River, China 557
yield optimization 585, 586, 591

Zimmermann's γ-family of operators 171–172, 192